Stem Cell Biology and Regenerative Medicine

Series Editor

Kursad Turksen

For further volumes:
http://www.springer.com/series/7896

Hossein Baharvand · Nasser Aghdami
Editors

Advances in Stem Cell Research

Humana Press

Editors
Hossein Baharvand
Royan Institute
University of Science and Culture
Tehran
Iran

Nasser Aghdami
Regenerative Medicine
Royan Institute for Stem Cell Biology
and Technology
Tehran
Iran

ISBN 978-1-61779-939-6 ISBN 978-1-61779-940-2 (eBook)
DOI 10.1007/978-1-61779-940-2
Springer New York Heidelberg Dordrecht London

Library of Congress Control Number: 2012937779

Printed on acid-free paper

Humana Press is a brand of Springer

Springer is part of Springer Science+Business Media (www.springer.com)

To the memory of Dr. Saeid Kazemi Ashtiani, a wonderful colleague, a great stem cell biologist, and an inspirational advocate of stem cell research in Iran
To our professors, teachers, students, and families

Preface

Keeping in line with the current focus that personalized medicine should be the future medicine, stem cells have received tremendous attention. In recognition of the growing excitement and potential of stem cells for both the advancement of biology and future clinical applications, we felt it timely to edit this book to address and discuss the current advances and topics on stem cells, covering aspects from stem cell nano-engineering, pluripotent stem cells, and cellular reprogramming to regenerative medicine. We are pleased to provide *Advances in Stem Cell Research* as a timely and comprehensive overview.

By itself, the field of stem cells has generated tremendous interest; however, development in this area will also pave the way for a breakthrough within regenerative medicine. In parallel, scientific developments in the post-genomic era and achievements in systems biology will allow for the necessary development of personalized medicine.

We want to sincerely thank all authors who have contributed to this book for their dedicated efforts and their excellent contributions. We hope that you, as a reader, will enjoy this volume.

We are also grateful to Dr. Hamid Gourabi, Dr. Abdolhossein Shahverdi, and Dr. Ahmad Vosough Dizaj for having faith in and supporting us throughout this project. We also wish to acknowledge the great support provided by many at Humana Press. A special thank you goes to our dedicated colleagues at Royan Institute for Stem Cell Biology and Technology who, with their tireless commitment toward stem cell research, have become crucial factors in encouraging us to edit this book. We are grateful to Zahra Maghari for her help with collecting the chapters and in the follow-up.

Contents

Contributors

Nasser Aghdami Cell Science Research Center, Royan Institute for Stem Cell Biology and Technology, ACECR, Tehran, Iran

Peter W. Andrews Department of Biomedical Science, Centre for Stem Cell Biology, The University of Sheffield, Sheffield, UK

Amir R. Aref Department of Biological Engineering, Massachusetts Institute of Technology, Cambridge, MA USA; Harvard Medical School, Dana-Farber Cancer Institute, Boston, MA, USA

Hossein Azizi Institute for Anatomy and Cell Biology, University of Heidelberg, Heidelberg, Germany

Hojae Bae Department of Medicine, Center for Biomedical Engineering, Harvard Medical School, Brigham and Women's Hospital, Cambridge, MA, USA; Harvard-MIT Division of Health Sciences and Technology, Massachusetts Institute of Technology, Cambridge, MA, USA

Duncan Baker Sheffield Diagnostic Genetic Services, Sheffield Children's NHS Trust, Sheffield, UK

Hossein Baharvand Department of Stem Cells and Developmental Biology, Cell Science Research Center, Royan Institute for Stem Cell Biology and Technology, ACECR, Tehran, Iran; Department of Developmental Biology, University of Science and Culture, ACECR, Tehran, Iran

Abbas Beh-Pajooh Stem Cell Biology, Cluster-of-Excellence, REBIRTH, Hannover Medical School, Hannover, Germany

Michele Boiani Max Planck Institute for Molecular Biomedicine, Münster, Germany

Kyunghee Byun Center for Genomics and Proteomics and Stem Cell Core Facility, Lee Gil Ya Cancer and Diabetes Institute, Gachon University of Medicine and Science, Incheon, Korea; Department of Anatomy and Cell Biology, Graduate School of Medicine, Gachon University of Medicine and Science, Incheon, Korea

Diana Campioni Hematology Section, University of Ferrara, Ferrara, Italy

Tobias Cantz Stem Cell Biology, Cluster-of-Excellence, REBIRTH, Hannover Medical School, Hannover, Germany

Trevor N. Collingwood Sigma-Aldrich, Inc., St. Louis, MO, USA

Sabine Conrad Insitute of Anatomy, University of Tübingen, Tübingen, Germany

Faramarz Edalat Department of Medicine, Center for Biomedical Engineering, Harvard Medical School, Brigham and Women's Hospital, Cambridge, MA, USA; Harvard-MIT Division of Health Sciences and Technology, Massachusetts Institute of Technology, Cambridge, MA, USA

Maria Francisca Eiriz Faculty of Medicine, Center for Neuroscience and Cell Biology, University of Coimbra, Coimbra, Portugal

James L. M. Ferrara Professor of Medicine and Pediatrics, Combined Bone Marrow Transplant Program, University of Michigan Cancer Center, Ann Arbor, MI, USA

Lino Ferreira CNC—Center for Neuroscience and Cell Biology, University of Coimbra, Coimbra, Portugal

Raquel Ferreira Faculty of Medicine, Center for Neuroscience and Cell Biology, University of Coimbra, Coimbra, Portugal

Mario Grãos Biocant—Center of Innovation in Biotechnology, Cantanhede, Portugal

Neil J. Harrison Department of Biomedical Science, Centre for Stem Cell Biology, The University of Sheffield, Sheffield, UK

Masaki Ieda Department of Clinical and Molecular Cardiovascular Research, Keio University School of Medicine, Shinjuku-ku, Tokyo, Japan; Department of Cardiology, Keio University School of Medicine, Shinjuku-ku, Tokyo, Japan

Anna Jauch Institute of Human Genetics, University Heidelberg, Heidelberg, Germany

Goo-Bo Jeong Department of Anatomy and Cell Biology, Graduate School of Medicine, Gachon University of Medicine and Science, Incheon, Korea

Ali Khademhosseini Wyss Institute for Biologically Inspired Engineering at Harvard University, Boston, MA, USA

Sang Bok Kim Department of Medicine, Center for Biomedical Engineering, Harvard Medical School, Brigham and Women's Hospital, Cambridge, MA, USA;

Harvard-MIT Division of Health Sciences and Technology, Massachusetts Institute of Technology, Cambridge, MA, USA

Francesco Lanza Hematology and Stem Cell Unit, Cremona, University of Ferrara, Ferrara, Italy

Bonghee Lee Center for Genomics and Proteomics and Stem Cell Core Facility, Lee Gil Ya Cancer and Diabetes Institute, Gachon University of Medicine and Science, Incheon, Korea; Department of Anatomy and Cell Biology, Graduate School of Medicine, Gachon University of Medicine and Science, Incheon, Korea

Xiaojun Lian Department of Chemical and Biological Engineering, University of Wisconsin, Madison, WI, USA

Shi-Lung Lin Division of Regenerative Medicine, WJWU & LYNN Institute for Stem Cell Research, Santa Fe Springs, CA, USA

Vincenzo Lionetti Laboratory of Medical Science, Institute of Life Sciences, Scuola Superiore Sant'Anna, Pisa, Italy; Laboratory of Molecular Biology and Stem Cell Engineering, National Institute of Biostructures and Biosystems, University of Bologna, Bologna, Italy; Department of Physiology, New York Medical College, Valhalla, NY, USA

João O. Malva Laboratory of Biochemistry and Cell Biology, Faculty of Medicine, University of Coimbra, Coimbra, Portugal

Sam Manoucheri Department of Medicine, Center for Biomedical Engineering, Harvard Medical School, Brigham and Women's Hospital, Cambridge, MA, USA; Harvard-MIT Division of Health Sciences and Technology, Massachusetts Institute of Technology, Cambridge, MA, USA

Endri Mauro Hematology and Stem Cell Unit, Cremona, University of Ferrara, Ferrara, Italy

Jae Min Cha Department of Medicine, Center for Biomedical Engineering, Harvard Medical School, Brigham and Women's Hospital, Cambridge, MA, USA; Harvard-MIT Division of Health Sciences and Technology, Massachusetts Institute of Technology, Cambridge, MA, USA

Ismael Neiva Faculty of Medicine, Center for Neuroscience and Cell Biology, University of Coimbra, Coimbra, Portugal

Sean P. Palecek Department of Chemical and Biological Engineering, University of Wisconsin, Madison, WI, USA

Annalisa Pasini Hematology and Stem Cell Unit, Cremona, University of Ferrara, Ferrara, Italy

Martin J. Pfeiffer Max Planck Institute for Molecular Biomedicine, Münster, Germany

Pavan Reddy Department of Internal Medicine, Bone Marrow Transplant Program, University of Michigan Cancer Center, Ann Arbor, MI, USA

Roberta Rizzo Microbiology Institute, University of Ferrara, Ferrara, Italy

Nadia Rosenthal European Molecular Biology Laboratory (EMBL), Mouse Biology Unit, Campus A. Buzzati-Traverso, Monterotondo, Italy; Australian Regenerative Medicine Institute (ARMI), Monash University, Melbourne, VIC, Australia

Tiago Santos Faculty of Medicine, Center for Neuroscience and Cell Biology, University of Coimbra, Coimbra, Portugal

Iman Saramipoor Behbahan Transplant Research Program, Department of Internal Medicine, University of California Davis Medical Center, Sacramento, CA, USA

Ali Seifinejad Department of Stem Cells and Developmental Biology, Cell Science Research Centre, Royan Institute for Stem Cell Biology and Technology, Academic Center for Education, Culture and Research (ACECR), Tehran, Iran

Malte Sgodda Stem Cell Biology, Cluster-of-Excellence, REBIRTH, Hannover Medical School, Hannover, Germany

Thomas Skutella Institute for Anatomy and Cell Biology, University of Heidelberg, Heidelberg, Germany

Martin Stehling Max Planck Institute for Molecular Biomedicine, Münster, Germany

Philippe Taupin School of Biotechnology, Dublin City University, Glasnevin, Dublin, Ireland

Jorge Valero Faculty of Medicine, Center for Neuroscience and Cell Biology, University of Coimbra, Coimbra, Portugal

Helena Vazão CNC—Center for Neuroscience and Cell Biology, University of Coimbra, Coimbra, PortugalBiocant—Center of Innovation in Biotechnology, Cantanhede, Portugal

Carlo Ventura Cardiovascular Department, S. Orsola—Malpighi Hospital, University of Bologna, Bologna, Italy

Manlio Vinciguerra Istituto EuroMEditerraneo di Scienza e Tecnologia (IEMEST), Palermo, Italy

Irma Virant-Klun Department of Obstetrics and Gynaecology, University Medical Centre Ljubljana, Ljubljana, Slovenia

Mark A. Zern Transplant Research Program, Department of Internal Medicine, University of California Davis Medical Center, Sacramento, CA, USA

Chapter 1
ES Cell Lines from Tetraploid Mouse Blastocysts

Martin J. Pfeiffer, Martin Stehling, Anna Jauch and Michele Boiani

Abstract Testing of mouse stem cells for pluripotency is mainly conducted with either of three different assays: embryoid body (EB) formation, teratoma formation, and tetraploid (4 N) embryo complementation. The complementation assay relies on the assumption that the 4 N embryo lacks in own pluripotency. Here we devised a new method for generating 4 N mouse embryos and, from these embryos, we derived functional 4 N embryonic stem (ES) cells. Our method uses the nuclear transplantation (NT) of two somatic cell nuclei (diploid, 2 N) into oocytes deprived of their metaphase II spindle (ooplasts), or the intracytoplasmic sperm injection (ICSI) of two sperm heads into intact oocytes. It follows that these embryos are 4 N from the beginning of embryonic life, in contrast to conventional methods that use the fusion of two 2 N embryo blastomeres. The derivation of 4 N ES cells proves that the inner cell mass (ICM) of a mouse 4 N blastocyst is pluripotent. Here we apply this new method to determine whether one mouse ooplasm can reprogram two somatic nuclei, and to separate the effects of chromosome duplication and centriole duplication from each other in the process of tetraploidization that can lead to tumor formation.

1.1 Introduction

Tetraploidy is the presence of twice the normal number of chromosomes in a cell. In mammals there are no confirmed tetraploid (4 N) species, and experimentally-induced 4 N embryos are incapable of significant if any development

M. J. Pfeiffer · M. Stehling · M. Boiani (✉)
Max Planck Institute for Molecular Biomedicine, Münster, Germany
e-mail: mboiani@mpi-muenster.mpg.de

A. Jauch
Institute of Human Genetics, University Heidelberg, Heidelberg, Germany

H. Baharvand and N. Aghdami (eds.), *Advances in Stem Cell Research*,
Stem Cell Biology and Regenerative Medicine, DOI: 10.1007/978-1-61779-940-2_1,

after implantation [1]. Yet these 4 N embryos have proved very useful in biological research: to assess pluripotency of stem cells, and to determine if gene mutations affect the differentiation ability of these cells. Most prominent members of the pluripotent stem cell family are the embryonic stem (ES) cells, which have been derived from the inner cell mass (ICM) of the diploid (2 N) blastocyst-stage embryo in the mouse as well as in other mammalian species. Of the precursor ICM, ES cells retain the pluripotency, that is, the ability to differentiate into cell lineages of the three embryonic germ layers, as well as germ cells. Pluripotency is tested by having ES cells differentiate in the permissive environment of an embryo, of an embryoid body, or even of an adult body.

The 4 N embryo provides an environment to assess 2 N ES cells for pluripotency, as follows: by aggregating ES cells with 4 N morulae, or by microinjecting ES cells inside the cavity of 4 N blastocysts. The ES cells replace the 4 N ICM but do not mingle with the 4 N trophectoderm (TE). As a result, a chimeric embryo arises in which the exogenous ES cells form the fetus but not the placenta, unless they suffer a mutation that affects the ICM's ability to produce normal embryonic lineages. This assay is referred to as 'tetraploid embryo complementation'. In fact, a sensitive transgenic method allowed to detect the cell progeny of 4 N blastomeres in the early postimplantation embryo up to the stage of gastrulation [2]. Thus, it is not clear if the ICM of 4 N blastocysts abstains from making an active contribution to the complementation assay, and if experimental tetraploidy has no other experimental application than supporting 2 N ES cells.

To date, three protocols are available to generate 4 N embryos. The first protocol is to microsurgically add a 2 N nucleus to a zygote [3, 4]. The second protocol used to produce 4 N embryos is the use of cytochalasin B to skip one cell division during 2 N embryo cleavage [5, 6]. The third protocol is to fuse the blastomeres of a 2 N 2-cell embryo with each other [7–9]. In the past the first method lagged far behind the other two in terms of applicability, as it required a micromanipulator, which in the meantime has become more and more common in the modern cell culture facilities.

In this chapter we introduce a new protocol to generate 4 N mouse embryos for purposes of research. Our protocol is based on the nuclear transplantation (NT) of two somatic cell nuclei into ooplasts, or the intracytoplasmic sperm injection (ICSI) of two sperm heads into intact metaphase II mouse oocytes, resulting in 4 N NT embryos and 4 N ICSI embryos, respectively. We show that these 4 N embryos have a functional ICM as measured by the derivation of 4 N NT ES cells. We characterized these cells for karyotypic stability and extent of pluripotency. Moreover, we provide applications of 4 N NT ES cells to solve important biological questions, for example: whether one ooplasm can reprogram two somatic nuclei, and how the effects of chromosome duplication and centriole duplication can be resolved from each other when cells become tetraploid.

Miyamoto and colleagues suggested that the mouse oocyte hosts reprogramming capacity sufficient for multiple somatic nuclei [10]. However, to date, nobody has challenged the reprogramming capacity of the mouse ooplasm by transplanting more than one somatic nucleus simultaneously in the same ooplasm.

The result of transplanting two 2 N nuclei in the same ooplasm is a 4 N embryo. As a measure of reprogramming, we examined whether this embryo can develop into a blastocysts with functional ICM.

In cancer formation, chromosome instability and aneuploidy is often preceded by transient tetraploidy [11]. However, when somatic cells become tetraploid, they also typically undergo centriole duplication. It is therefore of importance to cancer research to efficiently and effectively separate the effects of these two alterations—the duplication of chromosome number and that of centriole number—from each other. The mouse oocyte hosts a set of unique biological processes, such as the ability to haploidize its own DNA during meiosis, to remodel the sperm nucleus into a pronucleus and to destroy sperm's mitochondria after fertilization [12]. Yet another feature of the mouse oocyte is its ability to integrate a somatic nucleus while destroying its companion centriole(s) after somatic cell nuclear transfer (SCNT) [13]. These cell constructs are amenable to embryo production and ES cells derivation.

Therefore, by the transfer of two somatic nuclei into the mouse ooplasm combined with the derivation of ES cells, we tested the hypothesis that the mouse oocyte has reprogramming capacity in excess of one somatic nucleus, and that pure tetraploidy (i.e. twice the number of chromosomes without twice the number of centrioles) is an inherently unstable condition that exacerbates the tumorigenic properties of mouse ES cells with respect to teratoma formation.

1.2 Results

Ooplasts were transplanted with two cumulus cell nuclei (double SCNT, OG2 transgenic cells) and subjected to activation in the presence of cytochalasin B. These 4 N constructs went on cleaving and gave rise to blastocysts, which supported the derivation of cell lines under leukemia inhibitory factor (Lif) culture conditions on a layer of fibroblast feeder cells. Rates of cell line derivation were lower compared to ooplasts transplanted with one nucleus (single SCNT), with zygotes whose first cleavage had been prevented by cytochalasin B (4 N after fertilization) and with oocytes subjected to ICSI of two sperm heads and prevention of polar body extrusion (Table 1.1). We focused on the six cell lines that were derived after double SCNT. These cell lines are positive for alkaline phosphatase activity (Fig. 1.1) and express Oct4 promoter-driven GFP, both prominent markers for pluripotency. Therefore these cells qualify as 'ES cells', and given their origin, we call them '4 N NT ES cells'.

We asked if ES cell lines that have double the number of chromosomes suffered an increased rate of chromosome malsegregation and became more unstable, compared to 2 N counterparts. The 4 N NT ES cell lines presented a modal chromosome count of 80 at passage 20 (Fig. 1.3), determined by analyzing 103 chromosome spreads after Hoechst staining (Fig. 1.2). Using M-FISH probes we measured the chromosomal makeup of additional 117 individual 4 N NT ES cells,

Table 1.1 Derivation rates of 4N and 2N ES cells (NT, somatic cell nuclear transfer; F, fertilization)

ES cell	Outgrowths picked	ES cell lines derived	Rate	χ^2
Single NT[a]	47	16	0.34	1
Double NT	59	6	0.10	0.003
4N F (blastomere fusion)	26	8	0.31	0.776
Double ICSI (4N)	12	4	0.33	0.963

[a] used as control group to calculate χ^2

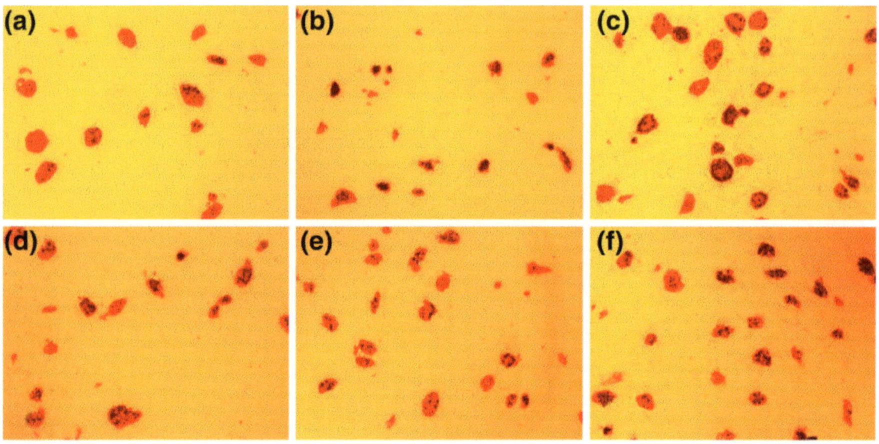

Fig. 1.1 AP-stained 4 N NT ES cell colonies. Lines 1–6 (**a**–**f**). *Red-colored* cells indicate the presence of alkaline phosphatase, a marker characteristic of pluripotent *Red-colored* stem cells. Original magnification 4X

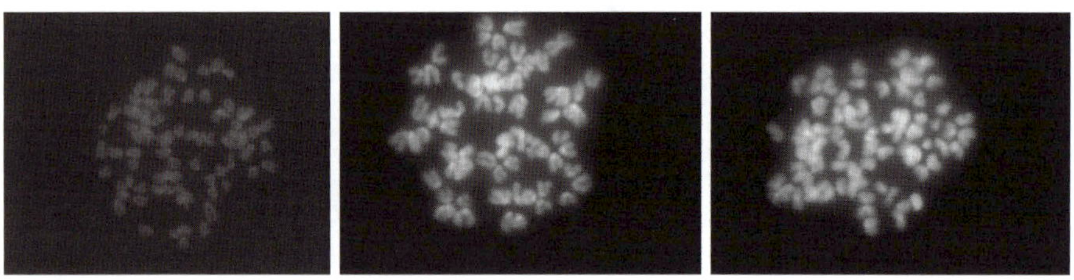

Fig. 1.2 Chromosome spreads of the 4 N NT ES cell line #4. The number of counted chromosomes were 77 (**a**), 80 (**b**), and 82 (**c**) Original magnification 100x

in comparison to that of 2 N ES cells derived from cloned and zygotic 2 N embryos (Fig. 1.3). In particular, it became possible to detect chromosome rearrangements (e.g. translocations, duplications, deletions) and chromosome

(a)

Number of chromosomes	Counted X times
Less than 75	31
75	3
76	6
77	10
78	7
79	12
80	22
81	8
82	2
83	0
84	2
85	0
More than 85	0

(b)

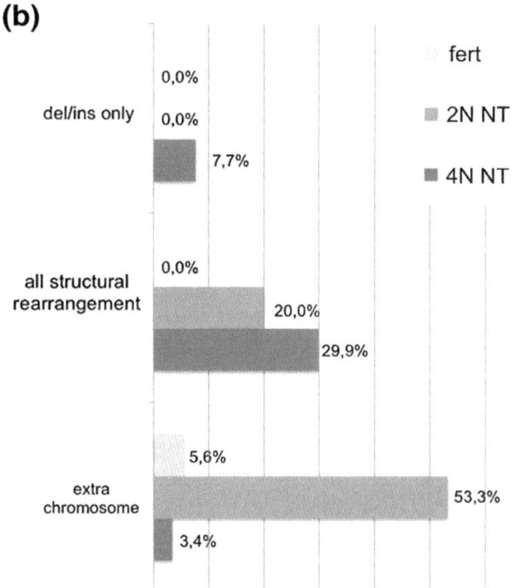

Fig. 1.3 Chromosome counts and M-FISH analysis of chromosome spreads (117 spreads of 4 N NT ES cells; 20 spreads of 2 N NT ES cells; 36 spreads of fert ES cells)

nondisjunction (leading to trisomy in 2 N cells, or to five copies of a chromosome in 4 N cells). The average nondisjunction rate per 4 N NT ES cell was 3.42% (4/117), that is, lower than that of 2 N NT ES cells analyzed in our previous study (53.33%; 8/15; $\chi^2 = 2.43\text{E}{-10}$) [13], and comparable to that of 2 N ES cells from naturally fertilized embryos (5.56%; 2/36; $\chi^2 = 0.56$). Chromosome rearrangements were detected exclusively in NT ES cells (Fig. 1.3).

ES cell pluripotency is typically tested by embryoid body (EB) formation in vitro, by teratoma formation in vivo, and by chimera formation after blastocyst injection and embryo transfer. Here, we tested the pluripotency of 4 N NT ES cells by the EB and the teratoma assay.

Three to five days after aggregation in hanging drops of medium without Lif, the 4 N NT ES cells formed spheroids. Differentiation was monitored live, which is made possible by the presence of the Oct4-GFP transgene. Compared to 2 N counterparts, GFP expression was sustained over a longer period of time in 4 N NT ES cells. Typically, 2 N EBs lose GFP expression after 3–4 days, whereas 4 N EBs retained GFP expression even longer than a week (Fig. 1.4).

ES cells have the ability to form solid tumors known as 'teratomas', originally described as testicular tumors in strain 129 mice [14]. This property allows to test if 4 N NT ES cells give rise to more aggressive tumors compared to diploid counterparts. Although none of the severe combined immunodeficient (SCID) mice injected with at least 1×10^6 4 N NT ES cells at early passage (pp. 6–7)

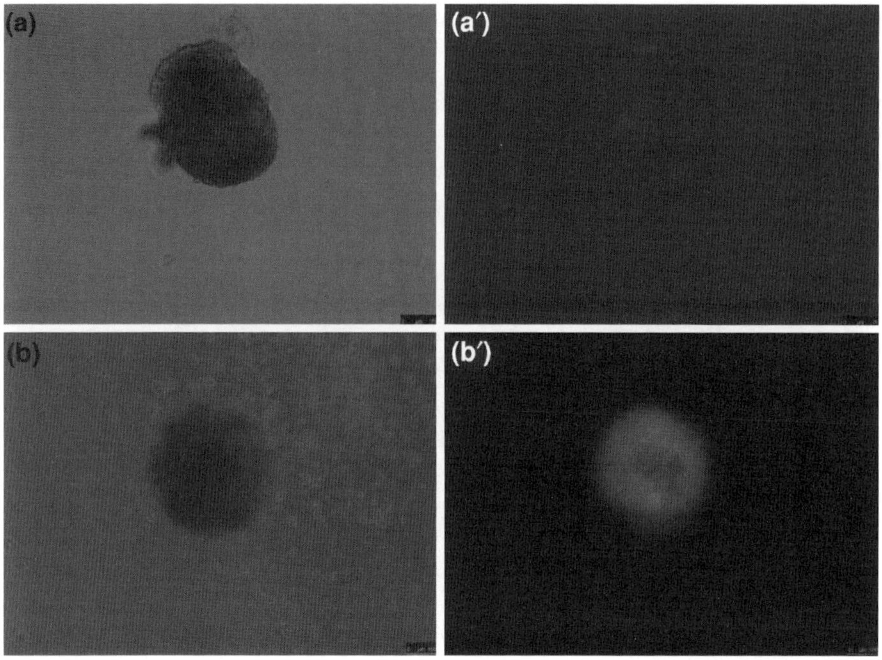

Fig. 1.4 Embryoid bodies of 2 N (a, a') and 4 N (b, b') NT ES cells (**a**, **b**, bright field; **a'**, **b'**, Oct4-GFP fluorescence)

developed teratomas after one month (unlike the case for 2 N counterparts), teratomas were detected after two months when cells at late passage (pp. 20–22) were used (Fig. 1.5). These data are summarized in Table 1.2. These cells were the siblings of those that were previously analyzed for their karyotype and found to be stable. Similar results were obtained with 4 N ES cells derived from fertilized embryos after inhibition of the first cleavage (no teratomas after one month, teratomas after two months). The 4 N teratomas grew slower and attained smaller size than the 2 N counterparts. Animal welfare considerations and animal protection laws prohibited us from letting the teratomas grow bigger than 10 mm (US) or 15 mm (Germany) in diameter over a period of time longer than two months. The important question of whether the 4 N NT ES cells remained tetraploid during and after teratoma formation was answered indirectly. We subjected the 4 N NT ES cells to in vitro differentiation into cardiomyocytes for ten days and then let the cardiomyocytes beat for another three weeks prior to performing a FACS analysis for ploidy. The cells presented the FACS profile typical of tetraploid cells (Fig. 1.6). Although we have not shown that the teratoma cells are 4 N in vivo, we have shown that 4 N NT ES cells retain the tetraploid karyotype when they differentiate over one month in vitro.

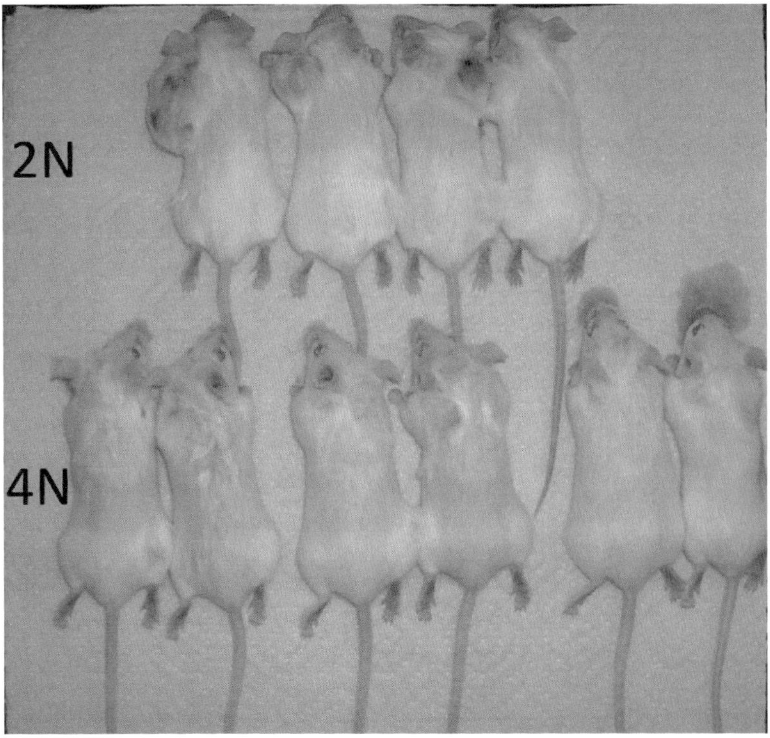

Fig. 1.5 SCID mice four weeks after subcutaneous injection of early-passage ES cells. Teratoma formation was only visible in mice injected with 2 N NT ES cells (*top*), but not after injection of 4 N NT ES cells (*bottom*)

Table 1.2 Teratoma formation after subcutaneous injection of ES cells in SCID mice (NT, somatic cell nuclear transfer; F, fertilization)

ES cell	Passage #	Days after injection	Teratoma/ mice injected	Days after injection	Teratoma/ mice injected	Total teratomas	χ^2
2N NT[a]	6–7	40	2/4	100	2/3	4/7	1
4N NT	6–7	34	0/7	100	0/3	0/10	0.006
4N NT	20–22	40	2/3	100	1/4	3/7	0.593
4N F	5	27	0/3	100	2/6	2/9	0.152

[a] used as control group to calculate χ^2

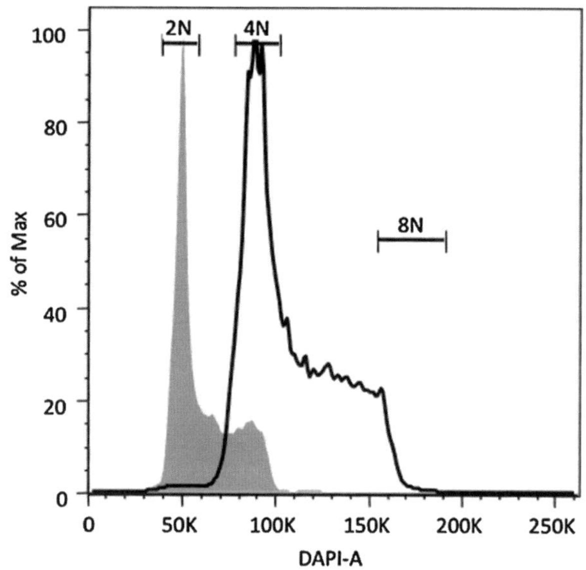

Fig. 1.6 FACS analysis of cardiomyocytes derived from 4 N (*black line*) and 2 N (*gray solid*) NT ES cells

1.3 Discussion

ES cells are pluripotent cell lines derived from the ICM of the blastocyst. The vast majority of mouse ES cell lines have been derived from 2 N blastocysts. In contrast, 4 N blastocysts are considered to lack in pluripotency, since on their own they are incapable of postimplantation development unless 2 N ES cells are added [15, 16]. Indeed, expression of Oct4, a pluripotency marker of the ICM, is lost by the 4 N expanded blastocysts [17]. Although 4 N embryos are incapable of full development, the presence of 4 N cells is normal within certain tissues of the adult body, such as neurons, muscle, and liver [18, 19]. 4 N ES cells have been obtained from existing cell lines after cell fusion [20] or after block of cytokinesis [21]. In a different approach, Yang and colleagues derived 4 N ES cells from metaphase II mouse oocytes transplanted with one somatic nucleus [22]. It should be noted that these 4 N cells contain one somatic and one gametic genome. Taken together, these finding indicates that although tetraploidy induction may be detrimental, tetraploid cells can have a normal life once the initial hurdles post-induction have been overcome. Applied to pluripotency, past studies showed that the maintenance of pluripotency is compatible with a 4 N state of the genome (as seen after cell fusion), but the induction of pluripotency may only be possible with support from an already pluripotent genome, such as that of an ES cell in case of fusion, or that of the oocyte in case of SCNT into intact (not enucleated) oocytes. Here we developed a unique approach to the derivation of 4 N ES cells from mouse embryos by simultaneously injecting two 2 N nuclei into a single, enucleated oocyte. With this tool in hand we were able to address two questions that pertain to the induction of pluripotency via oocytes, and to the chromosomal theory of tumorigenesis.

Miyamoto and colleagues suggested that the mouse oocyte hosts reprogramming capacity sufficient for multiple somatic nuclei, possibly up to 100 of them [10]. The ability to incorporate the somatic nuclei is supported by 10–15 microtubule organizing centers (MTOCs), which are present in the mouse ooplasm and build a microtubular spindle around the chromosomes in the absence of centrioles [23, 24]. This feature may allow multiple somatic nuclei to be transplanted in the same ooplasm without the formation of multipolar spindles and chaotic mitoses as a consequence. We observed that when two somatic nuclei were transplanted to the same ooplasm, they were both reprogrammed to pluripotency as measured by the derivation of 4 N ES cells. This achievement underscores the abundant reprogramming capacity of mammalian oocytes, which was suggested to be in excess of one nucleus [10], and the ability of a 4 N embryonic genome to support derivation of ES cells.

Yang and colleagues derived 4 N ES cells following NT into intact mouse oocytes [22]. Our 4 N ES cells differ from those of Wang and colleagues, in that the latter resulted from reprogramming of a single somatic nucleus transplanted into a metaphase II mouse oocyte, whereas our own reprogramming strategy relies solely on the reprogramming factors of the ooplasm, without the support of the maternal genome. Accordingly, our 4 N ES cells also have a balanced imprinting status (2/4 maternal, 2/4 paternal), those of Yang and colleagues have an unbalanced imprinting status (3/4 maternal, 1/4 paternal), at least at the initial passages of ES cell culture.

Thanks to the oocyte's ability to destroy somatic centrioles, we envisioned the use of double SCNT into ooplasts as an avenue to distinguish between chromosome duplication and centriole duplication as destabilizing factors that facilitate tumorigenesis. In a cell undergoing tetraploidization, the duplication of DNA content typically coexists with the duplication of centrioles, making it difficult to understand if the tetraploid genome per se, or rather the multipolar spindles, make cells prone to transformation.

We showed that 4 N NT ES cells not only can be derived, but they are chromosomally stable as measured up to passage 20. Interestingly, NT ES cells cloned from 1 and 2 nuclei presented structural rearrangements that were not detected in zygotic ES cells. The average rate of insertions and deletions (pooled) per ES cell was 7.7, 0 and 0% for 4 N NT, 2 N NT and for zygotic ES cells, respectively. Taking also *bona fide* chromosomal translocations into account, the average total rate of structural rearrangement rises to 29.9 and 20.0% for 4 N NT and 2 N NT ES cells, respectively, while it remains 0% for zygotic ES cells. These figures indicate that the cloning procedure leaves a genetic mark on derivative ES cells, in contrast to the prevalent view that cloned and zygotic ES cells are indistinguishable from each other [25, 26]. It is tempting to ascribe the chromosomal rearrangements to recombination events, since metaphase II oocytes have residual recombinogenic activity [27, 28]. In zygotes, the maternal and paternal genomes are physically separated from each other as DNA replication takes place during the first cell cycle, whereas in a cloned embryo, the two somatic genomes as well as the maternal and paternal chromosomes are mixed in the pronuclei that form after SCNT.

We derived six lines of 4 N NT ES cells, all of which were alkaline phosphatase-positive, suggestive of pluripotency. To test for actual pluripotency, we subjected the ES cells to EB and teratoma formation. Differences became apparent between 4 N NT ES cells and the conventional (2 N) ES cells. When ES cells are forced to aggregate in hanging drops of medium lacking Lif, they form spheroids whose cells undergo differentiation. The course of differentiation can be conveniently followed by Oct4 promoter-driven GFP expression without having to consume the EBs. Using this transgenic strategy, we observed that the 4 N EBs expressed GFP much longer compared to 2 N counterparts. This observation is in line with that of Luo and colleagues, who generated 4 N ES cells by colcemid treatment, and observed that these cells retained their pluripotency after LIF has been withdrawn [29]. Furthermore, we observed differences of teratoma formation after subcutaneous injection of 4 N NT ES cells in severe combined immunodeficient (SCID) mice. Teratomas are tumors composed of a variety of endodermal, mesodermal, and ectodermal tissues, arranged in a chaotic manner [30]. Hence, ES cells must differentiate in order to form teratomas. 4 N NT ES cells gave rise to teratomas, but it took longer and the penetrance was lower compared to 2 N NT ES cells.

At this stage, we cannot explain how the 4 N NT ES cells express the properties described above. It may be noted that 4 N NT ES cells are not simply endowed with double the amount of DNA, but have four alleles per gene instead of two. This feature is generally expected to lower the gene expression noise in many gene networks [31] and to make it more difficult to switch off pluripotency genes during cell differentiation. One of these genes is *Oct4* [32], located on mouse chromosome 17. It has been shown that Oct4 is a dose-dependent oncogenic fate determinant in mouse ES cells [33]. High Oct4 levels increase the teratogenic properties of ES cells, while Oct4 inactivation induces regression of the teratomas. Thus, the fact that 4 N NT ES cells are not more prone to form teratomas than the 2 N counterparts is puzzling.

The refractoriness of 4 N NT ES cell to form teratomas is interesting in the context of ES cell potential and its relationship to the moral status of ES cells. In principle, ES cells can reproduce the whole organism, whether via tetraploid embryo complementation [15, 16] or, in the future, maybe even by the formation of functional germ cells [34]. This is possible because ES cells are capable of 'epigenesis'—the ability to produce structured complexity. Hans-Werner Denker elaborated on this feature in great depth, remarking that ES cells "tend to autonomously create complexity by initiating pattern formation processes (self-organization in embryological terms) whenever they start cell differentiation, even in vitro or after transplantation to ectopic sites (teratoma formation)" [35]. Pending confirmation that our 4 N NT ES cells are refractory to form teratomas, these ES cells are envisaged to provide a solution—if only partial—to the ethical dilemma of ES cells, as follows: reproductive cloning would never be possible using 4 N NT ES cells, as tetraploidy is well known to be not permissive of full development in mammals. Reports of human gestations resulting in 4 N liveborn infants are extremely rare and do not indicate if all tissues, or just some (mosaicism), were 4 N [36].

Taken together, our data indicate that a single ooplasm reprograms two somatic cell nuclei to pluripotency. The 4 N ICM is vital and developmentally competent, since it gives rise to ES cells. These data challenge the notion that diploid ES cell transplanted to a 4 N blastocyst would form the fetus entirely on their own [15, 16]. We further learn from our data that a doubled number of chromosomes is not necessarily leading to 'catastrophic mitoses' [37, 38]. Since ES cells express a very high proliferation rate, the instances of mitotic errors would even be more frequent and thereby easier to observe. Twice the normal number of chromosomes in an ES cell does not per se increase the risk nor exacerbate the severity of tumor formation, as measured in the teratoma model.

In prospect, placing two nuclei in the same cytoplasmic environment allows to compare them on the same level - something that has not been possible so far in SCNT. This opportunity will help to clarify if the two genomes are processed equally by the reprogramming machinery, ultimately addressing the question of whether oocyte-mediated reprograming is a stochastic process or not. Pursuing this opportunity will undoubtedly rely on the method of double nuclear transfer, but also will require the development of new tools, for example to distinguish the two nuclei in such a way that it is possible to ascribe their origin to one genome or the other.

1.4 Materials and Methods

1.4.1 Mouse Handling and Gamete Recovery

Mice were maintained and used for experiments according to the ethical permit issued by the Landesamt für Natur, Umwelt und Verbraucherschutz (LANUV) of the State of North Rhine-Westphalia, Germany. Oocytes from B6C3F1 mice aged 6–8 weeks were used for somatic cell nuclear transfer (SCNT) and for intracytoplasmic sperm injection (ICSI). Cumulus cells from equally aged B6C3F1 or OG2 mice were used as nucleus donors for SCNT, the latter expressing an Oct4 promoter-driven GFP transgene (OG2; CBA-Tg(Pou5f1-EGFP)2Mnn/J, JAX stock number 004654; [39]). These mice were injected with 10 IU each of PMSG and hCG 48 h apart at 5 pm, and sacrificed by cervical dislocation 14 h after hCG injection to recover cumulus-oocyte complexes from the oviducts. Cumulus-oocyte complexes were handled in Hepes-buffered CZB (HCZB) medium, as previously described [13]. Oocytes were separated from cumulus cells by treatment with 50 IU/mL hyaluronidase in HCZB, aided by gentle pipetting with a mouth-operated micropipette. Oocytes were placed in embryo culture medium, consisting of α-modified Eagle's medium (α-MEM; M4526 Sigma) supplemented with 0.2% w/v BSA (Pentex; Serological Proteins Inc., Kankakee, IL, USA) and 50 μg/mL gentamicin sulfate, at 37°C and under 5% CO_2, until use. Cumulus cells were left in hyaluronidase and kept at 4°C until SCNT. Sperm was recovered from mature B6C3F1 males by swim-up in Whittingham medium, and incubated 2–3 h prior to ICSI. Severe combined immunodeficient (SCID) mice of either gender were bred in house and aged 6–7 weeks for use in the subcutaneous transplantation of ES cells.

1.4.2 Double Nuclear Transfer, ICSI, and Embryo Culture

Micromanipulations and embryo culture were performed at 30°C (room temperature), as previously described [40]. For SCNT, oocytes were first ‘enucleated’ (spindle-chromosome complex removed) before injection of one or two cumulus cell nuclei in the ooplasm, using a piezo-driven micropipette under Nomarski optics. The nucleus-transplanted oocytes were activated for 6 h in calcium-free α-MEM supplemented with 10 mM $SrCl_2$ and 5 μg/mL cytochalasin B. ICSI of one or two sperm heads was performed to generate 2 N and 4 N fertilized embryos, respectively; for the latter, extrusion of the second polar body was prevented by cytochalasin B. Embryos were placed in α-MEM and cultured at 37°C under 5% CO_2. On day 4 after activation, SCNT and ICSI blastocysts were processed for ES cell derivation.

1.4.3 Derivation and Culture of ES Cells

Derivation of ES cell lines from SCNT blastocysts was performed as described previously [41]. After the zonae pellucidae had been removed in Tyrode’s acidic solution (Sigma), blastocysts were transferred onto a feeder layer of γ-ray-inactivated mouse embryonic fibroblasts—MEFs (C3H background; 30 Gray irradiation dose) previously grown to confluency in 4-well dishes. At day 4 after plating, blastocysts had attached to the feeders and formed trophoblastic outgrowths. The cell clumps located at the center of the outgrowths were removed with a polished glass pipette, dissociated by trypsinization and transferred onto fresh feeder cells in a 96-well plate. Passage zero ES cells (after initial plating of the dissociated ICMs) were grown for six days in culture and further expanded by subsequent passaging (2–3 days) onto larger well sizes. ES cell culture medium consisted of knockout DMEM with 15% fetal bovine serum, glutamine and penicillin/streptomycin, nonessential amino acids, mercaptoethanol, 2000 Units/mL leukemia inhibitory factor (Lif), and 50 μM MEK1 inhibitor (PD98059). From passage 3 of ES cell culture the MEK1 inhibitor was omitted.

1.4.4 Alkaline Phosphatase (AP) Staining

The AP staining was performed following the protocol modified from Bernstine and colleagues [42]. Briefly, the culture plate with ES cell colonies on MEFs was thoroughly washed with 1× PBS and fixed with 4% paraformaldehyde in PBS 1X for 2 min. Fixative was removed and the plate rinsed with 1X rinse buffer (20 mM Tris–HCl, pH 7.4, 0.15 M NaCl, 0.05% Tween-20). The rinse buffer was removed and replaced with stain solution (0.4 mg/mL naphthol phosphate and 1 mg/mL Fast Red TR salt in Tris-maleate buffer 25 mM, pH 9), followed by incubation in the dark at room temperature for 15–30 min. AP positive colonies of cells stained red.

1.4.5 Karyotype Analysis

Karyotype analysis was performed with the air-drying preparation method for metaphase chromosome spreads [43] to determine the exact number of chromosomes of the ES cells. Briefly, the 4 N NT ES cells as well as a 2 N control cell lines were cultured in 6 cm dishes for three days. At this point, nocodazole was added to the culture medium (0.3 μg/ml) and the cells were incubated for 4–5 h to arrest in metaphase. The cells were washed, trypsinized, spun down and the pellet was resuspended in a 0.56 % hypotonic KCl solution. After 20 min incubation at room temperature, the cells were spun down again and the supernatant was gently removed. Pellets were fixed with ice-cold methanol:acetic acid (3:1) and kept at 4°C until further processing for chromosome spreads. The spreads were made by dropping the cell suspensions (30 μl drop) on glass slides and letting them dry out. The spreads were stained with Hoechst 33342 dye (1 μg/ml in PBS) and mounted in Vectashield. Pictures of the spreads were taken on a Nikon TE2000 fluorescence microscope.

M-FISH analysis was performed for greater chromosome detail. Mouse chromosome-specific painting probes were combinatorially labeled using 7 different fluorochromes and hybridized as described [44]. Metaphase spreads were examined using a Leica DM RXA epifluorescence microscope (Leica, Bensheim, Germany) equipped with a Sensys CCD camera (Photometrics, Tucson, AZ) and controlled by the Leica Q-FISH software (Leica Microsystems Imaging Solutions, Cambridge, UK). Image processing and karyotyping were performed using the Leica MCK software.

1.4.6 Ploidy Analysis by FACS

The ploidy status of ES cells was determined by flow cytometry. Cells were prepared and stained with DAPI using the CyStain DNA 2 step kit (Partec GmbH,

Münster, Germany) and analyzed on a FACSAria cell sorter equipped with a 375 nm laser (BD Biosciences). Data was analyzed using FlowJo software (Tree Star Inc.).

1.4.7 Embryoid Body (EB) Assay

Embryoid bodies were produced using the hanging drop method after feeder cell removal by sedimentation, using 500 ES cells per 20 μl medium without Lif. After three days in culture, EBs were plated on gelatinized plates for endoderm and mesoderm differentiation, and on matrigel-coated plates for ectoderm differentiation.

1.4.8 Teratoma Assay

The teratoma assay was performed to assess the ability of ES cells for tumor formation. Most ES cell lines form tumors when injected into immunodeficient mice [30]; these tumors are reminescent of the teratomas forming in the testes of strain 129 mice [14]. ES cells were cultured for three days under Lif conditions, washed, trypsinized, and collected. The cell suspension was transferred to a new gelatinized dish and the feeders were allowed to attach (sedimented) for one hour. After this time, most feeders have attached and the supernatant contains 90–95% ES cells. The ES cells were spun down and resuspended in a small volume of ES cell medium for subcutaneous injection into a mouse (10^6 cells in a volume of 200 μl). This cell suspension was injected in the dorsal neck region of SCID mice. The mice were returned to their cages and kept under normal conditions. After four weeks from the injection, mice were visually inspected for teratoma formation in the neck region.

1.4.9 Statistical Analysis

Statistical comparisons were performed using the χ^2 test.

References

1. Eakin GS, Behringer RR (2003) Tetraploid development in the mouse. Dev Dyn 228: 751–766
2. Eakin GS, Hadjantonakis AK, Papaioannou VE, Behringer RR (2005) Developmental potential and behavior of tetraploid cells in the mouse embryo. Dev Biol 288:150–159

3. Modlinski JA (1978) Transfer of embryonic nuclei to fertilised mouse eggs and development of tetraploid blastocysts. Nature 273:466–467
4. Modlinski JA (1981) The fate of inner cell mass and trophectoderm nuclei transplanted to fertilized mouse eggs. Nature 292:342–343
5. Snow MH (1975) Embryonic development of tetraploid mice during the second half of gestation. J Embryol Exp Morphol 34:707–721
6. Tarkowski AK, Witkowska A, Opas J (1977) Development of cytochalasin in B-induced tetraploid and diploid/tetraploid mosaic mouse embryos. J Embryol Exp Morphol 41:47–64
7. Graham C (1971) Virus assisted fusion of embryonic cells. Acta Endocrinol 68:S154–S167
8. Berg H (1982) Biological implications of electric field effects: Part V. Fusion of blastomeres and blastocysts of mouse embryos. Bioelectrochem Bioenerg 9:223–228
9. Kurischko A, Berg H (1986) Electrofusion of rat and mouse blastomeres. Bioelectrochem Bioenerg 15:513–519
10. Miyamoto K, Tsukiyama T, Yang Y, Li N, Minami N, Yamada M, Imai H (2009) Cell-free extracts from mammalian oocytes partially induce nuclear reprogramming in somatic cells. Biol Reprod 80:935–943
11. Margolis RL (2005) Tetraploidy and tumor development. Cancer Cell 8:353–354
12. Sutovsky P, Moreno R, Ramalho-Santos J, Dominko T, Simerly C, Schatten G (2000) Ubiquitinated sperm mitochondria, selective proteolysis and the regulation of mitochondrial inheritance in mammalian embryos. Biol Reprod 63:582–590
13. Balbach S, Jauch A, Böhm-Steuer B, Cavaleri F, Han YM, Boiani M (2007) Chromosome stability differs in cloned mouse embryos and derivative ES cells. Dev Biol 308:309–321
14. Stevens L, Little C (1954) Spontaneous testicular teratomas in an inbred strain of mice. Proc Natl Acad Sci U S A 40:1080–1087
15. Nagy A, Gócza E, Diaz EM, Prideaux VR, Iványi E, Markkula M, Rossant J (1990) Embryonic stem cells alone are able to support fetal development in the mouse. Development 110:815–821
16. Nagy A, Rossant J, Nagy R, Abramow-Newerly W, Roder JC (1993) Derivation of completely cell culture-derived mice from early-passage embryonic stem cells. Proc Natl Acad Sci U S A 90:8424–8428
17. Park MR, Lee AR, Bui HT, Park C, Park KK, Cho SG, Song H, Kim JH, Nguyen VT, Kim JH (2011) Chromosome remodeling and differentiation of tetraploid embryos during preimplantation development. Dev Dyn 240:1660–1669
18. Morillo SM, Escoll P, de la Hera A, Frade JM (2010) Somatic tetraploidy in specific chick retinal ganglion cells induced by nerve growth factor. Proc Natl Acad Sci U S A 107:109–114
19. Anatskaya OV, Vinogradov AE (2007) Genome multiplication as adaptation to tissue survival: evidence from gene expression in mammalian heart and liver. Genomics 89:70–80
20. Pralong D, Trounson AO, Verma PJ (2006) Cell fusion for reprogramming pluripotency: toward elimination of the pluripotent genome. Stem Cell Rev 2:331–340
21. Fujikawa-Yamamoto K, Miyagoshi M, Yamagishi H (2007) Establishment of a tetraploid cell line from mouse H-1 (ES) cells highly polyploidized with demecolcine. Cell Prolif 40: 327–337
22. Yang H, Shi L, Zhang S, Ling J, Jiang J, Li J (2010) High-efficiency somatic reprogramming induced by intact MII oocytes. Cell Res 20:1034–1042
23. Calarco-Gillam PD, Siebert MC, Hubble R, Mitchison T, Kirschner M (1983) Centrosome development in early mouse embryos as defined by an autoantibody against pericentriolar material. J Cell Biol 35:621–629
24. Combelles CM, Albertini DF (2001) Microtubule patterning during meiotic maturation in mouse oocytes is determined by cell cycle- specific sorting and redistribution of gamma-tubulin. Dev Biol 239:281–294
25. Brambrink T, Hochedlinger K, Bell G, Jaenisch R (2006) ES cells derived from cloned and fertilized blastocysts are transcriptionally and functionally indistinguishable. Proc Natl Acad Sci U S A 103:933–938

26. Wakayama S, Jakt ML, Suzuki M, Araki R, Hikichi T, Kishigami S, Ohta H, Van Thuan N, Mizutani E, Sakaide Y, Senda S, Tanaka S, Okada M, Miyake M, Abe M, Nishikawa S, Shiota K, Wakayama T (2006) Equivalency of nuclear transfer-derived embryonic stem cells to those derived from fertilized mouse blastocysts. Stem Cells 24:2023–2033
27. Perry AC, Wakayama T, Kishikawa H, Kasai T, Okabe M, Toyoda Y, Yanagimachi R (1999) Mammalian transgenesis by intracytoplasmic sperm injection. Science 284:1180–1183
28. Perry AC (2000) Hijacking oocyte DNA repair machinery in transgenesis? Mol Reprod Dev 56:319–324
29. Luo X, Fujikawa-Yamamoto K, Miyagoshi M, Yamagishi H (2010) Different responses between diploid and tetraploid H1 embryonic stem cells appeared during long-term culturing in L15F10 medium without leukemia inhibitory factor. Hum Cell 23:134–140
30. Pierce GB, Speers WC (1988) Tumors as caricatures of the process of tissue renewal: Prospects for therapy by directing differentiation. Cancer Res 48:1996–2004
31. Raj A, van Oudenaarden A (2008) Nature, nurture, or chance: stochastic gene expression and its consequences. Cell 135:216–226
32. Niwa H, Miyazaki J, Smith AG (2000) Quantitative expression of Oct-3/4 defines differentiation, dedifferentiation or self-renewal of ES cells. Nat Genet 24:372–376
33. Gidekel S, Pizov G, Bergman Y, Pikarsky E (2003) Oct-3/4 is a dose-dependent oncogenic fate determinant. Cancer Cell 4:361–370
34. Hübner K, Fuhrmann G, Christenson LK, Kehler J, Reinbold R, De La Fuente R, Wood J, Strauss JF 3rd, Boiani M, Schöler HR (2003) Derivation of oocytes from mouse embryonic stem cells. Science 300:1251–1256
35. Denker HW (2009) Induced pluripotent stem cells: how to deal with the developmental potential. Reprod Biomed Online 19:34–37
36. Guc-Scekic M, Milasin J, Stevanovic M, Stojanov LJ, Djordjevic M (2002) Tetraploidy in a 26-month-old girl (cytogenetic and molecular studies). Clin Genet 61:62–65
37. Storchova Z, Pellman D (2004) From polyploidy to aneuploidy, genome instability and cancer. Nat Rev Mol Cell Biol 5:45–54
38. Ganem NJ, Storchova Z, Pellman D (2007) Tetraploidy, aneuploidy and cancer. Curr Opin Genet Dev 17:157–162
39. Szabó PE, Hübner K, Schöler H, Mann JR (2002) Allele-specific expression of imprinted genes in mouse migratory primordial germ cells. Mech Dev 115:157–160
40. Pfeiffer MJ, Balbach ST, Esteves TC, Crosetto N, Boiani M (2010) Enhancing somatic nuclear reprogramming by Oct4 gain-of-function in cloned mouse embryos. Int J Dev Biol 54:1649–1657
41. Esteves TC, Balbach ST, Pfeiffer MJ, Araúzo-Bravo MJ, Klein DC, Sinn M, Boiani M (2011) Somatic cell nuclear reprogramming of mouse oocytes endures beyond reproductive decline. Aging Cell 10:80–95
42. Bernstine E, Hooper M, Grandchamp S, Ephrussi B (1973) Alkaline phosphatase activity in mouse teratoma. Proc Nat Acad Sci U S A 70:3899–3903
43. Rothfels K, Siminovitch L (1958) An air-Drying Technique for flattening chromosomes in mammalian cells grown in vitro. Biotechnic and Histochem 33:73–77
44. Jentsch I, Geigl J, Klein CA, Speicher MR (2003) Seven-fluorochrome mouse M-FISH for high-resolution analysis of interchromosomal rearrangements. Cytogenet Genome Res 103: 84–88

Chapter 2
The Significance of Culture Adaptation of Embryonic Stem Cells for Regenerative Medicine

Neil J. Harrison, Duncan Baker and Peter W Andrews

Abstract The promise that human embryonic stem (ES) cells hold for regenerative medicine has generated much excitement since their initial derivation. However, before the potential of these cells can be realised, efficient differentiation protocols must be devised, and the cells should be shown to pose no safety risk. Despite initial reports suggesting that human ES cells are karyotypically stable, during the last decade it has become apparent that they do acquire genetic and/or epigenetic changes during culture, reflecting an adaptation to life in vitro. This culture adaptation can affect ES cell growth and differentiation, but of particular concern is the potential link between adaptation and cancer, which would become an issue if the cells are to be used for transplantation. In this chapter we discuss the issues surrounding culture adaptation of ES cells, and the potential impacts, both positive and negative, it may have on the use of these cells for regenerative medicine.

2.1 The Principles of Culture Adaptation

> ..natural selection is daily and hourly scrutinising, throughout the world, the slightest variations; rejecting those that are bad, preserving and adding up all that are good; silently and insensibly working, whenever and wherever opportunity offers, at the improvement of each organic being in relation to its organic and inorganic conditions of life—Charles Darwin, *The Origin of Species* (1870).

N. J. Harrison · P. W. Andrews (✉)
Centre for Stem Cell Biology, Dept Biomedical Science,
The University of Sheffield, Sheffield, UK
e-mail: p.w.andrews@sheffield.ac.uk

N. J. Harrison
e-mail: N.J.Harrison@sheffield.ac.uk

D. Baker
Sheffield Diagnostic Genetic Services, Sheffield Children's NHS Trust, Sheffield, UK

H. Baharvand and N. Aghdami (eds.), *Advances in Stem Cell Research*,
Stem Cell Biology and Regenerative Medicine, DOI: 10.1007/978-1-61779-940-2_2,

The derivation of embryonic stem (ES) cells was achieved almost a century after Darwin's death, yet the evolutionary principles he described are inescapable in their maintenance. The ability to self-renew is one of the defining characteristics of ES cells, yet they are not resistant to mutation, and as such evolutionary pressures during culture will invariably lead to the generation of variant cells with improved growth capacity. The selection of cells is evident from the initial derivation of an ES cell line, since generation of an immortal cell line from the transient population of inner cell mass (ICM) cells in the blastocyst stage embryo requires some adaptation, albeit that, at least, initially such changes can be reversed to allow, in the laboratory mouse, the generation of germline chimaeras if ES cells are returned to a developing blastocyst. However, ES cells can continue to 'evolve' during their time in vitro, a process termed culture adaptation.

The culture adaptation of ES cells must then involve two, potentially independent, phenomena: mutation and selection. Without selection, and in the absence of a population bottleneck, such as cloning, a mutation would never reach sufficient levels in the population to be detected, and without mutation, no variants would exist. The selection pressure for advantageous mutations must be quite high since, assuming a routine passage ratio of 1:3 every 3–4 days, and a cell cycle time of 16 h, it is likely that upward of 90% of cells must be lost between passages. One study has used a Monte Carlo simulation to assess the relationships of these are key factors in ES cell growth [1]. In this study, the selective advantage of the variant cells that commonly appear in ES cultures was assessed by spiking cultures with cytogenetically abnormal cells and then comparing the rate at which they overgrew the culture with predictions from the simulation. Indeed the selective advantage of the variant cells was high and variant cells typically predominant in cultures within 10–15 passages, sometimes less. From this it was estimated that human ES cells acquire beneficial mutations at a rate of approximately 1×10^{-6}. This rate is similar to that reported for other cell types [2], though the short cell cycle time for ES cells perhaps makes them more prone to the acquisition of such mutations. Selection is more difficult to quantify, though it has the potential to be quite significant. With a large opportunity for selection, and seemingly frequent mutation, human ES cells are obviously strongly subject to adaptation.

2.2 Culture Adaptation and Cancer

The culture adaptation of ES cells increases their growth capacity in culture, which has indeed been observed during the prolonged passage of human ES cells [3]. For a stem cell, this increase in growth capacity must result from an increased propensity for self-renewal, which would increase the stem cell pool, over differentiation and/or death, which would decrease the stem cell pool. A similar shift in cell fate would benefit cancer cells, particularly cancer stem cells, since the evolution of a phenotype strongly biased towards self-renewal in a cancer stem cell would result in a highly aggressive tumour. With this in mind, it is notable that the genetic changes most frequently reported in culture adapted ES cells are also commonly observed in embryonal carcinoma (EC) cells, their malignant counterparts and the stem cells of teratocarcinomas [4, 5]. Teratocarcinomas are germ cell tumours (GCT) that occur most commonly as a testicular cancers (TGCT), and are perhaps the first examples of a stem cell-based cancer. Here, Kleinsmith and Pierce [6] showed that a single EC cell from a murine teratocarcinoma could recapitulate the original tumour when transplanted to a syngeneic recipient mouse. It should be noted that when culture adapted human ES cells are allowed to form teratomas in immunecompromised mice, the tumours tend to behave more like teratocarcinomas, retaining a stem cell component that can be excised and re-cultured [7, 8].

The karyotypic changes most frequently seen in ES and EC cells are the gains of material from chromosomes 12 and 17 (Fig. 2.1), particularly the p arm of chromosome 12 and the q arm of chromosome 17 [4]. The gain of material from chromosome 12p is also common in TGCT, so much so that it is considered diagnostic for this malignancy. In TGCT, the gain of material from chromosome 12p is mainly from 2 amplicons, 12p11.2-12 [9, 10] and 12p13 [11], and the specific amplification of 12p13 has also been seen in a culture adapted iPS cell line [12]. The potential candidate genes in the 12p13 band include GDF3, DPPA3, CCND2 and NANOG. NANOG and GDF3 were shown to be overexpressed in the culture adapted iPS cells with 12p13 amplification, and forced overexpression of NANOG has previously been shown to maintain human ES cells in an undifferentiated state [13]. However, not all culture adapted cells show an increase in NANOG [14, 15], including those which are trisomic for chromosome 12 (Alagaratnam et al., in preparation), suggesting that there may be a number of ways in which ES cells can adapt to culture.

The amplicons on chromosome 17 in TGCT were originally identified as 17q11-21 and 17q24qter [16], though recent work [17] suggested the region 17p11.2-q21.32 is most commonly amplified in EC cells. These data are in general agreement with the amplicon suggested for culture adapted human ES cells, approximated as 17q21-qter [4]. In particular, the terminal end of 17q (q25-ter) has been implicated in human ES cell culture adaptation, with reduced expression of Survivin (BIRCS) inducing apoptosis [18]. However, a small amplicon at 17p11.2 has also been identified in human ES cells. Amplification of this region has been reported in two independent human ES cell lines, and in one case as a homogeneous staining region (HSR), a cytogenetic representation of genetic amplification that is almost unique to cancer cells. The amplification of this region has also been reported in breast cancer [19] retinoblastoma [20], and frequently in osteosarcoma [21], suggesting this region may contain gene(s) which can impact on cell fate.

Aside from the changes on chromosomes 12 and 17, gains of material from chromosomes 1, 20 and X have also been seen with relative regularity. On chromosome 20, an amplicon at 20q11.21 has been identified through array-CGH by a number of groups [22–24] and may hold most promise for identifying specific genes involved in adaptation. The use of array-CGH technology has helped to identify and refine the common regions of genetic change in human ES cells, yet confirms the earlier karyological reports that chromosomal gain is far more

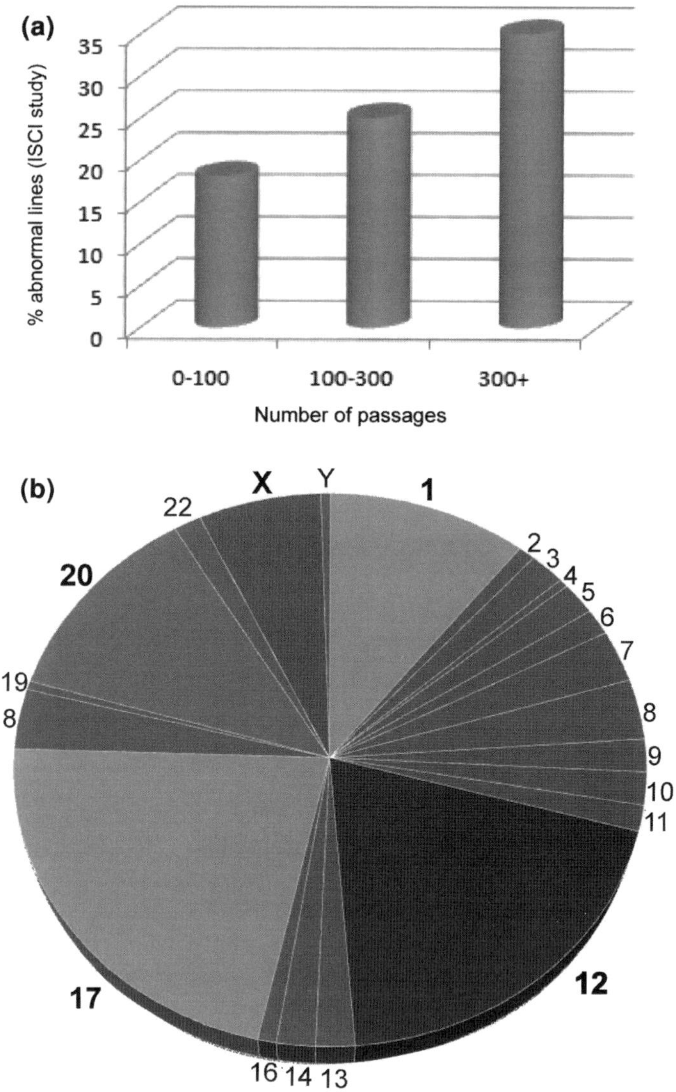

Fig. 2.1 Karyotypic abnormalities in hES cells. **a** The percentage of karyotypically abnormal cell lines (mosaic cultures are considered abnormal) compared against passage number. As passage number increases, so does the percentage of abnormal lines. **b** The percentage of abnormalities, per chromosome, reported for hES cells. The most common abnormalities are highlighted, and are associated with chromosomes 12, 17, 20 and X. Data shown is from the International Stem Cell Initiative (ISCI) study [39]

frequent than chromosomal loss in these cells. This may reflect a quirk in the DNA repair machinery of human ES cells, or suggest that increased gene expression may be more readily achieved through DNA amplification than epigenetic mechanisms in these cells.

2.3 Culture Adaptation and Regenerative Medicine

The use of stem cells in regenerative medicine will require their efficient differentiation into the cell type(s) of interest. Culture adaptation may change fate propensities in ES cells, biasing them not only against differentiation per se, but also potentially against differentiation along certain lineages. Human ES cells are not islands, these cells will be influenced by their neighbours, and the signals that they provide. As such, to survive as a stem cell, it would be beneficial to be surrounded by cells that provide signals that facilitate self-renewal, as opposed to those that encourage death or differentiation. Here, it is of interest that some culture adapted cells seem to be biased against endoderm differentiation [25], since endoderm cells are known to produce bone morphogenetic proteins (BMPs) which can cause human ES cell differentiation [26]. This offers an obvious mechanism by which variant cells could acquire a selective advantage, and would prove an obstacle to those trying to differentiate the cells to, e.g. pancreas or liver.

2.4 Identification of Culture Adapted ES Cells

The concerns associated with using culture adapted cells in regenerative medicine mean that methodologies to detect these cells, and/or mimimise the likelihood of their appearance, have become increasingly important. In this regard, defining what exactly constitutes ‘culture adapted’ is a fundamental question, yet one which is still without a definitive answer.

If one assumes that for a mutation to reach detectable levels it must bestow a growth advantage on the variant cell, then the identification of culture adapted ES cells could be achieved based on mutation. With karyotyping services readily available and array-CGH becoming more accessible, any change in DNA content can be observed. In addition, the advent of sequencing technologies may also allow identification of point mutations, or any novel splice variants. However, to demonstrate that a mutant has spread through a population requires regular monitoring, and the procedures to detect mutations are destructive to the cells. The use of CD30 as a marker for karyotypically abnormal cells was suggested [27], since this can be detected on live cells, yet subsequent results have shown that CD30 is not always associated with aneuploidy [28, 29]. In any case, altered gene expression patterns can also be acquired by epigenetic changes in the absence of detectable genetic changes. Although imprinting patterns tend to be fairly stable in human ES cells [30, 31] marked changes in DNA methylation patterns have been reported [32, 33]. A further example is the reported loss of X-chromosome inactivation upon adaptation leading to the functional overexpression of X-linked genes in the absence of any identifiable karyotypic change affecting that chromosome [14].

Functional tests likely present the best way to show that a cell has adapted, yet the assay(s) of choice have still to be decided. Across a number of studies, culture adapted cells have shown decreased apoptosis, increased cloning efficiency, lack of growth factor dependence, reduced differentiation and increased clonogenicity [14, 27, 34], yet sometimes there is discrepancy. For example, [29] showed no difference in apoptosis between a culture adapted subline and its karyotypically normal sister line. Perhaps a more simple way to show that a variant cell has a growth advantage is to mix it with an early passage sister line, and monitor its spread through the culture.

A further issue to be considered when assaying for adapted cells is their in vitro environment. Human ES cells are maintained in a number of different ways, with variation in substrate, media and passaging strategy, so that certain conditions are likely to promote/reduce the selection of certain mutants. Anecdotally, manually passaging human ES cells was assumed to guard against genetic change, and the work of [35] did suggest that when human ES cells are passaged manually they retain a normal karyotype, yet this study does not factor in population size. Manual passaging generally involves transfer of fewer cells on passage than bulk dissagregation techniques, so that the chances of propagating a relatively rare mutation are much decreased. Indeed, [1] showed that population size alone can have a major impact on the likelihood of acquiring a culture adapted cell line. Interestingly, this study also revealed that the maintenance of human ES cells in small cultures as opposed to large cultures reduced the probability of abnormal cell appearance. However, one must also remember that the most common changes observed in culture adapted ES cells (i.e. gain of material from chromosomes 12 and 17) are also seen in TGCT, which are maintained in very different conditions. Thus these particular mutations seem to provide stem cells with an intrinsic growth advantage, and may be very difficult to completely avoid. On the other hand, altered culture conditions might also affect the mutation rate.

2.5 Can Culture Adapted Cells be Utilised in Regenerative Medicine?

It is unlikely that mutation in ES cells will ever be completely prevented, and so eventual culture adaptation after a period in culture is perhaps unavoidable. Given this, it is pertinent to ask how hazardous these cells may be, and whether, in fact, culture adaptation actually presents any opportunities for research and regenerative medicine.

Inevitably, the initial derivation of immortal human ES cell lines must involve a degree of adaptation, although this might be expected to be initially reversible given the experience of using mouse ES cells to produce germ line chimeras and genetically altered mice. Thus, at no point is any ES cell line identical to those cells which exist in the ICM of the embryo. Further, although culture adapted cells may show tumourigenic behaviour when undifferentiated, there is no guarantee that the growth advantage which afforded their selection will persist when the cells are differentiated. In other words, if those pathways affected by adaptation are intrinsic to the stem cell state, then they will no longer be active when the cells differentiate, and as such these cells will likely behave as normal somatic cells. Since it is unlikely that undifferentiated cells would themselves be used in therapy, then these cells may not pose a safety risk, provided that there are adequate methods for ensuring that no undifferentiated cells persist in preparations of their differentiated derivatives for transplantation. However, since many of the genetic changes that underlie adaptation involve large chromosomal fragments, it may be that genes involved in driving stem cell adaptation might have other consequences for differentiated cells. It will therefore be essential that assays that assess the growth potential of differentiated culture adapted cells, and also the tissue stem cells formed from variant ES cells, are performed.

Apart from their applications in regenerative medicine, a more immediate use for differentiated cells derived from ES cells, and iPS cells, is in toxicology and drug screening in the pharmaceutical industry. Since culture adaptation has the effect of increasing the propensity of stem cells for self-renewal, it may then offer the practical advantage of making the cells easier to maintain. Toxicology screens using ES cells are likely to require large-scale productions of homogeneous cell populations, and these may be more easily achieved using culture adapted cells. Of course it would be essential to confirm experimentally that the adaptive changes in the stem cells do not significantly affect the behaviour of their differentiated derivatives in the particular application for which they are to be used.

2.6 Culture Adaptation as a Tool to Study Stem Cell Fate

To maximise the potential of ES cells, it is imperative that their basic biology is fully understood. Culture adaptation must change the fate of potential ES cells, and as such this process may provide an insight into pathways that affect cell growth. Further, considering the similarity between culture adapted ES cells and EC cells of TGCT, it seems plausible that this in vitro process may act as a model to study this cancer. The use of culture adapted cells may be particularly useful here, since these cells show only a small number of genetic changes compared to the highly aneuploid TGCT. Speculatively, bearing in mind the alleged transcriptional similarity between ES cells and cancer cells [36], it is possible that culture adaptation may also act as a paradigm for other cancers, particularly those with a stem cell component.

At present, there is little information relating to the molecular pathways affected by culture adaptation. From the common amplicons it is possible to pick potential candidate genes, but testing these genes, and their up- and down-stream targets, is not a trivial matter in human ES cells. Genetic manipulation will exert pressures on ES cells, which could promote culture adaptation, making results difficult to interpret. The issue is further complicated by the fact that adaptation could occur in many ways. A number of mechanisms appear to exist through which pluripotency pathways and/or survival pathways can be regulated, any of which could be the target of an adaptive change. This highlights the earlier point that culture adapted cells may have a range of phenotypes, making their identification challenging.

2.7 Summary

The culture adaptation of human ES cells is an inescapable truth. These cells are maintained in sub-optimal conditions, and over time variant cells will inevitably arise which show an increased growth capacity, and will overtake their normal neighbours. Certain passaging strategies, and of course reduced culture time, are likely to reduce the chances of generating abnormal cells, yet since it appears that at least some adaptive advantages are intrinsic to stem cell behaviour, such methods will only act to stem the tide for so long. Based on their similarity to human ES cells, the same principles will likely apply for iPS cells, and indeed a number of groups have reported mutations in these cells [37, 38]. It is then important to understand the potential problems culture adaptation may cause for regenerative medicine, and also to grasp the opportunities this process may provide.

One of the most concerning issues is the relationship between culture adaptation and oncogenesis. Culture adapted cells have been shown to acquire the same changes as those observed in EC cells, their malignant counterparts, and a number of their behaviours (e.g. increased clonogenicity, reduced apoptosis) are associated with transformation. However, at present, it is not known whether the growth changes observed in culture adapted cells will still manifest when the cells are fully differentiated. Since it is unlikely that undifferentiated cells would be used in therapy, it is possible that the adaptive changes active in the stem cells would have no impact on somatic cells. If this were true, then culture adapted cells may still be suited to therapy, particularly since they are much easier to maintain and expand. However, this would need to be assessed on a case-by-case basis. There are concerns regarding the differentiation potential of the adapted cells, since evidence suggests they may be biased towards, or against certain lineages. Although this may be the case, Melton and colleagues have already reported differing differentiation propensities between different 'normal' lines, making adapted lines no more problematic than those already available.

The culture adaptation of ES cells must bias their fate towards self-renewal, and away from differentiation or death. As such, adaptation provides an insight into the fundamental biology of these cells, and also potentially cancer stem cells. The relatively small number of changes observed in these cells should help hone the search for amplicons, and genes of interest, in TCGT in particular. The variety of manners through which adaptation can occur could in fact shed light on a number of pathways which could impact on stem cell fate, influencing both regenerative medicine and oncological stem cell research.

References

1. Olariu V, Harrison NJ, Coca D, Gokhale PJ, Baker D, Billings S, Kadirkamanathan V, Andrews PW (2010) Modeling the evolution of culture-adapted human embryonic stem cells. Stem Cell Res 4(1):50–56
2. Holliday R (1989) Chromosome error propagation and cancer. Trends Genet 5(2):42–45
3. Draper JS, Moore HD, Ruban LN, Gokhale PJ, Andrews PW (2004) Culture and characterization of human embryonic stem cells. Stem Cells Dev 13(4):325–336
4. Baker DE, Harrison NJ, Maltby E, Smith K, Moore HD, Shaw PJ, Heath PR, Holden H, Andrews PW (2007) Adaptation to culture of human embryonic stem cells and oncogenesis in vivo. Nat Biotechnol 25(2):207–215
5. Harrison NJ, Baker D, Andrews (2007) PWCulture adaptation of embryonic stem cells echoes germ cell malignancy. Int J Androl 30(4):275–281, discussion 281
6. Kleinsmith LJ, Pierce GB (1964) Multipotentiality of single embryonal carcinoma cells. Cancer Res 24:1544–1551
7. Andrews PW, Matin MM, Bahrami AR, Damjanov I, Gokhale P, Draper JS (2005) Embryonic stem (ES) cells and embryonal carcinoma (EC) cells: opposite sides of the same coin. Biochem Soc Trans 33(Pt 6):1526–1530
8. Blum B, Benvenisty N (2009) The tumorigenicity of diploid and aneuploid human pluripotent stem cells. Cell Cycle 8(23):3822–3830
9. Rodriguez S, Jafer O, Goker H, Summersgill BM, Zafarana G, Gillis AJ, van Gurp RJ, Oosterhuis JW, Lu YJ, Huddart R, Cooper CS, Clark J, Looijenga LH, Shipley JM (2003) Expression profile of genes from 12p in testicular germ cell tumors of adolescents and adults associated with i(12p) and amplification at 12p11.2–p12.1. Oncogene 22(12):1880–1891
10. Mostert MC, Verkerk AJ, van de Pol M, Heighway J, Marynen P, Rosenberg C, van Kessel AG, van Echten J, de Jong B, Oosterhuis JW, Looijenga LH (1998) Identification of the critical region of 12p over-representation in testicular germ cell tumors of adolescents and adults. Oncogene 16(20):2617–2627
11. Korkola JE, Houldsworth J, Chadalavada RS, Olshen AB, Dobrzynski D, Reuter VE, Bosl GJ, Chaganti RS (2006) Down-regulation of stem cell genes, including those in a 200-kb gene cluster at 12p13.31, is associated with in vivo differentiation of human male germ cell tumors. Cancer Res 66(2):820–827
12. Mayshar Y, Ben-David U, Lavon N, Biancotti JC, Yakir B, Clark AT, Plath K, Lowry WE, Benvenisty N (2010) Identification and classification of chromosomal aberrations in human induced pluripotent stem cells. Cell Stem Cell 7(4):521–531
13. Darr H, Mayshar Y, Benvenisty N (2006) Overexpression of NANOG in human ES cells enables feeder-free growth while inducing primitive ectoderm features. Development 133(6):1193–1201
14. Enver T, Soneji S, Joshi C, Brown J, Iborra F, Orntoft T, Thykjaer T, Maltby E, Smith K, Dawud RA, Jones M, Matin M, Gokhale P, Draper J, Andrews PW (2005) Cellular differentiation hierarchies in normal and culture-adapted human embryonic stem cells. Hum Mol Genet 14(21):3129–3140
15. Ji J, Werbowetski-Ogilvie TE, Zhong B, Hong SH, Bhatia M (2009) Pluripotent transcription factors possess distinct roles in normal versus transformed human stem cells. PLoS ONE 4(11):e8065
16. Kraggerud SM, Skotheim RI, Szymanska J, Eknaes M, Fossa SD, Stenwig AE, Peltomaki P, Lothe RA (2002) Genome profiles of familial/bilateral and sporadic testicular germ cell tumors. Genes Chromosomes Cancer 34(2):168–174

17. Korkola JE, Heck S, Olshen AB, Reuter VE, Bosl GJ, Houldsworth J, Chaganti RS (2008) In vivo differentiation and genomic evolution in adult male germ cell tumors. Genes Chromosomes Cancer 47(1):43–55. doi:10.1002/gcc.20504
18. Blum B, Bar-Nur O, Golan-Lev T, Benvenisty N (2009) The anti-apoptotic gene survivin contributes to teratoma formation by human embryonic stem cells. Nat Biotechnol 27(3): 281–287
19. Jung SH, Shin SH, Yim SH, Choi HS, Lee SH, Chung YJ (2009) Integrated analysis of copy number alteration and RNA expression profiles of cancer using a high-resolution whole-genome oligonucleotide array. Exp Mol Med 41(7):462–470
20. Ganguly A, Nichols KE, Grant G, Rappaport E, Shields C (2009) Molecular karyotype of sporadic unilateral retinoblastoma tumors. Retina 29(7):1002–1012. doi:10.1097/IAE.0b013 e3181a0be05
21. van Dartel M, Hulsebos TJ (2004) Amplification and overexpression of genes in 17p11.2 ~ p12 in osteosarcoma. Cancer Genet Cytogenet 153(1):77–80
22. Wu H, Kim KJ, Mehta K, Paxia S, Sundstrom A, Anantharaman T, Kuraishy AI, Doan T, Ghosh J, Pyle AD, Clark A, Lowry W, Fan G, Baxter T, Mishra B, Sun Y, Teitell MA (2008) Copy number variant analysis of human embryonic stem cells. Stem Cells 26(6):1484–1489
23. Spits C, Mateizel I, Geens M, Mertzanidou A, Staessen C, Vandeskelde Y, Van der Elst J, Liebaers I, Sermon K (2008) Recurrent chromosomal abnormalities in human embryonic stem cells. Nat Biotechnol 26(12):1361–1363
24. Lefort N, Feyeux M, Bas C, Feraud O, Bennaceur-Griscelli A, Tachdjian G, Peschanski M, Perrier AL (2008) Human embryonic stem cells reveal recurrent genomic instability at 20q11.21. Nat Biotechnol 26(12):1364–1366
25. Fazeli A, Liew CG, Matin MM, Elliott S, Jeanmeure LF, Wright PC, Moore H, Andrews PW (2011) Altered patterns of differentiation in karyotypically abnormal human embryonic stem cells. Int J Dev Biol 55(2):175–180
26. Pera MF, Andrade J, Houssami S, Reubinoff B, Trounson A, Stanley EG, Ward-van Oostwaard D, Mummery C (2004) Regulation of human embryonic stem cell differentiation by BMP-2 and its antagonist noggin. J Cell Sci 117(Pt 7):1269–1280
27. Herszfeld D, Wolvetang E, Langton-Bunker E, Chung TL, Filipczyk AA, Houssami S, Jamshidi P, Koh K, Laslett AL, Michalska A, Nguyen L, Reubinoff BE, Tellis I, Auerbach JM, Ording CJ, Looijenga LH, Pera MF (2006) CD30 is a survival factor and a biomarker for transformed human pluripotent stem cells. Nat Biotechnol 24(3):351–357
28. Thomson A, Wojtacha D, Hewitt Z, Priddle H, Sottile V, Di Domenico A, Fletcher J, Waterfall M, Corrales NL, Ansell R, McWhir J (2008) Human embryonic stem cells passaged using enzymatic methods retain a normal karyotype and express CD30. Cloning Stem Cells 10(1):89–106
29. Harrison NJ, Barnes J, Jones M, Baker D, Gokhale PJ, Andrews PW (2009) CD30 expression reveals that culture adaptation of human embryonic stem cells can occur through differing routes. Stem Cells 27(5):1057–1065
30. Rugg-Gunn PJ, Ferguson-Smith AC, Pedersen RA (2005) Epigenetic status of human embryonic stem cells. Nat Genet 37(6):585–587
31. Adewumi O, Aflatoonian B, Ahrlund-Richter L, Amit M, Andrews PW, Beighton G, Bello PA, Benvenisty N, Berry LS, Bevan S, Blum B, Brooking J, Chen KG, Choo AB, Churchill GA, Corbel M, Damjanov I, Draper JS, Dvorak P, Emanuelsson K, Fleck RA, Ford A, Gertow K, Gertsenstein M, Gokhale PJ, Hamilton RS, Hampl A, Healy LE, Hovatta O, Hyllner J, Imreh MP, Itskovitz-Eldor J, Jackson J, Johnson JL, Jones M, Kee K, King BL, Knowles BB, Lako M, Lebrin F, Mallon BS, Manning D, Mayshar Y, McKay RD, Michalska AE, Mikkola M, Mileikovsky M, Minger SL, Moore HD, Mummery CL, Nagy A, Nakatsuji N, O'Brien CM, Oh SK, Olsson C, Otonkoski T, Park KY, Passier R, Patel H, Patel M, Pedersen R, Pera MF, Piekarczyk MS, Pera RA, Reubinoff BE, Robins AJ, Rossant J, Rugg-Gunn P, Schulz TC, Semb H, Sherrer ES, Siemen H, Stacey GN, Stojkovic M, Suemori H, Szatkiewicz J, Turetsky T, Tuuri T, van den Brink S, Vintersten K, Vuoristo S, Ward D, Weaver TA, Young LA, Zhang W (2007) Characterization of human

embryonic stem cell lines by the international stem cell initiative. Nat Biotechnol 25(7):803–816

32. Allegrucci C, Young LE (2007) Differences between human embryonic stem cell lines. Hum Reprod Update 13(2):103–120
33. Calvanese V, Horrillo A, Hmadcha A, Suarez-Alvarez B, Fernandez AF, Lara E, Casado S, Menendez P, Bueno C, Garcia-Castro J, Rubio R, Lapunzina P, Alaminos M, Borghese L, Terstegge S, Harrison NJ, Moore HD, Brustle O, Lopez-Larrea C, Andrews PW, Soria B, Esteller M, Fraga MF (2008) Cancer genes hypermethylated in human embryonic stem cells. PLoS ONE 3(9):e3294
34. Werbowetski-Ogilvie TE, Bosse M, Stewart M, Schnerch A, Ramos-Mejia V, Rouleau A, Wynder T, Smith MJ, Dingwall S, Carter T, Williams C, Harris C, Dolling J, Wynder C, Boreham D, Bhatia M (2009) Characterization of human embryonic stem cells with features of neoplastic progression. Nat Biotechnol 27(1):91–97
35. Mitalipova MM, Rao RR, Hoyer DM, Johnson JA, Meisner LF, Jones KL, Dalton S, Stice SL (2005) Preserving the genetic integrity of human embryonic stem cells. Nat Biotechnol 23(1):19–20
36. Ben-Porath I, Thomson MW, Carey VJ, Ge R, Bell GW, Regev A, Weinberg RA (2008) An embryonic stem cell-like gene expression signature in poorly differentiated aggressive human tumors. Nat Genet 40(5):499–507
37. Gore A, Li Z, Fung HL, Young JE, Agarwal S, Antosiewicz-Bourget J, Canto I, Giorgetti A, Israel MA, Kiskinis E, Lee JH, Loh YH, Manos PD, Montserrat N, Panopoulos AD, Ruiz S, Wilbert ML, Yu J, Kirkness EF, Izpisua Belmonte JC, Rossi DJ, Thomson JA, Eggan K, Daley GQ, Goldstein LS, Zhang K (2011) Somatic coding mutations in human induced pluripotent stem cells. Nature 471(7336):63–67
38. Hussein SM, Batada NN, Vuoristo S, Ching RW, Autio R, Narva E, Ng S, Sourour M, Hamalainen R, Olsson C, Lundin K, Mikkola M, Trokovic R, Peitz M, Brustle O, Bazett-Jones DP, Alitalo K, Lahesmaa R, Nagy A, Otonkoski T (2011) Copy number variation and selection during reprogramming to pluripotency. Nature 471(7336):58–62
39. Amps K et al (2011) Screening a large, ethnically diverse population of human embryonic stem cells identifies a chromosome 20 minimal amplicon that confers a growth advantage. In preparation

Chapter 3
Biomanufacturing Human Pluripotent Stem Cells for Therapeutic Applications

Xiaojun Lian and Sean P. Palecek

Abstract Human embryonic stem cells (hESCs) and human induced pluripotent stem cells (iPSCs) can be propagated indefinitely and have the potential to differentiate to all cell types of the body. In addition, patient-specific human iPSCs and hESCs containing point mutations in genes that cause disease can be used to study diseases that have no adequate human in vitro or animal models. Harnessing the potential of these cells holds promise for future applications in cell therapy and regenerative medicine. Common methods of expanding and differentiating human pluripotent stem cells (hPSCs), including hESC and iPSC, require serum, mouse or human feeder cells, or feeder-conditioned medium. These methods are labor intensive and hard to scale, and sources of variability including growth factor fluctuations during preparation and culture complicate large-scale hPSC bioprocesses. Biomanufacturing cells from hPSCs requires development of fully defined, xeno-free culture medium and substrates formulated with human-derived, human recombinant proteins or chemically synthetic substrates under cGMP and improved processes for monitoring cell status and genomic stability during expansion and differentiation. In addition, robust and scalable differentiation methods must be developed. This chapter discusses recent progress and remaining challenges facing production of hPSC-derived cells.

X. Lian · S. P. Palecek (✉)
Department of Chemical and Biological Engineering,
University of Wisconsin, Madison, WI, USA
e-mail: palecek@engr.wisc.edu

H. Baharvand and N. Aghdami (eds.), *Advances in Stem Cell Research*,
Stem Cell Biology and Regenerative Medicine, DOI: 10.1007/978-1-61779-940-2_3,

3.1 Development of Defined Culture Systems

3.1.1 Defined Media for hPSC Expansion

3.1.1.1 Identification of Soluble Factors Regulating hPSC Self-Renewal

The first isolated human embryonic stem cell lines [1] were cocultured with mouse embryonic fibroblast (MEF) feeder cells in medium containing fetal bovine serum (FBS). This initial culture system requires manual selection of individual compact undifferentiated colonies by micropipette selection during each passage. This protocol is extremely labor-intensive and infeasible to scale. Patient-specific human iPSCs [2, 3] and hESCs containing point mutations [4] in genes that cause disease require similar culture conditions as normal hESCs. In the early stages of hESC culture, the critical factors important for maintaining undifferentiated states of hESCs were gradually clarified. These factors have led to development of robust, defined media for hPSC culture which are replacing the MEF-coculture system as the preferred hPSC culture platform.

In most hESC culture conditions, BMP-like activity in the medium is the dominating differentiation-inducing factor [5, 6]. FGF2 (bFGF) and activin/Nodal/TGFβ1, which inhibit or counteract BMP-like activity, are commonly supplemented to hESC culture medium to repress spontaneous differentiation [7]. Activin/Nodal/TGFβ signals directly regulate expression of pluripotency gene Nanog [8, 9] to facilitate hPSC self-renewal.

3.1.1.2 Development of Fully Defined Media for hPSC Expansion

Fully-defined expansion conditions are crucial for robust expansion of hPSCs for therapeutic applications (Table 3.1). Vallier et al. reported an IMDM/F12-based chemically defined medium (CDM), which contains bFGF and Activin A/Nodal, and is used in conjunction with a FBS-derived extracellular matrix (ECM) [10]. The use of FBS as a matrix, which is undefined and variable in composition, is not ideal for large-scale biomanufacturing applications. bFGF and Activin A are thought to promote self-renewal synergistically by repressing the BMP-like activity [10]. However, 20 ng/ml bFGF itself was found to be sufficient for sustaining hESC self-renewal in another basal medium, N2B27, on Matrigel [11]. The animal substrate Matrigel, a complex undefined mixture of extracellular matrix proteins and growth factors derived from Engelberth-Holm-Swarm tumor basement membrane [12], is often used as an hPSC culture substrate, but it is undefined and must be subject to costly testing to ensure performance of supporting hPSC self-renewal and lack from pathogens.

In order to achieve hESC expansion in fully defined culture conditions, hESCs have been cultured on human fibronectin in defined HESCO medium containing bFGF, insulin, transferrin, Wnt3a, April or BAFF, albumin, and 2.5× cholesterol lipid supplement [13]. This fully defined system supports hESC self-renewal and

Table 3.1 Defined media for long-term growth of hESC and iPSC

Basal medium	Active components	Long-term self-renewal potential	Karyotypic stability	Reference
IMDM/F12	Activin A, FGF2	10 passages	Stable	[10]
DMEM/F12 (Defined HESCO)	Wnt3a, FGF, insulin, transferrin, April/BAFF, cholesterol, and albumin	24 passages	Normal after 23 passages	[13]
DMEM/F12	N2, B27, FGF2, BSA, BME	27 passages	Normal after 22 passages	[11]
ESF medium	LIF, FGF2, Ascorbic acid, heparin sulfate	25 passages on type I collagen	Normal	[14]
DMEM/F12 (StemPro)	Heregulin-1β EGF domain, Activin A, LR3-IGF1, FGF2	9 months	Normal	[15]
DMEM/F12 (TeSR)	bFGF, TGFβ1, LiCl, GABA, and pipecolic acid	7 months	Normal after 13 passages; however, new derived lines are abnormal	[16]
DMEM/F12 (E8)	Ascorbic acid, selenium, FGF2, insulin, NaHCO3, transferring, TGFβ1, or Nodal	3 months	Normal after 3 months in culture	[19]

normal karyotype for at least 24 passages. hESCs can also be expanded on collagen I in an ESF medium containing LIF, FGF2, ascorbic acid, heparin sulfate for at least 25 passages [14].

Insulin-like growth factors also play a key role in maintaining pluripotency. Wang et al. revealed that conditioned medium (CM) from MEFs mediates hESC proliferation by triggering IGF1R/IR and ERBB-family tyrosine phosphorylation which they showed was indispensible for hESC self-renewal [15]. Thus, they included IGF1R and ERBB2 ligands, LR-IGF1 and the EGF domain of heregulin-1β, along with FGF2 and Activin A in their defined medium, called DC-HAIF in the article and later commercialized as StemPro. Using this medium and fibronectin as a matrix, the authors were able to maintain hESCs in an undifferentiated state for 9 months and cells cultured for 6 months could form teratomas when injected in mice. These studies indicate that hESCs can be maintained on different types of defined matrix in conjunction with corresponding defined media.

Ludwig et al. developed a DMEM/F12-based defined medium, named TeSR, that contains bFGF, TGFβ1, LiCl, GABA, and pipecolic acid for long-term self-renewal and derivation of hESCs [16]. These conditions completely eliminated the use of animal products such as nonhuman sialic acid, Neu5Gc, which is a significant step forward in developing a robust and scalable biomanufacturing process for hPSC expansion. Furthermore, they used the defined medium and matrix to derive two new cell lines from just five blastocysts. However, both hESC lines generated were karyotypically abnormal; this may be attributable to the enzymatic dissociation used to passage the cells [17].

Although TeSR could be used in the complete absence of animal proteins, the inclusion of human serum albumin and human-sourced matrix proteins makes those conditions extraordinary expensive, impractical for routine use, and not completely defined because of albumin's ability to bind lipids and other impurities [18]. Chen et al. [19] examined pairwise interactions between each factor present in TeSR and developed a β-mercaptoethanol (BME) and serum albumin-free medium. By using this new medium (E8) and vitronectin-coated substrates, improved generation of human iPSCs has been shown. However, derivation of karyotypically normal new human embryonic stem cell lines has not yet been demonstrated in this system.

3.1.2 Defined Substrates for hPSC Expansion

3.1.2.1 Extracellular Matrix-Based Substrates

While Matrigel and other extracellular matrix proteins remain the gold standard for hPSC culture, substantial progress has been made in engineering matrices to support hPSC self-renewal in recent years. However, many of these defined matrices require the use of a particular complementary culture medium, suggesting interplay between matrix and soluble factors in regulating cell pluripotency.

A better understanding of this interplay is needed to develop robust matrices compatible with multiple culture media. One ECM protein, human vitronectin, which binds $\alpha_v\beta_5$ integrin on the undifferentiated hESC surface, has been shown to support hESC long-term self-renewal with MEF conditioned medium and mTeSR1 [20]. Truncated forms of vitronectin, including N-terminal and C-terminal truncated VTN-NC and N-terminal truncated VTN-N, were found to support hPSC attachment and survival better than wild type vitronectin [19]. Furthermore, VTN-NC in conjunction with E8-based medium supported improved iPSC reprogramming efficiencies compared to culture with mouse fibroblast feeder cells, in FBS-containing medium, or in TeSR [19].

Another ECM protein, human laminin-511 (LN-511) promotes hESC, and hiPSC self-renewal in several chemically defined media, including TeSR1 and StemPro [21]. Adhesion of hESCs to LN-511 is dependent on the $\alpha_6\beta_1$ integrin. In addition, LN-511 has been shown to support blastocyst inner cell mass attachment and spreading, suggesting that this matrix may be suitable for derivation of new hESC and iPSC lines as well as expansion of existing lines.

The ability of vitronectin and laminin-derived proteins to support hPSC self-renewal highlights the importance of cell–matrix interactions in hPSC expansion. Cell–cell contact mediated by E-cadherin may also play a key role in self-renewal of hESCs. Nagaoka et al. showed that a plate coated with a fusion protein of recombinant human E-cadherin and the IgG Fc domain performed almost identically to Matrigel for hESC culture, based on the criteria of cell morphology, doubling time, expression of pluripotent markers, and differentiation ability [22]. hE-cadherin-Fc was shown to support hESC self-renewal for 53 passages in defined mTeSR1 medium and these cells were able to generate teratomas in severe combined immunodeficiency (SCID) mice. The mechanisms by which these hE-cadherin-Fc surfaces stimulate intercellular communication pathways to regulate pluripotency are not yet clear, but they provide a promising platform for hPSC expansion and differentiation.

3.1.2.2 Synthetic Substrates

Synthetic non-biological matrices would likely offer lower cost and better reproducibility and scalability than protein-based substrates. Walker et al. reported that poly-D-lysine surfaces maintained hESCs and iPSCs when used in conjunction with mTeSR1 medium supplemented with blebbistatin, a muscle myosin II inhibitor. The cells were expanded in this system for more than 20 passages with pluripotency and genomic integrity intact [23]. Although a poly-D-lysine coated surface provides a simple, easy to fabricate surface, the effects of long-term blebbistatin exposure on hPSC differentiation capacity are not known.

Klim et al. screened FGF receptor-binding peptides, integrin-binding peptides, and peptides derived from fibronectin, laminin, bone sialoprotein and vitronectin for their ability to bind hESCs cells as well as maintain hESCs in a pluripotent state [24]. They used a high-throughput array of self-assembled monolayers (SAMs) to identify the best peptides and combination of peptides at various concentrations. They reported that the heparin-binding peptide GKKQRFRHRNRKG derived from vitronectin can mediate hESC binding at a relatively low concentration (0.5% peptide surface density) and maintain self-renewal. The synthetic peptide was modified for functionalization of a variety of surfaces by SAM assembly, covalent conjugation of reactive bromoacetamide groups to glass slides, and binding of biotinylated peptides to streptavidin-coated polystyrene. Substrates containing the heparin-binding peptide interacted with cell-surface glycosaminoglycans and could be used with the defined medium mTeSR1 to culture multiple hESC and iPSC lines for more than 3 months, during which pluripotency and genome integrity were maintained. However, the presence of ROCK inhibitors in the mTeSR1 medium was required to maintain hPSC attachment and growth. The effects of long-term inclusion of ROCK inhibitors in the culture medium on hPSC expansion and subsequent differentiation are not yet known. Perhaps substrates comprised of multiple peptides that activate distinct cell surface receptors might alleviate the need for ROCK inhibitors in the culture medium.

Melkoumian et al. described synthetic peptide acrylate surfaces (PAS) for supporting hESC self-renewal [25]. The peptide candidates for hESC binding were derived from active domains of the ECM proteins, such as bone sialoprotein (BSP), vitronectin (VN), fibronectin, and laminin. They showed that either BSP-PAS or VN-PAS exhibited a similar ability as Matrigel in mediating hESC binding and proliferation in three different media, X-VIVO with growth factors, mTeSR1, and medium supplemented with knockout serum replacer (KOSR). VN-PAS and X-VIVO medium supported hESC self-renewal for more than 10 passages after which the hESCs maintained a normal karyotype and the ability to differentiate to cells in each of the three germ layers. This synthetic surface does not require ROCK inhibitor for hPSC attachment and further expansion. Importantly, the VN-PAS substrate was also suitable for differentiation of hPSCs to cardiomyocytes, suggesting that this substrate may permit development of defined expansion and differentiation processes for generating large quantities of cells in clinically relevant lineages.

hESCs and iPSCs do not efficiently attach to uncoated tissue culture treated polystyrene surfaces. However, it is possible to fabricate polymers that bind hESCs on tissue culture polystyrene dishes to facilitate hESC culture. Villa-Diaz et al. identified that poly[2-(methacryloyloxy)ethyl dimethyl-(3-sulfopropyl)ammonium hydroxide] (PMEDSAH) was able to maintain various hESC and hiPSC lines [26]. hESCs grown on PMEDSAH-coated plates in MEF-conditioned medium remained in an undifferentiated state, maintain pluripotency, and did not acquire an abnormal karyotype. However, this substrate was not effective with all cell lines and media formulations tested. H9, but not BG01, hESCs could be expanded on PMEDSAH substrates in StemPro medium for 10 passages with confirmation of pluripotency and stable karyotype. However, neither cell line could be maintained on PMEDSAG in mTeSR1 medium. The variability of effectiveness of this substrate between cell lines and media formulations might result from differential adsorption of media factors or cell-secreted proteins to the PMEDSAH.

3.1.2.3 Mechanisms of Substrate Maintenance of Self-Renewal

The culture substrate must support initial hPSC attachment, but its long-term role in maintaining pluripotency is unclear. hPSCs secrete extracellular matrix proteins, including laminin-511 and nidogen-1 [27]. Laminin-511 and nidogen-1 support assembly of singularized hESCs into multicellular complexes, which may be important in their survival and expansion. Decellularized matrices secreted by human fibroblasts have also been shown to facilitate hESC and iPSC expansion and pluripotency over 15 passages [28]. Additional analysis of how hPSCs modify their substrate and the subsequent effects on cell proliferation and differentiation is needed.

In conclusion, chemically defined, synthetic, xeno-free, scalable media, and substrates that support the self-renewal and directed differentiation of hESCs and iPSCs will be useful for both research purposes and development of cell therapies. Development of these systems has been in part driven by a deeper understanding of molecular mechanisms regulating self-renewal, pluripotency, and differentiation, but also by hypothesis directed and high-throughput screening approaches. As pathways that control hPSC behavior become better understood, more robust and perhaps simpler media and matrices will be defined, and the field may come to a consensus culture system that is effective for both research and biomanufacturing applications.

3.2 Large-Scale Production of Somatic Cells from hPSCs

Biomanufacturing of somatic cells from hPSCs faces many of the same challenges as expansion of undifferentiated cells (Fig. 3.1). However, the differentiation process is more complex since appropriate developmental cues must be presented to guide cells through a differentiation process which often takes days to weeks. In addition, undesired cells often comprise a substantial fraction of the total population and must be removed, and desired cells must be profiled for expression of appropriate markers and phenotypes.

3.2.1 Embryoid Body Differentiation Processes

Many differentiation protocols utilize formation of embryoid bodies (EBs), aggregate of hPSC-derived cells. EBs can be formed by suspension culture of hESCs in the absence of self-renewal factors and the presence of serum or defined induction cues, and generally contain a wide variety of differentiated cell types. EBs typically recapitulate events that naturally occur in embryonic development, although in a very spatially disorganized fashion. EBs are amenable to large-scale processes because they can be cultured in suspension and contain many different cell types of interest. Challenges facing use of EBs in biomanufacturing include the generally low yield of desired cell types, the need to purify the desired population, and the lack of defined EB culture systems.

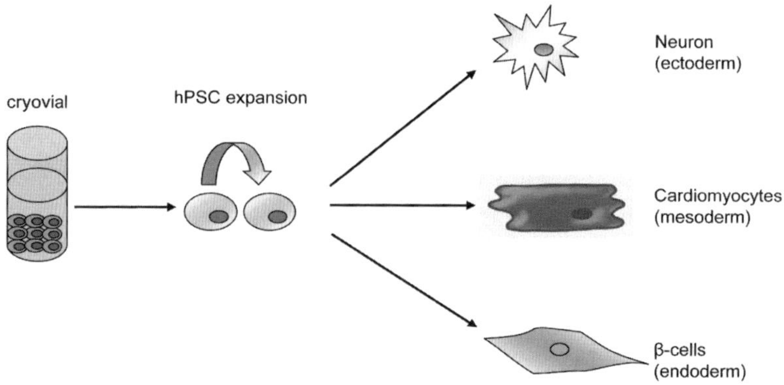

Fig. 3.1 Overview of biomanufacturing human pluripotent stem cells for therapeutic applications. hPSCs in cryogenic storage must be thawed, expanded under defined conditions, and further differentiated into clinical applicable lineages, including but not restricted to neurons, cardiomyocytes, and β-cells

Burridge et al. developed a protocol to generate contracting cardiomyocytes from hPSC-derived EBs by systematic optimization of 45 process variables, including concentration and timing of cardiac inductive factors, culture medium composition, cell density, and colony morphology [29]. This protocol generated 65–89% pure cardiomyocytes in 11 different hESC and iPSC lines, suggesting the potential of EB-based methods in efficient and scalable cell differentiation processes.

3.2.2 Directed Differentiation

While EBs are often effective at recapitulating developmental events, directed differentiation protocols which apply differentiation cues to cells at various stages of development are more amenable to biomanufacturing processes because of the tighter control over differentiation, which typically results in higher yields and more robust differentiation. As a better understanding of developmental cues is obtained and systems to apply these cues in the appropriate context are engineered, defined, and scalable processes to generate differentiated cells from hPSCs are becoming more common. Some examples of such defined culture protocols are discussed here.

3.2.2.1 Neuroepithelial Cells

Inhibition of SMAD signaling was found to direct greater than 80% of hESCs to a neural progenitor fate [30]. In another study, treatment of hESCs with inhibitors of glycogen synthase 3, TGFβ, and Notch pathways generated primitive neuroepithelial cells, which were able to be converted to midbrain and hindbrain neuronal subtypes by subsequent treatment with protein growth and differentiation factors [31]. A defined system for generating approximately 30% pure populations of dopaminergic neurons from neuroepithelial progenitors by treatment with FGF8 and sonic hedgehog has also been reported [32].

Neural progenitors derived from hESCs have also been directed to retinal pigment epithelial (RPE) cells by treatment with nicotinamide and TGF superfamily ligands [33]. These RPE cells can rescue defects in animal retinal degeneration models. RPE and photoreceptor cells have also been differentiated from hPSCs using small molecules, including a casein I kinase inhibitor, an ALK4 inhibitor, a Rho-associated kinase inhibitor, retinoic acid, and taurine [34].

3.2.2.2 Cardiomyocytes

Substantial progress has been made in generating functional cardiomyocytes from hPSCs by directed differentiation protocols. TGFβ superfamily ligands Activin A and BMP4, and Wnt3a induce mesendoderm development of hPSCs cultured in a monolayer [35]. A substantial fraction of these cells, up to 30% depending on the cell line, develop into contractile cardiomyocytes. Another protocol uses Activin A, BMP4, bFGF, and VEGF to induce mesoderm formation, then inhibits canonical Wnt signaling via Dkk1 to specify cardiac lineages [36]. Sorting Flk-1^{+}/c-kit^{+} cardiomyocyte progenitors substantially enhanced the purity of cardiomyocytes generated by this protocol, but may be difficult to implement in a scalable biomanufacturing process [36]. More robust processes that efficiently generate cardiomyocytes from many different hPSC lines and methods to mature the embryonic-like cardiomyocytes that result from these protocols are needed.

3.2.2.3 Islet β Cells

Another somatic cell type of significant therapeutic potential is the pancreatic β-cell. Cells with the ability to monitor and respond to changes in glucose by releasing insulin to regulate blood sugar levels would offer diabetics an option in managing their disease. One protocol to generate insulin-producing β-cells from hESCs relies on a complex mixture of proteins and small molecules to guide hPSCs through definitive endoderm, foregut, pancreatic endoderm, and endocrine progenitors to pancreatic endocrine cells [37]. Another protocol generated definitive endoderm by treating hPSCs with Activin A and sodium butyrate, then EGF, bFGF, Noggin, nicotinamide, and IGF-II were added at various stages to specify pancreatic endoderm [38].

Small molecules which guide hPSC differentiation toward β-cells have also been identified. The compounds IDE1 and IDE2 have been shown to induce definitive endoderm differentiation via activation of TGFβ/Nodal signaling [39]. Stauprimide downregulates c-MYC expression and causes hPSCs to preferentially differentiate to definitive endoderm [40]. Definitive endoderm cells can be directed to pancreatic endoderm by treatment with (-)-Indolactam, an agonist of protein kinase C signaling [41].

3.2.2.4 Other Cells and Tissues

Defined methods have also been developed to generate epidermal keratinocytes and 3D skin tissue constructs from hESCs and iPSCs via treatment with retinoic acid and BMP4 [42–44]. Defined differentiation systems for generating hematopoietic progenitors and blood cells, including erythroid cells, dendritic cells, macrophages, neutrophils, and megakaryocytes have been developed [45, 46], raising the interesting possibility of reconstituting blood from a stem cell source. Chondrocytes expressing SOX9 have been generated from hESCs using a three stage defined protocol that guides cells through mesendoderm, mesoderm, and chondrocyte lineages by treatment with BMP4 and other TGF superfamily ligands [47]. Hepatocytes have been derived from hESCs using Activin A, FGF2, BMP4, and a PI3 K inhibitor to guide cells to endodermal progenitors, then FGF10, retinoic acid, and an Activin/Nodal inhibitor to generate hepatocytes [48]. These and future defined systems that yield pure populations of these cells will facilitate evaluation of these cells in animal models and clinical trials, and be an important part of developing hPSC-based cellular therapies.

3.3 Scalable 3D Culture Systems for hPSC Expansion and Differentiation

Due to the large number of cells required for hESC and iPSC therapeutic applications, the development of large-scale culture systems for cell expansion and differentiation is critical. Scaling hPSC culture for biomanufacturing applications faces many challenges. Cell density must be increased by orders of magnitude compared to lab scale culture. This is typically achieved by culture in suspension or on suspended microparticles. However, hPSCs are prone to aggregation in suspension and are sensitive to shear stresses. Cell aggregates are heterogeneous in size and may severely limit diffusion of nutrients, growth factors, metabolic byproducts, and dissolved gases within the aggregate, which may in turn affect expansion potential, lead to spontaneous differentiation, or cause or acquisition of abnormal karyotypes. In addition, loss of matrix attachment in large-scale cultures may induce undesired differentiation, leading to culture contamination. Furthermore, many of the hPSC culture systems described earlier in this chapter often require undefined or expensive chemical component, which limit process development and implementation. The following sections will review application of bioreactors and scalable culture systems to hPSC expansion and differentiation, and will discuss challenges facing future culture system improvements as related to biomanufacturing.

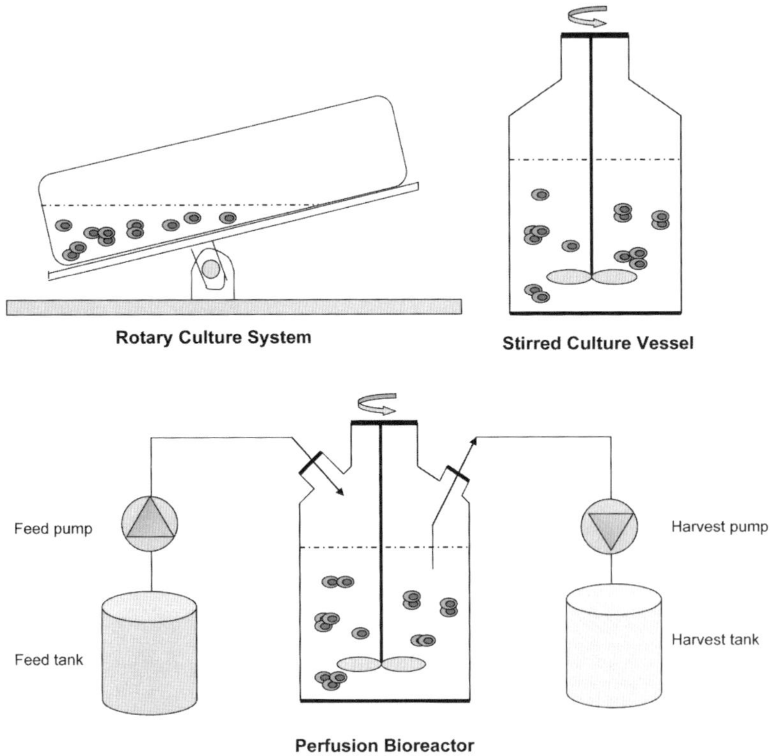

Fig. 3.2 Scalable suspension culture systems for expansion and differentiation of human pluripotent stem cells

3.3.1 Rotary and Stirred Culture Systems

Rotary cell culture systems provide a well-mixed environment and efficient gas transfer through a permeable silicon membrane. Appropriate inoculation density of hESCs and iPSCs is important for rotary cell culture systems. No or only partial formation of aggregates occurred with too low inoculation density which reduce the cell vitality, while high density will cause extensive aggregate fusion and formation of heterogenous clusters which could induce cell differentiation. Scale-up of low inoculation density, 0.33×10^5 cells/ml, hPSC cultures have been shown in ROCK inhibitor-supplemented mTeSR1 medium in Erlenmeyer flasks rotated at 50 rpm, with up to a 6-fold expansion achieved in 4 days [49]. However, high-density (1×10^6 cells/ml) cultures were shown to be better suited to a rotating Petri dish platform, resulting in a 2- to 3-fold expansion within 7 days [49]. Although rotary culture systems do not typically generate high shear forces, which can induce death or differentiation of hESCs [50], compared to stirred culture vessels, they are limited in size and thus may not be capable of producing large quantities of cells required in biomanufacturing applications (Fig. 3.2).

Stirred culture systems such as spinner flasks and stirred-tank bioreactors have been widely used in the scale-up of mammalian cell culture. These systems provide a well-mixed, homogeneous culture environment with robust control, and monitoring of culture parameters, including dissolved gases, nutrient and metabolite concentrations, perfusion rates, and mixing speeds. Four hESC and two iPSC lines cultured in spinner flasks in medium containing Knockout Serum Replacer supplemented with the IL6RIL6 chimera (interleukin-6 receptor fused to interleukin-6) and basic fibroblast growth factor have been shown to expand by approximately 25-fold over 10 days, and maintained pluripotency and differentiation capacity after 20 passages [49, 51].

Cameron et al. reported a greater than 3-fold improvement in hESC-derived EB expansion in spinner flasks as compared to static cultures [52]. The EBs in the spinner flasks were also more homogeneous in size and morphology. Yirme et al. demonstrated that hydrodynamic conditions during seeding affect cell metabolic and growth rates in hESC-derived EBs in a glass ball impeller spinner flask [53]. Sargent et al. further showed that mixing speed in a rotary culture affects murine EB formation efficiency, EB size and morphology, and cell differentiation fate [54]. EBs cultured in a rotary suspension system were enriched in cardiomyocytes compared to EBs maintained in static culture [55]. Together, these studies highlight the importance of considering mixing on expansion and differentiation of hPSCs.

3.3.2 Perfusion Bioreactors

Perfusion bioreactors, in which medium flows through the culture vessel, have also been used to expand and differentiate hPSCs. A study of the effects of flow on hESC growth in microfluidic bioreactors illustrated that at low flow rates cell expansion is limited by nutrient depletion and waste accumulation while cells at high flow rates undergo detachment and/or spontaneous differentiation [56]. Thus, an optimum perfusion rate exists to expand undifferentiated cells. One approach to addressing the shear sensitivity of hPSCs in a perfusion reactor is to culture cells in grooves on a surface, so that convection supplies nutrients to the cells but does not expose them to high shear stresses. Culturing fibroblasts in microchannels and microwells enabled a perfusion rate an order of magnitude higher than that which supported culture on a planar substrate [57]. Microwell culture of hESCs has also been demonstrated [58–60], and would also likely allow for higher perfusion rates than have been used in other systems. Intermittent perfusion, short periods of flow followed by longer periods of static culture, have also been shown to reduce shear sensitivity of hESCs cultured in microchannels [61].

Perfusion has been shown to enhance the density of hESCs expanded in MEF-conditioned medium by 70% as compared to static culture [62]. A comparison of hESC expansion in stirred tank and perfusion systems demonstrated higher cell densities in a perfusion system when oxygen was controlled to support aerobic metabolism [63].

A scalable bioprocess was developed to generate cardiomyocytes from murine ESCs in a perfusion bioreactor [64]. In this system, hypoxia and encapsulation of cell aggregates were found to induce cardiomyocyte differentiation. Cardiomyocytes were selected based on G418 resistance, and optimization of reactor operation yielded greater than 3 cardiomyocytes per ESC after 2 weeks of differentiation. Niebruegge et al. generated a pure population of 4.6×10^9 cardiomyocytes from murine ESCs in a 2 l perfusion bioreactor, also using antibiotic selection [65]. These studies demonstrate the feasibility of scalable pluripotent stem cell differentiation processes, which will be improved upon implementation of new, defined differentiation protocols discussed in Sect. 2.2.

3.3.3 Expansion and Differentiation of hPSCs on Microcarriers

While rotary, stirred culture, and perfusion bioreactors are best suited to expand mammalian cells that can grow in suspension, microcarriers permit culture of adhesion-dependent cells at a high density in these bioreactors. Microcarriers are small, on the order of 100 s of microns, particles that provide support for high density growth of anchorage-dependent cells. These particles provide a high surface area to volume ratio and can be suspended in bioreactors, including rotary bioreactors, spinner flasks, and fluidized beds. Microcarriers can be fabricated from a variety of materials, including dextrans, glass, polymers, and extracellular matrix proteins, and can be functionalized to present desired surface properties. Several studies have demonstrated the feasibility and challenges of hPSC culture on microcarriers. A proof-of-principle study demonstrated that hESCs could be cultured in suspension with MEF-conditioned medium on trimethyl ammonium-coated polystyrene microcarriers in low attachment 6-well plates [66]. Recently, hESCs were expanded as undifferentiated cells in 3D suspension cultures on Matrigel-coated cellulose microcarriers in a stirred culture vessel [67, 68]. The microcarrier system achieved 2- to 4-fold higher cell densities than 2D colony cultures. Stable, continuous propagation of two hESC lines on cellulose microcarriers in spinner flask has been demonstrated in MEF-conditioned media for 6 months [67]. More importantly these cellulose microcarrier spinner flask cultures were also compatible with two serum-free defined media, StemPro and mTeSR1 [67]. Instead of Matrigel, laminin coated microcarriers are also capable of supporting the long-term propagation and pluripotency of multiple hESC lines, permitting culture on defined matrices [69]. Microcarriers have also been used to produce feeder fibroblasts. Human fibroblasts have been expanded on microcarriers and these human fibroblasts are suitable for ex vivo expansion of clinical-grade hESCs [70]. However, given the developments of defined culture media, it is unlikely that expansion of feeder cells will play a significant role in hPSC biomanufacturing applications.

Microcarrier size and composition were found to affect expansion and differentiation fates of hPSCs. Positively-charged cylindrical cellulose microcarriers and large positively charged spherical dextran microcarriers exhibited high cell expansion potential and levels of pluripotency [69]. Lower cell yields were obtained using smaller diameter spherical or macroporous beads [69]. Lecina et al. investigated effects of microcarrier composition and morphology on cardiac myocyte differentiation efficiency from hESCs cultured in serum-free medium [71]. Cells cultured on TOSOH-10 microcarriers produced 80% beating EBs containing 20% cardiomyocytes. Lock and Tzanakakis induced definitive endoderm differentiation using Activin A, Wnt3a, and serum in 80% of cells hESCs cultured on microcarriers [72]. Better insight into how microcarrier geometry and surface chemistry affect cell expansion and differentiation is needed to develop efficient, scalable microcarrier processes.

3.3.4 Carrier-Free Suspension Culture of hPSCs

While microcarriers represent a logical transition from lab scale plate culture to bioreactors, it is not clear that hPSCs are only capable of anchorage-dependent expansion. hPSCs have been expanded in suspension in spinner flasks, unattached to any support structure, time periods of approximately 6 weeks [66]. Steiner et al. discovered that when hESC clusters in suspension were cultured in Neurobasal medium, they did not differentiate to neural lineages. In fact, a high fraction of the cells in the clusters remained undifferentiated. They subsequently optimized the culture medium by including KOSR, laminin, gelatin, fibronectin, Nutridoma-CS, and the neurotrophins BDNF, NT3, and NT4 to the Neurobasal medium. Within this defined suspension culture system, hESCs displayed a doubling time and apoptosis activity comparable to cells cultured on standard MEF feeders and were confirmed to be pluripotent and karyotypically normal after prolonged culture [73]. Furthermore, new hESC lines could be derived and maintained in suspension, suggesting a new feeder-free defined system for derivation of new hESC lines. When the hESCs cultured in suspension returned to 2D culture, they displayed similar differentiation ability to neural lineages. This demonstration of expanding hESCs in suspension culture without a solid support suggests the promise of high density hPSC culture for clinical and industrial applications.

3.4 Future Perspectives and Challenges

Substantial progress has been made in developing laboratory scale defined culture systems for hPSCs. While these defined media and matrices represent a major step forward in large-scale production of hPSCs and their progeny, substantial efforts in bioprocess development are necessary. Most published bioreactor studies are limited in scale and scope, in part resulting from difficulty and cost associated with these projects. Logically, the foundation of hPSC bioreactor development has been prior work in mammalian cell culture (e.g., CHO cells), but unique properties of hPSCs will result in added complexity of hPSC biomanufacturing over current methodologies.

As the field of hPSC biomanufacturing matures toward generating cells for human therapies, substantial effort will be needed to ensure the quality of the desired product during the production process. Cell viability and density must be monitored, most likely through automated cell counting. Cell differentiation status must also be monitored. Cell surface markers provide easy but better differentiation markers are needed to distinguish cells at various differentiation stages. Depending on the particular cell type generated and its specific application, functional benchmarks must also be met. For example, cardiomyocytes must be able to electromechanically integrate with native heart tissue and islet cells must secrete insulin in a glucose-dependent manner.

Product safety is of utmost concern in cell-based therapeutics. Culture contamination by bacteria, fungi, viruses, and endotoxins must be monitored by genetic and/or chemical testing. hPSCs are prone to acquiring chromosomal abnormalities, typically gains of chromosome arms or full chromosomes, that increase expansion rates and affect differentiation potential [74]. Safe, high-throughput passaging protocols and improved identification and removal of cells with abnormal genomes, such as by high resolution single nucleotide polymorphism (SNP) or comparative genomic hybridization (CGH) analyses, are needed [75, 76]. Undifferentiated hPSCs also pose a safety concern in which they have the potential to form teratomas [77]. Residual undifferentiated cells can be removed from a mixed population of cells using immunodepletion with antibodies against pluripotency surface markers, including SSEA-5, CD9, CD30, CD50, CD90, and CD200 [78].

Finally, defined and efficient methods for generating functional somatic cells from hPSCs are needed. This will require a better mechanistic understanding of how cells make fate choices, engineered systems that provide microenvironmental cues in the appropriate spatial and temporal fashion, and methods to characterize cell differentiation state and phenotype.

Acknowledgments The authors thank members of the Palecek lab for helpful discussions of the topic matter. Support for this chapter was provided by the National Science Foundation grant EFRI-0735903 and National Institute of Biomedical Imaging and Bioengineering grant R01EB007534.

References

1. Thomson JA, Itskovitz-Eldor J, Shapiro SS, Waknitz MA, Swiergiel JJ, Marshall VS, Jones JM (1998) Embryonic stem cell lines derived from human blastocysts. Science 282(5391): 1145–1147
2. Yu J, Vodyanik MA, Smuga-Otto K, Antosiewicz-Bourget J, Frane JL, Tian S, Nie J, Jonsdottir GA, Ruotti V, Stewart R, Slukvin II, Thomson JA (2007) Induced pluripotent stem cell lines derived from human somatic cells. Science 318(5858):1917–1920

3. Takahashi K, Tanabe K, Ohnuki M, Narita M, Ichisaka T, Tomoda K, Yamanaka S (2007) Induction of pluripotent stem cells from adult human fibroblasts by defined factors. Cell 131(5):861–872
4. Soldner F, Laganiere J, Cheng AW, Hockemeyer D, Gao Q, Alagappan R, Khurana V, Golbe LI, Myers RH, Lindquist S, Zhang L, Guschin D, Fong LK, Vu BJ, Meng X, Urnov FD, Rebar EJ, Gregory PD, Zhang HS, Jaenisch R (2011) Generation of isogenic pluripotent stem cells differing exclusively at two early onset Parkinson point mutations. Cell 146(2):318–331
5. Xu RH, Chen X, Li DS, Li R, Addicks GC, Glennon C, Zwaka TP, Thomson JA (2002) BMP4 initiates human embryonic stem cell differentiation to trophoblast. Nat Biotechnol 20(12):1261–1264
6. Xu RH, Peck RM, Li DS, Feng X, Ludwig T, Thomson JA (2005) Basic FGF and suppression of BMP signaling sustain undifferentiated proliferation of human ES cells. Nat Methods 2(3):185–190
7. Levenstein ME, Ludwig TE, Xu RH, Llanas RA, VanDenHeuvel-Kramer K, Manning D, Thomson JA (2006) Basic fibroblast growth factor support of human embryonic stem cell self-renewal. Stem Cells 24(3):568–574
8. Vallier L, Mendjan S, Brown S, Chng Z, Teo A, Smithers LE, Trotter MW, Cho CH, Martinez A, Rugg-Gunn P, Brons G, Pedersen RA (2009) Activin/Nodal signalling maintains pluripotency by controlling Nanog expression. Development 136(8):1339–1349
9. Xu RH, Sampsell-Barron TL, Gu F, Root S, Peck RM, Pan G, Yu J, Antosiewicz-Bourget J, Tian S, Stewart R, Thomson JA (2008) NANOG is a direct target of TGFbeta/activin-mediated SMAD signaling in human ESCs. Cell Stem Cell 3(2):196–206
10. Vallier L, Alexander M, Pedersen RA (2005) Activin/Nodal and FGF pathways cooperate to maintain pluripotency of human embryonic stem cells. J Cell Sci 118(Pt 19):4495–4509
11. Yao S, Chen S, Clark J, Hao E, Beattie GM, Hayek A, Ding S (2006) Long-term self-renewal and directed differentiation of human embryonic stem cells in chemically defined conditions. Proc Natl Acad Sci U S A 103(18):6907–6912
12. Xu C, Inokuma MS, Denham J, Golds K, Kundu P, Gold JD, Carpenter MK (2001) Feeder-free growth of undifferentiated human embryonic stem cells. Nat Biotechnol 19(10):971–974
13. Lu J, Hou R, Booth CJ, Yang SH, Snyder M (2006) Defined culture conditions of human embryonic stem cells. Proc Natl Acad Sci U S A 103(15):5688–5693
14. Furue MK, Na J, Jackson JP, Okamoto T, Jones M, Baker D, Hata R, Moore HD, Sato JD, Andrews PW (2008) Heparin promotes the growth of human embryonic stem cells in a defined serum-free medium. Proc Natl Acad Sci U S A 105(36):13409–13414
15. Wang L, Schulz TC, Sherrer ES, Dauphin DS, Shin S, Nelson AM, Ware CB, Zhan M, Song CZ, Chen X, Brimble SN, McLean A, Galeano MJ, Uhl EW, D'Amour KA, Chesnut JD, Rao MS, Blau CA, Robins AJ (2007) Self-renewal of human embryonic stem cells requires insulin-like growth factor-1 receptor and ERBB2 receptor signaling. Blood 110(12):4111–4119
16. Ludwig TE, Levenstein ME, Jones JM, Berggren WT, Mitchen ER, Frane JL, Crandall LJ, Daigh CA, Conard KR, Piekarczyk MS, Llanas RA, Thomson JA (2006) Derivation of human embryonic stem cells in defined conditions. Nat Biotechnol 24(2):185–187
17. Draper JS, Smith K, Gokhale P, Moore HD, Maltby E, Johnson J, Meisner L, Zwaka TP, Thomson JA, Andrews PW (2004) Recurrent gain of chromosomes 17q and 12 in cultured human embryonic stem cells. Nat Biotechnol 22(1):53–54
18. Garcia-Gonzalo FR, Izpisua Belmonte JC (2008) Albumin-associated lipids regulate human embryonic stem cell self-renewal. PLoS ONE 3(1):e1384
19. Chen G, Gulbranson DR, Hou Z, Bolin JM, Ruotti V, Probasco MD, Smuga-Otto K, Howden SE, Diol NR, Propson NE, Wagner R, Lee GO, Antosiewicz-Bourget J, Teng JM, Thomson JA (2011) Chemically defined conditions for human iPSC derivation and culture. Nat Methods 8(5):424–429
20. Mummery CL, Braam SR, Zeinstra L, Litjens S, Ward-van Oostwaard D, van den Brink S, van Laake L, Lebrin F, Kats P, Hochstenbach R, Passier R, Sonnenberg A (2008) Recombinant vitronectin is a functionally defined substrate that supports human embryonic stem cell self-renewal via alpha V beta 5 integrin. Stem Cells 26(9):2257–2265

21. Rodin S, Domogatskaya A, Strom S, Hansson EM, Chien KR, Inzunza J, Hovatta O, Tryggvason K (2010) Long-term self-renewal of human pluripotent stem cells on human recombinant laminin-511. Nat Biotechnol 28(6):611–615
22. Duncan SA, Nagaoka M, Si-Tayeb K, Akaike T (2010) Culture of human pluripotent stem cells using completely defined conditions on a recombinant E-cadherin substratum. Bmc Developmental Biology 10, 60
23. Sato N, Walker A, Su H, Conti MA, Harb N, Adelstein RS (2010) Non-muscle myosin II regulates survival threshold of pluripotent stem cells. Nature Communications 1, 71
24. Klim JR, Li L, Wrighton PJ, Piekarczyk MS, Kiessling LL (2010) A defined glycosaminoglycan-binding substratum for human pluripotent stem cells. Nat Methods 7(12):989–994
25. Melkoumian Z, Weber JL, Weber DM, Fadeev AG, Zhou Y, Dolley-Sonneville P, Yang J, Qiu L, Priest CA, Shogbon C, Martin AW, Nelson J, West P, Beltzer JP, Pal S, Brandenberger R (2010) Synthetic peptide-acrylate surfaces for long-term self-renewal and cardiomyocyte differentiation of human embryonic stem cells. Nat Biotechnol 28(6):606–610
26. Villa-Diaz LG, Nandivada H, Ding J, Nogueira-de-Souza NC, Krebsbach PH, O'Shea KS, Lahann J, Smith GD (2010) Synthetic polymer coatings for long-term growth of human embryonic stem cells. Nat Biotechnol 28(6):581–583
27. Evseenko D, Schenke-Layland K, Dravid G, Zhu Y, Hao Q-L, Scholes J, Wang XC, Maclellan WR, Crooks GM (2009) Identification of the critical extracellular matrix proteins that promote human embryonic stem cell assembly. Stem Cells Dev 18(6):919–928
28. Abraham S, Sheridan SD, Miller B, Rao RR (2010) Stable propagation of human embryonic and induced pluripotent stem cells on decellularized human substrates. Biotechnol Progr 26(4):1126–1134
29. Burridge PW, Thompson S, Millrod MA, Weinberg S, Yuan X, Peters A, Mahairaki V, Koliatsos VE, Tung L, Zambidis ET (2011) A universal system for highly efficient cardiac differentiation of human induced pluripotent stem cells that eliminates interline variability. PLoS ONE 6(4):e18293
30. Chambers SM, Fasano CA, Papapetrou EP, Tomishima M, Sadelain M, Studer L (2009) Highly efficient neural conversion of human ES and iPS cells by dual inhibition of SMAD signaling. Nat Biotechnol 27(3):275–280. doi:10.1038/nbt.1529
31. Li W, Sun W, Zhang Y, Wei W, Ambasudhan R, Xia P, Talantova M, Lin T, Kim J, Wang X, Kim WR, Lipton SA, Zhang K, Ding S (2011) Rapid induction and long-term self-renewal of primitive neural precursors from human embryonic stem cells by small molecule inhibitors. Proc Natl Acad Sci U S A 108(20):8299–8304
32. Ma L, Liu Y, Zhang SC (2011) Directed differentiation of dopamine neurons from human pluripotent stem cells. Methods Mol Biol 767:411–418
33. Idelson M, Alper R, Obolensky A, Ben-Shushan E, Hemo I, Yachimovich-Cohen N, Khaner H, Smith Y, Wiser O, Gropp M, Cohen MA, Even-Ram S, Berman-Zaken Y, Matzrafi L, Rechavi G, Banin E, Reubinoff B (2009) Directed differentiation of human embryonic stem cells into functional retinal pigment epithelium cells. Cell Stem Cell 5(4):396–408
34. Osakada F, Jin ZB, Hirami Y, Ikeda H, Danjyo T, Watanabe K, Sasai Y, Takahashi M (2009) In vitro differentiation of retinal cells from human pluripotent stem cells by small-molecule induction. J Cell Sci 122(Pt 17):3169–3179
35. Laflamme MA, Chen KY, Naumova AV, Muskheli V, Fugate JA, Dupras SK, Reinecke H, Xu C, Hassanipour M, Police S, O'Sullivan C, Collins L, Chen Y, Minami E, Gill EA, Ueno S, Yuan C, Gold J, Murry CE (2007) Cardiomyocytes derived from human embryonic stem cells in pro-survival factors enhance function of infarcted rat hearts. Nat Biotechnol 25(9):1015–1024
36. Yang L, Soonpaa MH, Adler ED, Roepke TK, Kattman SJ, Kennedy M, Henckaerts E, Bonham K, Abbott GW, Linden RM, Field LJ, Keller GM (2008) Human cardiovascular progenitor cells develop from a KDR+ embryonic-stem-cell-derived population. Nature 453(7194):524–528

37. D'Amour KA, Bang AG, Eliazer S, Kelly OG, Agulnick AD, Smart NG, Moorman MA, Kroon E, Carpenter MK, Baetge EE (2006) Production of pancreatic hormone-expressing endocrine cells from human embryonic stem cells. Nat Biotechnol 24(11):1392–1401
38. Jiang J, Au M, Lu K, Eshpeter A, Korbutt G, Fisk G, Majumdar AS (2007) Generation of insulin-producing islet-like clusters from human embryonic stem cells. Stem Cells 25(8):1940–1953
39. Borowiak M, Maehr R, Chen S, Chen AE, Tang W, Fox JL, Schreiber SL, Melton DA (2009) Small molecules efficiently direct endodermal differentiation of mouse and human embryonic stem cells. Cell Stem Cell 4(4):348–358
40. Zhu S, Wurdak H, Wang J, Lyssiotis CA, Peters EC, Cho CY, Wu X, Schultz PG (2009) A small molecule primes embryonic stem cells for differentiation. Cell Stem Cell 4(5):416–426
41. Chen S, Borowiak M, Fox JL, Maehr R, Osafune K, Davidow L, Lam K, Peng LF, Schreiber SL, Rubin LL, Melton D (2009) A small molecule that directs differentiation of human ESCs into the pancreatic lineage. Nat Chem Biol 5(4):258–265
42. Metallo CM, Ji L, de Pablo JJ, Palecek SP (2008) Retinoic acid and bone morphogenetic protein signaling synergize to efficiently direct epithelial differentiation of human embryonic stem cells. Stem Cells 26(2):372–380
43. Metallo CM, Azarin SM, Moses LE, Ji L, de Pablo JJ, Palecek SP (2010) Human embryonic stem cell-derived keratinocytes exhibit an epidermal transcription program and undergo epithelial morphogenesis in engineered tissue constructs. Tissue Eng Part A 16(1):213–223
44. Itoh M, Kiuru M, Cairo MS, Christiano AM (2011) Generation of keratinocytes from normal and recessive dystrophic epidermolysis bullosa-induced pluripotent stem cells. Proc Natl Acad Sci U S A 108(21):8797–8802
45. Salvagiotto G, Burton S, Daigh CA, Rajesh D, Slukvin II, Seay NJ (2011) A defined, feeder-free, serum-free system to generate in vitro hematopoietic progenitors and differentiated blood cells from hESCs and hiPSCs. PLoS ONE 6(3):e17829
46. Chicha L, Feki A, Boni A, Irion O, Hovatta O, Jaconi M (2011) Human pluripotent stem cells differentiated in fully defined medium generate hematopoietic CD34- and CD34+ progenitors with distinct characteristics. PLoS ONE 6(2):e14733
47. Oldershaw RA, Baxter MA, Lowe ET, Bates N, Grady LM, Soncin F, Brison DR, Hardingham TE, Kimber SJ (2010) Directed differentiation of human embryonic stem cells toward chondrocytes. Nat Biotechnol 28(11):1187–1194
48. Touboul T, Hannan NR, Corbineau S, Martinez A, Martinet C, Branchereau S, Mainot S, Strick-Marchand H, Pedersen R, Di Santo J, Weber A, Vallier L (2010) Generation of functional hepatocytes from human embryonic stem cells under chemically defined conditions that recapitulate liver development. Hepatology 51(5):1754–1765. doi:10.1002/hep.23506
49. Zweigerdt R, Olmer R, Singh H, Haverich A, Martin U (2011) Scalable expansion of human pluripotent stem cells in suspension culture. Nat Protoc 6(5):689–700
50. Oh SKW, Leung HW, Chen A, Choo ABH, Reuveny S (2011) Agitation can induce differentiation of human pluripotent stem cells in microcarrier cultures. Tissue Eng Part C-Me 17(2):165–172
51. Amit M, Laevsky I, Miropolsky Y, Shariki K, Peri M, Itskovitz-Eldor J (2011) Dynamic suspension culture for scalable expansion of undifferentiated human pluripotent stem cells. Nat Protoc 6(5):572–579
52. Cameron CM, Hu WS, Kaufman DS (2006) Improved development of human embryonic stem cell-derived embryoid bodies by stirred vessel cultivation. Biotechnol Bioeng 94(5):938–948
53. Yirme G, Amit M, Laevsky I, Osenberg S, Itskovitz-Eldor J (2008) Establishing a dynamic process for the formation, propagation, and differentiation of human embryoid bodies. Stem Cells Dev 17(6):1227–1241
54. Sargent CY, Berguig GY, Kinney MA, Hiatt LA, Carpenedo RL, Berson RE, McDevitt TC (2010) Hydrodynamic modulation of embryonic stem cell differentiation by rotary orbital suspension culture. Biotechnol Bioeng 105(3):611–626

55. Sargent CY, Berguig GY, McDevitt TC (2009) Cardiomyogenic differentiation of embryoid bodies is promoted by rotary orbital suspension culture. Tissue Eng Pt A 15(2):331–342
56. Titmarsh D, Hidalgo A, Turner J, Wolvetang E, Cooper-White J (2011) Optimization of flowrate for expansion of human embryonic stem cells in perfusion microbioreactors. Biotechnol Bioeng 108(12):2894–2904
57. Korin N, Bransky A, Khoury M, Dinnar U, Levenberg S (2009) Design of well and groove microchannel bioreactors for cell culture. Biotechnol Bioeng 102(4):1222–1230
58. Mohr JC, Zhang J, Azarin SM, Soerens AG, de Pablo JJ, Thomson JA, Lyons GE, Palecek SP, Kamp TJ (2010) The microwell control of embryoid body size in order to regulate cardiac differentiation of human embryonic stem cells. Biomaterials 31(7):1885–1893
59. Mohr JC, de Pablo JJ, Palecek SP (2006) 3-D microwell culture of human embryonic stem cells. Biomaterials 27(36):6032–6042
60. Khademhosseini A, Ferreira L, Blumling J 3rd, Yeh J, Karp JM, Fukuda J, Langer R (2006) Co-culture of human embryonic stem cells with murine embryonic fibroblasts on microwell-patterned substrates. Biomaterials 27(36):5968–5977
61. Korin N, Bransky A, Dinnar U, Levenberg S (2009) Periodic "flow-stop" perfusion microchannel bioreactors for mammalian and human embryonic stem cell long-term culture. Biomed Microdevices 11(1):87–94
62. Fong WJ, Tan HL, Choo A, Oh SK (2005) Perfusion cultures of human embryonic stem cells. Bioprocess Biosyst Eng 27(6):381–387
63. Serra M, Brito C, Sousa MF, Jensen J, Tostoes R, Clemente J, Strehl R, Hyllner J, Carrondo MJ, Alves PM (2010) Improving expansion of pluripotent human embryonic stem cells in perfused bioreactors through oxygen control. J Biotechnol 148(4):208–215
64. Bauwens C, Yin T, Dang S, Peerani R, Zandstra PW (2005) Development of a perfusion fed bioreactor for embryonic stem cell-derived cardiomyocyte generation: oxygen-mediated enhancement of cardiomyocyte output. Biotechnol Bioeng 90(4):452–461
65. Niebruegge S, Nehring A, Bar H, Schroeder M, Zweigerdt R, Lehmann J (2008) Cardiomyocyte production in mass suspension culture: embryonic stem cells as a source for great amounts of functional cardiomyocytes. Tissue Eng Part A 14(10):1591–1601
66. Phillips BW, Horne R, Lay TS, Rust WL, Teck TT, Crook JM (2008) Attachment and growth of human embryonic stem cells on microcarriers. J Biotechnol 138(1–2):24–32
67. Oh SK, Chen AK, Mok Y, Chen X, Lim UM, Chin A, Choo AB, Reuveny S (2009) Long-term microcarrier suspension cultures of human embryonic stem cells. Stem Cell Res 2(3):219–230
68. Nie Y, Bergendahl V, Hei DJ, Jones JM, Palecek SP (2009) Scalable culture and cryopreservation of human embryonic stem cells on microcarriers. Biotechnol Prog 25(1):20–31
69. Chen AK, Chen X, Choo AB, Reuveny S, Oh SK (2011) Critical microcarrier properties affecting the expansion of undifferentiated human embryonic stem cells. Stem Cell Res 7(2):97–111
70. Phillips BW, Lim RYM, Tan TT, Rust WL, Crook JM (2008) Efficient expansion of clinical-grade human fibroblasts on microcarriers: Cells suitable for ex vivo expansion of clinical-grade hESCs. J Biotechnol 134(1–2):79–87
71. Lecina M, Ting S, Choo A, Reuveny S, Oh S (2010) Scalable platform for human embryonic stem cell differentiation to cardiomyocytes in suspended microcarrier cultures. Tissue Eng Part C Methods 16(6):1609–1619
72. Lock LT, Tzanakakis ES (2009) Expansion and differentiation of human embryonic stem cells to endoderm progeny in a microcarrier stirred-suspension culture. Tissue Eng Part A 15(8):2051–2063
73. Steiner D, Khaner H, Cohen M, Even-Ram S, Gil Y, Itsykson P, Turetsky T, Idelson M, Aizenman E, Ram R, Berman-Zaken Y, Reubinoff B (2010) Derivation, propagation and controlled differentiation of human embryonic stem cells in suspension. Nat Biotechnol 28(4):361–364

74. Taapken SM, Nisler BS, Newton MA, Sampsell-Barron TL, Leonhard KA, McIntire EM, Montgomery KD (2011) Karotypic abnormalities in human induced pluripotent stem cells and embryonic stem cells. Nat Biotechnol 29(4):313–314
75. Elliott AM, Elliott KA, Kammesheidt A (2010) High resolution array-CGH characterization of human stem cells using a stem cell focused microarray. Mol Biotechnol 46(3):234–242
76. Laurent LC, Ulitsky I, Slavin I, Tran H, Schork A, Morey R, Lynch C, Harness JV, Lee S, Barrero MJ, Ku S, Martynova M, Semechkin R, Galat V, Gottesfeld J, Izpisua Belmonte JC, Murry C, Keirstead HS, Park HS, Schmidt U, Laslett AL, Muller FJ, Nievergelt CM, Shamir R, Loring JF (2011) Dynamic changes in the copy number of pluripotency and cell proliferation genes in human ESCs and iPSCs during reprogramming and time in culture. Cell Stem Cell 8(1):106–118
77. Gutierrez-Aranda I, Ramos-Mejia V, Bueno C, Munoz-Lopez M, Real PJ, Macia A, Sanchez L, Ligero G, Garcia-Parez JL, Menendez P (2010) Human induced pluripotent stem cells develop teratoma more efficiently and faster than human embryonic stem cells regardless the site of injection. Stem Cells (Dayton, Ohio) 28(9):1568–1570
78. Tang C, Lee AS, Volkmer JP, Sahoo D, Nag D, Mosley AR, Inlay MA, Ardehali R, Chavez SL, Pera RR, Behr B, Wu JC, Weissman IL, Drukker M (2011) An antibody against SSEA-5 glycan on human pluripotent stem cells enables removal of teratoma-forming cells. Nat Biotechnol 29(9):829–834

Chapter 4
hESC-Derived Hepatocytes

Iman Saramipoor Behbahan and Mark A. Zern

Abstract Liver transplantation is the most effective treatment for end-stage liver diseases. However, shortage of donor livers has limited its clinical application. Transfusion of primary hepatocyte from cadaveric livers is a potential effective alternative. Since donor livers are scarce, and adult hepatocytes cannot be expanded in culture, hepatocyte transplantation has not become a viable substitute to whole organ transplantation. Since hESCs replicate virtually indefinitely in culture, an infinite source of hepatocytes will therefore be available for therapeutics and for toxicology and pharmacology studies. This review will provide scientists and physicians with the latest achievements in the rapidly growing field of embryonic stem cell biology as it pertains to liver cell differentiation.

4.1 Introduction

Patients with life-threatening liver disease can be effectively treated with liver transplantation. However, this requires a major surgical procedure that is associated with considerable morbidity and mortality. In addition, most recipients of a donor liver require lifelong immune suppression that frequently leads to complications. Most importantly, the long-standing shortage of donor livers has rendered this treatment unavailable to most patients. Consequently, thousands of patients with end-stage liver disease die each year while on a waiting list for liver transplantation, and tens of thousands are never put on this list. Infusion of hepatocytes

I. S. Behbahan (✉) · M. A. Zern
Transplant Research Program, Department of Internal Medicine,
University of California Davis Medical Center, Sacramento, CA, USA
e-mail: saramipoor@yahoo.com

H. Baharvand and N. Aghdami (eds.), *Advances in Stem Cell Research*,
Stem Cell Biology and Regenerative Medicine, DOI: 10.1007/978-1-61779-940-2_4,

isolated from cadaveric livers has previously been shown to be effective in correcting liver diseases in humans, providing proof-of-principle for the efficacy and safety of cell therapy [1–3]. However, since donor livers are scarce, and adult hepatocytes cannot be expanded in culture, hepatocyte transplantation has not become a viable alternative to whole organ transplantation. Clearly, an abundant source of hepatocytes is needed to realize the potential of cell therapy for severe liver diseases.

Our research group and others have recently made significant progress in guiding the differentiation of human embryonic stem cells (hESCs) into hepatocytes. Since hESCs replicate virtually indefinitely in culture, an infinite source of hepatocytes will therefore be available for therapeutics and for toxicology and pharmacology studies. This review will provide scientists and physicians with the latest achievements in the rapidly growing field of embryonic stem cell biology as it pertains to liver cell differentiation.

4.2 Human Embryonic Stem Cells

Embryonic stem cells offer outstanding potential for regenerative medicine. After the first report of the derivation of human ESC lines in 1998 by James Thomson and his group [4], numerous hESC lines have been created with more than a hundred in the National Institutes of Health registry (http://stemcells.nih.gov/research/registry/). Moreover, hESCs possess a normal karyotype [5] with high levels of telomerase activity, and they demonstrate a continuous renewal capability. They can be propagated in an undifferentiated state in vitro and have the potential to differentiate into all specialized cell types [6, 7]. Therefore, these cells provide a powerful tool for the investigation of cellular identity and early mammalian development, as well as having great potential for cellular therapeutics.

4.2.1 Differentiation of Human Embryonic Stem Cells Toward the Hepatic Lineage

The capacity of endodermal differentiation in mouse ESC (mESC) was reported by Abe et al. and Levinson-Dushnik and Benvenistry [8, 9]. Later, Duncan et al. demonstrated that mECS have the ability to differentiate to visceral endoderm that expresses liver-specific proteins [10]. Finally, hepatocyte-like cells were generated by Hamsaki et al. applying specific growth factors in mESC [11]. The differentiation of hESCs into hepatocytes was first reported by Rambhatla and colleagues [12]. Since then, many studies have focused on enhancing the culture conditions to obtain a more homogeneous cell population. Recently, most of the protocols follow stepwise differentiation through definitive endoderm imitating embryonic liver development.

4.3 Spontaneous Differentiation

ESCs differentiate spontaneously into the three germ layers, among them endodermal cells expressing AFP and albumin [13, 14]. Lavon et al. demonstrated that hESC differentiated into hepatic-like cells through the intermediates step of EB formation. They showed that albumin-expressing cells isolated by fluorescence-activated cell sorting were capable of growing in vitro for a few weeks [15, 16].

Our group, among others, extended the original finding by optimizing growth factors, ECM, and medium selection to enhance the purity of the differentiation process using EBs [17–21]. In addition, it was suggested by Baharvand et al. that plating EB in a 3D collagen type I scaffold with the sequential addition of growth factors is more efficient than employing 2D cultures. They also reported the migration of cells out of aggregates and the formation of multilayer cord-like structures in a 3D culture system [19]. However, hepatocyte differentiation through EBs is no longer considered the most effective approach. The major problem is the relatively low differentiation rate and nonhomogeneous population of cells after differentiation.

4.4 Stepwise Differentiation

4.4.1 Differentiation Toward Definitive Endoderm by Nodal Signaling Induction

Definitive endoderm is formed by inward migration of the epiblast from the primitive streak during gastrulation [22, 23]. The liver, lung, pancreas, thyroid, and gastrointestinal tract generate from developing endoderm [24].

Nodal signaling pathway, a member of the transforming growth factor-β (TGF-β) superfamily, is involved in endoderm specification during embryonic development. Activin A is a protein agonist for the Nodal signaling pathway by binding to the cell surface transmembrane type II receptor. This signaling induces a response which eventually results in phosphorylation of SMAD2 and SMAD3, the key components of the Nodal signaling pathway [23]. The seminal study employing this approach was reported by D'Amour et al. who described differentiation of hESCs to definitive endoderm with more than 80% purity using high levels (100 ng/ml) of Activin A and low serum concentrations [25]. Since then, the most effective hepatocyte differentiation protocols have employed this approach of definitive endoderm formation as the first step.

Ishii et al. compared endodermal differentiation through EB or without forming EBs on different ECMs. They found the best result by employing Matrigel-coated dishes with 100 ng/ml Activin A without employing EBs. In addition, they showed that HGF, BMP4, and FGF4 have no significant effect on endodermal differentiation, and they suggested an inhibitory effect of all-trans-retinoic acid on endodermal differentiation. The presence of BMP4 and FGF4 reduced the effects of Activin A and HGF on endodermal differentiation [26]. Brolén et al. showed that using Activin A and FGF2 together enhanced the expression of endodermal markers (Sox17, HNF3β, and CXCR4) in comparison with using Activin A alone [27].

Hay and colleagues examined Wnt3a expression in liver at different times during human liver development. They reported that in the first trimester Wnt3a activity developed in the portal system with no parenchymal expression. In contrast, in the second trimester Wnt3a expression was observed both in the portal region and in the liver parenchyma suggesting a potential role in early hepatogenesis. Furthermore, they found enhanced endodermal differentiation of hESC using Activin A plus Wnt3a in comparison with Activin A plus sodium butyrate. They postulated that Activin A plus Wnt3a-induced a rapid and more homogeneous hepatic endoderm production and led to more functional hepatocyte-like cells [28].

4.4.2 Hepatic Induction

FGFs and BMPs appear to play significant roles in liver specification. Gouon-Evans et al. demonstrated that hESCs differentiated in the presence of Activin A in serum-free conditions to efficiently generate an endoderm progenitor population. Hepatic specification of these progenitors occur by BMP4 in combination with bFGF and Activin A [29].

Then, Cai et al. reported that FGF4 or BMP2 alone in the culture medium had little effect on liver specification from definitive endoderm. However, their combination lead to a significant increase in albumin expression in vitro, and this effect was confirmed by the FGFR1 inhibitor Su5402 or the BMP inhibitor Noggin. Also, they showed definitive endoderm induction is essential for the effectiveness of FGF4 and BMP2 on hepatic induction. They reported similar results for the combinations of aFGF and BMP4, bFGF and BMP4, or FGF4 and BMP4 [30]. DMSO, which modifies histone acetylation, appears to have an inductive effect toward hepatocyte lineage specification and has been utilized in some protocols [31–33].

A combination of BMP2/4 and FGF1/2/4 was applied by Brolén et al. for hepatic induction. They observed multiple EpCAM-expressing cell clusters with adjacent groups of CK7 and CK19 positive cells and some CD54 positive areas. In the control group which was not induced by BMPs and FGFs, although EpCAM, α1-antitrypsin, AFP, CK7/8/18/19, and CD54 positive cells appeared, there was a considerable cell morphology difference with a significant decrease in liver-specific gene expression [27].

Johannesson et al. found that Activin A-treated hESCs tend toward the hepatic lineage instead of toward pancreatic progenitor cells if they are not treated with FGF4 and retinoic acid (RA) [34]. On the other hand, Ameri et al. reported a dose-related response to FGF2 regarding cell fate selection. They demonstrated that low levels of FGF2 enhanced hepatic induction while intermediate doses of FGF2 resulted in pancreatic fate selection [35]. In summary, many factors have been shown to affect hepatic induction including the timing, concentration and specific variety of growth factors, as well as the methods of generating definitive endoderm.

4.4.3 Maturation

In this step, differentiated cells present a polygonal hepatocyte morphology with distinct round nuclei that have cytoplasmic vacuoles [28, 36]. Hepatocyte maturation strategies are designed to replicate the role of growth factors in liver development. In general, sequential and different levels of HGF and/or, OSM or EGF with glucocorticoids and insulin, in a hepatocyte-optimized culture medium are used (Table 4.1).

4.5 Differentiation Toward Liver Precursor Cells

The isolation of a hepatic precursor cell line derived from hESC has drawn the attention of some researchers because the establishment of a method to drive hESC or iPS cells to a stable bipotent liver stem cell line which has renewal capability would bypass many expensive and difficult differentiation steps and would enhance clinical applications that require a large quantity of cells. Haiyun Pei and colleagues reported the isolation of liver progenitor-like cells which were isolated from portal zone 1 that expressed stromal progenitor markers, such as CD29, CD44, CD71, CD90, and CD105, and were negative for CD34 and CD45 [37]. Zhao et al. studied N-cadherin-expressing hepatic endoderm that was derived from hESCs by induction of definitive endoderm and then treatment with BMP2 and FGF4 for 5 days. These cells were maintained and passaged on mouse embryonic stromal feeders cells for more than 100 days. They also reported that these cells can be differentiated toward hepatocyte- and cholangiocyte-like cells [38].

Li F et al. differentiated mouse ES cells through a stepwise protocol to a cell population containing hepatic progenitor cells. They separated epithelial cell adhesion molecule (EpCAM) positive and c-Kit negative cells from this population and reported that hepatic progenitor colonies from single c-Kit$^-$EpCAM$^+$ cells could be expanded and maintained hepatoblast-like characteristics for an extended period. Then, enriched c-Kit$^-$EpCAM$^+$ cells were transplanted into the livers of Fah-deficient mice. They showed that c-Kit$^-$EpCAM$^+$ cells matured into functional hepatocytes in vivo and repopulated mouse livers without inducing tumorigenesis [39].

Table 4.1 Step-wise differentiation of human embryonic stem cell through definitive endoderm with Activin A

Cell line	Before definitive endoderm	Definitive endoderm induction	Hepatic specification	Hepatic differentiation	Hepatic maturation	Special evaluation	Reference
H1, H9	———	Activin A 3 days	FGF4 BMP2 5 days	HGF 5 day	OSM, Dex 5 days	Evaluation of entry of HIV-HCV Pseudotype Viruses into hESC-Derived Hepatic Cells. In vivo study in SCID mice	[30]
H1, H9	———	Activin A 100 ng/ml Wnt3a 50 ng/ml, RPMI 1640, B27 3 days	1% DMSO, KO DMEM, 20% SR, 4 days	———	HGF, OSM, L-15, 10% FBS, 7 days	Intrasplenic injection of cells in NOD-SCID mice	[28]
H1, H9	———	Activin A 0.5% FBS + RPMI 3 days 2% KOSR instead of FBS in Days 3–5	FGF-4, HGF, 2% KOSR, RPMI media on Cllagen I Days 5–8	MM media, BSA, FGF-4, HGF + OSM + Dex Days 8–10	HCM Media, FGF-4 HGF, OSM Dex Day 11–20	Portal injection of definitive endoderm cells in NOD-SCID mice	[36]
H1	Embryoid body formation 2 days	Activin A bFGF 3 days	HGF DMSO 8 days	———	Dex 3 days	Splenic injection of ASGPR 1 positive cells to uPA-SCID mice, Nagase- analbuminemic-rat, NOD-SCID mice	[47]

(continued)

Table 4.1 (continued)

Cell line	Before definitive endoderm	Definitive endoderm induction	Hepatic specification	Hepatic differentiation	Hepatic maturation	Special evaluation	Reference
SA001, SA002, SA002.5, SA167	———	Activin A FGF-2 1–5 days	BMP 2/4 FGF 1/2/4 6–17 days	———	EGF, Insulin, Transferri, Hydrocortion, Ascorbic Acid 18–45 days	Differentiation in three cell lines was successful	[27]
H9	———	No Serum- Activin A 2 days Low serum, Activin A, SB, in RPMI 1640 3–6 days	FGF4, HGF, Dex, BMP2/ 4,Insulin ,DMSO, IMDM 10–14 days	———	FGF4, HGF, Dex, DMSO, OSM, HCM Until use	Drug metabolism evaluation by LC/MS/MS	[47]
H9	Activin A10 ng/ml FGF2 12 ng/ml 2 days	100 ng/mL Activin A, 20 ng/mL FGF2, 10 ng/mL BMP4, 10 μM LY294003 3 days	50 ng/mL FGF10, 10^{-7} M retinoic acid, 10 μM SB431542 3 + 2 days	———	30 ng/mLFGF4, 50 ng/mL HGF, 50 ng/mL EGF. 10 days	uPAxrag2gammac $^{-/-}$ mice Injection	[88]

4.6 Oxygen Pressure and Mitochondrial Function in Hepatocyte Differentiation

Early embryonic development mostly occurs in low oxygen concentration. In addition, hypoxic microenvironments in adults provide adult stem cell niches that control cellular differentiation [40, 41]. The effect of low oxygen on embryonic stem cell culture is controversial [42]. Low oxygen seems to decrease the chance of spontaneous differentiation in mESC culture conditions in some reports; however, other studies suggest that it reduces the expression of pluripotency genes. Low oxygen concentration enhances the production of some lineages, such as neurons, cardiomyocytes, hematopoietic progenitors, endothelial cells, and chondrocytes (reviewed in Refs. [40] and [42]). Multiple O2-sensitive intracellular mechanisms such as hypoxia-inducible transcription factors (HIFs), the environmental sensing of mammalian target of rapamycin (mTOR), and the endoplasmic reticulum (ER) stress response have been described [41]. It appears that hypoxia increases the mitochondrial membrane potential and subsequently reactive oxygen species which activate HIF-1 expression [43].

Si-Tayeb et al. used different oxygen pressures during the differentiation of hESC and hiPSC to hepatocytes. They maintained undifferentiated hESC and hiPSC at low oxygen pressure, then generated the definitive endoderm at a normal oxygen pressure, and finally hepatic induction and specification was performed in a hypoxic environment. A high percentage of hepatic differentiation is reported. However, the advantage of the different oxygen pressures is not discussed in comparison to a control group [44].

4.7 Characterization and Functional Evaluation of Differentiated Hepatocytes

There is no consensus for the proper characterization of hepatocyte-like cells derived from stem cells. Generally, hepatocyte-like cells are recognized by their morphology, liver-specific mRNAs, protein markers, and their functional abilities during differentiation. Each phase of differentiation is delineated by specific markers (See Table 4.2 and Refs. [45] and [46]).

4.8 In Vivo Studies

Clinical transplantation is one of the main potential goals of hepatocyte generation from hESC; however, only a few studies have evaluated these cells in vivo. Jun Cai et al. showed that differentiated hESCs were able to engraft in the livers of SCID mice injured by CCl4 and that they expressed human alpha-1 antitrypsin for

Table 4.2 Characterization and functional evaluation of differentiated hepatocytes and drug metabolism

Goal of test	
Definitive endoderm	SOX17, GSC, FOXA2
Primary hepatic differentiation	HNF3b, AFP, and transthyretin (TTR)
Intermediate phase of hepatogenesis	HNF1α, HNF4α, Albumin, and CK18
Mature hepatocytes	Tryptophan-oxygenase, tyrosine amino-transferase, C/EBPα, specific CYPs, asialoglycoprotein receptor 1(AGPR1)
Functional assay	Evaluation of glycogen accumulation by PAS staining
	Hepatocyte morphology under electron microscopy
	Urea synthesis, albumin production
	Indocyanine Green uptake
	Measurement of coagulation factor VII activity, entry of HIV-HCV pseudotype viruses into hESC-derived hepatic cells
Drug metabolism ability of hepatocyte like-cells	Pentoxyresorufin-O-dealkylase (PROD) assay for the assessment of the cytochrome p450 system
	CYP3A4 induction of hESC-derived hepatocytes by rifampin and midazolam by LC–MS/MS
	Hepatic CYP 1A1/1A2-mediated Ethoxyresorufin-O-deethylase (EROD) activities

at least 2 months [30]. Our group transduced hESC-differentiated hepatocytes using a lentiviral vector containing a triple fusion protein and transplanted transduced hESC-differentiated hepatocytes into the livers of NOD-SCID mice. Then we investigated the luciferase-induced bioluminescence in the animal livers by a charge-coupled device camera. We showed that the transplanted cells had engrafted into the liver and were functional [21]. Agarwal et al. transplanted definitive endoderm cells derived from hESCs into the portal vein of NOD-SCID mice that were previously treated with CCl4 and retrosine, and they detected that the cells expressed human-specific mitochondrial antigen, as well as CD26 and alpha-1-antitrypsin in the livers of the injected mice over time [36]. Hay et al. showed that hepatocyte-like cells could engraft in the NOD-SCID mouse liver. They reported that Wnt3a-treated cells expressed higher levels of human albumin in the serum than did the control cells [28]. Basma et al. isolated the ASGPR positive cells from differentiated hepatocyte-like cells by flow cytometery. Then, they transplanted these cells into the spleen of Alb-uPA SCID mice. Very high levels of human albumin (1,000–2,000 ng/ml) and alpha-1-antitrypsin were detected in the animals' serum 75 days after transplantation. In histological evaluation, small clusters of human CD18-expressing cells without any tumor formation were observed. Then intrasplenic injection of one million sorted cells was performed in an FK506-immune suppressed Nagase analbuminemic rat or NOD-SCID mice after 70 or 50% partial hepatectomy and retrosine treatment. They reported 20,000 ng/ml human albumin in the transplanted animal's serum with large clusters of human engrafted cells in histological evaluation. However, a well-differentiated adenocarcinoma was detected as well [47].

4.9 Small Molecules

A major challenge in stem cell therapy is the development of defined culture and differentiation systems for the maintenance of GMP-compatible cells which are not contaminated by pathogens or xenogens. Replacing animal sera and mouse embryonic fibroblast (MEF) feeder cells with defined components of human origin has been a common strategy [48–52]. Other studies attempted to eliminate the need for MEF and growth factors by application of small molecules to 'chemically direct' maintenance and differentiation of hESCs as a powerful approach to this problem.

Recently Tsutsui et al. showed that combination of three small molecule inhibitors against Rho-associated kinase (ROCK), glycogen synthase kinase (GSK), and mitogen-activated protein kinase (MEK) maintained hESC on a fibronectin-coated surface for 20 passages. However, the addition of bFGF was necessary for preventing the loss of OCT4 over time [53].

The application of small molecules in hepatic differentiation has also been reported. Bone HK et al. demonstrated the inhibition of (GSK-3) in hESC by a specific small molecule (1 M) results in transient Nodal expression. They also reported that longer induction generated a cell population which expressed α-fetoprotein and HNF4α. Although, they demonstrated that these cells differentiated into albumin-producing hepatocyte-like cells, their functional capacities, and engraftement in vivo have not been investigated [54].

4.10 Epigenetics

Epigenetic control mechanisms play critical roles in molecular regulation of embryonic stem cells differentiation. Undifferentiated stem cells undergo substantial alterations in the epigenetic and coding regions during differentiation to specific cell fates [55]. These mechanisms include DNA methylation and histone modifications, and ATP-dependent chromatin remodeling which modulate the chromatin structure and regulate the accessibility of transcription factors. Gene expression is regulated by various histone modifications including acetylation, methylation, sumoylation, etc.

Kim et al. investigated DNA methylation at three levels of in vitro hepatocyte differentiation (hESC, definitive endoderm, and hepatocyte). They reported dynamic methylation changes in intergenic regions of the human genome during differentiation [56].

Xu et al. isolated mouse embryonic endoderm cells by cell sorting and assessed histone modifications at regulatory elements of silent genes that are activated during liver or pancreas specification. They suggested that BMP signaling which is mediated by SMAD4 results in histone acetylation at liver target elements. It further enhances liver gene activation by histone acetyltransferase P300 recruitment which results in enhancement of liver bud formation. They also propose that histone modifications of P300 mainly have a regulatory role rather than absolute induction of liver specification. In addition, histone methyltransferase Ezh2 was shown to have modulatory roles in liver and pancreas specification [57].

Kim SW et al. investigated genomic targets for SMAD2/3, SMAD3, SMAD4, FOXH1 and the active and repressive chromatin marks, H3K4me3 and H3K27me3, in hESCs and derived endoderm using ChIP-seq. They reported a highly dynamic pattern in DNA in association with SMAD2/3, SMAD4, and FOXH1. However, they discovered an optimal bivalent signature at 32 gene loci for driving endoderm commitment. Initially, this signature is marked by both H3K4me3 and H3K27me3 as a very broad bivalent domain in hESCs. After 24 h, in SMAD2/3, an H3K27me3 reduction occurred and these loci become monovalent marked by H3K4me3 [58].

4.11 The Tumorigenicity of Human Embryonic Stem Cells

When hESCs are injected into immunodeficient mice, they form teratomas [5, 59]. Moreover, normal diploid hESCs can potentially undergo culture-adaptation and after transplantation transform to aggressive tumors, which resemble teratocarcinomas [59, 60]. Thus there are concerns that transplanted hESC can cause malignant teratocarcinomas in humans. Even benign teratomas are not an acceptable complication of embryonic stem cell therapy in humans. Hence, tumorigenicity of hESC is one of the main obstacles for their clinical application which must be addressed before safe translation of stem cell therapy into the clinic.

Some molecular characteristics of tumor cells and ESC, such as rapid proliferation rate, lack of contact inhibition, and a tendency for genomic instability, are similar [5, 59, 61–66]. They both express high telomerase activity and oncogenes such as MYC9 and KLF4 [67–69]. In addition, a noticeable similarity in gene expression patterns, microRNA (miRNA) signatures, and epigenetic status between ESC and cancer cells have been reported [70–74].

In some reports, transplantation of even small number of hESCs are sufficient for tumor formation [75, 76]. Complete differentiation and total destruction of semi-differentiated cells in culture systems as well as rapid detection of tumors in the transplanted patient's body have been suggested to address this problem. Several attempts have been reported, describing the removal of undifferentiated hESCs, such as applying magnetic-activated cell sorting and fluorescence-activated cell sorting [59, 77]. This is an active area of research that is critical for the safe use of these cells.

4.12 MicroRNAs During Hepatocyte Differentiation

MicroRNAs (miRNAs) are crucial in eukaryotic gene expression regulation, and they have been shown to modulate development, infection, immunity, and carcinogenesis [78, 79]. These non-coding RNAs also play a key role in the differentiation of stem cell populations. Kim et al. evaluated the alteration of miRNAs from hESCs to a hepatocyte lineage through definitive endoderm. They detected enriched expression of miR-512-3p, miR-512-5p and miR-520c-3p in hESCs, and miR-9, miR-205 and miR-375 in hESC-derived DE cells. Levels of miR-10a, miR-122, and miR-2 expression were enhanced in hESC-derived hepatocytes which was similar to the miRNAs patterns in liver tissue [80]. Tzur et al. discovered sodium butyrate induction of hESCs upregulated miR-24 and miR-10a as well as some liver-enriched miRNAs, including miR-122 and miR-192 [81].

4.13 Future Directions and Challenges

The differentiation of hESC and other stem cells along a hepatocyte lineage has undergone enormous changes over the last decade. The progress has been substantial, with the greatest paradigm shift being the emergence of hiPSC which resolve many of the ethical issues surrounding hESC. Although the progress toward the development of improved protocols is impressive, a number of problems exist before the cells can be used for cell transplantation trials in humans or for toxicology or pharmacology studies. For example, xeno-free and feeder-free growth and differentiation conditions must be established which are effective, reproducible, robust, and relatively inexpensive. This may include the development and use of small molecules and synthetic biocompatible ECMs which can substitute for the highly expensive growth factors and xeno-derived ECMs.

New approaches may be developed in the near future that may be even more effective than our present technology of differentiating hepatocytes. For example, recent reports have shown the transdifferentiation of fibroblasts to neuronal cells and cardiomyocytes [82–85] and very recently mouse fibroblasts to hepatocytes [86, 87].

Despite continual improvements in vitro function, the engraftment and proliferation of hepatocyte-like cells from hESC has in general been disappointing in rodents. One likely explanation is that the differentiated human cells may not be compatible with the extracellular matrix and growth factor niche of rodent livers. Thus it may be that nonhuman primates will provide a better environment for hepatocyte engraftment, differentiation, and proliferation. Moreover, studies conducted in nonhuman primates will obviously be better indicators of results in humans.

Tumorogenisity is still one of the main obstacles in the clinical application of hESC, and a major question is how safe must the cells be before they can be used? How does one define lack of tumorogenisity? What is an acceptable risk, and how does one determine that a differentiated cell population is free from the presence of any early progenitor cells? Another major issue is the loss of proliferation of cells when they become significantly differentiated. This leads to the question of whether a sufficient number of mature hepatocyte-like cells that have been derived

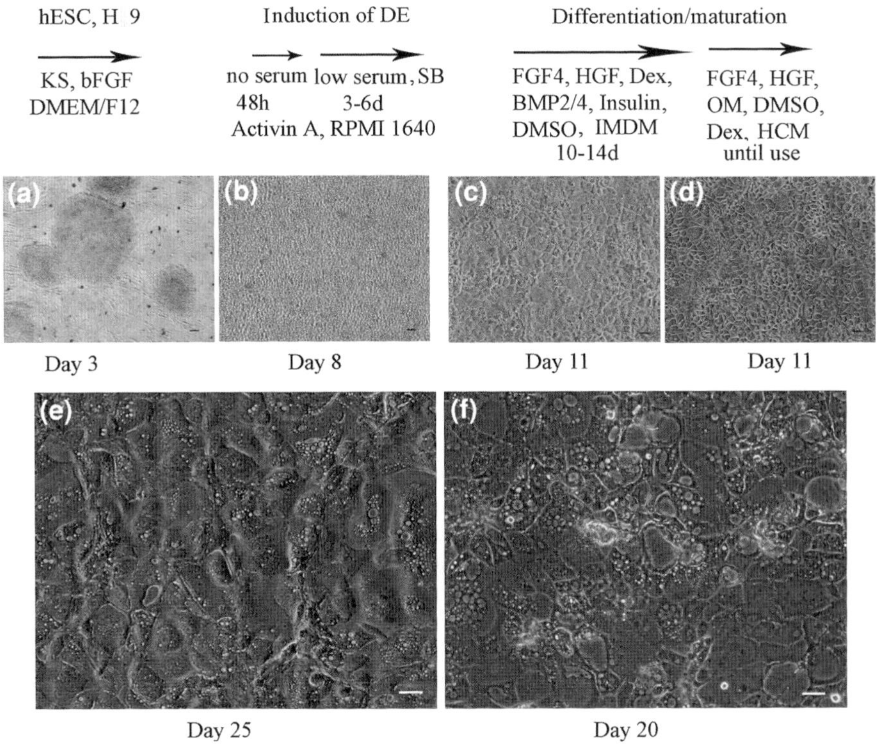

Fig. 4.1 A stepwise protocol for differentiation of hESC toward hepatocytes (from Ref. [31]). The *top panel* shows schematic illustration representing the differentiation strategy; the *middle* and the *bottom* panels show the morphological changes from undifferentiated hESC to hepatocytes during differentiation. **a** Undifferentiated hESC. **b** DE cells at day 8 after induction. **c** Differentiated cells at day 11 cultured on coated-plates after splitting from DE. **d** Differentiated cells at day 11 directly differentiated from DE without splitting. **e** Differentiated cells at day 25 cultured on coated-plates after splitting from DE. **f** Differentiated cells at day 20 directly differentiated from DE without splitting. Scale bar—200 μm (**a** and **b**), 100 μm (**c**–**f**). Magnification: (**a**, **b**), × 40; (**c**–**f**), × 100. Abbreviations: *KS* knockout serum replacer, *DE* definitive endoderm, *SB* sodium butyrate, *Dex* dexamethasone, *OM* Oncostatin M, *HCM* hepatocyte culture medium

from hESC can be obtained so that cell transplantation can be clinically effective. Another variant of this question is at what stage should cells that are undergoing differentiation be transplanted? Should they be transplanted at an early progenitor stage when they are rapidly proliferating, yet not mature, or wait until they are fully differentiated yet much less proliferative? These are but a few of the many technical questions that must be addressed in the coming years by investigators in the field as our new-found success makes technical questions take the forefront of research endeavors, and basic research yields to translational approaches (Fig. 4.1).

References

1. Dhawan A, Puppi J, Hughes RD, Mitry RR (2010) Human hepatocyte transplantation: current experience and future challenges. Nat Rev Gastroenterol Hepatol 7:288–298
2. Fisher RA, Strom SC (2006) Human hepatocyte transplantation: worldwide results. Transplantation 82:441–449
3. Ito M, Nagata H, Miyakawa S, Fox IJ (2009) Review of hepatocyte transplantation. J Hepatobiliary Pancreat Surg 16:97–100
4. Thomson JA, Itskovitz-Eldor J, Shapiro SS, Waknitz MA, Swiergiel JJ, Marshall VS et al (1998) Embryonic stem cell lines derived from human blastocysts. Science 282:1145–1147
5. Buzzard JJ, Gough NM, Crook JM, Colman A (2004) Karyotype of human ES cells during extended culture. Nat Biotechnol 22:381–382
6. Heins N, Englund MC, Sjoblom C, Dahl U, Tonning A, Bergh C et al (2004) Derivation, characterization, and differentiation of human embryonic stem cells. Stem Cells 22:367–376
7. Reubinoff BE, Pera MF, Fong CY, Trounson A, Bongso A (2000) Embryonic stem cell lines from human blastocysts: somatic differentiation in vitro. Nat Biotechnol 18:399–404
8. Abe K, Niwa H, Iwase K, Takiguchi M, Mori M, Abe SI et al (1996) Endoderm-specific gene expression in embryonic stem cells differentiated to embryoid bodies. Exp Cell Res 229: 27–34
9. Levinson-Dushnik M, Benvenisty N (1997) Involvement of hepatocyte nuclear factor 3 in endoderm differentiation of embryonic stem cells. Mol Cell Biol 17:3817–3822
10. Duncan SA, Nagy A, Chan W (1997) Murine gastrulation requires HNF-4 regulated gene expression in the visceral endoderm: tetraploid rescue of Hnf-4(-/-) embryos. Development 124:279–287
11. Hamazaki T, Iiboshi Y, Oka M, Papst PJ, Meacham AM, Zon LI et al (2001) Hepatic maturation in differentiating embryonic stem cells in vitro. FEBS Lett 497:15–19
12. Rambhatla L, Chiu CP, Kundu P, Peng Y, Carpenter MK (2003) Generation of hepatocyte-like cells from human embryonic stem cells. Cell Transplant 12:1–11
13. Itskovitz-Eldor J, Schuldiner M, Karsenti D, Eden A, Yanuka O, Amit M et al (2000) Differentiation of human embryonic stem cells into embryoid bodies compromising the three embryonic germ layers. Mol Med 6:88–95
14. Schuldiner M, Yanuka O, Itskovitz-Eldor J, Melton DA, Benvenisty N (2000) Effects of eight growth factors on the differentiation of cells derived from human embryonic stem cells. Proc Natl Acad Sci U S A 97:11307–11312
15. Lavon N, Yanuka O, Benvenisty N (2004) Differentiation and isolation of hepatic-like cells from human embryonic stem cells. Differentiation 72:230–238
16. Lavon N, Benvenisty N (2005) Study of hepatocyte differentiation using embryonic stem cells. J Cell Biochem 96:1193–1202
17. Shirahashi H, Wu J, Yamamoto N, Catana A, Wege H, Wager B et al (2004) Differentiation of human and mouse embryonic stem cells along a hepatocyte lineage. Cell Transplant 13: 197–211
18. Schwartz RE, Linehan JL, Painschab MS, Hu WS, Verfaillie CM, Kaufman DS (2005) Defined conditions for development of functional hepatic cells from human embryonic stem cells. Stem Cells Dev 14:643–655
19. Baharvand H, Hashemi SM, Kazemi Ashtiani S, Farrokhi A (2006) Differentiation of human embryonic stem cells into hepatocytes in 2D and 3D culture systems in vitro. Int J Dev Biol 50:645–652
20. Baharvand H, Hashemi SM, Shahsavani M (2008) Differentiation of human embryonic stem cells into functional hepatocyte-like cells in a serum-free adherent culture condition. Differentiation 76:465–477
21. Duan Y, Catana A, Meng Y, Yamamoto N, He S, Gupta S et al (2007) Differentiation and enrichment of hepatocyte-like cells from human embryonic stem cells in vitro and in vivo. Stem Cells 25:3058–3068

22. Wodarz A, Nusse R (1998) Mechanisms of Wnt signaling in development. Annu Rev Cell Dev Biol 14:59–88
23. Sulzbacher S, Schroeder IS, Truong TT, Wobus AM (2009) Activin A-induced differentiation of embryonic stem cells into endoderm and pancreatic progenitors-the influence of differentiation factors and culture conditions. Stem Cell Rev 5:159–173
24. Jung J, Zheng M, Goldfarb M, Zaret KS (1999) Initiation of mammalian liver development from endoderm by fibroblast growth factors. Science 284:1998–2003
25. D'Amour KA, Agulnick AD, Eliazer S, Kelly OG, Kroon E, Baetge EE (2005) Efficient differentiation of human embryonic stem cells to definitive endoderm. Nat Biotechnol 23: 1534–1541
26. Ishii T, Fukumitsu K, Yasuchika K, Adachi K, Kawase E, Suemori H et al (2008) Effects of extracellular matrixes and growth factors on the hepatic differentiation of human embryonic stem cells. Am J Physiol Gastrointest Liver Physiol 295:G313–G321
27. Brolen G, Sivertsson L, Bjorquist P, Eriksson G, Ek M, Semb H et al (2010) Hepatocyte-like cells derived from human embryonic stem cells specifically via definitive endoderm and a progenitor stage. J Biotechnol 145:284–294
28. Hay DC, Fletcher J, Payne C, Terrace JD, Gallagher RC, Snoeys J et al (2008) Highly efficient differentiation of hESCs to functional hepatic endoderm requires Activin A and Wnt3a signaling. Proc Natl Acad Sci U S A 105:12301–12306
29. Gouon-Evans V, Boussemart L, Gadue P, Nierhoff D, Koehler CI, Kubo A, Shafritz DA, Keller G (2006) BMP-4 is required for hepatic specification of mouse embryonic stem cell-derived definitive endoderm. Nat Biotechnol 24:1402–1411
30. Cai J, Zhao Y, Liu Y, Ye F, Song Z, Qin H et al (2007) Directed differentiation of human embryonic stem cells into functional hepatic cells. Hepatology 45:1229–1239
31. Duan Y, Ma X, Zou W, Wang C, Bahbahan IS, Ahuja TP, Tolstikov V, Zern MA (2010) Differentiation and characterization of metabolically functioning hepatocytes from human embryonic stem cells. Stem Cells 28:674–686
32. Hay DC, Zhao D, Ross A, Mandalam R, Lebkowski J, Cui W (2007) Direct differentiation of human embryonic stem cells to hepatocyte-like cells exhibiting functional activities. Cloning Stem Cells 9:51–62
33. Hay DC, Fletcher J, Payne C, Terrace JD, Gallagher RC, Snoeys J et al (2008) Highly efficient differentiation of hESCs to functional hepatic endoderm requires Activin A and Wnt3a signaling. Proc Natl Acad Sci U S A 105:12301–12306
34. Johannesson M, Stahlberg A, Ameri J, Sand FW, Norrman K, Semb H (2009) FGF4 and retinoic acid direct differentiation of hESCs into PDX1-expressing foregut endoderm in a time. and concentration-dependent manner. PLoS One 4:e4794
35. Ameri J, Stahlberg A, Pedersen J, Johansson JK, Johannesson MM, Artner I et al (2010) FGF2 specifies hESC-derived definitive endoderm into foregut/midgut cell lineages in a concentration-dependent manner. Stem Cells 28:45–56
36. Agarwal S, Holton KL, Lanza R (2008) Efficient differentiation of functional hepatocytes from human embryonic stem cells. Stem Cells 26:1117–1127
37. Pei H, Yang Y, Xi J, Bai Z, Yue W, Nan X et al (2009) Lineage restriction and differentiation of human embryonic stem cells into hepatic progenitors and zone 1 hepatocytes. Tissue Eng Part C Methods 15:95–104
38. Zhao D, Chen S, Cai J, Guo Y, Song Z, Che J et al (2009) Derivation and characterization of hepatic progenitor cells from human embryonic stem cells. PLoS One 4:e6468
39. Li F, Liu P, Liu C, Xiang D, Deng L, Li W et al (2010) Hepatoblast-like progenitor cells derived from embryonic stem cells can repopulate livers of mice. Gastroenterology. 139:2158–2169.e8
40. Simon MC, Keith B (2008) The role of oxygen availability in embryonic development and stem cell function. Nat Rev Mol Cell Biol 9:285–296
41. Wion D, Christen T, Barbier EL, Coles JA (2009) PO(2) matters in stem cell culture. Cell Stem Cell 5:242–243
42. Millman JR, Tan JH, Colton CK (2009) The effects of low oxygen on self-renewal and differentiation of embryonic stem cells. Curr Opin Organ Transplant 14:694–700

43. Hannoun Z, Filippi C, Sullivan G, Hay DC, Iredale JP (2010) Hepatic endoderm differentiation from human embryonic stem cells. Curr Stem Cell Res Ther 5:233–244
44. Si-Tayeb K, Noto FK, Nagaoka M, Li J, Battle MA, Duris C et al (2010) Highly efficient generation of human hepatocyte-like cells from induced pluripotent stem cells. Hepatology 51:297–305
45. Snykers S, De Kock J, Rogiers V, Vanhaecke T (2009) In vitro differentiation of embryonic and adult stem cells into hepatocytes: state of the art. Stem Cells 27:577–605
46. Behbahan IS, Duan Y, Lam A, Khoobyari S, Ma X, Ahuja TP, Zern MA (2011) New Approaches in the Differentiation of Human Embryonic Stem Cells and Induced Pluripotent Stem Cells toward Hepatocytes. Stem Cell Rev 7:748–759
47. Basma H, Soto-Gutierrez A, Yannam GR, Liu L, Ito R, Yamamoto T et al (2009) Differentiation and transplantation of human embryonic stem cell-derived hepatocytes. Gastroenterology 136:990–999
48. Hovatta O, Mikkola M, Gertow K, Strömberg AM, Inzunza J, Hreinsson J et al (2003) A culture system using human foreskin fibroblasts as feeder cells allows production of human embryonic stem cells. Hum Reprod 18:1404–1409
49. Richards M, Fong CY, Chan WK, Wong PC, Bongso A (2002) Human feeders support prolonged undifferentiated growth of human inner cell masses and embryonic stem cells. Nat Biotechnol 20:933–936
50. Richards M, Tan S, Fong CY, Biswas A, Chan WK, Bongso A et al (2003) Comparative evaluation of various human feeders for prolonged undifferentiated growth of human embryonic stem cells. Stem Cells 21:546–556
51. Stojkovic P, Lako M, Stewart R, Przyborski S, Armstrong L, Evans J, Murdoch A, Strachan T, Stojkovic M (2005) An autogeneic feeder cell system that efficiently supports growth of undifferentiated human embryonic stem cells. Stem Cells 23:306–314
52. Wang Q, Fang ZF, Jin F, Lu Y, Gai H, Sheng HZ (2005) Derivation and growing human embryonic stem cells on feeders derived from themselves. Stem Cells 23:1221–1227
53. Tsutsui H, Valamehr B, Hindoyan A, Qiao R, Ding X, Guo S, Witte ON, Liu X, Ho CM, Wu H (2011) An optimized small molecule inhibitor cocktail supports long-term maintenance of human embryonic stem cells. Nat Commun 25:167
54. Bone HK, Nelson AS, Goldring CE, Tosh D, Welham MJ (2011) A novel chemically directed route for the generation of definitive endoderm from human embryonic stem cells based on inhibition of GSK-3. J Cell Sci 124:1992–2000
55. Christophersen NS, Helin K (2010) Epigenetic control of embryonic stem cell fate. J Exp Med 207:2287–2295
56. Kim M, Kang TW, Lee HC, Han YM, Kim H, Shin HD et al (2011) Identification of DNA methylation markers for lineage commitment of in vitro hepatogenesis. Hum Mol Genet [Epub ahead of print]
57. Xu CR, Cole PA, Meyers DJ, Kormish J, Dent S, Zaret KS (2011) Chromatin "prepattern" and histone modifiers in a fate choice for liver and pancreas. Science 332:963–966
58. Kim SW, Yoon SJ, Chuong E, Oyolu C, Wills A, Gupta R, Baker J (2011) Chromatin and transcriptional signatures for nodal signaling during endoderm formation in hESCs. Dev Biol [Epub ahead of print]
59. Ben-David U, Benvenisty N (2011) The tumorigenicity of human embryonic and induced pluripotent stem cells. Nat Rev Cancer 11:268–277
60. Shih CC, Forman SJ, Chu P, Slovak M (2007) Human embryonic stem cells are prone to generate primitive, undifferentiated tumors in engrafted human fetal tissues in severe combined immunodeficient mice. Stem Cells Dev 16:893–902
61. Dreesen O, Brivanlou AH (2007) Signaling pathways in cancer and embryonic stem cells. Stem Cell Rev 3:7–17
62. Knoepfler PS (2009) Deconstructing stem cell tumorigenicity: a roadmap to safe regenerative medicine. Stem Cells 27:1050–1056
63. Blum B, Benvenisty N (2008) The tumorigenicity of human embryonic stem cells. Adv Cancer Res 100:133–158

64. Amit M, Carpenter MK, Inokuma MS, Chiu CP, Harris CP, Waknitz MA, Itskovitz-Eldor J, Thomson JA (2000) Clonally derived human embryonic stem cell lines maintain pluripotency and proliferative potential for prolonged periods of culture. Dev Biol 227:271–278
65. Baker DE, Harrison NJ, Maltby E, Smith K, Moore HD, Shaw PJ, Heath PR, Holden H, Andrews PW (2007) Adaptation to culture of human embryonic stem cells and oncogenesis in vivo. Nature Biotech 25:207–215
66. Harrison NJ, Baker D, Andrews PW (2007) Culture adaptation of embryonic stem cells echoes germ cell malignancy. Int J Androl 30:275–281
67. Hiyama E, Hiyama K (2007) Telomere and telomerase in stem cells. Br J Cancer 96: 1020–1024
68. Eilers M, Eisenman RN (2008) Myc's broad reach. Genes Dev 22:2755–2766
69. Evans PM, Liu C (2008) Roles of Krupel-like factor 4 in normal homeostasis, cancer and stem cells. Acta Biochim Biophys Sin 40:554–564
70. Sperger JM, Chen X, Draper JS, Antosiewicz JE, Chon CH, Jones SB, Brooks JD, Andrews PW, Brown PO, Thomson JA (2003) Gene expression patterns in human embryonic stem cells and human pluripotent germ cell tumors. Proc Natl Acad Sci U S A 100: 13350–13355
71. Ben-Porath I, Thomson MW, Carey VJ, Ge R, Bell GW, Regev A, Weinberg RA (2008) An embryonic stem cell-like gene expression signature in poorly differentiated aggressive human tumors. Nature Gene 40:499–507
72. Wong DJ, Liu H, Ridky TW, Cassarino D, Segal E, Chang HY (2008) Module map of stem cell genes guides creation of epithelial cancer stem cells. Cell Stem Cell 2:333–344
73. Neveu P, Kye MJ, Qi S, Buchholz DE, Clegg DO, Sahin M, Park IH, Kim KS, Daley GQ, Kornblum HI, Shraiman BI, Kosik KS (2010) MicroRNA profiling reveals two distinct p53-related human pluripotent stem cell states. Cell Stem Cell 7:671–681
74. Calvanese V, Horrillo A, Hmadcha A, Suarez-Alvarez B, Fernandez AF, Lara E et al (2008) Cancer genes hypermethylated in human embryonic stem cells. PLoS ONE 3:e3294
75. Lee AS, Tang C, Cao F, Xie X, van der Bogt K, Hwang A, Connolly AJ, Robbins RC, Wu JC (2009) Effects of cell number on teratoma formation by human embryonic stem cells. Cell Cycle 8:2608–2612
76. Hentze H, Soong PL, Wang ST, Phillips BW, Putti TC, Dunn NR (2009) Teratoma formation by human embryonic stem cells: evaluation of essential parameters for future safety studies. Stem Cell Res 2:198–210
77. Fong CY, Peh GS, Gauthaman K, Bongso A (2009) Separation of SSEA-4 and TRA-1-60 labelled undifferentiated human embryonic stem cells from a heterogeneous cell population using magnetic-activated cell sorting (MACS) and fluorescence-activated cell sorting (FACS). Stem Cell Rev 5:72–80
78. Griffiths-Jones S, Saini HK, van Dongen S, Enright AJ (2008) miRBase: tools for microRNA genomics. Nucleic Acids Res 36:D154–D158
79. Ambros V (2004) The functions of animal microRNAs. Nature 431:350–355
80. Kim N, Kim H, Jung I, Kim Y, Kim D, Han YM (2011) Expression profiles of miRNAs in human embryonic stem cells during hepatocyte differentiation. Hepatol Res 41:170–183
81. Tzur G, Levy A, Meiri E, Barad O, Spector Y, Bentwich Z, Mizrahi L, Katzenellenbogen M, Ben-Shushan E, Reubinoff BE, Galun E (2008) MicroRNA expression patterns and function in endodermal differentiation of human embryonic stem cells. PLoS One 3:e3726
82. Manohar R, Lagasse E (2009) Transdetermination: a new trend in cellular reprogramming. Mol Ther 17:936–938
83. Vierbuchen T, Ostermeier A, Pang ZP, Kokubu Y, Sudhof TC, Wernig M (2010) Direct conversion of fibroblasts to functional neurons by defined factors. Nature 463:1035–1041
84. Ieda M, Fu JD, Delgado-Olguin P, Vedantham V, Hayashi Y, Bruneau BG et al (2010) Direct reprogramming of fibroblasts into functional cardiomyocytes by defined factors. Cell 142: 375–386
85. Nicholas CR, Kriegstein AR (2010) Regenerative medicine: cell reprogramming gets direct. Nature 463:1031–1032

86. Sekiya S, Suzuki A (2011) Direct conversion of mouse fibroblasts to hepatocyte-like cells by defined factors. Nature [Epub ahead of print]
87. Huang P, He Z, Ji S, Sun H, Xiang D, Liu C, Hu Y, Wang X, Hui L (2011) Induction of functional hepatocyte-like cells from mouse fibroblasts by defined factors. Nature [Epub ahead of print]
88. Touboul T, Hannan NR, Corbineau S, Martinez A, Martinet C, Branchereau S, et al (2010) Generation of functional hepatocytes from human embryonic stem cells under chemically defined conditions that recapitulate liver development. Hepatology 51:1754–1765

Chapter 5
Advances in Induced Pluripotent Stem Cell Biology

Ali Seifinejad

Abstract Conversion of somatic cells into a pluripotent state is possible by the introduction of reprogramming transcription factors (Oct4, Sox2, c-Myc, and Klf4) into somatic cells. Induced pluripotent stem cells (iPSCs), which hold great promise for the future of regenerative medicine, disease modeling, and drug discovery are embryonic stem cell (ESC)-like cells that display most characteristics of ESCs. There are some challenges regarding the establishment of iPSCs, their similarity with ESCs, and their biomedical applications. Improvements in gene delivery methods, establishment of iPSC lines without the use of Yamanaka factors, and in vitro modeling of complex diseases are among hot topics of this field to be discussed in this chapter.

5.1 Introduction

After the establishment of mouse embryonic stem cells (ESCs) in 1981 by Martin Evans [1] and human ESCs in 1998 by James Thomson [2], the field of regenerative medicine received much attention. The scientific community had access to new, potent cells that could be used in the clinic to save patients who suffered from diseases in which cells of the diseased tissues were either lost or abnormal. Since then, most researchers have attempted to generate patient compatible pluripotent stem cells for clinic use that do not cause immune rejection after transplantation and, on the other hand, lack ethical issues related to ESCs. Utilization of routes

A. Seifinejad (✉)
Department of Stem Cells and Developmental Biology, Cell Science Research Center, Academic Center for Education, Culture and Research (ACECR), Royan Institute for Stem Cell Biology and Technology, Tehran, Iran
e-mail: aliseify@yahoo.com

H. Baharvand and N. Aghdami (eds.), *Advances in Stem Cell Research*, Stem Cell Biology and Regenerative Medicine, DOI: 10.1007/978-1-61779-940-2_5,

such as somatic cell nuclear transfer (SCNT) [3–5] as well as nuclear fusion of somatic cells with pluripotent cells [6, 7] (Fig. 5.1a) are among the attempts that have made it possible to generate patient-specific pluripotent stem cells, but each has its own ethical and technical issues as well as difficulties. The accumulated understandings of the mechanisms that underlie the above techniques, as well as ESCs, lead to a breakthrough. Groundbreaking, innovative research by Shinya Yamanaka's group at Kyoto University in 2006 introduced the generation of induced pluripotent stem cells (iPSCs) [8]. Their research has made it possible to easily create patient-specific stem cells in vitro. In this chapter, some of the issues related to iPSC generation and establishment, as well as problems in their basic biology will be discussed.

The story of iPSCs began with 24 genes: pluripotency related, growth related, and others with unknown functions that were upregulated in stem cells. Subsequently, these genes were introduced into mouse embryonic fibroblasts (MEF). Of these 24 genes, 20 were eliminated, thus resulting in four genes that were screened and examined in different combinations. The four genes (Oct4, Sox2, c-Myc, and Klf4: OSMK), according to Yamanaka's lab, had enough power to convert MEFs into a stem cell-like state [8]. Soon after this achievement, James Thomson's laboratory in Wisconsin successfully converted human fetal and postnatal fibroblasts into human iPSCs (hiPSCs) by a combination of four factors. Two factors were the same as used by Yamanaka's group (Oct4 and Sox2) and two were different (Lin28 and Nanog) [9]. This work was coincidentally repeated by the technology's pioneer group in Kyoto, but with their own original cocktail (OSMK) [10], after which most laboratories began to establish their own hiPSC and mouse iPSC lines with different goals and applications [11–15]. There have been several modifications and improvements in both the basic biology and application of iPSC lines established thus far, as discussed in the next sections.

5.2 The Magic Transcription Factors

Early researches on iPSCs have demonstrated that these four transcription factors could fully reprogram somatic cells into an ESC-like state. How this magic cocktail performs this conversion is still elusive. According to Liu et al. [16], the numbers of target genes for c-Myc, Klf4, Oct4, and Sox2 are 3869, 1505, 904, and 864, respectively in mouse ESCs. When Oct4 is artificially overexpressed or repressed in ESCs, it differentiates into primitive endoderm-like cells and a trophectodermal lineage. For Sox2, its depletion in mouse ESCs leads to differentiation into multiple lineages, including the trophectoderm, which shows the importance of its role in maintaining pluripotency [17]. It has been shown that along with Nanog, Oct4 and Sox2 together occupy at least 353 genes within human ESCs [18]. Forced expression of Klf4 in cultured cells results in the inhibition of DNA synthesis and cell cycle progression. c-Myc has a potent role in most human cancers when its expression is deregulated [19, 20]. Considering this

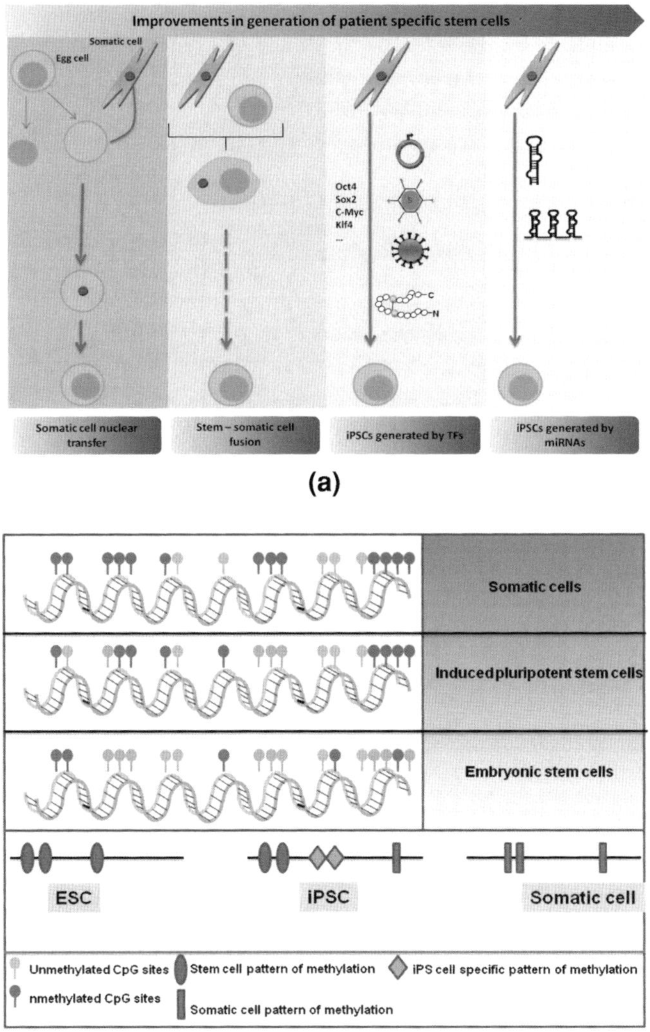

Fig. 5.1 **a** Inducing reprogramming of somatic cells into a pluripotent state and the generation of patient-specific stem cells. From *left* to *right*. SCNT, stem—somatic cell fusion, induced pluripotency by transcription factors, and induced pluripotency only with miRNAs. It seems that with better understanding of basic cellular and molecular biology it would be possible to generate fully reprogrammed cells easily and safely. **b** Observation of differentially methylated regions between iPSC lines and ESC lines according to some reports. They may be due to incomplete reprogramming of somatic cell methylation patterns known as epigenetic memory, or due to reprogramming techniques that establish new regions of methylation with a type of pattern different from the original somatic cells as well as ESCs, which are called iPSC-specific differential methylation regions

complication, a strong stoichiometry should be established between these transcription factors with the intent to achieve a stable conversion of somatic cells to pluripotent stem cells. Variations in the relative ratios of transcription factor expression might affect efficiency of hiPSC induction. It has been shown that the target genes of the four factors strongly overlap in ESCs and iPSCs [21, 22]. In iPSCs, the promoters of 565 genes are co-bound by the four Yamanaka factors [23]. It seems that Oct3/4, Sox2, and Klf4 might function together to induce target gene expression, thus establishing a stable stemness network, whereas c-Myc might enhance the accessibility of target genes by stimulating DNA replication. c-Myc greatly enhances the initial steps of reprogramming, therefore its major contribution has been demonstrated to be in the initial reprogramming events [17, 21]. The screening process is ongoing to discover new transcription factors and genes for reprogramming. However, there are new methods and approaches that reprogram somatic cells to an embryonic state without the use of transcription factors, which will be discussed.

5.3 Technical Advances in Methods for Establishment of iPSCs

5.3.1 Retroviruses and Lentiviruses

Retro- and lentiviruses were the first vehicles to deliver reprogramming factors into somatic cells. Scientists have used these viruses either as individual viruses to carry one gene or as single expression cassettes carrying multiple genes, where the number of integrations is decreased [15, 24–26]. Although these vehicles are still used in reprogramming trials, the safety issue related to integration of the viral genome into the host cell genome has raised concerns about oncogene activation [27, 28]. This is one of the most important issues that has led scientists to move toward the use of safer, integration-free methods. However, excision of integrated sequences by the Cre-loxP system in which the vector contains a loxP site at the 3′ long terminal repeat that is excised after expression of the Cre-recombinase gene is one approach that ameliorates the disadvantages of the integration method. Another approach involves transgene removal by using a transposase in the PB (piggyBac transposition) system [29–31].

5.3.2 Adenoviruses, Sendai Virus Vectors, and Vector Transfection

As mentioned above, the goal of researchers in the iPSC community is to generate safer iPSC lines. In this regard, there are reports in which the integration of exogenous nucleic acid has not been detected in the genome [27, 32, 33]. At Harvard University, Hochedlinger's group has used adenoviral vectors to convert somatic cells to pluripotent cells. Since the expressions by adenoviruses are transient and viral DNA does not integrate into the host genome, they could not detect any vector integration upon genomic analysis. On the other hand, other groups have generated cells by transfection of naked vectors into target cells; upon analysis, these researchers did not detect vector integration into the genome. In these methods, since transgenes are not integrated during cell proliferation those cells harboring transgenes are diluted, therefore their efficiency is very low relative to the integration methods.

In a recent advancement in the generation of iPSC lines without integration, a Japanese group has used temperature-sensitive Sendai virus vectors to deliver reprogramming factors into host cells. By using these RNA virus vectors and by introducing temperature-sensitive mutations in them, the researchers were able to establish iPSCs from human fibroblasts and $CD34^+$ cord blood cells. Vectors were easily removable at non-permissive temperatures [34]. Although all of these techniques have provided us with safer iPSCs, each struggles with their own difficulties and problems, particularly concerning efficiency.

5.3.3 Delivery of Protein and mRNA Version of Reprogramming Transcription Factors

It is believed that delivery of transcripted or translated versions of genes into host cells will result in their accelerated functions. In addition, their function without integration into the host genome is undoubtedly a reasonable expectation that they could be used for generating safer iPSCs. To support this idea, two groups have successfully established iPSCs using just the protein version of Yamanaka's factors (Fig. 5.1a) [35, 36]. They have transduced their Tat-fused transcription factors into target cells after production of the proteins in bacterial [35] or eukaryotic cells [36]. This method, as with the methods mentioned above, suffers from the problem of low efficiency and needs optimization. Technical issues, such as in vitro post-translational modifications of recombinant proteins, as well as purification of recombinant proteins and their half-lives, are among the most important technical issues that should be considered.

Recent breakthroughs in which scientists have successfully generated iPSCs by transfection of the mRNA of Yamanaka's factors into somatic cells at high efficiencies have brought these cells one step closer to the clinic setting [37, 38]. It has previously been shown that transcriptome transfer of one cell type to another can produce a predictable cellular phenotype. In one study, the total transcriptome of astrocytes which contained RNAs in the appropriate relative amounts that gave rise to regulatory RNA and translated proteins has been transferred into post-mitotic neurons. The astrocyte RNA population converted 44% of the neuronal host cells into the destination astrocyte-like phenotype [39]. In a recent study, in order to obtain a highly efficient generation of iPSCs by mRNA transfection, researchers have in vitro produced mRNA that incorporated modified ribonucleoside bases, which exhibited a reduced immunogenic profile and increased sustainability of RNA-mediated protein expression. With contribution from mRNAs of five transcription factors (Oct4, Sox2, c-Myc, Klf4, and Lin28), and transfection of these synthetic mRNAs into human fibroblasts under reduced oxygen culture conditions, the researchers succeeded in generating iPSCs with an efficiency of 4.4%. Although this method seemed to be very efficient and safe, the protocols used were multi-step and complex. This issue should be considered in experimental designs [37].

5.3.4 Safer Approaches that Eliminate the Application of Exogenous Reprogramming Transcription Factors

The main challenge in the field of iPSCs is the removal of integrating viruses, and whether there are possibilities that can replace well-known reprogramming factors with the intent to generate iPSCs at a higher efficiency. To reach these goals scientists have examined two hypotheses: application of small molecules and generation of iPSCs using miRNAs.

5.3.4.1 Small Molecules

The role of small molecules in stem cell biology has been intensively investigated by different groups [40–42]. It has long been hypothesized that small molecules can modulate cellular processes through their involvement in cellular signaling pathways [40, 43]. In direct cell reprogramming trials that have been carried out by the introduction of four transcription factors into somatic cells, the important role of small molecules has been established. It can be said that small molecules play a role in iPSC biology by enhancing both the efficiency of iPSC generation and replacing exogenous transcription factors [44]. Small molecules in iPSC biology function as DNA methyltransferase and histone deacetylase (HDAC) inhibitors, signaling pathway inhibitors, and inducers [42, 43, 45, 46]. By administration of DNA methyltransferase and HDAC inhibitors, Huangfu et al. [45] have improved the efficiency of mouse iPSCs more than 100-fold and successfully removed Klf4 and c-Myc from their cocktail, thus generating hiPSCs with only two factors (Oct4 and Sox2) [46]. There are studies in which small molecules have specifically replaced one transcription factor [47]. It seems that small molecules have the potential to fulfill scientists' dreams of generating patient-specific cells by using these molecules without any genetic materials. There are reports regarding the use of only small molecules to generate partially reprogrammed cells [41].

5.3.4.2 MicroRNAs as Another Approach to Generate Safer iPSCs

The role of miRNAs in the cellular process and stem cell biology has been intensively investigated by different groups. Since ESCs have distinct miRNA signatures [48, 49], logically it seems that they could play a role in reprogramming a cell's fate. Thus, it could be possible to develop alternatives for transgenes that are currently used in direct reprogramming. The work by Judson et al. [50] has proven this hypothesis to some extent in which they have shown that ESC-specific miRNAs (miR-291-3p, miR-294, and miR-295) promote induced pluripotency. By transfecting miR-291-3p, miR-294, and miR-295 along with Oct4, Sox2, and Klf4, they have observed an increase in the efficiency of generating iPSCs. This increase was noted to occur without c-Myc in the cocktail. After addition of c-Myc, the three factors along with used miRNAs could not increase the efficiency of generating iPSCs. It is believed that c-Myc promotes early stages of reprogramming through activating the miR-290 family and, as mentioned above, these miRNAs are downstream effectors of c-Myc during reprogramming. After activation of the miR-290 family, the need for c-Myc expression at later stages of reprogramming is dispensable [21, 51]. It can be said that expression of the miR-290 family at the initial stages of reprogramming could replace c-Myc to promote reprogramming [50]. There is a report concerning c-Myc repression of MEF-enriched miRNAs such as miR-21, let-7a, and miR-29a during reprogramming. It has been shown that after introduction of c-Myc into MEFs for the generation of iPSCs, the expression level of MEF-enriched miRNAs decreased and depletion of miR-21 and miR-29a enhanced reprogramming efficiency in MEF [51].

The power of miRNAs in cell fate reprogramming has been shown when two groups reported that, without the need for transcription factors, miRNAs could convert somatic cells into iPSCs per se (Fig. 5.1a) [52, 53]. By expressing the miR302/367 cluster in mouse and human somatic cells, Anokye-Danso et al. [53] could successfully convert these cells to iPSCs without the need for transcription factors, both over a shorter time period and more efficiently than the transcription factor approach. These researchers have observed greater than two orders of magnitude efficiency when compared to the standard Oct4/Sox2/Klf4/Myc-mediated methods. Although the method is highly efficient, problems concerning the application of viral vectors to deliver miRNAs into cells is one challenge which again raises safety issues. However, in a recent work, Miyoshi et al. [52] have transfected mature double-stranded miRNAs from a combination of the mir-200c, mir-302s,and mir-369s family converted mouse and human cells to pluripotency. Since there is no viral vector to be used in this method and mature miRNAs are directly introduced into cells, the method holds great promise for regenerative medicine and biomedical research; however, this method suffers from the problem of low efficiency. The mechanisms behind microRNA functions in cell fate reprogramming have largely remained unknown. For some, such as miR302, its role in cell fate reprogramming has previously been shown by Lin et al. [54, 55] in which they converted human skin cancer cells and hair follicle cells (hHFCs) into a pluripotent state. It seems that, by inhibiting translation of some epigenetic regulators (AOF1/2, MECP1/2 and DNMT1), global DNA demethylation is induced and finally activation of core pluripotency factors (Oct4, Sox2, and Nanog) complete the reprogramming process [56]. There are some miRNAs that enhance reprogramming efficiency by specifically targeting one transcription factor. As reported by Pfaff et al. [57], miRNA family miR-130/301/721 enhances iPSC generation via repression of Meox2.

Although the role of miRNAs in induced pluripotency has strongly been established, there are some miRNAs that act as barriers to reprogramming. It has been demonstrated that the miR-34 family, particularly miR-34a, exhibits p53-dependent induction during reprogramming and their depletion in mice somatic cells significantly increases the efficiency of reprogramming. Repression of the pluripotency genes Nanog, Sox2, and N-myc by miR34a is one reason for reprogramming suppression [58].

5.4 iPSCs and ESCs: The Level of Similarity

After establishment of the first iPSC colonies from mice, followed by human somatic cells, scientists had to confirm their cells' pluripotency and stemness characteristics. Scientists could only evaluate their established cells by comparing them with ESCs as a gold standard for analyzing iPSCs. In consideration of ESCs as an index for evaluating iPSCs, the most important question that arises is to what extend these two cells are equivalent. In 2006, when the first iPSC colonies were reported by Japanese scientists, their global gene expression patterns and DNA methylation status were analyzed. The researchers concluded that these iPSCs were not identical to ESCs [8]. In another study they found 1,267 genes differentially expressed between hiPSCs and human ESCs [10]. These results were repeated by other groups, all of who have reached the same conclusion that iPSCs and ESCs are not identical in view of their total gene expression profiles [9, 15, 59]. In 2009, UCLA scientists reported that iPSCs "share a gene expression signature that defines the iPSC state as unique from that of ESCs" and "should be considered a unique subtype of pluripotent cells" [60]. Carrying out a detailed genome-wide expression analysis between human ESC and iPSC clones at different passages, they have concluded that iPSCs are highly similar to ESCs, but they are more similar to each other than to ESCs and late passage iPSCs are closer to ESCs than early passage iPSCs. The study has shown that "(1) at early passage hiPSC lines are incompletely reset to a hESC-like expression pattern and (2) even at late passage differences between hESCs and hiPSCs persist and reflect an imperfect resetting of somatic cell expression to an ESC-like state." The feature seen in early passage hiPSCs is not a common feature of low passage pluripotent stem cells but is specific to hiPSCs. However, the hallmark of the study was when the researchers compared the transcriptome data from four labs and observed that 15 genes are consistently expressed differentially between early passage iPSC and

ESC lines. The researchers also found that their iPSC lines have established the H3K27 methylation state in the same manner as ESCs, which suggests that early and late hiPSC gene expression signatures probably do not arise as a result of faulty resetting of H3K27 trimethylation during reprogramming. Nevertheless, they detected few miRNAs that consistently expressed differently between human ESCs and late hiPSCs.

The UCLA scientists have reported the presence of 15 genes that consistently express differently between iPSC and ESC lines. Two other studies, however, after analyzing their own materials in addition to the data of Chin et al. [60] by using improved statistical methods have claimed that there were no consistent gene expression differences between iPSC and ESC lines [61, 62]. The researchers determined that the observed differences were lab-specific variances and not due to ESC/iPSC identity. Guenther et al. [62] have analyzed histone modifications for H3K4me3 and H3K23me3 and scanned the entire genome for identification of regions with significant differences in H3K4me3 and H3K27me3 occupancy between ESCs and iPSCs. They found very small genomic rejoins that had different H3K4me3 and H3K23me3 occupancies. Other studies have investigated whole genome DNA methylation, histone modification, and gene expression status between iPSC and ESCs, and came to the conclusion that these two types of pluripotent stem cells have different features [63–66]. With regard to some residual gene expression from the original cell type which might be due to epigenetic memory retained from the original cells and iPSC-specific differentially methylated regions (iDMRs) not observed in progenitor somatic cells and ESCs (Fig. 5.1b), which might be a product of aberrant reprogramming [64], it seems that observation of these types of differences is not uncommon [67–69]. It is believed that long-term passaging may erase this memory; however, it may cause some unwanted outcomes such as "culture adaptation" and contribute to tumorigenic differences between early and late passage cells of the same hiPSC line [70]. Regarding the report showing consistent differentially expressed genes among ESCs [71], it is undoubtedly true that variations exist among iPSC lines regarding their cells of origin, genetic background, method of derivation, culture condition, passage number, and most importantly the statistical method used. However, there is no certainty regarding the consistent expression in these lines. Therefore, it is possible to eliminate this variation by using more standard methods, and the same materials and methods for analysis.

5.5 Applications

Ease of derivation of iPSCs brings to mind the question of what would be the most important use for these cells. Perhaps the main purpose would be regenerative medicine, which is a distant goal. However, modeling of human diseases and the generation of tissue-specific cell types for drug discovery could be immediate uses for this technology.

5.5.1 *Application of Patient-Specific Cells for Generation of Immune-Compatible Tissue Cell Types*

It is generally believed that the primary, final goal in generating iPSCs is regenerative medicine. Therefore, these cells should be studied in clinical trials and used for transplantation of patients' own cells as cures for diseases. However, in order to reach this goal some important issues need to be overcome. Problems such as integration of transgenes into the genome, the application of certain oncogenes (c-Myc and Klf4) for the generation of iPSCs in addition to other problems such as whether the established cells are fully reprogrammed and cell tumorigenicity are considered to be the most important barriers for clinical application of iPSCs as created by the current methods.

5.5.2 *Disease Modeling*

The ultimate goal of scientists is to study disease development in vitro, which has somewhat been fulfilled by the establishment of iPSCs. The generation of hiPSCs provides a new method for elucidating the mechanisms underlying human disease. There are some established ESC lines that have been derived from diseased embryos diagnosed by PGD (Preimplantation genetic diagnosis) analysis [72, 73]. The wide range of diseases that could be covered by ESCs is very low since most diseases cannot be detected by PGD analysis. Additionally, it is difficult to find embryos that suffer from a wide range of diseases.

The generation of iPSCs provides a very promising approach for the establishment of pluripotent cells from a wide range of diseases in vitro. Thus far, there are many iPSC lines from different diseases that have been produced by different groups [74–78]. In 2008, a group of scientist at Harvard University succeeded to establish iPSC lines from patients with a variety of genetic diseases from either Mendelian or complex inheritance [75]. In their study, the authors only produced iPSC lines from patients who suffered from adenosine deaminase deficiency-related severe combined immunodeficiency, Shwachman-Bodian-Diamond syndrome, Gaucher disease type III, Duchenne and Becker muscular dystrophy, Parkinson disease, Huntington disease, juvenile-onset type 1 diabetes mellitus, Down syndrome/trisomy 21, and the carrier state of Lesch-Nyhan syndrome. The researchers characterized these established cell lines according to their mutations and pluripotency characteristics. Our group has also succeeded in generating iPSCs from a number of liver diseases that include tyrosinemia type I, GSD1, progressive familial hereditary cholestasis, and Crigler-Najjar syndrome. We have differentiated these iPSCs into hepatocyte-like lineages to test albumin secretion and LDL uptake [26]. However, there are reports in which the authors have analyzed the phenotype of disease-derived iPSC lines for their potential to differentiate to cell types of interest, i.e., certain diseases. Additionally, in some cases, they have shown disease-specific changes in cell survival and function. Examples of modeled and functionally analyzed diseases include the following neurological and neurodegenerative conditions: amyotrophic lateral sclerosis, spinal muscular atrophy, Parkinson disease, familial dysautonomia, fragile X (FX) syndrome, and Rett syndrome. Hematopoietic disorders include beta-thalassemia, Fanconi anaemia, and dyskeratosis congenita. In patients who suffer from cardiovascular disease, such examples include long QT syndrome types 1 and 2, and leopard syndrome, whereas in metabolic disorders, modeled diseases include familial hypercholesterolemia and glycogen storage disease type 1a (GSD1a) [29, 76–87], (review: [44, 88–90]).

Although modeling of diseases with defects in a single gene by using iPSC technology seems to be promising, there are beliefs that the application of iPSC technology to model more complex diseases such as psychiatric diseases would be more challenging. One of the more surprising reports about the application of iPSCs to model complex diseases is a model of the debilitating psychiatric disorder, schizophrenia [74]. By obtaining fibroblasts from childhood-onset schizophrenia from both schizophrenic children and adults, the researchers have succeeded in establishing iPSC lines for this disease. It seems that studying those more complex diseases, such as psychiatric disorders (schizophrenia) at the cellular and molecular levels in vitro would not be more informative. However, after reprogramming patient-derived fibroblast cells to iPSCs and their subsequent differentiation toward affected, neuronal cells, they found that neurons derived from schizophrenia hiPSCs showed diminished neuronal connectivity in conjunction with decreased neurite numbers, PSD95-protein levels, and glutamate receptor expression [74]. The study has also shown alterations in the gene expression profile of schizophrenia hiPSC neurons for components of the cyclic AMP and WNT signaling pathways. This result resembled, to some extent, a study by Wang et al. [91] in which the researchers have detected some alterations in gene expression in fibroblasts derived from schizophrenia patients. It could be assumed that these alterations have shown their effects after differentiation toward neuronal cells, which was accomplished after the generation of iPSCs from those fibroblast cells. Since some aspects of schizophrenia are manifested at the peripheral level in proliferating cell types, it seems that application of iPSCs for investigation of this disease at the molecular and cellular levels would be more informative. Modeling more complex disease such as schizophrenia in the lab, by using iPSC technology, paves the way toward investigation of other diseases in vitro with the use of this technology. Although the specificity of each disease may be problematic during investigations, thus it should be noted that such studies be complemented by using animal models to study these diseases in order to better understand the molecular mechanisms of the disease.

5.5.3 Generation of Desired Cell Types for Drug Discovery and Screening

Another application of patient-specific iPSCs, in addition to generating disease models, is the potential they have as an unlimited cell source for drug screening and discovery for various diseases. The availability of hiPSCs provides very exciting opportunities for general and patient-specific high throughput drug screening. In drug discovery trials, in addition to development of different animal models for a vast variety of diseases, pharmaceutical companies have long been using diverse cell culture models that range from cell cultures derived from tumor cell lines or cells immortalized through genetic transformation to primary cell cultures from tissues, but recently robust ESCs and iPSCs have eliminated many drawbacks of previous in vitro models, even they have certain difficulties.

In addition to the application of iPSCs for the study of basic biology and disease pathologies [92], there are reports about the application of iPSCs to achieve biological products. Our group has succeeded in generating iPSCs from an individual with Bombay blood type [93]. Currently, we are attempting to generate a universal donor red blood cell from this iPSC line. There is another group that has already succeeded in generating red blood cells from iPSC lines [94]. Given the fact that, having more than approximately 81 million units of donated whole blood and 20 million of plasma annually, we are still faced with blood shortages. Therefore, these products would be of tremendous help for patients who suffer from these shortages. It is believed that the application of iPSC technology to study basic biology, developmental biology, disease modeling, regenerative medicine, and the creation of biological products is increasing daily.

5.6 Concluding Remarks

Although there are debates on the different aspects of iPSCs including efficiency, safety, similarity to ESCs, the remaining epigenetic signature of the original cell, immunogenicity, and many important questions related to the technical aspect of these cells, it seems iPSCs can play an important role in disease modeling and drug screening studies. It is believed that establishing iPSCs without any remaining epigenetic memory from the original cell type and without any single nucleotide variations in comparison with the parental cell type is the primary focus for scientists working on iPSC biology. Creating iPSCs that acquire the ground state of pluripotency characteristics mandates established, defined culture conditions and convenient high efficient protocols. More studies need to uncover the molecular mechanisms of direct reprogramming which may help scientists to design new establishment protocols. Prior to their clinical use, although iPSCs have some specific challenges, they share some bottlenecks with ESCs including the cell status after additional passages and the establishment of more efficient differentiation protocols. Another challenging issue regards the decreased efficiency of iPSCs compared with ESCs according to some reports, which raises additional questions about the differentiation potential of iPSCs [95]. Direct conversion of one cell type to another one, or "transdifferentiation", is another approach that may help researchers clearly understand cellular plasticity and its application in regenerative and personalized medicine.

References

1. Evans MJ, Kaufman MH (1981) Establishment in culture of pluripotential cells from mouse embryos. Nature 292(5819):154–156
2. Thomson JA, Itskovitz-Eldor J, Shapiro SS, Waknitz MA, Swiergiel JJ, Marshall VS, Jones JM (1998) Embryonic stem cell lines derived from human blastocysts. Science 282(5391):1145–1147
3. Gurdon JB, Wilmut I (2011) Nuclear transfer to eggs and oocytes. Cold Spring Harb Perspect Biol 3:a002659
4. Byrne JA, Pedersen DA, Clepper LL, Nelson M, Sanger WG, Gokhale S, Wolf DP, Mitalipov SM (2007) Producing primate embryonic stem cells by somatic cell nuclear transfer. Nature 450(7169):497–502
5. Simerly C, Dominko T, Navara C, Payne C, Capuano S, Gosman G, Chong KY, Takahashi D, Chace C, Compton D, Hewitson L, Schatten G (2003) Molecular correlates of primate nuclear transfer failures. Science 300(5617):297
6. Tada M, Takahama Y, Abe K, Nakatsuji N, Tada T (2001) Nuclear reprogramming of somatic cells by in vitro hybridization with ES cells. Curr Biol 11(19):1553–1558
7. Cowan CA, Atienza J, Melton DA, Eggan K (2005) Nuclear reprogramming of somatic cells after fusion with human embryonic stem cells. Science 309(5739):1369–1373
8. Takahashi K, Yamanaka S (2006) Induction of pluripotent stem cells from mouse embryonic and adult fibroblast cultures by defined factors. Cell 126(4):663–676
9. Yu J, Vodyanik MA, Smuga-Otto K, Antosiewicz-Bourget J, Frane JL, Tian S, Nie J, Jonsdottir GA, Ruotti V, Stewart R, Slukvin II, Thomson JA (2007) Induced pluripotent stem cell lines derived from human somatic cells. Science 318(5858):1917–1920
10. Takahashi K, Tanabe K, Ohnuki M, Narita M, Ichisaka T, Tomoda K, Yamanaka S (2007) Induction of pluripotent stem cells from adult human fibroblasts by defined factors. Cell 131(5):861–872
11. Okita K, Ichisaka T, Yamanaka S (2007) Generation of germline-competent induced pluripotent stem cells. Nature 448(7151):313–317
12. Wernig M, Meissner A, Foreman R, Brambrink T, Ku M, Hochedlinger K, Bernstein BE, Jaenisch R (2007) In vitro reprogramming of fibroblasts into a pluripotent ES-cell-like state. Nature 448(7151):318–324
13. Maherali N, Sridharan R, Xie W, Utikal J, Eminli S, Arnold K, Stadtfeld M, Yachechko R, Tchieu J, Jaenisch R, Plath K, Hochedlinger K (2007) Directly reprogrammed fibroblasts show global epigenetic remodeling and widespread tissue contribution. Cell Stem Cell 1(1):55–70
14. Totonchi M, Taei A, Seifinejad A, Tabebordbar M, Rassouli H, Farrokhi A, Gourabi H, Aghdami N, Hosseini-Salekdeh G, Baharvand H (2009) Feeder- and serum-free establishment and expansion of human induced pluripotent stem cells. Int J Dev Biol 54:877–886
15. Lowry WE, Richter L, Yachechko R, Pyle AD, Tchieu J, Sridharan R, Clark AT, Plath K (2008) Generation of human induced pluripotent stem cells from dermal fibroblasts. Proc Natl Acad Sci U S A 105(8):2883–2888

16. Liu X, Huang J, Chen T, Wang Y, Xin S, Li J, Pei G, Kang J (2008) Yamanaka factors critically regulate the developmental signaling network in mouse embryonic stem cells. Cell Res 18(12):1177–1189
17. Niwa H (2007) How is pluripotency determined and maintained? Development 134(4): 635–646
18. Boyer LA, Lee TI, Cole MF, Johnstone SE, Levine SS, Zucker JP, Guenther MG, Kumar RM, Murray HL, Jenner RG, Gifford DK, Melton DA, Jaenisch R, Young RA (2005) Core transcriptional regulatory circuitry in human embryonic stem cells. Cell 122(6):947–956
19. Yamanaka S (2007) Strategies and new developments in the generation of patient-specific pluripotent stem cells. Cell Stem Cell 1(1):39–49
20. Knoepfler PS (2008) Why myc? An unexpected ingredient in the stem cell cocktail. Cell Stem Cell 2(1):18–21
21. Sridharan R, Tchieu J, Mason MJ, Yachechko R, Kuoy E, Horvath S, Zhou Q, Plath K (2009) Role of the murine reprogramming factors in the induction of pluripotency. Cell 136(2): 364–377
22. Papapetrou EP, Tomishima MJ, Chambers SM, Mica Y, Reed E, Menon J, Tabar V, Mo Q, Studer L, Sadelain M (2009) Stoichiometric and temporal requirements of Oct4, Sox2, Klf4, and c-Myc expression for efficient human iPSC induction and differentiation. Proc Natl Acad Sci U S A 106(31):12759–12764
23. Huang J, Chen T, Liu X, Jiang J, Li J, Li D, Liu XS, Li W, Kang J, Pei G (2009) More synergetic cooperation of Yamanaka factors in induced pluripotent stem cells than in embryonic stem cells. Cell Res 19(10):1127–1138
24. Carey BW, Markoulaki S, Hanna J, Saha K, Gao Q, Mitalipova M, Jaenisch R (2009) Reprogramming of murine and human somatic cells using a single polycistronic vector. Proc Natl Acad Sci U S A 106(1):157–162
25. Sommer CA, Stadtfeld M, Murphy GJ, Hochedlinger K, Kotton DN, Mostoslavsky G (2008) iPS cell generation using a single lentiviral stem cell cassette. Stem Cells 27:543–549
26. Ghodsizadeh A, Taei A, Totonchi M, Seifinejad A, Gourabi H, Pournasr B, Aghdami N, Malekzadeh R, Almadani N, Salekdeh GH, Baharvand H (2010) Generation of liver disease-specific induced pluripotent stem cells along with efficient differentiation to functional hepatocyte-like cells. Stem Cell Rev 6(4):622–632
27. Okita K, Nakagawa M, Hyenjong H, Ichisaka T, Yamanaka S (2008) Generation of mouse induced pluripotent stem cells without viral vectors. Science 322(5903):949–953
28. Yamanaka S (2009) A fresh look at iPS cells. Cell 137(1):13–17
29. Soldner F, Hockemeyer D, Beard C, Gao Q, Bell GW, Cook EG, Hargus G, Blak A, Cooper O, Mitalipova M, Isacson O, Jaenisch R (2009) Parkinson's disease patient-derived induced pluripotent stem cells free of viral reprogramming factors. Cell 136(5):964–977
30. Woltjen K, Michael IP, Mohseni P, Desai R, Mileikovsky M, Hamalainen R, Cowling R, Wang W, Liu P, Gertsenstein M, Kaji K, Sung HK, Nagy A (2009) piggyBac transposition reprograms fibroblasts to induced pluripotent stem cells. Nature 458:766–770
31. Yusa K, Rad R, Takeda J, Bradley A (2009) Generation of transgene-free induced pluripotent mouse stem cells by the piggyBac transposon. Nat Methods 6:363–369
32. Yu J, Hu K, Smuga-Otto K, Tian S, Stewart R, Slukvin II, Thomson JA (2009) Human Induced pluripotent stem cells free of vector and transgene sequences. Science 324:797–801
33. Stadtfeld M, Nagaya M, Utikal J, Weir G, Hochedlinger K (2008) Induced pluripotent stem cells generated without viral integration. Science 322(5903):945–949
34. Ban H, Nishishita N, Fusaki N, Tabata T, Saeki K, Shikamura M, Takada N, Inoue M, Hasegawa M, Kawamata S, Nishikawa S (2011) Efficient generation of transgene-free human induced pluripotent stem cells (iPSCs) by temperature-sensitive Sendai virus vectors. Proc Natl Acad Sci U S A 108(34):14234–14239
35. Zhou H, Wu S, Joo JY, Zhu S, Han DW, Lin T, Trauger S, Bien G, Yao S, Zhu Y, Siuzdak G, Scholer HR, Duan L, Ding S (2009) Generation of induced pluripotent stem cells using recombinant proteins. Cell Stem Cell 4:381–384

36. Kim D, Kim CH, Moon JI, Chung YG, Chang MY, Han BS, Ko S, Yang E, Cha KY, Lanza R, Kim KS (2009) Generation of human induced pluripotent stem cells by direct delivery of reprogramming proteins. Cell Stem Cell 4(6):472–476
37. Warren L, Manos PD, Ahfeldt T, Loh YH, Li H, Lau F, Ebina W, Mandal PK, Smith ZD, Meissner A, Daley GQ, Brack AS, Collins JJ, Cowan C, Schlaeger TM, Rossi DJ (2010) Highly efficient reprogramming to pluripotency and directed differentiation of human cells with synthetic modified mRNA. Cell Stem Cell 7(5):618–630
38. Yakubov E, Rechavi G, Rozenblatt S, Givol D (2010) Reprogramming of human fibroblasts to pluripotent stem cells using mRNA of four transcription factors. Biochem Biophys Res Commun 394(1):189–193
39. Sul JY, Wu CW, Zeng F, Jochems J, Lee MT, Kim TK, Peritz T, Buckley P, Cappelleri DJ, Maronski M, Kim M, Kumar V, Meaney D, Kim J, Eberwine J (2009) Transcriptome transfer produces a predictable cellular phenotype. Proc Natl Acad Sci U S A 106(18):7624–7629
40. Xu Y, Shi Y, Ding S (2008) A chemical approach to stem-cell biology and regenerative medicine. Nature 453(7193):338–344
41. Park J, Kim C, Tang Y, Amano T, Lin CJ, Tian XC (2011) Reprogramming of mouse fibroblasts to an intermediate state of differentiation by chemical induction. Cell Reprogram 13(2):121–131
42. Li Y, Zhang Q, Yin X, Yang W, Du Y, Hou P, Ge J, Liu C, Zhang W, Zhang X, Wu Y, Li H, Liu K, Wu C, Song Z, Zhao Y, Shi Y, Deng H (2011) Generation of iPSCs from mouse fibroblasts with a single gene, Oct4, and small molecules. Cell Res 21(1):196–204
43. Ichida JK, Blanchard J, Lam K, Son EY, Chung JE, Egli D, Loh KM, Carter AC, Di Giorgio FP, Koszka K, Huangfu D, Akutsu H, Liu DR, Rubin LL, Eggan K (2009) A small-molecule inhibitor of tgf-Beta signaling replaces sox2 in reprogramming by inducing nanog. Cell Stem Cell 5(5):491–503
44. Seifinejad A, Tabebordbar M, Baharvand H, Boyer LA, Salekdeh GH (2010) Progress and promise towards safe induced pluripotent stem cells for therapy. Stem Cell Rev 6(2):297–306
45. Huangfu D, Maehr R, Guo W, Eijkelenboom A, Snitow M, Chen AE, Melton DA (2008) Induction of pluripotent stem cells by defined factors is greatly improved by small-molecule compounds. Nat Biotechnol 26(7):795–797
46. Huangfu D, Osafune K, Maehr R, Guo W, Eijkelenboom A, Chen S, Muhlestein W, Melton DA (2008) Induction of pluripotent stem cells from primary human fibroblasts with only Oct4 and Sox2. Nat Biotechnol 26(11):1269–1275
47. Lyssiotis CA, Foreman RK, Staerk J, Garcia M, Mathur D, Markoulaki S, Hanna J, Lairson LL, Charette BD, Bouchez LC, Bollong M, Kunick C, Brinker A, Cho CY, Schultz PG, Jaenisch R (2009) Reprogramming of murine fibroblasts to induced pluripotent stem cells with chemical complementation of Klf4. Proc Natl Acad Sci U S A 106(22):8912–8917
48. Gangaraju VK, Lin H (2009) MicroRNAs: key regulators of stem cells. Nat Rev Mol Cell Biol 10(2):116–125
49. Marson A, Levine SS, Cole MF, Frampton GM, Brambrink T, Johnstone S, Guenther MG, Johnston WK, Wernig M, Newman J, Calabrese JM, Dennis LM, Volkert TL, Gupta S, Love J, Hannett N, Sharp PA, Bartel DP, Jaenisch R, Young RA (2008) Connecting microRNA genes to the core transcriptional regulatory circuitry of embryonic stem cells. Cell 134(3):521–533
50. Judson RL, Babiarz JE, Venere M, Blelloch R (2009) Embryonic stem cell-specific microRNAs promote induced pluripotency. Nat Biotechnol 27:459–461
51. Yang CS, Li Z, Rana TM (2011) microRNAs modulate iPS cell generation. RNA 17: 1451–1460
52. Miyoshi N, Ishii H, Nagano H, Haraguchi N, Dewi DL, Kano Y, Nishikawa S, Tanemura M, Mimori K, Tanaka F, Saito T, Nishimura J, Takemasa I, Mizushima T, Ikeda M, Yamamoto H, Sekimoto M, Doki Y, Mori M (2011) Reprogramming of mouse and human cells to pluripotency using mature MicroRNAs. Cell Stem Cell 8(6):633–638
53. Anokye-Danso F, Trivedi CM, Juhr D, Gupta M, Cui Z, Tian Y, Zhang Y, Yang W, Gruber PJ, Epstein JA, Morrisey EE (2011) Highly efficient miRNA-mediated reprogramming of mouse and human somatic cells to pluripotency. Cell Stem Cell 8(4):376–388

54. Lin SL, Chang DC, Lin CH, Ying SY, Leu D, Wu DT (2011) Regulation of somatic cell reprogramming through inducible mir-302 expression. Nucleic Acids Res 39(3):1054–1065
55. Lin SL, Chang DC, Chang-Lin S, Lin CH, Wu DT, Chen DT, Ying SY (2008) Mir-302 reprograms human skin cancer cells into a pluripotent ES-cell-like state. RNA 14(10):2115–2124
56. Lin SL (2011) Concise review: deciphering the mechanism behind induced pluripotent stem cell generation. Stem Cells 29(11):1645–1649
57. Pfaff N, Fiedler J, Holzmann A, Schambach A, Moritz T, Cantz T, Thum T (2011) miRNA screening reveals a new miRNA family stimulating iPS cell generation via regulation of Meox2. EMBO Rep 12(11):1153–1159
58. Choi YJ, Lin CP, Ho JJ, He X, Okada N, Bu P, Zhong Y, Kim SY, Bennett MJ, Chen C, Ozturk A, Hicks GG, Hannon GJ, He L (2011) miR-34 miRNAs provide a barrier for somatic cell reprogramming. Nat Cell Biol 13(11):1353–1360
59. Park IH, Zhao R, West JA, Yabuuchi A, Huo H, Ince TA, Lerou PH, Lensch MW, Daley GQ (2008) Reprogramming of human somatic cells to pluripotency with defined factors. Nature 451(7175):141–146
60. Chin MH, Mason MJ, Xie W, Volinia S, Singer M, Peterson C, Ambartsumyan G, Aimiuwu O, Richter L, Zhang J, Khvorostov I, Ott V, Grunstein M, Lavon N, Benvenisty N, Croce CM, Clark AT, Baxter T, Pyle AD, Teitell MA, Pelegrini M, Plath K, Lowry WE (2009) Induced pluripotent stem cells and embryonic stem cells are distinguished by gene expression signatures. Cell Stem Cell 5(1):111–123
61. Newman AM, Cooper JB (2010) Lab-specific gene expression signatures in pluripotent stem cells. Cell Stem Cell 7(2):258–262
62. Guenther MG, Frampton GM, Soldner F, Hockemeyer D, Mitalipova M, Jaenisch R, Young RA (2010) Chromatin structure and gene expression programs of human embryonic and induced pluripotent stem cells. Cell Stem Cell 7(2):249–257
63. Doi A, Park IH, Wen B, Murakami P, Aryee MJ, Irizarry R, Herb B, Ladd-Acosta C, Rho J, Loewer S, Miller J, Schlaeger T, Daley GQ, Feinberg AP (2009) Differential methylation of tissue- and cancer-specific CpG island shores distinguishes human induced pluripotent stem cells, embryonic stem cells and fibroblasts. Nat Genet 41(12):1350–1353
64. Lister R, Pelizzola M, Kida YS, Hawkins RD, Nery JR, Hon G, Antosiewicz-Bourget J, O'Malley R, Castanon R, Klugman S, Downes M, Yu R, Stewart R, Ren B, Thomson JA, Evans RM, Ecker JR (2011) Hotspots of aberrant epigenomic reprogramming in human induced pluripotent stem cells. Nature 471(7336):68–73
65. Bock C, Kiskinis E, Verstappen G, Gu H, Boulting G, Smith ZD, Ziller M, Croft GF, Amoroso MW, Oakley DH, Gnirke A, Eggan K, Meissner A (2011) Reference maps of human ES and iPS cell variation enable high-throughput characterization of pluripotent cell lines. Cell 144(3):439–452
66. Ohi Y, Qin H, Hong C, Blouin L, Polo JM, Guo T, Qi Z, Downey SL, Manos PD, Rossi DJ, Yu J, Hebrok M, Hochedlinger K, Costello JF, Song JS, Ramalho-Santos M (2011) Incomplete DNA methylation underlies a transcriptional memory of somatic cells in human iPS cells. Nat Cell Biol 13(5):541–549
67. Kim K, Doi A, Wen B, Ng K, Zhao R, Cahan P, Kim J, Aryee MJ, Ji H, Ehrlich LI, Yabuuchi A, Takeuchi A, Cunniff KC, Hongguang H, McKinney-Freeman S, Naveiras O, Yoon TJ, Irizarry RA, Jung N, Seita J, Hanna J, Murakami P, Jaenisch R, Weissleder R, Orkin SH, Weissman IL, Feinberg AP, Daley GQ (2010) Epigenetic memory in induced pluripotent stem cells. Nature 467(7313):285–290
68. Polo JM, Liu S, Figueroa ME, Kulalert W, Eminli S, Tan KY, Apostolou E, Stadtfeld M, Li Y, Shioda T, Natesan S, Wagers AJ, Melnick A, Evans T, Hochedlinger K (2010) Cell type of origin influences the molecular and functional properties of mouse induced pluripotent stem cells. Nat Biotechnol 28(8):848–855
69. Marchetto MC, Yeo GW, Kainohana O, Marsala M, Gage FH, Muotri AR (2009) Transcriptional signature and memory retention of human-induced pluripotent stem cells. PLoS ONE 4(9):e7076

70. Ben-David U, Benvenisty N (2011) The tumorigenicity of human embryonic and induced pluripotent stem cells. Nat Rev Cancer 11(4):268–277
71. Abeyta MJ, Clark AT, Rodriguez RT, Bodnar MS, Pera RA, Firpo MT (2004) Unique gene expression signatures of independently-derived human embryonic stem cell lines. Hum Mol Genet 13(6):601–608
72. Verlinsky Y, Strelchenko N, Kukharenko V, Rechitsky S, Verlinsky O, Galat V, Kuliev A (2005) Human embryonic stem cell lines with genetic disorders. Reprod Biomed Online 10(1):105–110
73. Taei A, Gourabi H, Seifinejad A, Totonchi M, Shahbazi E, Valojerdi MR, Eftekhari P, Karimian L, Baharvand H (2010) Derivation of new human embryonic stem cell lines from preimplantation genetic screening and diagnosis-analyzed embryos. In Vitro Cell Dev Biol Anim 46(3–4):395–402
74. Brennand KJ, Simone A, Jou J, Gelboin-Burkhart C, Tran N, Sangar S, Li Y, Mu Y, Chen G, Yu D, McCarthy S, Sebat J, Gage FH (2011) Modelling schizophrenia using human induced pluripotent stem cells. Nature 473(7346):221–225
75. Park IH, Arora N, Huo H, Maherali N, Ahfeldt T, Shimamura A, Lensch MW, Cowan C, Hochedlinger K, Daley GQ (2008) Disease-specific induced pluripotent stem cells. Cell 134(5):877–886
76. Ebert AD, Yu J, Rose FF Jr, Mattis VB, Lorson CL, Thomson JA, Svendsen CN (2009) Induced pluripotent stem cells from a spinal muscular atrophy patient. Nature 457(7227): 277–280
77. Dimos JT, Rodolfa KT, Niakan KK, Weisenthal LM, Mitsumoto H, Chung W, Croft GF, Saphier G, Leibel R, Goland R, Wichterle H, Henderson CE, Eggan K (2008) Induced pluripotent stem cells generated from patients with ALS can be differentiated into motor neurons. Science 321(5893):1218–1221
78. Moretti A, Bellin M, Welling A, Jung CB, Lam JT, Bott-Flugel L, Dorn T, Goedel A, Hohnke C, Hofmann F, Seyfarth M, Sinnecker D, Schomig A, Laugwitz KL (2010) Patient-specific induced pluripotent stem-cell models for long-QT syndrome. N Engl J Med 363(15): 1397–1409
79. Hargus G, Cooper O, Deleidi M, Levy A, Lee K, Marlow E, Yow A, Soldner F, Hockemeyer D, Hallett PJ, Osborn T, Jaenisch R, Isacson O (2010) Differentiated Parkinson patient-derived induced pluripotent stem cells grow in the adult rodent brain and reduce motor asymmetry in Parkinsonian rats. Proc Natl Acad Sci U S A 107(36):15921–15926
80. Raya A, Rodriguez-Piza I, Guenechea G, Vassena R, Navarro S, Barrero MJ, Consiglio A, Castella M, Rio P, Sleep E, Gonzalez F, Tiscornia G, Garreta E, Aasen T, Veiga A, Verma IM, Surralles J, Bueren J, Izpisua Belmonte JC (2009) Disease-corrected haematopoietic progenitors from Fanconi anaemia induced pluripotent stem cells. Nature 460(7251):53–59
81. Ye L, Chang JC, Lin C, Sun X, Yu J, Kan YW (2009) Induced pluripotent stem cells offer new approach to therapy in thalassemia and sickle cell anemia and option in prenatal diagnosis in genetic diseases. Proc Natl Acad Sci U S A 106(24):9826–9830
82. Itzhaki I, Maizels L, Huber I, Zwi-Dantsis L, Caspi O, Winterstern A, Feldman O, Gepstein A, Arbel G, Hammerman H, Boulos M, Gepstein L (2011) Modelling the long QT syndrome with induced pluripotent stem cells. Nature 471(7337):225–229
83. Urbach A, Bar-Nur O, Daley GQ, Benvenisty N (2010) Differential modeling of fragile X syndrome by human embryonic stem cells and induced pluripotent stem cells. Cell Stem Cell 6(5):407–411
84. Marchetto MC, Carromeu C, Acab A, Yu D, Yeo GW, Mu Y, Chen G, Gage FH, Muotri AR (2010) A model for neural development and treatment of Rett syndrome using human induced pluripotent stem cells. Cell 143(4):527–539
85. Batista LF, Pech MF, Zhong FL, Nguyen HN, Xie KT, Zaug AJ, Crary SM, Choi J, Sebastiano V, Cherry A, Giri N, Wernig M, Alter BP, Cech TR, Savage SA, Reijo Pera RA, Artandi SE (2011) Telomere shortening and loss of self-renewal in dyskeratosis congenita induced pluripotent stem cells. Nature 474:399–402

86. Carvajal-Vergara X, Sevilla A, D'Souza SL, Ang YS, Schaniel C, Lee DF, Yang L, Kaplan AD, Adler ED, Rozov R, Ge Y, Cohen N, Edelmann LJ, Chang B, Waghray A, Su J, Pardo S, Lichtenbelt KD, Tartaglia M, Gelb BD, Lemischka IR (2010) Patient-specific induced pluripotent stem-cell-derived models of LEOPARD syndrome. Nature 465(7299):808–812
87. Rashid ST, Corbineau S, Hannan N, Marciniak SJ, Miranda E, Alexander G, Huang-Doran I, Griffin J, Ahrlund-Richter L, Skepper J, Semple R, Weber A, Lomas DA, Vallier L (2010) Modeling inherited metabolic disorders of the liver using human induced pluripotent stem cells. J Clin Invest 120(9):3127–3136
88. Saha K, Jaenisch R (2009) Technical challenges in using human induced pluripotent stem cells to model disease. Cell Stem Cell 5(6):584–595
89. Unternaehrer JJ, Daley GQ (2011) Induced pluripotent stem cells for modelling human diseases. Philos Trans R Soc Lond B Biol Sci 366(1575):2274–2285
90. Zhu H, Lensch MW, Cahan P, Daley GQ (2011) Investigating monogenic and complex diseases with pluripotent stem cells. Nat Rev Genet 12(4):266–275
91. Wang L, Lockstone HE, Guest PC, Levin Y, Palotas A, Pietsch S, Schwarz E, Rahmoune H, Harris LW, Ma D, Bahn S (2010) Expression profiling of fibroblasts identifies cell cycle abnormalities in schizophrenia. J Proteome Res 9(1):521–527
92. Baek KH, Zaslavsky A, Lynch RC, Britt C, Okada Y, Siarey RJ, Lensch MW, Park IH, Yoon SS, Minami T, Korenberg JR, Folkman J, Daley GQ, Aird WC, Galdzicki Z, Ryeom S (2009) Down's syndrome suppression of tumour growth and the role of the calcineurin inhibitor DSCR1. Nature 459(7250):1126–1130
93. Seifinejad A, Taei A, Totonchi M, Vazirinasab H, Hassani SN, Aghdami N, Shahbazi E, Yazdi RS, Salekdeh GH, Baharvand H (2010) Generation of human induced pluripotent stem cells from a Bombay individual: moving towards "universal-donor" red blood cells. Biochem Biophys Res Commun 391(1):329–334
94. Dias J, Gumenyuk M, Kang H, Vodyanik M, Yu J, Thomson JA, Slukvin II (2011) Generation of red blood cells from human induced pluripotent stem cells. Stem Cells Dev 20:1639–1647
95. Hu BY, Weick JP, Yu J, Ma LX, Zhang XQ, Thomson JA, Zhang SC (2010) Neural differentiation of human induced pluripotent stem cells follows developmental principles but with variable potency. Proc Natl Acad Sci U S A 107(9):4335–4340

Chapter 6
iPS Cells: New Applications for Metabolic Liver Diseases

Tobias Cantz, Abbas Beh-Pajooh and Malte Sgodda

Abstract Gaining insight into the biology of human stem cells and the early embryonic development helps us to generate new ways for generation of tissue-specific cells from embryonic stem cells or other pluripotent stem cell resources. The landmark studies from Shinya Yamanaka and others opened a new era in stem cell biology, by providing a technology that allows the generation of pluripotent stem cells from patients' cells and, thus, allowing the generation of disease-specific stem cells. But in their pluripotent stage those stem cells cannot be readily used for studies of the disease phenotype and improved differentiation protocols need to be applied to get a disease-specific cell phenotype. Combining these two major advances in the field of stem cell biology, new approaches for studying metabolic liver diseases in correlation to each patient's symptoms are available, which provide a breakthrough in new drug target research or individual gene therapy.

6.1 Introduction: Stem Cells and Hepatic Cell Transplantation

The liver might be an ideal target organ for cell-based therapy for genetic diseases. More than 200 liver-based defects are identified, which can cause hepatic and extrahepatic diseases and such inborn errors of liver metabolism account for 15–20% of pediatric liver transplantations. However, not all patients with metabolic diseases are urgent candidates for full organ transplantation and might

T. Cantz (✉) · A. Beh-Pajooh · M. Sgodda
Stem Cell Biology, Cluster-of-Excellence, REBIRTH,
Hannover Medical School, Hannover, Germany
e-mail: cantz.tobias@mh-hannover.de

H. Baharvand and N. Aghdami (eds.), *Advances in Stem Cell Research*,
Stem Cell Biology and Regenerative Medicine, DOI: 10.1007/978-1-61779-940-2_6,

not be selected for transplantation due to the limited number of available donor organs. But these patients experience poor quality of life due to specific treatments or diet restrictions. Since adult hepatocytes represent the smallest and most pivotal functional entity of the liver, hepatocyte transplantation has already been proven to restore liver function. Hereby, hepatocytes engraft in the recipient liver and functionally connect with the surrounding cells [1, 2]. In case of an immense liver injury and/or selective growth advantage, engrafted cells respond to growth stimuli and repopulate recipient's liver [3, 4]. Hence, the elucidation of the molecular pathways of liver regeneration and extensive preclinical cell transplantation experiments in animals have led to the application of hepatocyte transplantation in a limited number of patients with hereditary metabolic liver diseases and acute liver failure [5–7]. Meanwhile, some clinical centers are currently evaluating cell transplantation approaches in phase I and II clinical trials using primary human hepatocytes from non-transplantable donor organs [8–10].

As a preferable source for hepatic cell transplants, patient-derived stem cells were extensively studied [11–13] and much enthusiasm arose from patient-derived induced pluripotent stem (iPS) cells. The generation of iPS cells from adult somatic cells by retrovirus-mediated expression of pluripotency-associated genes in mice [14–17] and in humans [18–21] offers a unique tool to generate patient-specific iPS cells. In a different approach, James Thomson identified a slightly modified approach using a lentiviral expression system to deliver the transcription factors of Oct4, Sox2, Nanog, and Lin28 to human dermal fibroblasts [19] for iPS cell generation. A lot of progress is reported with respect to safer strategies for the generation of transgene-free iPS cells and alternative approaches to generate clinically applicable iPS cells have been reported using conventional plasmids [22], non-integrating adenoviral [23], Sendai viral [24], and episomal vectors [25] as well as protein transduction [26, 27]. Very recently, efficient reprogramming of human cells was shown through multiple transfection with synthetic mRNAs [28]. However, some recent publications questioned the clinical applicability of human iPS cells by describing substantial genomic abnormalities, such as copy number variations or point mutations [29–31], and altered epigenetic modifications, such as DNA methylation patterns [32]. Furthermore, the immunogenicity of patient-derived iPS cells might be still an issue, as it was demonstrated that murine iPS cells could be rejected upon transplantation into syngenic recipients [33]. Based on recent improvements applying polycistronic, excisable, and improved lentiviral constructs [34, 35] that prevent unfavorable reprogramming factor stoichiometries [36, 37], we argue that an efficient lentiviral reprogramming system might be the most suitable approach to generate human iPS cells for current pre-clinical studies.

The use of stem cell-derived hepatic cells for transplantation will not only need efficient in vitro differentiation protocols to obtain a definitive hepatic phenotype, but also requires in vivo repopulation and differentiation capabilities upon transplantation. In previous studies we were able to demonstrate that fetal liver cells exhibit these capabilities and efficiently engraft into host livers and acquire a fully mature phenotype within 2 weeks after transplantation [38]. However, in the context of murine and human embryonic stem cell-derived hepatic cells, we observed only limited repopulation capabilities of such stem cell-derived hepatic cells [39]. In a recent study, Haridass et al. [40] compared adult fetal and ES-derived hepatic cells from murine and human origin as transplants in mice, which allowed a semi-quantitative in vivo analysis. In mice expressing the toxic uroplasminogen activator (uPA) under the control of the albumin promoter from only one allele, severe hepatic failure occurs within the first weeks after birth. However, if the transgene is eliminated in rare events, such spontaneously "repaired" cells can repopulate the liver and the mice can develop normally. Transplanted hepatic cells have a competitive growth advantage and therefore can reliably be studied in this particular system. When these mice were cross-bred to mice from an immuno-deficient Rag2$^{-/-}\gamma c^{-/-}$ background, also human cells could be transplanted and such xenografts were accepted without signs of immunorejection. Interestingly, the ability of repopulating these uPA-mouse livers increased with the developmental stage of the transplanted hepatic cells, which means that the size and number of regeneration nodules was biggest in mice transplanted with adult hepatocytes and lowest in mice transplanted with murine ES-derived cells. Furthermore, murine cells showed more efficient engraftment and repopulation than human cells. Concluding from these findings it is not too surprising that even in this highly selective model for xenogenic cell transplantation, human ES-derived hepatic cells failed to repopulate uPA-mouse livers. To date, no unequivocal report has described the functional restoration of liver diseases using human ES-derived cells. Therefore, we speculate that the existing hepatic differentiation protocols need to be optimized to generate more mature hepatic cells that resemble advanced human fetal hepatoblasts. Taken into account that human iPS cells might be less susceptible for specific differentiation stimuli, even the best available ES-differentiation protocols might fail to generate hepatic cells that are mature enough to exhibit a functional phenotype in vitro. Therefore, in these days a lot of emphasis is laid on comparisons of various human iPS cells and ES cells as well as on defining crucial steps for successful generation of human iPS cell-derived hepatic cells.

But, besides the direct use of iPS cell derivatives as therapeutic cell transplants, patient-specific, or disease-specific iPS cells are a unique resource to study the pathophysiology and new therapeutic strategies of the respective disorder. The implications of genetic defects during the specification of the affected cell type can be investigated and the severity of the defects can be correlated to the individual course of the disease. Most importantly, derivatives of disease-specific iPS cells offer an unlimited cell resource for in vitro studies and high throughput screenings of potentially new drug targets.

6.2 Modeling of Metabolic Liver Diseases Using iPS Cells

In order to use iPS cells for pathophysiologic studies of the disease, the relevant cell type must be generated in vitro. To this end, suitable differentiation protocols need to be applied and a mature cell phenotype need to be acquired to adequately model the respective disease. In the past, several groups including ours succeeded in the generation of hepatic precursor cells from human and mouse embryonic stem cell lines [41–43]. With the existing differentiation protocols a primitive hepatic phenotype with the fetal gene expression patterns can be induced in the majority of the embryonic stem cells [39, 44]. Transplantation of these cells, however, have so far been resulted only in scattered formation of hepatocytes or were reported to form small hepatocyte clusters in major urinary protein promoter driven urokinase-type plasminogen activator (uPA) and $FAH^{-/-}$ mice [45, 46]. Nevertheless, in more recent publications, hepatic cell differentiation of human ES cells was refined [47, 48] and hepatic cells suitable for pharmacological testing have been described [49]. Fortunately, recent insights into various signaling pathways that are involved in hepatocyte differentiation of cells have increased our knowledge to induce a mature hepatic phenotype more efficiently [46, 50, 51]. Therefore, more advanced protocols for murine cells [45, 52] could be adapted for the use in human ESCs as recently described [48, 53–56].

Currently, those improved hepatic differentiation protocols seem to be appropriate for studies of disease models, because they give rise to hepatic cells, displaying the respective disease phenotype. This was demonstrated for a murine model of Wilson's disease (toxic milk mice), a model of acute tyrosinamia type I (FAH mice), and for a model of α1-antitrypsin deficiency (PiZ mice), where in all three experimental models the respective metabolic defect could be observed in hepatic cells derived from the respective iPS cells [57]. Moreover, various human liver disease-specific iPS cells were described covering patients suffering from α1-antitrypsin deficiency, Glucose-6-phosphate deficiency, LDL-receptor mutations, Crigler-Najjar-Syndrome, Tyrosinemia type I, Progressive Familial Hereditary Cholestasis, as well as from Wilson's disease [58–60]. For the latter, a gene correction approach has already been applied, which suggests that an efficient rescue of the Wilson's disease phenotype could be achieved by a lentiviral expression of an intact *ATP7b* gene copy.

6.3 Genetic Correction of Disease-Specific iPS Cells

If iPS cell–based transplantation approaches are considered for genetic disorders, gene therapy must also be considered to rescue the diseased phenotype [61]. The genetic engineering of ES and iPS cells appears to be rather inefficient, but recently published results on a targeted genetic engineering approach showed great promise [62]. As a proof-of-principle, it was already demonstrated that iPS cells from a murine sickle cell anemia model could be repaired by homologous recombination and were successfully used as a transplant for the correction of the disease [63]. However, this approach is rather inefficient and is difficult to be applied to human ES or iPS cells. The use of tailored zinc fingers attached to an endonuclease for site-specific double-strand breaks substantially increases the efficiency of recombination, but the flanking regions of the repair constructs are only 0.5–1 kbp long, and therefore specific zinc fingers need to be designed for each targeted locus [62]. Furthermore, the classic approach of electroporation to introduce BACs and deliver more than 100 kbp of gene flanking regions is accompanied by severe DNA damage (double-strand breaks) and by offsite integrations of part of the transferred DNA [64, 65]. Considering orphan diseases, gene correction by zinc finger–based homologous recombination approaches cannot not be easily used, unless new targeting strategies are designed for each new mutation. Therefore, third-generation self-inactivating lentiviral vectors, which are much safer in terms of oncogenic transformation than conventional gamma-retroviral vectors [66] are supposed to be a reasonable option. In fact, lentiviral vectors have already been used in clinical setting as part of the therapeutic intervention for a variety of disease states, including advanced forms of HIV infection, inherited disorders affecting hematopoietic cells, and Parkinson's disease [67]. Such improved vector architecture was used in a recent gene therapy trial for X-linked adrenoleukodystrophy [68] and is also currently used in an ongoing multicenter trial for Wiskott-Aldrich syndrome, coordinated by the San Raffaele Telethon Institute for Gene Therapy [69]. A very well studied example for the efficacy of a lentiviral gene correction for the originating fibroblast population has been publised for Fanconi anemia—specific iPS cells [70]. Nevertheless, significant silencing of the transduced transgene has been observed during in vitro differentiation into hematopoietic cells. In a recent report from our laboratory, we also observed significant silencing of a lentiviral trangene upon in vivo differentiation of the gene-corrected iPS cells in a tetraploid embryo aggregation experiment [71]. In this study, we generated iPS cells from a murine model of tyrosinemia type 1 (fumarylacetoacetate hydrolase deficiency; $FAH^{-/-}$ mice) and rescued the diseased phenotype by lentiviral transduction of an *FAH* transgene, which resulted in a constitutive expression of *FAH* in the gene-corrected iPS cells (Fig. 6.1). We then applied tetraploid embryo aggregations to generate fully iPS cell-derived mice and analyzed the capability of the gene-corrected iPS cells to form functional hepatic cells. Interestingly, we obtained healthy mice from the gene-corrected iPS cells, but less than 50% of the hepatocytes showed FAH expression in the liver. With this set of data we confirmed that lentiviral gene correction strategies face issues of transgene silencing upon in vivo differentiation as well. Nevertheless, those cells that actively express the gene correction transgene do not acquire functional restrictions and might be applied toward cell-based therapeutic applications.

6.4 Outlook

Combining two recent breakthroughs in stem cell biology, that is the generation of patient-derived pluripotent stem cells and the generation of hepatic cells from uncommitted pluripotent stem cells, new approaches might emerge treating various metabolic liver diseases.

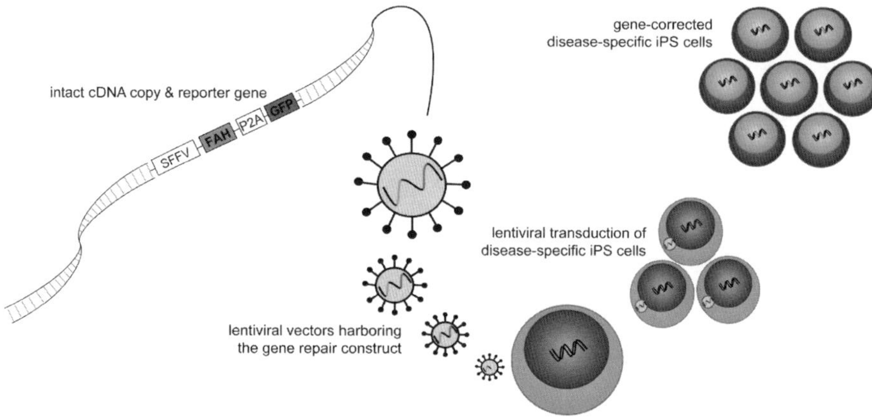

Fig. 6.1 Strategy for lentiviral gene therapy in iPS cells. Integrating lentiviral vectors combine the preferable capabilities of sustained transgene expression and efficient stable transfection rates. For the genetic correction of a inherited disease a constitutively active promoter, such as the spleen focus-forming virus (SFFV) promoter, can be used to drive the expression of an intact cDNA copy encoding the respective gene, e.g., *FAH*. Also, a co-expression of a fluorescent reporter (GFP) can be achieved by linking that cDNA using a 2a-peptidase motif. After lentiviral vector production, the viral cargo can be delivered to the disease-specific iPS cells and the repair constructs get integrated into the host genome. Finally, the intact gene copy is expressed along with the fluorescent reporter

Although this approach holds great promise, it lacks feasibility as long as suitable automated cell culture systems are missing, due to their enormous personnel workload. Furthermore, safety analyses ruling out major risks due to extended cell proliferation in vitro and cell manipulations during iPS generation and gene correction of the cells, and hepatic specification prior to transplantation should be considered. Moreover, chromosomal aberrations, which also deregulate epigenetic modifications, might be a potential risk for cellular misfunctioning or tumorigenesis.

With respect to individualized regenerative medicine, patient-specific cell lines might be generated and gene corrected prior to transplantation of in vitro-differentiated cells. To this end, safe and sustained gene correction techniques should be elaborated as the currently available technologies still harbor risks of either oncogenic transformation and severe DNA damage, or substantial silencing of the rescuing transgene during differentiation processes.

Because such issues are not relevant for in vitro studies on the diseased phenotypes of the respective disease, patient-specific iPS cells will be of high interest for studying those metabolic liver diseases lacking a suitable animal model or to evaluate new drug targets on patients' cells. These cells not only offer a unique opportunity for studies in pathophysiology, but may also serve as an unlimited resource for studying new drug targets using small compound screens.

References

1. Gupta S, Rajvanshi P, Lee CD (1995) Integration of transplanted hepatocytes into host liver plates demonstrated with dipeptidyl peptidase IV-deficient rats. Proc Natl Acad Sci USA 92(13):5860–5864
2. Gupta S, Rajvanshi P, Sokhi R, Slehria S, Yam A, Kerr A, Novikoff PM (1999) Entry and integration of transplanted hepatocytes in rat liver plates occur by disruption of hepatic sinusoidal endothelium. Hepatology 29(2):509–519
3. Weglarz TC, Degen JL, Sandgren EP (2000) Hepatocyte transplantation into diseased mouse liver: kinetics of parenchymal repopulation and identification of the proliferative capacity of tetraploid and octaploid hepatocytes. Am J Pathol 157(6):1963–1974
4. Grompe M (2006) Principles of therapeutic liver repopulation. J Inherit Metab Dis 29(2–3):421–425
5. Fox IJ, Chowdhury JR, Kaufman SS, Goertzen TC, Chowdhury NR, Warkentin PI, Dorko K, Sauter BV, Strom SC (1998) Treatment of the Crigler-Najjar syndrome type I with hepatocyte transplantation. N Engl J Med 338(20):1422–1426
6. Fisher RA, Strom SC (2006) Human hepatocyte transplantation: worldwide results. Transplantation 82(4):441–449
7. Schneider A, Attaran M, Meier PN, Strassburg C, Manns MP, Ott M, Barthold M, Arseniev L, Becker T, Panning B (2006) Hepatocyte transplantation in an acute liver failure due to mushroom poisoning. Transplantation 82(8):1115–1116
8. Dhawan A, Mitry RR, Hughes RD (2006) Hepatocyte transplantation for liver-based metabolic disorders. J Inherit Metab Dis 29(2–3):431–435
9. Meyburg J, Das AM, Hoerster F, Lindner M, Kriegbaum H, Engelmann G, Schmidt J, Ott M, Pettenazzo A, Luecke T, Bertram H, Hoffmann GF, Burlina A (2009) One liver for four children: first clinical series of liver cell transplantation for severe neonatal urea cycle defects. Transplantation 87(5):636–641
10. Puppi J, Strom SC, Hughes RD, Bansal S, Castell JV, Dagher I, Ellis EC, Nowak G, Ericzon BG, Fox IJ, Gomez-Lechon MJ, Guha C, Gupta S, Mitry RR, Ohashi K, Ott M, Reid LM, Roy-Chowdhury J, Sokal E, Weber A, Dhawan A (2012) Improving the techniques for human hepatocyte transplantation: report from a consensus meeting in London. Cell Transplant 21(1):1–10
11. Asgari S, Pournasr B, Salekdeh GH, Ghodsizadeh A, Ott M, Baharvand H (2010) Induced pluripotent stem cells: a new era for hepatology. J Hepatol 53(4):738–751
12. Cantz T, Manns MP, Ott M (2008) Stem cells in liver regeneration and therapy. Cell Tissue Res 331(1):271–282
13. Sancho-Bru P, Najimi M, Caruso M, Pauwelyn K, Cantz T, Forbes S, Roskams T, Ott M, Gehling U, Sokal E, Verfaillie CM, Muraca M (2009) Stem and progenitor cells for liver repopulation: can we standardise the process from bench to bedside? Gut 58(4):594–603
14. Takahashi K, Yamanaka S (2006) Induction of pluripotent stem cells from mouse embryonic and adult fibroblast cultures by defined factors. Cell 126(4):663–676
15. Okita K, Ichisaka T, Yamanaka S (2007) Generation of germline-competent induced pluripotent stem cells. Nature 448(7151):313–317
16. Meissner A, Wernig M, Jaenisch R (2007) Direct reprograming of genetically unmodified fibroblasts into pluripotent stem cells. Nat Biotechnol 25(10):1177–1181
17. Maherali N, Sridhara R, Wei X, Utikal J, Eminli S, Arnold K, Stadtfeld M, Yachechko R, Tchieu J, Jaenisch R, Plath K, Hochedlinger K (2007) Directly reprogrammed fibroblasts show global epigenetic remodeling and widespread tissue contribution. Cell Stem Cell 1(1):55–70
18. Takahashi K, Tanabe K, Ohnuki M, Narita M, Ichisaka T, Tomoda K, Yamanaka S (2007) Induction of pluripotent stem cells from adult human fibroblasts by defined factors. Cell 131(5):861–872

19. Yu J, Vodyanik MA, Smuga-Otto K, Antosiewicz-Bourget J, Frane JL, Tian S, Nie J, Jonsdottir GA, Ruotti V, Stewart R, Slukvin II, Thomson JA (2007) Induced pluripotent stem cell lines derived from human somatic cells. Science 318(5858):1917–1920
20. Park IH, Zhao R, West JA, Yabuuchi A, Huo H, Ince TA, Lerou PH, Lensch MW, Daley GQ (2008) Reprograming of human somatic cells to pluripotency with defined factors. Nature 451(7175):141–146
21. Zaehres H, Kogler G, Arauzo-Bravo MJ, Bleidissel M, Santourlidis S, Weinhold S, Greber B, Kim JB, Buchheiser A, Liedtke S, Eilken HM, Graffmann N, Zhao X, Meyer J, Reinhardt P, Burr B, Waclawczyk S, Ortmeier C, Uhrberg M, Schöler HR, Cantz T, Wernet P (2010) Induction of pluripotency in human cord blood unrestricted somatic stem cells. Exp Hematol 38(9):809–818
22. Okita K, Nakagawa M, Hyenjong H, Ichisaka T, Yamanaka S (2008) Generation of mouse induced pluripotent stem cells without viral vectors. Science 322(5903):949–953
23. Stadtfeld M, Nagaya M, Utikal J, Weir G, Hochedlinger K (2008) Induced pluripotent stem cells generated without viral integration. Science 322(5903):945–949
24. Fusaki N, Ban H, Nishiyama A, Saeki K, Hasegawa M (2009) Efficient induction of transgene-free human pluripotent stem cells using a vector based on Sendai virus, an RNA virus that does not integrate into the host genome. Proc Jpn Acad Ser B Phys Biol Sci 85(8):348–362
25. Yu J, Hu K, Smuga-Otto K, Tian S, Stewart R, Slukvin II, Thomson JA (2009) Human induced pluripotent stem cells free of vector and transgene sequences. Science 324(5928): 797–801
26. Kim D, Kim CH, Moon JI, Chung YG, Chang MY, Han BS, Ko S, Yang E, Cha KY, Lanza R, Kim KS (2009) Generation of human induced pluripotent stem cells by direct delivery of reprograming proteins. Cell Stem Cell 4(6):472–476
27. Utikal J, Polo JM, Stadtfeld M, Maherali N, Kulalert W, Walsh RM, Khalil A, Rheinwald JG, Hochedlinger K (2009) Immortalization eliminates a roadblock during cellular reprograming into iPS cells. Nature 460(7259):1145–1148
28. Warren L, Manos PD, Ahfeldt T, Loh YH, Li H, Lau F, Ebina W, Mandal PK, Smith ZD, Meissner A, Daley GQ, Brack AS, Collins JJ, Cowan C, Schlaeger TM, Rossi DJ (2010) Highly efficient reprograming to pluripotency and directed differentiation of human cells with synthetic modified mRNA. Cell Stem Cell 7(5):618–630
29. Mayshar Y, Ben-David U, Lavon N, Biancotti JC, Yakir B, Clark AT, Plath K, Lowry WE, Benvenisty N (2010) Identification and classification of chromosomal aberrations in human induced pluripotent stem cells. Cell Stem Cell 7(4):521–531
30. Hussein SM, Batada NN, Vuoristo S, Ching RW, Autio R, Narva E, Ng S, Sourour M, Hamalainen R, Olsson C, Lundin K, Mikkola M, Trokovic R, Peitz M, Brustle O, Bazett-Jones DP, Alitalo K, Lahesmaa R, Nagy A, Otonkoski T (2011) Copy number variation and selection during reprograming to pluripotency. Nature 471(7336):58–62
31. Gore A, Li Z, Fung HL, Young JE, Agarwal S, Antosiewicz-Bourget J, Canto I, Giorgetti A, Israel MA, Kiskinis E, Lee JH, Loh YH, Manos PD, Montserrat N, Panopoulos AD, Ruiz S, Wilbert ML, Yu J, Kirkness EF, Izpisua Belmonte JC, Rossi DJ, Thomson JA, Eggan K, Daley GQ, Goldstein LS, Zhang K (2011) Somatic coding mutations in human induced pluripotent stem cells. Nature 471(7336):63–67
32. Lister R, Pelizzola M, Kida YS, Hawkins RD, Nery JR, Hon G, Antosiewicz-Bourget J, O'Malley R, Castanon R, Klugman S, Downes M, Yu R, Stewart R, Ren B, Thomson JA, Evans RM, Ecker JR (2011) Hotspots of aberrant epigenomic reprograming in human induced pluripotent stem cells. Nature 471(7336):68–73
33. Zhao T, Zhang ZN, Rong Z, Xu Y (2011) Immunogenicity of induced pluripotent stem cells. Nature 474(7350):212–215. doi:10.1038/nature10135
34. Sommer CA, Stadtfeld M, Murphy GJ, Hochedlinger K, Kotton DN, Mostoslavsky G (2009) Induced pluripotent stem cell generation using a single lentiviral stem cell cassette. Stem Cells 27(3):543–549

35. Warlich E, Kühle J, Cantz T, Brugman MH, Mätzig T, Galla M, Filipczyk A, Halle S, Klump H, Schöler HR, Baum C, Schroeder T, Schambach A (2011) Lentiviral vector design and imaging approaches to visualize the early stages of cellular reprograming. Mol Ther 19(4):782–789
36. Papapetrou EP, Tomishima MJ, Chambers SM, Mica Y, Reed E, Menon J, Tabar V, Mo Q, Studer L, Sadelain M (2009) Stoichiometric and temporal requirements of Oct4, Sox2, Klf4, and c-Myc expression for efficient human iPSC induction and differentiation. Proc Natl Acad Sci USA 106(31):12759–12764
37. Tiemann U, Sgodda M, Warlich E, Ballmaier M, Schöler HR, Schambach A, Cantz T (2011) Optimal reprograming factor stoichiometry increases colony numbers and affects molecular characteristics of murine induced pluripotent stem cells. Cytometry Part A 79A:426–435
38. Cantz T, Zuckerman DM, Burda MR, Dandri M, Goricke B, Thalhammer S, Heckl WM, Manns MP, Petersen J, Ott M (2003) Quantitative gene expression analysis reveals transition of fetal liver progenitor cells to mature hepatocytes after transplantation in uPA/RAG-2 mice. Am J Pathol 162(1):37–45
39. Sharma AD, Cantz T, Vogel A, Schambach A, Haridass D, Iken M, Bleidissel M, Manns MP, Scholer HR, Ott M (2008) Murine embryonic stem cell-derived hepatic progenitor cells engraft in recipient livers with limited capacity of liver tissue formation. Cell Transplant 17(3):313–323
40. Haridass D, Yuan Q, Becker PD, Cantz T, Iken M, Rothe M, Narain N, Bock M, Norder M, Legrand N, Wedemeyer H, Weijer K, Spits H, Manns MP, Cai J, Deng H, Di Santo JP, Guzman CA, Ott M (2009) Repopulation efficiencies of adult hepatocytes, fetal liver progenitor cells, and embryonic stem cell-derived hepatic cells in albumin-promoter-enhancer urokinase-type plasminogen activator mice. Am J Pathol 175(4):1483–1492
41. Kania G, Blyszczuk P, Jochheim A, Ott M, Wobus AM (2004) Generation of glycogen- and albumin-producing hepatocyte-like cells from embryonic stem cells. Biol Chem 385(10): 943–953
42. Hamazaki T, Iiboshi Y, Oka M, Papst PJ, Meacham AM, Zon LI, Terada N (2001) Hepatic maturation in differentiating embryonic stem cells in vitro. FEBS Lett 497(1):15–19
43. Agarwal S, Holton KL, Lanza R (2008) Efficient differentiation of functional hepatocytes from human embryonic stem cells. Stem Cells 26(5):1117–1127
44. Jochheim A, Hillemann T, Kania G, Scharf J, Attaran M, Manns MP, Wobus AM, Ott M (2004) Quantitative gene expression profiling reveals a fetal hepatic phenotype of murine ES-derived hepatocytes. Int J Dev Biol 48(1):23–29
45. Heo J, Factor VM, Uren T, Takahama Y, Lee JS, Major M, Feinstone SM, Thorgeirsson SS (2006) Hepatic precursors derived from murine embryonic stem cells contribute to regeneration of injured liver. Hepatology 44(6):1478–1486
46. Gouon-Evans V, Boussemart L, Gadue P, Nierhoff D, Koehler CI, Kubo A, Shafritz DA, Keller G (2006) BMP-4 is required for hepatic specification of mouse embryonic stem cell-derived definitive endoderm. Nat Biotechnol 24(11):1402–1411
47. Hay DC, Zhao D, Ross A, Mandalam R, Lebkowski J, Cui W (2007) Direct differentiation of human embryonic stem cells to hepatocyte-like cells exhibiting functional activities. Cloning Stem Cells 9(1):51–62
48. Cai J, Zhao Y, Liu Y, Ye F, Song Z, Qin H, Meng S, Chen Y, Zhou R, Song X, Guo Y, Ding M, Deng H (2007) Directed differentiation of human embryonic stem cells into functional hepatic cells. Hepatology 45(5):1229–1239
49. Ek M, Soderdahl T, Kuppers-Munther B, Edsbagge J, Andersson TB, Bjorquist P, Cotgreave I, Jernstrom B, Ingelman-Sundberg M, Johansson I (2007) Expression of drug metabolizing enzymes in hepatocyte-like cells derived from human embryonic stem cells. Biochem Pharmacol 74(3):496–503
50. Zaret KS, Grompe M (2008) Generation and regeneration of cells of the liver and pancreas. Science 322(5907):1490–1494
51. McLean AB, D'Amour KA, Jones KL, Krishnamoorthy M, Kulik MJ, Reynolds DM, Sheppard AM, Liu H, Xu Y, Baetge EE, Dalton S (2007) Activin a efficiently specifies

definitive endoderm from human embryonic stem cells only when phosphatidylinositol 3-kinase signaling is suppressed. Stem Cells 25(1):29–38

52. Loya K, Eggenschwiler R, Ko K, Sgodda M, Andre F, Bleidissel M, Scholer HR, Cantz T (2009) Hepatic differentiation of pluripotent stem cells. Biol Chem 390(10):1047–1055
53. Baharvand H, Hashemi SM, Shahsavani M (2008) Differentiation of human embryonic stem cells into functional hepatocyte-like cells in a serum-free adherent culture condition. Differentiation 76(5):465–477
54. Si-Tayeb K, Noto FK, Nagaoka M, Li J, Battle MA, Duris C, North PE, Dalton S, Duncan SA (2010) Highly efficient generation of human hepatocyte-like cells from induced pluripotent stem cells. Hepatology 51(1):297–305
55. Sullivan GJ, Hay DC, Park IH, Fletcher J, Hannoun Z, Payne CM, Dalgetty D, Black JR, Ross JA, Samuel K, Wang G, Daley GQ, Lee JH, Church GM, Forbes SJ, Iredale JP, Wilmut I (2010) Generation of functional human hepatic endoderm from human induced pluripotent stem cells. Hepatology 51(1):329–335
56. Touboul T, Hannan NR, Corbineau S, Martinez A, Martinet C, Branchereau S, Mainot S, Strick-Marchand H, Pedersen R, Di Santo J, Weber A, Vallier L (2010) Generation of functional hepatocytes from human embryonic stem cells under chemically defined conditions that recapitulate liver development. Hepatology 51(5):1754–1765
57. Eggenschwiler R, Loya K, Sgodda M, Andre F, Cantz T (2011) Hepatic differentiation of murine disease-specific induced pluripotent stem cells allows disease modelling in vitro. Stem Cells Int 2011:924782
58. Ghodsizadeh A, Taei A, Totonchi M, Seifinejad A, Gourabi H, Pournasr B, Aghdami N, Malekzadeh R, Almadani N, Salekdeh GH, Baharvand H (2010) Generation of liver disease-specific induced pluripotent stem cells along with efficient differentiation to functional hepatocyte-like cells. Stem Cell Rev 6(4):622–632
59. Rashid ST, Corbineau S, Hannan N, Marciniak SJ, Miranda E, Alexander G, Huang-Doran I, Griffin J, Ahrlund-Richter L, Skepper J, Semple R, Weber A, Lomas DA, Vallier L (2010) Modeling inherited metabolic disorders of the liver using human induced pluripotent stem cells. J Clin Invest 120(9):3127–3136
60. Zhang S, Chen S, Li W, Guo X, Zhao P, Xu J, Chen Y, Pan Q, Liu X, Zychlinski D, Lu H, Tortorella MD, Schambach A, Wang Y, Pei D, Esteban MA (2011) Rescue of ATP7B function in hepatocyte-like cells from Wilson's disease induced pluripotent stem cells using gene therapy or the chaperone drug curcumin. Hum Mol Genet 20(16):3176–3187
61. Kiskinis E, Eggan K (2010) Progress toward the clinical application of patient-specific pluripotent stem cells. J Clin Invest 120(1):51–59
62. Zou J, Maeder ML, Mali P, Pruett-Miller SM, Thibodeau-Beganny S, Chou BK, Chen G, Ye Z, Park IH, Daley GQ, Porteus MH, Joung JK, Cheng L (2009) Gene targeting of a disease-related gene in human induced pluripotent stem and embryonic stem cells. Cell Stem Cell 5(1):97–110
63. Hanna J, Wernig M, Markoulaki S, Sun CW, Meissner A, Cassady JP, Beard C, Brambrink T, Wu LC, Townes TM, Jaenisch R (2007) Treatment of sickle cell anemia mouse model with iPS cells generated from autologous skin. Science 318(5858):1920–1923
64. Meaking WS, Edgerton J, Wharton CW, Meldrum RA (1995) Electroporation-induced damage in mammalian cell DNA. Biochim Biophys Acta 1264(3):357–362
65. Lakshmipathy U, Buckley S, Verfaillie C (2007) Gene transfer via nucleofection into adult and embryonic stem cells. Methods Mol Biol 407:115–126
66. Montini E, Cesana D, Schmidt M, Sanvito F, Bartholomae CC, Ranzani M, Benedicenti F, Sergi LS, Ambrosi A, Ponzoni M, Doglioni C, Di Serio C, von Kalle C, Naldini L (2009) The genotoxic potential of retroviral vectors is strongly modulated by vector design and integration site selection in a mouse model of HSC gene therapy. J Clin Invest 119(4): 964–975
67. Schambach A, Baum C (2008) Clinical application of lentiviral vectors—concepts and practice. Curr Gene Ther 8(6):474–482

68. Cartier N, Hacein-Bey-Abina S, Bartholomae CC, Veres G, Schmidt M, Kutschera I, Vidaud M, Abel U, Dal-Cortivo L, Caccavelli L, Mahlaoui N, Kiermer V, Mittelstaedt D, Bellesme C, Lahlou N, Lefrere F, Blanche S, Audit M, Payen E, Leboulch P, l'Homme B, Bougneres P, Von Kalle C, Fischer A, Cavazzana-Calvo M, Aubourg P (2009) Hematopoietic stem cell gene therapy with a lentiviral vector in X-linked adrenoleukodystrophy. Science 326(5954):818–823
69. Aiuti A, Ferrua F, Scaramuzza S, Bosticardo M, Castiello C, Evangelio C, Casiraghi M, Ripamonti A, Vallanti G, Radrizzani M, Salomoni M, Galy A, Finocchi A, Cicalese MP, Biffi A, Ciceri F, Naldini L, Villa A, Roncarolo MG (2010) Gene therapy trial with lentiviral vector transduced CD34+ cells for the treatment of Wiskott-Aldrich Syndrome. Hum Gene Ther 21(10):1439–1440
70. Raya A, Rodriguez-Piza I, Guenechea G, Vassena R, Navarro S, Barrero MJ, Consiglio A, Castella M, Rio P, Sleep E, Gonzalez F, Tiscornia G, Garreta E, Aasen T, Veiga A, Verma IM, Surralles J, Bueren J, Izpisua Belmonte JC (2009) Disease-corrected haematopoietic progenitors from Fanconi anaemia induced pluripotent stem cells. Nature 460(7251):53–59
71. Wu G, Liu N, Rittelmeyer I, Sharma AD, Sgodda M, Zaehres H, Bleidissel M, Greber B, Gentile L, Han DW, Rudolph C, Steinemamm D, Schambach A, Ott M, Schöler HR, Cantz T (2011) Generation of healthy mice from gene-corrected disease-specific induced pluripotent stem cells. PLoS Biol 9(7):e1001099

Chapter 7
Vascular Differentiation of Human Pluripotent Stem Cells

Helena Vazão, Mário Grãos and Lino Ferreira

Abstract Human pluripotent stem cells (hPSCs) are an unlimited source of endothelial and smooth muscle cells for regenerative medicine, drug screening, and toxicology assays. Significant advances have been made in the last 10 years in driving the differentiation of human embryonic stem cells into vascular cells and some of the approaches have been implemented recently in the differentiation of induced pluripotent stem cells. This chapter reviews the progresses done so far in the vascular differentiation of hPSCs focusing in the efficiency and kinetics of the differentiation protocols, inductive signals, maturation, and functionality of the hPSC-derived cells and their therapeutic potential.

7.1 Introduction

Human pluripotent stem cells (hPSCs) including human embryonic stem cells (hESCs) and human-induced pluripotent stem cells (hiPSCs) are promising sources of vascular cells for regenerative medicine, drug discovery, and toxicology assays. hESCs derived from the inner cell mass of blastocysts were originally isolated by James Thomson and co-workers [1]. hiPSCs derived by the transfection of adult skin cells (fibroblasts) with four pluripotency genes were isolated and characterized for the first time by Shinya Yamanaka and James Thompson labs in 2007 [2, 3].

H. Vazão · L. Ferreira (✉)
CNC—Center for Neuroscience and Cell Biology,
University of Coimbra, Coimbra, Portugal
e-mail: lino@biocant.pt

H. Vazão · M. Grãos · L. Ferreira
Biocant—Center of Innovation in Biotechnology, Cantanhede, Portugal

H. Baharvand and N. Aghdami (eds.), *Advances in Stem Cell Research*,
Stem Cell Biology and Regenerative Medicine, DOI: 10.1007/978-1-61779-940-2_7,

hiPSCs have properties similar to hESCs including similar morphology, karyotype, high level of telomerase, expression of characteristic genes and proteins, ability to give rise to cells of the three germ layers; endoderm, ectoderm and mesoderm, and can be maintained in vitro for several months without losing their properties [1, 4].

hPSCs are an important system to study embryonic vasculogenesis, being vasculogenesis the process by which blood vessels are formed de novo. In mammals, vasculogenesis occurs in parallel with hematopoiesis, the formation of blood cells [5]. In addition, hPSCs represent an unlimited source of vascular cells for regenerative medicine. hPSC-derived vascular cells can be useful for both initiating vasculogenesis and angiogenesis in clinical ischemic conditions [6–10]. Angiogenesis stimulation by hPSC-derived vascular cells can be induced from their secretion of pro-angiogenic and pro-survival growth factors and cytokines as well as from the physical contact with pericytes or smooth muscle cells [11]. Finally, although less explored, hPSC-derived vascular cells can be an important model for drug screening and toxicology assays [12]. The establishment of hPSC cell lines from patients with inherited diseases presenting vascular abnormality should enable the identification of new drugs to treat the disease.

Several methodologies have been reported in the differentiation of human pluripotent stem cells (both human embryonic stem cells and induced pluripotent stem cells) into vascular cells and these achievements have been recently reviewed elsewhere [11, 13–15]. A major challenge in this field is to develop more efficient protocols (in terms of differentiation yield and kinetics) for the differentiation of hPSCs, and in conditions that might be suitable for clinical translation (i.e., in serum-free conditions). Another challenge is related to the development of methodologies that are conducive to the full maturation of vascular cells.

This chapter reviews the different methodologies used so far to derive vascular cells from human pluripotent stem cells, specifically taking into account the yields, kinetics of the differentiation process, and maturation of the vascular cells (Fig. 7.1). The differences between vascular cells derived from hESCs and hiPSCs will be analyzed. In the second section of this chapter we will describe aspects related to developmental cues conducive to the vascular differentiation of human pluripotent stem cells. In the third and fourth sections we will describe the vascular potential of hESCs and hiPSCs, respectively. Finally in the fifth section we will refer examples related to the application of the hPSC-derived vascular cells for tissue engineering.

7.2 Embryonic Vascular Differentiation

During embryonic development, there are four key steps in vasculogenesis: (i) the establishment of the angioblasts from mesoderm, (ii) the assembly of angioblasts into vascular structures, (iii) the formation of vascular lumens, and (iv) the organization of continuous vascular networks. Vasculogenesis occurs at two distinct embryonic locations during development: the extraembryonic and

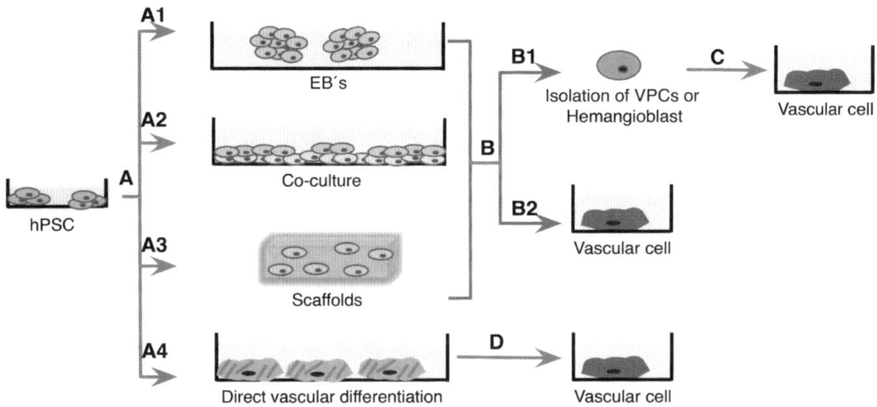

Fig. 7.1 Schematic diagram of the differentiation protocols used in the vascular differentiation of human pluripotent stem cells. (*A1*): Differentiation through an EBs stage; (*A2*): differentiation in a co-culture system with feeder cells; (*A3*): differentiation in 3D scaffolds; (*A4*): differentiation in 2D cultures using biomolecules. The differentiation process can include (*B1*) isolation of VPCs or hemanglioblats or the direct differentiation of the cells into vascular cells (*B2*, *C*, *D*). *hPSC*: human pluripotent stem cells; *EBs*: embryoid bodies; *VPCs*: vascular progenitor cells

intraembryonic tissues [5, 16]. Extraembryonic vasculogenesis precedes intraembryonic vascular development, and in mammals appear as blood islands assembling within the mesodermal layer of the yolk sac. Endocardium and great vessels are the first intraembryonic endothelial structures formed during development [17].

Some of the steps in vasculogenesis have been reproduced and identified in the hESC differentiation system [18, 19]. It has been shown that hESC-derived hemangioblast (a precursor of endothelial and hematopoietic cells) expresses KDR and develops within 72–96 h of EB differentiation, the stage during which KDR and CD117 are expressed on distinct populations, prior to the expression of CD31 and CD34. These human hemangioblasts generate distinct blast colonies that display hematopoietic and endothelial potential [18].

The molecular regulation in the vascular development has been clarified by several studies in the last decade (reviewed in Ref. [16]). Fibroblast growth factor (FGF), bone morphogenetic protein (BMP), and Wnt signaling pathways seem to be key components required for hematovascular differentiation in differentiating embryonic stem cells [16, 20]. The vascular endothelial growth factor (VEGF) is the first secreted molecule with specificity for the endothelium during development. Cells that respond to VEGF must first express its receptors, Flk-1/KDR and Flt-1 (VEGFR2 and VEGFR1) [21]. Therefore, the appearance of Flk-1/KDR expression is a marker for hemangioblasts during early development [18]. In addition to the early events, there are regulatory factors required for blood vessels to assemble. For example, some paracrine factors have an important role in endothelial cell survival and remodeling of the capillary complex (e.g., angiopoietin) [22].

After the vasculogeneis and angiogenesis, primordial vascular smooth muscle cells (VSMCs) are recruited to the endothelium to form a multilayered vessel wall. Platelet-derived growth factors are required for recruitment of pericytes and smooth muscle cells to invest in developing arteries and establish vasomotor tone [23]. As VSMCs become associated with the vascular endothelium, they produce and organize extracellular matrix (ECM) molecules within the vessel wall [24]. Primordial VSMCs also proliferate shortly after association with the nascent endothelium [25]. As the vasculature matures, the presence of VSMCs provides for vascular tone, control of peripheral resistance, and for distribution of blood flow throughout the developing organism [25].

Most of the VSMCs derive from mesoderm, with the exception of VSMCs from the great vessels (ascending aorta and branchial arches) that are derived from the cranial neural crest and VSMCs of the coronary artery, which are derived from cells that have migrated from the epicardium (reviewed in Ref. [26]). Although the molecular basis for phenotypic modulation of VSMCs is poorly understood, specific cytoskeletal and contractile proteins such as alpha-smooth muscle actin (α-SMA), calponin, caldesmon, smooth muscle myosin heavy chain (SM-MHC), and smoothelin have been found to characterize VSMC differentiation [27, 28]. In contrast to α-SMA, which may also be expressed by other cells, the smooth muscle-specific isoforms of myosin heavy chains, SM-1 and SM-2, appear to be specific markers for VSMCs [29]. Smoothelin, a protein identified from adult chicken gizzard smooth muscle, is expressed in all vertebrates and is purported to be a very late marker for differentiated, contractile SMCs [29].

7.3 Vascular Differentiation of hESCs

7.3.1 EC Differentiation

Differentiation methodologies: yields and kinetics. Several methodologies have been used for the vascular differentiation of hESCs such as embryoid bodies (EBs) which recapitulates in vivo embryogenesis [9, 30], a mixture of EBs with 2D or 3D culture systems [7, 8, 10, 31] and co-culture with cell lines [32–35]. Typically, EB differentiation protocols, where cells are allowed to spontaneously differentiate into all the germinative cell layers, are relatively inefficient in driving the vascular differentiation of the cells (yields below 5%) [8, 9]. Although, recent studies have used different approaches to ameliorate the efficiency of the system by using inductive agents (e.g., TGFβ-1 inhibitor [36] and $VEGF_{165}$ [37]), bioengineering approaches (including scaffolds [38] or micro- and nanoparticles [39]), or specific culture conditions (e.g., hypoxia) [40], the yield (with few exceptions, the yield is generally below 5–10%) is relatively low. Methodologies using co-culture systems might be more efficient, in terms of yield, in the vascular differentiation of hESCs. Several types of feeder layers have been used including mouse embryonic fibroblasts [34], OP9 stromal cells [35, 41] M2-10B4 stromal cells [42], and S17 stromal cells [33].

Endothelial cells (ECs) can be isolated from a heterogeneous cell population by fluorescence-activated cell sorting or magnetic-activated cell sorting. A marker that has often been used to isolate ECs is CD31 [9, 43]. In some cases (either in methodologies comprising EB or co-culture with cell lines), the protocols involve the isolation of hemangioblasts (cells that have the capacity to give rise to endothelial and hematopoietic cells; e.g. $CD31^+$/KDR/VE-cad^+ cells isolated at day 10 of EB differentiation [30], KDR^+ cells isolated at day 4 of EB differentiation [18], or $CD34^{bright}CD31^+KDR^+$ cells isolated at day 14 of a co-culture differentiation system [33]), or vascular progenitor cells (progenitor cells that have the ability to give rise to endothelial and smooth muscle cells; e.g. $CD34^+$ cells isolated from EBs at day 10 [8], $CD34^+$ cells isolated from hESCs co-cultured with M2-10B4 stromal cells or Wnt1 expressing M2-10B4 for 13–15 days [42], KDR^+/TRA1-60^- cells isolated at day 8 of a co-culture differentiation system [32]). In both cases, hemangioblasts or vascular progenitor cells are then cultured in medium conditions that drive their differentiation into ECs. Although the isolation of these progenitor cells are important issues to understand related to developmental biology and their therapeutic effect, the percentage of these cells is relatively low (below 5%). A recent study has reported an efficient methodology to derive functional hESC-derived $CD34^+$ progenitor cells (vascular progenitor cells) by both inhibition of MEK/ERK signaling and activation of BMP4 signaling [44]. hESC-derived $CD34^+$ cells were differentiated into functional endothelial cells and contributed to blood perfusion and limb salvage through neovasculogenesis in hind limb ischemic mice.

Few of the protocols reported so far have been carried out in serum-free conditions [31, 36, 45]. Kane and co-workers have expanded hESCs in feeder-free conditions using pluripotency medium that was then replaced by endothelial growth medium [45]. In these conditions, 82% of the cells co-express vascular endothelial-cadherin (VE-cad) and CD31 at day 21 of differentiation. Although it is necessary to show that the same efficacy can be obtained for several cell lines of hESCs and the phenotype is preserved after several cell passages, the results are very encouraging for the potential clinical application of these cells. In a separate study, Rafii and co-workers developed a methodology for the long-term expansion of hESC-derived ECs in serum-free conditions [36]. They have identified TGFβ inhibitor SB431542 as a key molecule for human endothelial differentiation and proliferation in serum-free conditions.

Inductive signals. The differentiation of hESCs into vascular progenitor cells and then into ECs is largely dependent on the modulation of several intracellular signaling pathways including BMP4, Wnt, and VEGF. BMP4 (and also BMP2 or 7, but not BMP9), VEGF, and FGF2 were shown to be critical for the induction of hESCs into vascular progenitor cells ($CD34^+CD31^+$ cells) [46]. Using a 12-day protocol relying on a stromal cell co-culture and defined media formulations, Bai and colleagues achieved 10% efficiency and showed that BMP4, acting through the Smad1/5/8 pathway, is required at an early stage of differentiation (days 1–6), whereas VEGF and FGF2 are necessary at later stages (day 4 up to day 12). The early requirement for BMP4 is thought to be related with the blocking of pluripotency and the induction into mesodermal lineages. ECs could be further differentiated from the $CD34^{+}CD31^{+}$ subpopulation with good efficiency, since approximately 1.3×10^4 ECs were obtained from 1×10^4 $CD34^{+}CD31^{+}$ cells, which in turn had been achieved from 1×10^5 hESCs.

Wnt signaling has also been investigated to enhance the differentiation of hESCs into vascular progenitor cells with the capacity to further differentiate into ECs. Wnt1 (and Wnt5, to a lesser extent), expressed and putatively secreted by Wnt1- (or Wnt5-) overexpressing stromal cells used in a stromal-hESC co-culture system, increased the efficiency of hESC differentiation into $CD34^{+}$ cells between three- and fourfold, as compared to the effect of the unmodified feeder cells [33, 42]. Canonical (Wnt1) signaling pathway seems to be the most important for the endothelial differentiation of hESCs [33]; however, it is unclear the signaling events occurring downstream of Wnt.

The differentiation of vascular progenitor cells into ECs is mediated by VEGF signaling. Vascular progenitor cells ($CD34^{+}$ cells) can be differentiated into cells with mature and functional features of ECs (both in vitro and in vivo), by stimulation with VEGF [8]. It was later shown that the presence of VEGF at the EB differentiation stage increases the presence of vascular progenitor cells with potential to differentiate into ECs [37], suggesting that VEGF plays an inductive role at various stages of hESC-derived ECs differentiation.

Maturation and functionality of hESC-derived ECs. Several studies have shown that hESC-derived ECs are functional since they are able to synthesize nitric oxide, form vascular network in vitro when cultured on top of Matrigel, metabolize acetylated low-density lipopoprotein (DiI-Ac-LDL), and produce extracellular matrix for developing a basal lamina [7–10, 30, 31, 36]. Furthermore, in vivo studies have shown that hESC-derived ECs form microvessels that anastomose with animal vasculature [9, 10, 42, 47], enhancing blood flow [7, 9, 32, 44]. Unfortunately, to the best of our knowledge, there is no report showing the specification of the human hPSC-derived ECs into arterial, venous, lymphatic, or brain-like phenotype either in vitro or after transplantation in animal models. A recent study has shown that mouse ESC-derived ECs have a preferential venous phenotype [48], and this might be due to the fact that venous is the default EC phenotype during development [49]. During embryonic development, specification into arterial-, venous-, lymphatic-, or brain-derived ECs is defined at gene level and is mediated by several signaling pathways including VEGF, Notch, and ephrin before circulation begins [50, 51]. Studies in mouse have shown that ephrinB2 and its receptor EphB4 are differentially expressed in arterial and venous endothelial cells, respectively, before the onset of circulation in the developing embryo [52]. After the onset of the circulation, the distinct hemodynamic forces found in arteries and veins, such as blood flow rate, direction, and pressure, can be a major driver in the specification of the endothelial cells [50, 51]. Further research is needed to evaluate the effect of these parameters in the phenotype of hESC-derived ECs.

Therapeutic potential of hESC-derived ECs. The therapeutic potential of hESC-derived ECs has been evaluated in animal models of hind limb [7, 31, 32, 41, 45], myocardial infarction [6, 10, 31], and after subcutaneous transplantation in immunodefficient rodents [8, 9]. Most studies have demonstrated that the transplanted hESC-derived ECs could be successfully incorporated into the host vasculature and significantly accelerate local blood flow [7, 32, 41, 53]. Importantly, mature ECs (not derived from hESCs) had not the same therapeutic effect when transplanted in the same animal models and thus suggesting that the hESC-derived ECs have high pro-angiogenic properties [32, 45]. Indeed, these cells express several angiogenic growth factors including VEGF, bFGF, HGF, and PDGF-BB [32]. Recent studies indicate that a key player for the enhanced therapeutic effect of hESC-derived ECs is Sirt1, a nicotinamide adenine dinucleotide (NAD^+)-dependent histone deacetylase [54]. Inhibition of Sirt1 attenuates the differences in functionality between hESC-derived ECs and adult ECs.

Several variables affect the final therapeutic outcome of the ECs derived from hPSCs. For example, the combined transplantation of hESC-derived ECs with hESC-derived SMCs has higher therapeutic effect than each cell type administered alone [41]. Both cells injected intra-arterially can augment neovascularization in an animal model of hindlimb ischemia. In addition, co-implantation of hESC-derived ECs with mouse embryonic fibroblasts form durable blood vessels in vivo [34]. At the moment, it is unclear whether the improvement is due to cell–cell interactions or soluble factors released by the mural cells. The differentiation stage of hESC-derived ECs also affects their therapeutic outcome. Highly differentiated [32] or very immature (KDR^+/TRA1-60^-) [32] hESC-derived ECs do not improve local blood flow in murine hindlimb ischemia model. However, hESC-derived ECs after moderate differentiation (10 days) produce nitric oxide, express angiogenesis-associated mature miRNAS (miR-126, 130a, 1331 and b, and 210), and augment neovascularization in hindlimb ischemia [45].

7.3.2 SMC Differentiation

Differentiation methodologies: yields and kinetics. Different strategies have been reported to differentiate hESCs into smooth muscle like-cells (SMLCs) by exposing a monolayer of undifferentiated hESCs to retinoic acid [55] or a combination of cell culture medium and extracellular matrix environment [8, 12, 56] either in single-hESC- [57], embryoid bodies (EBs)- [8] or stromal cell- [42] culture conditions. Most of the protocols reported so far have high efficiencies (i.e., high yields) in the differentiation of hESCs into SMLCs. For example, more than 90% of hESCs treated with basal medium supplemented with retinoic acid (10 μM) for 15 days express α-SMA. However, only 65% of the cells express mature markers such as SM-MHC. We [8, 58] and others [41] have also reported efficient (above 90% of the cells expressed α-SMA) differentiation protocols comprising the isolation of vascular progenitor cells ($CD34^+$ [8, 58]) or KDR^+/

$TRA1^-$ [41] followed by their differentiation in specific medium. We found that vascular progenitor cells ($CD34^+$ cells) have higher SMC potential than $CD34^-$ cells and $PDGF_{BB}$ and RA were the most effective agents to drive the differentiation of hESCs into SMLCs [58]. We have further demonstrated that the SMLCs contract and relax in response to SMC agonists or inhibitors, respectively, and the effect is mediated by Rho A/Rho kinase- and Ca^{2+}/CaM/MLCK-dependent pathways.

Inductive signals. Two types of differentiation protocols have been adopted so far for the differentiation of hESCs into SMCs: (i) single-step protocols characterized by the direct differentiation of undifferentiated hESCs into SMCs; and (ii) two-step protocols characterized by the isolation of vascular progenitor cells and then the differentiation of the progenitor cells into SMCs. Relative to the first type of protocols, it has been shown that retinoic acid [55], PDGF-BB [12, 57] and TGF-β1 [12, 57] signaling pathways are very important for the direct differentiation of hESCs into SMCs. Relative to the second type of protocols, the inductive signals to enhance the generation of vascular progenitor cells are described in Sect. 7.3.1, while the signals to induce the differentiation of vascular progenitor cells into SMCs include PDGF-BB [8, 41] and retinoic acid [58]. We have shown that vascular progenitor cells ($CD34^+$ cells) differentiate into vascular smooth muscle-like cells in culture medium supplemented with PDGF-BB [8, 58]. The cells exhibited spindle-shape morphology, expressed several SMC markers (such as α-SMA, SM-MHC, calponin, caldesmon and SMα-22), had the ability to contract and relax in vitro in response to carbachol and atropine, and when transplanted subcutaneously in an animal model contributed to the formation of microvessels. We have also demonstrated that TGF-β1 had low inductive effect in the differentiation of vascular progenitor cells into SMCs [58]. In line with these studies, others have demonstrated that hESC-derived vascular progenitor cells (presenting a $Tra^-VEGF\text{-}R2^+PDGF\text{-}R^+$ phenotype) required PDGF-BB to differentiate into vascular smooth muscle-like cells [41]. These cells could enhance blood flow after transplantation into a mouse hindlimb ischemia model.

Maturation and functionality of SMCs. The development of mature contractile SMCs from stem cells occurs in multiple steps comprising (i) the commitment to the SMC lineage, (ii) the differentiation into early immature, and (iii) the maturation into the mature contractile phenotype [59]. We and others have reported that vascular progenitor cells (CD34+ cells [8], $KDR^+/TRA1\text{-}61^-$ cells [41]) could give rise to SMLCs when cultured in medium supplemented with $PDGF_{BB}$; however, the differentiation of SMLCs was not complete since the assembly of α-SMA or SM-MHC proteins into filaments was not observed [8], and the presence of some SMC markers was not detected [41]. Recently, we have shown that the co-culture of SMLCs with fully differentiated hVSMCs induces the assembly of α-SMA and calponin proteins into individualized filaments [58]. This indicates that the cells are able to maturate into a fully contractile phenotype. We have further shown that the maturation of SMLCs can be induced by endothelin-1, an agonist of RhoA pathway (Fig. 7.2).

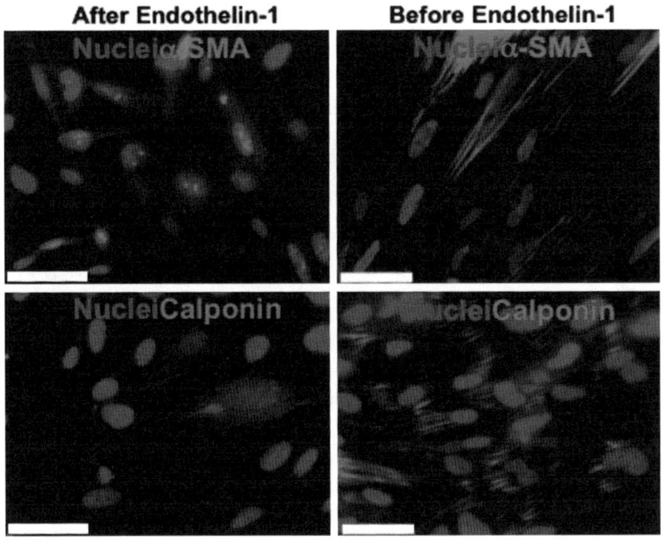

Fig. 7.2 Expression and organization of SMC proteins in hESC-derived smooth muscle progenitor cells before and after treatment with endothelin-1 for 3 days. Bar corresponds to 50 μm. Figure adapted from Ref. [58] with permission from PLoS One

Therapeutic potential of SMCs. Very few studies have evaluated the therapeutic potential and functionality of SMCs transplanted in vivo. It has been shown that SMLCs can contribute for the formation of functional blood microvessels in vivo [8, 41] and improve blood flow after transplantation in an ischemic hind limb mouse model [41]. In this case, further studies are needed to understand whether the regenerative potential of SMCs is mediated by their ability to contribute structurally for the formation of blood vessels or their ability to secrete paracrine factors. It is also unclear the regenerative potential of mature hESC-derived SMCs [58].

7.4 Vascular Differentiation of hiPSCs

7.4.1 EC Differentiation

Differentiation methodologies: yields and kinetics. Several methodologies have been reported to induce the vascular differentiation of hiPSCs, including co-culture with OP9 cells [53], and EBs. In general, the vascular differentiation protocols of hESCs have been successfully implemented in hiPSCs. However, studies indicate that the vascular potential of hiPSCs is, in some aspects, different from hESCs. For example, TRA1-60^{-}/KDR^{+}/VE-cadherin^{+} give rise to ECs while

TRA1-60^{-}/KDR^{+}/VE-cadherin^{-} give rise to SMCs [53]. Importantly, the differentiation efficiency was comparable to hESCs [53]. However, according to some studies, hiPSC-derived ECs had limited expansion capacity in vitro, and exhibited a gradual loss of CD31 marker expression over a period of 2 weeks [43, 60]. This contrasts with the capacity of hESC-derived endothelial cells to proliferate robustly in vitro with minimal loss of CD31 marker (85% for hESC-derived ECs vs 28% for hiPSC-derived ECs, at day 21) [60]. The variation in gene expression between hiPSCs and hESCs might account for the differences observed. Comparison between single hESCs and single hiPSCs revealed more heterogeneity in gene expression levels in the hiPSCs [60]. In addition, the variation in gene expression between hiPSC-derived ECs and hESC-derived ECs might contribute significantly for the biological differences among cells [43]. Several studies have revealed that somatic memory does exist [61–63]. That may explain the slower growth and fast loss of endothelial phenotype, depending on the donor somatic cells.

Therapeutic potential of hiPSC-derived ECs. Few studies have documented the differentiation of hiPSCs into ECs [43, 44, 53, 54] and thus it is unclear the therapeutic potential of hiPSC-derived ECs for the treatment of ischemic diseases and how different is that potential from the one observed for hESC-derived ECs. So far, studies have shown that hiPSC-derived ECs have similar functional activity as normal ECs, including the ability to produce endothelial nitric oxide synthase [53] and the ability to form capillary-like structures when seeded in Matrigel [53, 64]. Furthermore, the transplantation of hiPSC-derived cells in ischemic hind limb mice improved blood perfusion and limb salvage [44]. Further research is needed to investigate the therapeutic potential of these cells in other disease settings.

7.4.2 SMC Differentiation

Differentiation methodologies: yields and kinetics. The generation of hiPS cells from human aortic vascular smooth muscle cells (hAVSMCs) and their differentiation into SMCs was recently reported [65]. hiPSCs from hAVSMCs were obtained using lentiviral transduction of defined transcription factors and differentiated back into SMCs. hiPS cell-derived SMCs were very similar to parental hAVSMCs in gene expression patterns, epigenetic modifications of pluripotency related genes, and in vitro functional properties.

Therapeutic potential of hiPSC-derived SMCs. Currently there are no studies documenting the in vivo functionality of hiPSC-derived SMCs. So far, the in vitro results have shown that hiPSC-derived SMCs were very similar to somatic SMCs in gene expression and in vitro functional properties [53, 65]. The hiPSC-derived SMCs exhibited increased intracellular calcium influx in response to vasoconstrictive carbachol and KCl-induced membrane depolarization, and contract after treatment with carbachol or angiotensin II [65].

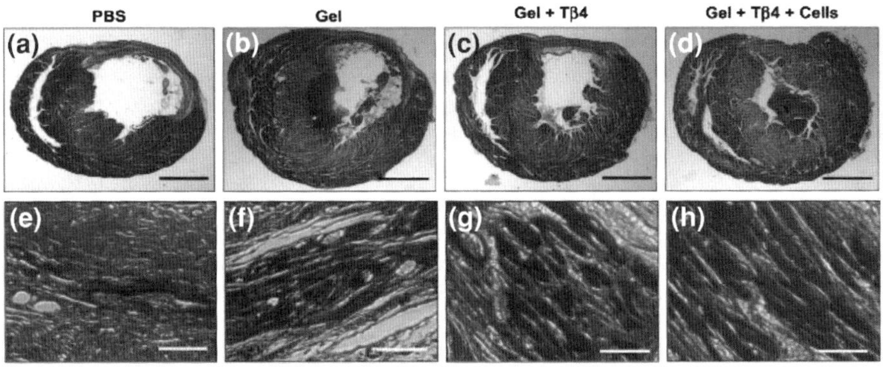

Fig. 7.3 Preservation and alignment of cardiomyocytes in infarcted myocardium treated with hESC-derived vascular cells encapsulated in a metalloprotease-sensitive gel with a pro-survival peptide. **a–d** Representative trichrome stains of transverse heart sections. Collagen in the infarct areas is shown in *blue*, whereas myocytes are in *red*. Animals were treated with **a** PBS, **b** gel, **c** gel + thymosin $\beta 4$ (T$\beta 4$), and **d** gel + T$\beta 4$ + cells. The cardiomyocytes appeared to be better preserved and aligned in bundles in hearts treated with the combined T$\beta 4$ and cell delivery, as compared to the rats treated with PBS, gel alone or T$\beta 4$ delivery alone. Low (**a–d**) and high (**e–h**) magnification. Bar corresponds to 2.5 mm (**a–d**) and 100 μm (**e–h**). Figure adapted from Ref. [6] with permission from Elsevier

7.5 Tissue engineering Applications of Vascular Cells Derived From hPSCs

hPSC are promising for the revascularization of ischemic tissues such as ischemic leg, chronic wounds, and infarcted heart [6, 7, 10]. Yet, there is a demand for scaffolds to allow efficient cellular transplantation in vivo, retaining them at the site they are needed, while improving their engraftment. The scaffold can be engineered to support migration, proliferation, and differentiation of vascular progenitor cells or differentiated cells and to help in the organization of these cells at three-dimensional level [66–68]. Examples of scaffolds that can support EC migration and their 3D organization include MMP-sensitive hydrogels incorporating RGD epitopes to promote cell adhesion [6, 69], fibrin gels [70], collagen gels [37], PLGA matrices [9], poly(hydroxyethyl methacrylate) gels [37], among others.

Pro-survival factors can be used to increase the survival of vascular cells after transplantation. Recently, we used a pro-angiogenic and pro-survival factor (thymosin $\beta 4$) along with hESC-derived vascular cells encapsulated in a MMP-sensitive gel for the treatment of myocardial infarction [6]. The gel, acting as scaffold, substituted the degrading extracellular matrix in the infarcted myocardium of rats and promoted structural organization of native ECs, while some of the delivered hESC-derived vascular cells formed de novo capillaries in the infarct zone (Fig. 7.3).

Table 7.1 Derivation of vascular cells from hESCs and hiPSCs

hPSC*s*	Models of differentiation	Pathways/molecules	Differentiation in vascular cells	Characterization and in vivo studies	Reference
hESC	Collagen IV/$Flk1^+$/E-$cadherin^-$	VEGF; PDGF-BB	EC/MC	EC and SMC markers; Tube formation; Contribution to the development of vasculature in chick embryos	[47]
hESC	VEGF-R2-positive but tumor rejection antigen 1–60 (TRA1-60)-negative cell isolation		EC/SMC	New blood vessels and improved blood flow	[76]
hESC	12 days of EB formation and 1.5 mg/ml of collagen I gel/ EGM-2 medium with 5% knockout serum replacement	Wnt2	EC	Functional vessel formation, improved cardiac function	[10]
hESC	10 days of EB formation; $CD34^+$ cell isolation	VEGF; PDGF-BB	EC/SMC	Formation of a functional microvasculature	[8]
hESC	PECAM-1+ cell isolation		EC	Tube formation; in vivo new vessels formation	[9]
hESC	$VEGFR2^+TRA1^-$VE-cadherin+ cells isolation	PDGF-BB	EC/MC	Blood flow recovery of ischemic limbs, long-term vascular integrity	[41]
hESC/hiPSC	TRA1–60-Flk1+ cells; VE-cadherin–positive and –negative populations	VEGF, PDGF-BB	EC/SMC	EC and SMC markers	[53]
hESC	vWF^+ cells isolation		EC	Improves blood perfusion and limb salvage by facilitating postnatal neovascularization in a mouse model of hindlimb ischemia	[7]

(continued)

Table 7.1 (continued)

hPSC*s*	Models of differentiation	Pathways/molecules	Differentiation in vascular cells	Characterization and in vivo studies	Reference
hESC/hiPSC	$Flk1^+$/VE-$cadherin^+$ cells isolation; Sirt1	VEGF	EC	EC markers	[54]
hESC/hiPSC	$CD34^+$ cells isolation	Inactivation of MEK/ERK signaling and activation of BMP4;	EC	Blood perfusion and limb salvage through the neovasculogenesis in hind limb ischemic mice	[44]
hESC	M2-10B4 stromal cells or Wnt1 expressing M2-10B4 for 13–15 days; CD34+ cells isolation	Wnt1	EC/SMC	EC and SMC markers; tube formation	[42]
hESC/hiPSC	EB formation; $CD31^+$ cell isolation	VEGF	EC	EC markers	[43]
hESC	$CD31^+CD34^+$ cell isolation Smad1/5/8	BMPs; TGFβ; VEGF	EC	EC markers	[46]
hESC	$CD34^+$	VEGF; RA	SMC	SMC markers and functionality	[58]
hESC	2D culture	PDGF-BB; TGFβ	SMC	SMC markers	[57]
hESC	CD144 promoter screen/inhibition of TGFβ	Id1	EC	$hCD144p^-GFP^+$ cells	[36]

Another approach to induce the three-dimensional organization of the vascular cells is by the use of prefabricate scaffolds that include channels that later may be coated with vascular cells. Nanofabrication techniques can be used to engineer network structures that might mimic the capillary network [71, 72]. Also, inkjet printing can be used to pattern cells into tubular structures [73] and be used in the future for the assembly of complex vascularized tissues obtained from hPSC.

7.6 Future Directions

Both hESCs and hiPSCs are promising sources of vascular cells for the treatment of ischemic diseases [74] and for drug discovery and toxicology assays. Similarities exist between hESCs and hiPSCs regarding their potential to differentiate into vascular cells, but whether the regulatory pathways that govern their differentiation are similar remains to be determined. Studies in both hESCs and hiPSCs suggest that factors implicated in EC differentiation are expressed in both cell systems but in slightly different temporal pattern [44]. Although vascular tissue engineering is promising for therapeutic applications, additional studies are needed to gain further insights about the long-term functionality of engrafted hPSC-derived vascular cells. Furthermore, there is a need to develop cell differentiation and maturation protocols in the absence of animal products in order to enhance the knowledge about cellular differentiation circuitry and potentially for the clinical application of these cell-based therapies. As compared to other stem cells, human pluripotent stem cells can generate higher number of vascular cells unaffected by disease or aging processes. Currently, there are several clinical trials using stem cells for the treatment of ischemic diseases. Most of them use mesenchymal stem cells, progenitor cells, and adult autologous hematopoietic stem cells [75]. Human pluripotent stem cells, specifically human embryonic stem cells, have been recently used in clinical trials to treat spinal cord injury, Stargardt′s Macular Dystrophy, and Dry Age-Related Macular Degeneration. The demonstration of the safety and efficacy of hESC-derived cells in these clinical trials might open new horizons for their clinical application in other disease settings (Table 7.1).

Acknowledgments This work was supported in part by a Marie Curie-Reintegration Grant (FP7-People-2007-4-3-IRG; contract n° 230929), the MIT-Portugal program, the Portuguese Foundation for Science and Technology (FCT; PTDC/SA-BEB/098468/2008; PTDC/CTM/099659/2008), and Crioestaminal (project n° CENTRO-01-0202-FEDER-005476 "INJECT-CORD", project headed by Crioestaminal S.A. and supported by the Portuguese program QREN/Mais Centro and co-funded by the European Regional Development Fund) grants to LF. The authors would also like to thank FCT for the fellowship to HV (SFRH/BD/40077/2007).

References

1. Thomson JA, Itskovitz-Eldor J, Shapiro SS, Waknitz MA, Swiergiel JJ, Marshall VS, Jones JM (1998) Embryonic stem cell lines derived from human blastocysts. Science 282(5391):1145–1147
2. Takahashi K, Tanabe K, Ohnuki M, Narita M, Ichisaka T, Tomoda K, Yamanaka S (2007) Induction of pluripotent stem cells from adult human fibroblasts by defined factors. Cell 131(5):861–872
3. Yu J, Vodyanik MA, Smuga-Otto K, Antosiewicz-Bourget J, Frane JL, Tian S, Nie J, Jonsdottir GA, Ruotti V, Stewart R, Slukvin II, Thomson JA (2007) Induced pluripotent stem cell lines derived from human somatic cells. Science 318(5858):1917–1920
4. Takahashi K, Yamanaka S (2006) Induction of pluripotent stem cells from mouse embryonic and adult fibroblast cultures by defined factors. Cell 126(4):663–676
5. Schmidt A, Brixius K, Bloch W (2007) Endothelial precursor cell migration during vasculogenesis. Circ Res 101(2):125–136
6. Kraehenbuehl TP, Ferreira LS, Hayward AM, Nahrendorf M, van der Vlies AJ, Vasile E, Weissleder R, Langer R, Hubbell JA (2011) Human embryonic stem cell-derived microvascular grafts for cardiac tissue preservation after myocardial infarction. Biomaterials 32(4):1102–1109
7. Cho S-W, Moon S-H, Lee S-H, Kang S-W, Kim J, Lim JM, Kim H-S, Kim B-S, Chung H-M (2007) Improvement of postnatal neovascularization by human embryonic stem cell-derived endothelial-like cell transplantation in a mouse model of hindlimb ischemia. Circulation 116(21):2409–2419
8. Ferreira LS, Gerecht S, Shieh HF, Watson N, Rupnick MA, Dallabrida SM, Vunjak-Novakovic G, Langer R (2007) Vascular progenitor cells isolated from human embryonic stem cells give rise to endothelial and smooth muscle like cells and form vascular networks in vivo. Circ Res 101(3):286–294
9. Levenberg S, Golub JS, Amit M, Itskovitz-Eldor J, Langer R (2002) Endothelial cells derived from human embryonic stem cells. Proc Natl Acad Sci USA 99(7):4391–4396
10. Li Z, Wilson KD, Smith B, Kraft DL, Jia F, Huang M, Xie X, Robbins RC, Gambhir SS, Weissman IL, Wu JC (2009) Functional and transcriptional characterization of human embryonic stem cell-derived endothelial cells for treatment of myocardial infarction. PLoS ONE 4(12):e8443
11. Kane NM, Xiao Q, Baker AH, Luo Z, Xu Q, Emanueli C (2011) Pluripotent stem cell differentiation into vascular cells: a novel technology with promises for vascular re(generation). Pharmacol Ther 129(1):29–49
12. Gerecht-Nir S, Ziskind A, Cohen S, Itskovitz-Eldor J (2003) Human embryonic stem cells as an in vitro model for human vascular development and the induction of vascular differentiation. Lab Invest 83(12):1811–1820
13. Bai H, Wang ZZ (2008) Directing human embryonic stem cells to generate vascular progenitor cells. Gene Ther 15(2):89–95
14. Levenberg S, Zoldan J, Basevitch Y, Langer R (2007) Endothelial potential of human embryonic stem cells. Blood 110:806–814
15. Iacobas I, Vats A, Hirschi KK (2010) Vascular potential of human pluripotent stem cells. Arterioscler Thromb Vasc Biol 30(6):1110–1117
16. Ferguson JE, Kelley RW, Patterson C (2005) Mechanisms of endothelial differentiation in embryonic vasculogenesis. Arterioscler Thromb Vasc Biol 25(11):2246–2254
17. Drake CJ, Fleming PA (2000) Vasculogenesis in the day 6.5 to 9.5 mouse embryo. Blood 95(5):1671–1679
18. Kennedy M, D'Souza SL, Lynch-Kattman M, Schwantz S, Keller G (2007) Development of the hemangioblast defines the onset of hematopoiesis in human ES cell differentiation cultures. Blood 109(7):2679–2687

19. Jakobsson L, Kreuger J, Claesson-Welsh L (2007) Building blood vessels–stem cell models in vascular biology. J Cell Biol 177(5):751–755
20. Flamme I, Risau W (1992) Induction of vasculogenesis and hematopoiesis in vitro. Development 116(2):435–439
21. Motoike T, Markham DW, Rossant J, Sato TN (2003) Evidence for novel fate of Flk1+ progenitor: contribution to muscle lineage. Genesis 35(3):153–159
22. Suri C, Jones PF, Patan S, Bartunkova S, Maisonpierre PC, Davis S, Sato TN, Yancopoulos GD (1996) Requisite role of angiopoietin-1, a ligand for the TIE2 receptor, during embryonic angiogenesis. Cell 87(7):1171–1180
23. Peppel K, Zhang L, Orman ES, Hagen P-O, Amalfitano A, Brian L, Freedman NJ (2005) Activation of vascular smooth muscle cells by TNF and PDGF: overlapping and complementary signal transduction mechanisms. Cardiovasc Res 65(3):674–682
24. Rzucidlo EM, Martin KA, Powell RJ (2007) Regulation of vascular smooth muscle cell differentiation. J Vasc Surg 45(Suppl A):A25–A32
25. Lee SH, Hungerford JE, Little CD, Iruela-Arispe ML (1997) Proliferation and differentiation of smooth muscle cell precursors occurs simultaneously during the development of the vessel wall. Dev Dyn 209(4):342–352
26. Hungerford JE, Little CD (1999) Developmental biology of the vascular smooth muscle cell: building a multilayered vessel wall. J Vasc Res 36(1):2–27
27. Rensen SSM, Doevendans PAFM, van Eys GJJM (2007) Regulation and characteristics of vascular smooth muscle cell phenotypic diversity. Neth Heart J Mon J Neth Soc Cardiol Neth Heart Found 15(3):100–108
28. Frid M, Shekhonin B, Koteliansky V, Glukhova M (1992) Phenotypic changes of human smooth muscle cells during development: late expression of heavy caldesmon and calponin. Dev Biol (Print) 153(2):185–193
29. Yoshida T, Owens GK (2005) Molecular determinants of vascular smooth muscle cell diversity. Circ Res 96(3):280–291
30. Wang L, Li L, Shojaei F, Levac K, Cerdan C, Menendez P, Martin T, Rouleau A, Bhatia M (2004) Endothelial and hematopoietic cell fate of human embryonic stem cells Originates from primitive endothelium with hemangioblastic properties. Immunity 21(1):31–41
31. Lu S-J, Feng Q, Caballero S, Chen Y, Moore MAS, Grant MB, Lanza R (2007) Generation of functional hemangioblasts from human embryonic stem cells. Nat Methods 4(6):501–509
32. Sone M, Itoh H, Yamahara K, Yamashita JK, Yurugi-Kobayashi T, Nonoguchi A, Suzuki Y, Chao T-H, Sawada N, Fukunaga Y, Miyashita K, Park K, Oyamada N, Sawada N, Taura D, Tamura N, Kondo Y, Nito S, Suemori H, Nakatsuji N, Nishikawa S-i, Nakao K (2007) Pathway for differentiation of human embryonic stem cells to vascular cell components and their potential for vascular regeneration. Arterioscler Thromb Vasc Biol 27(10):2127–2134
33. Woll PS, Morris JK, Painschab MS, Marcus RK, Kohn AD, Biechele TL, Moon RT, Kaufman DS (2008) Wnt signaling promotes hematoendothelial cell development from human embryonic stem cells. Blood 111(1):122–131
34. Wang ZZ, Au P, Chen T, Shao Y, Daheron LM, Bai H, Arzigian M, Fukumura D, Jain RK, Scadden DT (2007) Endothelial cells derived from human embryonic stem cells form durable blood vessels in vivo. Nat Biotechnol 25(3):317–318
35. Vodyanik MA, Thomson JA, Slukvin II (2006) Leukosialin (CD43) defines hematopoietic progenitors in human embryonic stem cell differentiation cultures. Blood 108(6):2095–2105
36. James D, Nam H-s, Seandel M, Nolan D, Janovitz T, Tomishima M, Studer L, Lee G, Lyden D, Benezra R, Zaninovic N, Rosenwaks Z, Rabbany SY, Rafii S (2010) Expansion and maintenance of human embryonic stem cell-derived endothelial cells by TGFbeta inhibition is Id1 dependent. Nat Biotechnol 28(2):161–166
37. Nourse MB, Halpin DE, Scatena M, Mortisen DJ, Tulloch NL, Hauch KD, Torok-Storb B, Ratner BD, Pabon L, Murry CE (2010) VEGF induces differentiation of functional endothelium from human embryonic stem cells: implications for tissue engineering. Arterioscler Thromb Vasc Biol 30(1):80–89

38. Ferreira LS, Gerecht S, Fuller J, Shieh HF, Vunjak-Novakovic G, Langer R (2007) Bioactive hydrogel scaffolds for controllable vascular differentiation of human embryonic stem cells. Biomaterials 28(17):2706–2717
39. Ferreira L, Squier T, Park H, Choe H, Kohane DS, Langer R (2008) Human embryoid bodies containing nano- and microparticulate delivery vehicles. Adv Mater 20(12):2285
40. Prado-Lopez S, Conesa A, Arminan A, Martinez-Losa M, Escobedo-Lucea C, Gandia C, Tarazona S, Melguizo D, Blesa D, Montaner D, Sanz-Gonzalez S, Sepulveda P, Gotz S, O'Connor JE, Moreno R, Dopazo J, Burks DJ, Stojkovic M (2010) Hypoxia promotes efficient differentiation of human embryonic stem cells to functional endothelium. Stem Cells 28(3):407–418
41. Yamahara K, Sone M, Itoh H, Yamashita J, Yurugi-Kobayashi T, Homma K, ChaoT-H, Miyashita K, Park K, Oyamada N, Sawada N, Taura D, Fukunaga Y, Tamura N, Nakao K (2008) Augmentation of neovascularizaiton in hindlimb ischemia by combined transplantation of human embryonic stem cells-derived endothelial and mural cells. PLOS one 3(2):e1666
42. Hill KL, Obrtlikova P, Alvarez DF, King JA, Keirstead SA, Allred JR, Kaufman DS (2010) Human embryonic stem cell-derived vascular progenitor cells capable of endothelial and smooth muscle cell function. Exp Hematol 38(3):246–257 e241
43. Li Z, Hu S, Ghosh Z, Han Z, Wu JC (2011) Functional characterization and expression profiling of human induced pluripotent stem cell- and embryonic stem cell-derived endothelial cells. Stem Cells Dev 20:1701–1710
44. Park S-W, Jun Koh Y, Jeon J, Cho Y-H, Jang M-J, Kang Y, Kim M-J, Choi C, Sook Cho Y, Chung H-M, Young Koh G, Han Y-M (2010) Efficient differentiation of human pluripotent stem cells into functional CD34+ progenitor cells by combined modulation of the MEK/ERK and BMP4 signaling pathways. Blood 116(25):5762–5772
45. Kane NM, Meloni M, Spencer HL, Craig MA, Strehl R, Milligan G, Houslay MD, Mountford JC, Emanueli C, Baker AH (2010) Derivation of endothelial cells from human embryonic stem cells by directed differentiation: analysis of microRNA and angiogenesis in vitro and in vivo. Arterioscler Thromb Vasc Biol 30(7):1389–1397
46. Bai H, Gao Y, Arzigian M, Wojchowski DM, Wu W-S, Wang ZZ (2010) BMP4 regulates vascular progenitor development in human embryonic stem cells through a smad-dependent pathway. J Cell Biochem 109(2):363–374
47. Yamashita J, Itoh H, Hirashima M, Ogawa M, Nishikawa S, Yurugi T, Naito M, Nakao K (2000) Flk1-positive cells derived from embryonic stem cells serve as vascular progenitors. Nature 408(6808):92–96
48. Glaser DE, Gower RM, Lauer NE, Tam K, Blancas AA, Shih AJ, Simon SI, McCloskey KE (2011) Functional characterization of embryonic stem cell-derived endothelial cells. J Vasc Res 48(5):415–428
49. Red-Horse K, Ueno H, Weissman IL, Krasnow MA (2010) Coronary arteries form by developmental reprogramming of venous cells. Nature 464(7288):549–553
50. Kume T (2010) Specification of arterial, venous, and lymphatic endothelial cells during embryonic development. Histol Histopathol 25(5):637–646
51. Swift MR, Weinstein BM (2009) Arterial-venous specification during development. Circ Res 104(5):576–588
52. Wang HU, Chen ZF, Anderson DJ (1998) Molecular distinction and angiogenic interaction between embryonic arteries and veins revealed by ephrin-B2 and its receptor Eph-B4. Cell 93(5):741–753
53. Taura D, Sone M, Homma K, Oyamada N, Takahashi K, Tamura N, Yamanaka S, Nakao K (2009) Induction and isolation of vascular cells from human induced pluripotent stem cells–brief report. Arterioscler Thromb Vasc Biol 29(7):1100–1103
54. Homma K, Sone M, Taura D, Yamahara K, Suzuki Y, Takahashi K, Sonoyama T, Inuzuka M, Fukunaga Y, Tamura N, Itoh H, Yamanaka S, Nakao K (2010) Sirt1 plays an important role in mediating greater functionality of human ES/iPS-derived vascular endothelial cells. Atherosclerosis 212(1):42–47

55. Huang H, Zhao X, Chen L, Xu C, Yao X, Lu Y, Dai L, Zhang M (2006) Differentiation of human embryonic stem cells into smooth muscle cells in adherent monolayer culture. Biochem Biophys Res Commun 351(2):321–327
56. Xie C-Q, Zhang J, Villacorta L, Cui T, Huang H, Chen YE (2007) A highly efficient method to differentiate smooth muscle cells from human embryonic stem cells. Arterioscler Thromb Vasc Biol 27(12):e311–e312
57. Vo E, Hanjaya-Putra D, Zha Y, Kusuma S, Gerecht S (2010) Smooth-muscle-like cells derived from human embryonic stem cells support and augment cord-like structures in Vitro. Stem Cell Rev Rep 6(2):237–247
58. Vazão H, das Neves RP, Grãos M, Ferreira L (2011) Towards the maturation and characterization of smooth muscle cells derived from human embryonic stem cells. PLoS ONE 6(3):e17771
59. Owens GK (1995) Regulation of differentiation of vascular smooth muscle cells. Physiol Rev 75(3):487–517
60. Narsinh KH, Sun N, Sanchez-Freire V, Lee AS, Almeida P, Hu S, Jan T, Wilson KD, Leong D, Rosenberg J, Yao M, Robbins RC, Wu JC (2011) Single cell transcriptional profiling reveals heterogeneity of human induced pluripotent stem cells. J Clin Invest 121(3): 1217–1221
61. Doi A, Park I-H, Wen B, Murakami P, Aryee MJ, Irizarry R, Herb B, Ladd-Acosta C, Rho J, Loewer S, Miller J, Schlaeger T, Daley GQ, Feinberg AP (2009) Differential methylation of tissue- and cancer-specific CpG island shores distinguishes human induced pluripotent stem cells, embryonic stem cells and fibroblasts. Nat Genet 41(12):1350–1353
62. Stadtfeld M, Apostolou E, Akutsu H, Fukuda A, Follett P, Natesan S, Kono T, Shioda T, Hochedlinger K (2010) Aberrant silencing of imprinted genes on chromosome 12qF1 in mouse induced pluripotent stem cells. Nature 465(7295):175–181
63. Urbach A, Bar-Nur O, Daley GQ, Benvenisty N (2010) Differential modeling of fragile X syndrome by human embryonic stem cells and induced pluripotent stem cells. Cell Stem Cell 6(5):407–411
64. Choi K, Yu J, Smuga-Otto K, Salvagiotto G, Rehrauer W, Vodyanik M, Thomson J, Slukvin I (2009) Hematopoietic and endothelial differentiation of human induced pluripotent stem cells. Stem Cells 27(3):559
65. Lee T-H, Song S-H, Kim KL, Yi J-Y, Shin G-H, Kim JY, Kim J, Han Y-M, Lee SH, Lee S-H, Shim SH, Suh W (2010) Functional recapitulation of smooth muscle cells via induced pluripotent stem cells from human aortic smooth muscle cells. Circ Res 106(1):120–128
66. Kraehenbuehl TP, Aday S, Ferreira LS (2010) Scaffolding for three-dimensional embryonic vasculogenesis biological and medical physics, Biomed Eng 49–67
67. Ferreira L, Pedroso DCS, Vazão H, Gomes RSM (2010) Stem cell-based therapies for heart regeneration: what did the bench teach us? Cardiovasc Hematol Disord Drug Targets 10(3): 173–185
68. Gerecht S, Ferreira LS, Langer R (2010) Vascular differentiation of human embryonic stem cells in bioactive hydrogel-based scaffolds. Methods Mol Biol 584:333–354
69. Kraehenbuehl TP, Ferreira LS, Zammaretti P, Hubbell JA, Langer R (2009) Cell-responsive hydrogel for encapsulation of vascular cells. Biomaterials 30(26):4318–4324
70. Pedroso DC, Tellechea A, Moura L, Fidalgo-Carvalho I, Duarte J, Carvalho E, Ferreira L (2011) Improved survival, vascular differentiation and wound healing potential of stem cells co-cultured with endothelial cells. PLoS ONE 6(1):e16114
71. Fidkowski C, Kaazempur-Mofrad MR, Borenstein J, Vacanti JP, Langer R, Wang Y (2005) Endothelialized microvasculature based on a biodegradable elastomer. Tissue Eng 11(1–2):302–309. doi:10.1089/ten.2005.11.302
72. Kaihara S, Borenstein J, Koka R, Lalan S, Ochoa ER, Ravens M, Pien H, Cunningham B, Vacanti JP (2000) Silicon micromachining to tissue engineer branched vascular channels for liver fabrication. Tissue Eng 6(2):105–117
73. Xu T, Jin J, Gregory C, Hickman JJJ, Boland T (2005) Inkjet printing of viable mammalian cells. Biomaterials 26(1):93–99

74. Narazaki G, Uosaki H, Teranishi M, Okita K, Kim B, Matsuoka S, Yamanaka S, Yamashita JK (2008) Directed and systematic differentiation of cardiovascular cells from mouse induced pluripotent stem cells. Circulation 118(5):498–506
75. Lasala GP, Minguell JJ (2011) Vascular disease and stem cell therapies. Br Med Bull 98:187–197
76. Sone M, Itoh H, Yamahara K, Yamashita JK, Yurugi-Kobayashi T, Nonoguchi A, Suzuki Y, Chao T-H, Sawada N, Fukunaga Y, Miyashita K, Park K, Oyamada N, Sawada N, Taura D, Tamura N, Kondo Y, Nito S, Suemori H, Nakatsuji N, Nishikawa S-i, Nakao K (2007) Pathway for differentiation of human embryonic stem cells to vascular cell components and their potential for vascular regeneration. Arterioscler Thromb Vasc Biol 27(10):2127–2134

Chapter 8
Mechanism of MicroRNA-Mediated Global DNA Demethylation in Human iPS Cells

Shi-Lung Lin

Abstract Global DNA demethylation is the most essential epigenetic mechanism for reprogramming somatic cells to induced pluripotent stem cells (iPS cells or iPSCs). However, none of Yamanaka factors (Oct4, Sox2, Klf4 and c-Myc) or Thomson factors (Oct4, Sox2, Nanog and Lin28) can define such an important mechanism. Our studies for the first time reveal that a unique family of microRNA (miRNA), namely miR-302, is responsible for inducing not only global DNA demethylation but also chromatin modification in human iPSCs and embryonic stem cells (ESCs). miR-302 functions to concurrently silence multiple key epigenetic regulators, such as AOF1 (LSD2, KDM1B), AOF2 (LSD1, KDM1), MECP1/2, DNMT1, and HDAC2, to reset genomic DNA methylation patterns and hence reactivate the full spectrum of ESC-specific transcriptome expression. Moreover, silencing AOF2 further destabilizes DNMT1 activity and prevents replication-dependent DNA remethylation during iPSC division. Inhibition of both AOF1 and AOF2 also increases methylation of histone 3 on lysine 4 (H3K4me2/3), a chromatin mark for ESCs and reprogrammed iPSCs. These epigenetic events highly resemble zygotic reprogramming during early embryonic development. In this review, through deciphering the function of miR302, we explain how nuclear reprogramming is initiated in iPSCs, of which the knowledge is useful for improving the efficiency of current iPSC generation technologies.

S.-L. Lin (✉)
Division of Regenerative Medicine, WJWU and LYNN Institute for Stem Cell Research, Santa Fe Springs, CA, USA
e-mail: shilungl@mirps.org

H. Baharvand and N. Aghdami (eds.), *Advances in Stem Cell Research*, Stem Cell Biology and Regenerative Medicine, DOI: 10.1007/978-1-61779-940-2_8,

8.1 Introduction

Global demethylation of genomic DNAs occurs only in two natural embryonic development events; one is during migration of primordial germ cells (PGCs) into the embryonic gonads (approximately embryonic day E10.5–E13.5) and the other is in the 2−8-cell-stage zygotic cells after fertilization [1–4]. Parental imprints are erased and reestablished in germline PGCs but largely preserved in postfertilized zygotic cells [5, 6], indicating that the germline and zygotic demethylation mechanisms are not completely identical. However, many aspects of these natural demethylation mechanisms remain unclear and controversial. The recent discovery of somatic cell reprogramming (SCR) to form induced pluripotent stem cells (iPSCs) introduces a new kind of man-made global demethylation mechanism comparable to the natural ones [7]. Such SCR-mediated demethylation triggers massive erasure of genomic methylation sites but preserves parental imprints similar to zygotic demethylation [8], nevertheless, occurring in somatic cells rather than postfertilized zygotes (Fig. 8.1). Hence, SCR demethylation seems more like a forced zygotic demethylation mechanism in somatic cells. As a result, some zygotic components required for the completion of epigenetic reprogramming during early embryonic development may be missing in SCR. This observation is further supported by recent evidence that reprogramming by somatic cell nuclear transfer (SCNT) results in more similar epigenome and transcriptome patterns to those of embryonic stem cells (ESCs) than does the direct reprogramming by four Yamanaka factors [9].

To provide effective SCR for efficient iPSC generation, we need to understand the zygotic components responsible for global demethylation. Global demethylation marks the first sign of both zygotic reprogramming and SCR and is necessary for the epigenetic reprogramming of cell nuclei [10]. Erasure of prior methylation sites and resetting them to an ESC-specific methylation pattern is important to activate reprogramming-associated genes required for pluripotent stem cell formation. In the 2−8-cell-stage mammalian zygotes, DNA (cytosine-5-)-methyltransferase 1 (DNMT1) inherited from oocytes is excluded from the nucleus by an unknown mechanism, whereas zygotic expression is very limited [11–13]. DNMT1 functions to maintain inherited CpG methylation patterns by methylating the newly replicated DNA during the S-phase cell cycle; thus, loss of DNMT1 activity leads to passive global DNA demethylation following the zygotic cleavage divisions before the morula stage [3, 4, 14, 15]. Nevertheless, demethylation triggered by this passive mechanism will generate two hemimethylated zygotic cells, which has never be seen in an embryo. In addition, the demethylation of paternal DNA has been found to start before the first zygotic cleavage division [3, 4, 16], suggesting another active demethylation mechanism responsible for enzymatically removing the paternal and hemimethylation sites. Several theories have been proposed for this active mechanism, involving 5-methylcytosine (5mC)/5-hydroxymethylcytosine (5hmC) deamination and mismatch/excision repair pathways; however, the details of these pathways remain elusive and controversial [17–19]. One other component

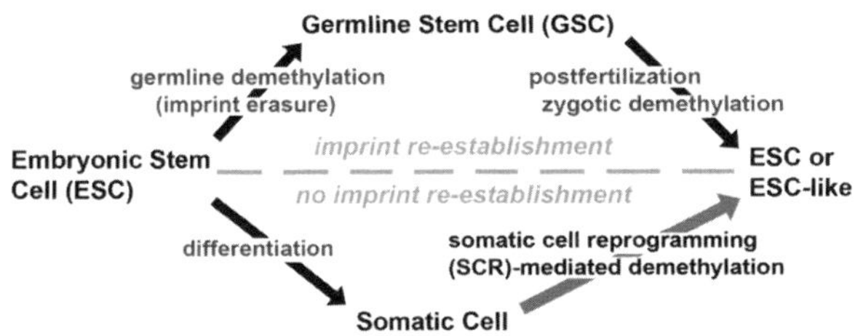

Fig. 8.1 Comparison of natural germline-zygote and manually forced SCR demethylation pathways. Global demethylation occurs naturally in two developmental stages, one during formation and migration of PGCs into the embryonic gonads (germline demethylation) and the other in 2–8-cell-stage zygotes before the morula stage (zygotic demethylation). Parental imprints are erased and reestablished during germline demethylation but not in zygotic demethylation, resulting in different genomic imprinting patterns from the parents. On the other hand, SCR demethylation can be forced in somatic cells by manually introducing the expression of either miR-302 or four defined reprogramming factors (Oct4-Sox2-Klf4-cMyc or Oct4-Sox2-Nanog-Lin28). Notably, Oct4 and Sox2 can also induce miR-302 expression. Due to the effects of miR-302-targeted epigenetic gene silencing, the somatic cell genome is forcedly demethylated without the erasure of parental imprints. As a result, expression of parental germline elements is still inactivated by the imprints. Given that parental elements are essential for normal zygotic development, their deficiency may cause developmental defects in SCR-induced iPSCs

important for zygotic demethylation is the simultaneous downregulation of methyl-CpG binding protein 1 and 2 (MECP1/MECP2) and methyl-CpG binding domain 2 (MBD2), whose functions are all highly involved in the maintenance of genomic methylation loci [20–22]. Taken together, and despite the unknown mechanisms, all the above evidences indicate that zygotic demethylation occurs in a condition lacking DNMT1, MECP1/2, and MBD2 activities. Then, what zygotic factor(s) can inhibit these epigenetic targets? To this, we found miR-302, which has been shown to concurrently and specifically silence these targeted epigenetic genes in human iPSCs [23–25]. By this means, miR-302 is able to create an intracellular environment suitable for SCR initiation.

MicroRNA (miRNA) is a class of small non-coding RNAs (ncRNAs) that function to suppress the translation of their target genes through complementary binding and formation of RNA-induced silencing complexes (RISCs) in the 3′-untranslated regions (3′-UTRs) of the targeted gene transcripts. Because this kind of miRNA-target interaction does not require perfect complementarity, miRNA can simultaneously silence multiple genes through a partially matched targeting mechanism to coordinate the functions and downstream interactions of these targeted genes. miR-302 is the most abundant miRNA in hESCs and iPSCs but not found in differentiated somatic cells [26, 27]. Recently, mouse iPSCs were also found to express an elevated level of miR-302 rather than miR-291/294/295, the regular marker for mouse ESCs [28]. According to this feature, miR-302 serves as a major gene silencer in hESCs and early zygotes. Based on the analytic results of online miRNA-target prediction programs TARGETSCAN (http://www.targetscan.org/) and PICTAR-VERT (http://pictar.mdc-berlin.de/), the majority of miR-302-targeted genes are transcripts of differentiation-associated genes and developmental signals, such as members of the RAS-MAPK, TGFß-SMAD and LEFTY pathways, indicating its important role in preventing stem cell differentiation. The genomic sequence encoding miR-302 is located in the 4q25 locus of human chromosome 4, a conserved region frequently associated with longevity [29]. More precisely, miR-302 is encoded in the intronic region of the *La ribonucleoprotein domain family member 7* (*LARP7*, *PIP7S*) gene and expressed via an intronic miRNA biogenesis mechanism [23, 30–32]. Native miR-302 consists of four familial sense homologues (miR-302b, c, a and d) and three distinct antisense members (miR-302b*, c* and a*), all of which are transcribed together as a polycistronic RNA cluster along with another miRNA, miR-367 [26]. Although the overall function of these miR-302 members is largely unknown, their unique expression pattern and abundance in hESCs and iPSCs infers a pivotal role in regulating stem cell pluripotency.

Species-differential and concentration-dependent targeting are two major factors determining the function of a miRNA. miRNA function may vary in different animal species or at different cellular concentrations. For example, studies in mice have indicated that miR-302 may silence cyclin-dependent kinase inhibitor 1A (CDKN1A, p21Cip1) to accelerate cell proliferation, whereas human *CDKN1A* gene does not contain any valid target site for miR-302 [24]. Also, *BMI1 polycomb ring finger oncogene* (*BMI-1*), a cancer stem cell marker gene, is a validated target for miR-302 in humans but not in mice [24]. This kind of species differences suggests that mouse is not an appropriate model for evaluating miR-302 function in humans. In fact, mouse embryos express the miR-290 family rather than miR-302. On the other hand, because of the miRNA-target binding through partial complementarity, a higher miRNA concentration often renders a stronger silencing effect on target genes. Using this dose-dependent gene silencing effect, miRNA can either fine-tune or completely alter the expression rates and functions of its targeted genes [25, 33]. Here, we exemplify the interaction between miR-302 and cyclin-dependent kinase 2 (CDK2) to show this concentration-dependent effect. When the cellular miR-302 expression is the same or lower than the level found in hESCs, miR-302 silences only large tumor suppressor homolog 2 (LATS2) but not CDK2 to promote cell proliferation, whereas at a higher concentration over 1.1–1.3 folds of the hESC level, both CDK2 and LATS2 are silenced and hence cell cycle is attenuated at the G1-phase check point [23, 24, 31]. Given that LATS2 inhibits the cell cycle by blocking the cyclin E-CDK2 pathway [34], the miR-302-mediated silencing of CDK2 can actually compensate this LATS2 function to inhibit cell cycle progression. Moreover, silencing CDK2 may also counteract the suppressive effect of miR-367 on CDKN1C (p57, Kip2), an inhibitor against both CDK2 and CDK4, subsequently resulting in a reduced cell cycle rate. In view of these species-differential and concentration-dependent target variations identified so far, there may be many more to be found. To prevent any confusion due to these variations, this review will focus on the miR-302 function in humans and at a concentration comparable to or beyond the normal hESC level.

8.2 Intronic Expression of miR-302

To reiterate the natural expression mechanism of miR-302, we have developed a *pSpRNAi-RGFP* vector mimicking the process of miRNA biogenesis from mammalian introns [30, 35, 36]. As shown in Fig. 8.2, primary miRNA precursors (pri-miRNAs) are first transcribed by type-II RNA polymerases (Pol-II) and excised by spliceosomal components and/or Drosha-like endoribonuclease to form miRNA precursors (pre-miRNAs), which are then exported out of the nucleus by Ran-GTP and Exportin-5 and further processed by Dicer-like RNaseIII endoribonucleases in cytoplasm to form mature miRNAs [30, 37–40]. However, the role of Drosha may not be required in this mechanism because depletion of >85% Drosha by siRNA reduces only almost a half of the miR-302 expression, indicating that some other endoribonucleases may replace Drosha for processing intronic pre-miRNAs [38]. Also, because mammalian intron often contains nonsense (i.e., translational stop) codons recognized by the nonsense-mediated decay (NMD) system, a cellular RNA surveillance mechanism [41, 42], the non-hairpin structures of an intron can be quickly degraded by NMD to prevent excessive RNA accumulation. RNA oversaturation has been reported to be a major problem for the direct (exonic) expression of short hairpin RNAs (shRNAs) and miRNAs in mammalian cells [43]. Under tight regulation by the cellular Pol-II transcription, RNA splicing and NMD systems, intronic miRNA biogenesis has an evolutional advantage in safety control. Using this mechanism, we have successfully identified the gene silencing effects of various intronic shRNAs/miRNAs in mouse and human cells in vitro as well as in mouse skins, chicken embryos, and zebrafish in vivo [30, 35, 37, 44–46]. Similar approaches have also been shown to express polycistronic shRNAs/miRNAs for multiple gene silencing [47, 48]. Based on these previous studies and efforts, we have established a means to express the intronic miR-302 familial cluster using its own natural biogenesis mechanism.

Prior attempts at investigating the "miR-302" function all failed to show the evidence of complete expression of all miR-302 familial members. Our approach using intronic miR-302 expression has overcome this drawback and successfully demonstrated a full spectrum of miR-302a, b, c, and d expression in human iPSCs similar to that in hESCs, as determined by both microarray and northern blot analyses [23, 25]. However, there are still several minor differences compared to the natural expression pattern of miR-302. First, we purposely redesigned the miR302 cluster in order to effectively insert it into the *SpRNAi* intron. Second, while the expressions of other miR-302 members were abundantly detected in human iPSCs after transfection, the antisense-strand miR-302b (miR-302b*) was not observed (Fig. 8.3). Last, the miR-302 expression was driven by a constitutive *cytomegaloviral* (*CMV*) or a *tetracycline-inducible* (*Tet-On*) promoter rather than its native promoter. Notably, the utilization of *Tet-On*-inducible expression offers a great advantage in studying the dose-dependent miR-302 effects on gene targeting. In view of these differences, the miR-302 function discussed here and in our previous studies represents the interactive activities of miR-302a, a*, b, c, c*, and d

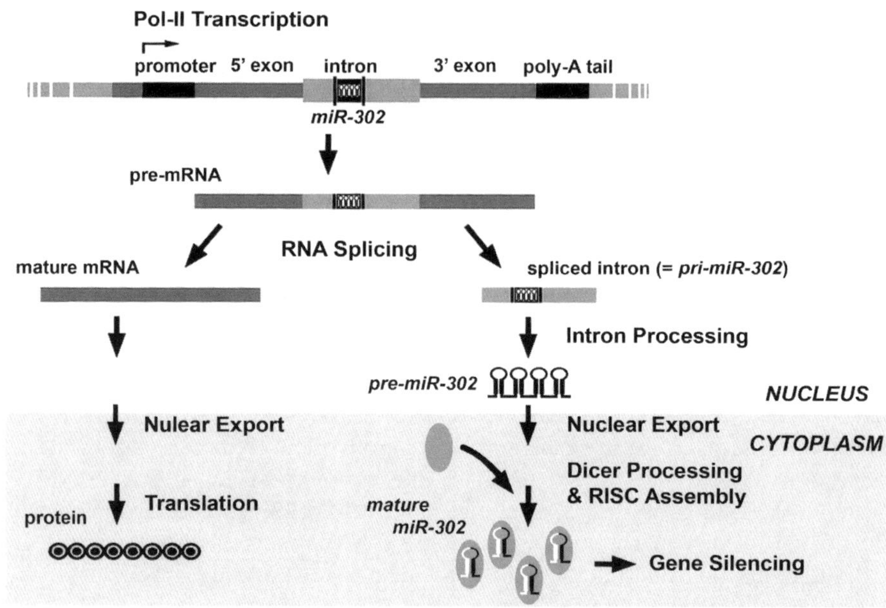

Fig. 8.2 Mechanism of natural miR-302 biogenesis. The intronic miR-302 familial cluster is transcribed together with a host gene by Pol-II and spliced out of the gene transcript by spliceosomal components and/or Drosha to form precursor miR-302 (pre-miR-302), which is then exported out of the nucleus by Ran-GTP and Exportin-5 and further processed by Dicer-like RNaseIII endoribonuclease in cytoplasm to form mature miR-302. For gene silencing, mature miR-302 is assembled into a RNA-induced silencing complex (RISC) with argonaute proteins and Dicer endoribonucleases and then functions to suppress the translation or cause direct degradation of targeted messenger RNAs (mRNAs)

but not miR-302b*. As the miRNA-targeting program PICTAR-VERT has predicted that Dicer1 is a target of miR-302b*, miR-302b* may silence Dicer1 to reduce the overall expression levels of the miR-302 family and many other ESC-specific miRNAs and hence hinders the reprogramming efficiency of iPSC generation. Due to lack of miR-302b* expression, our approach is able to bypass this self-regulation mechanism of miR-302 and successfully delivers a sufficient sense miR-302 concentration for reprogramming somatic cells to iPSCs.

8.3 Induction of Human iPSC Formation

SCR is an epigenetic process to reprogram differentiated somatic cells into an ESC-like pluripotent state. These reprogrammed pluripotent cells are named iPS cells or iPSCs. Global DNA demethylation is the first sign of SCR, yet our understanding to it is limited by the introduction of either Yamanaka (Oct4, Sox2,

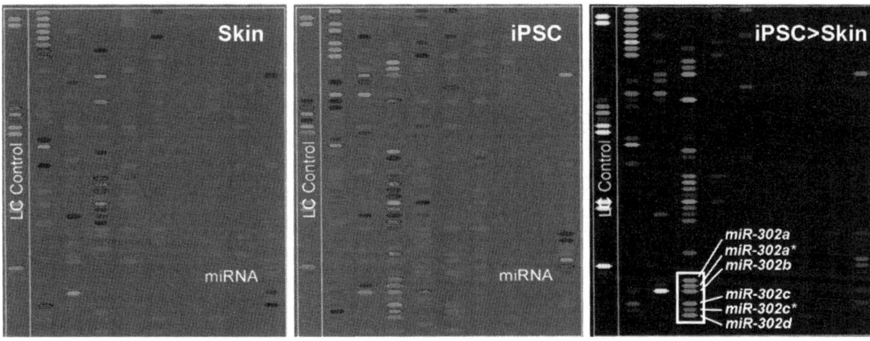

Call list (differentially expressed transcripts with p-value < 0.01)

No.	Probe_ID	Skin cell Signal	iPSC Signal	log2 (iPSC / Skin cell)
1	hsa-miR-302b	67.18	28,423.66	8.78
2	hsa-miR-302d	55.69	18,679.25	8.38
3	hsa-miR-302a	49.21	14,840.17	8.20
4	hsa-miR-302a*	42.91	13,875.22	8.16
5	hsa-miR-302c	33.38	8,641.02	7.99
6	hsa-miR-9	1.00	250.28	7.97
7	hsa-miR-346	4,392.06	27.61	-7.25
8	hsa-miR-374	17.34	2,720.12	7.22
9	hsa-miR-612	1,630.33	13.11	-6.94
10	hsa-miR-20b	42.99	5,270.39	6.94
11	hsa-miR-363	7.29	760.52	6.83
12	hsa-miR-542-5p	136.52	1.63	-6.50
13	hsa-miR-376a	5.77	494.89	6.45
14	hsa-miR-517a	10.20	737.14	6.24
15	hsa-miR-20a	95.17	7,200.26	6.21
16	hsa-miR-516-5p	8.43	601.21	6.11
17	hsa-miR-517b	8.44	573.21	6.06
18	hsa-miR-526a	4.22	278.40	6.05
19	hsa-miR-17-5p	97.58	6,656.73	6.04
24	hsa-miR-302c*	29.83	1,894.16	5.93

Fig. 8.3 Microarray confirmation of ectopically introduced miR-302 expression in iPSCs. miR-302 functions to reprogram both human normal and cancerous skin/hair follicle cells (i.e., keratinocytes and melanocytes) to ESC-like iPSCs, namely mirPSCs. Results of miRNA microarray analyses reveal that these mirPSCs express abundant miR-302a, a*, b, c, c*, and d, but not miR-302b*. *Upper panels* demonstrate the results of miR-302 expression in mirPSCs derived from human melanoma Colo-829 cells. The call list of differentially expressed miRNAs

Klf4 and c-Myc) or Thomson (Oct4, Sox2, Nanog and Lin28) factors [7, 49, 50]. Nevertheless, none of these reprogramming factors defined so far can explain how and why global demethylation occurs in iPSCs. Not until the identification of another key reprogramming factor, miR-302, have we finally seen the first insight into this unsolved mechanism. Using ectopic transfection, we have shown that the forced expression of miR-302 in human somatic cells can induce strong Oct4, Sox2, Nanog, and Lin28 co-expression and hence results in a reprogramming mechanism similar to the iPSC generation using Thomson factors [23, 25]. By deciphering the miR-302-targeted genes and their interactions, we found that miR-302 is able to functionally replace all previously defined reprogramming factors to reprogram both human normal and cancerous cells to iPSCs [23–25]. We referred these miR-302-induced iPSCs as mirPSCs.

mirPSCs proliferate at a rate approximately 20–24 h/cycle [23, 24], which is identical to that of zygotic cells before the morula stage but slower than the fast proliferation rate of blastocyst-derived hESCs and Yamanaka-factor-induced iPSCs (12–16 h/cycle) [7, 50]. Naturally, early mammalian zygotes before the morula (8–16-cell) stage present a slower cell cycle rate (20–24 h/cycle) than that of late blastocyst-derived hESCs (15–16 h/cycle) [51, 52]. Although both mammalian zygotes and hESCs are known to display a short G1 phase, the G1 phase of a 2−8-cell-stage zygote is still significantly longer than that of hESCs by 4 ± 1 h. Moreover, the mirPSC colonies are morphologically grown like three-dimensional cell aggregations and highly sensitive to the stimulation of leukemia inhibitory factor (LIF; Fig. 8.4); both of the features are frequently observed in mouse ESCs, which are known to be isolated at an earlier embryonic stage (E3–E3.5) than that of hESCs (E5–E6) [53, 54]. Based on these natural differences during embryonic development, it suggests that mirPSCs may represent an earlier developmental stage than do blastocyst-derived hESCs. However, iPSC generation has been proposed to involve either a cell cycle-dependent (with Klf4) or cell cycle-independent (with Nanog) reprogramming process [55], in which the former seemingly cannot explain the slower proliferation rate of mirPSCs. In fact, this hypothesis overlooked that Klf4, an upstream transcription factor of Nanog, induces not only Nanog but also many other oncogenes, which promote fast cell proliferation as a result unrelated to the reprogramming function of Nanog [56–59]. Therefore, fast cell cycle, although can increase iPSC number, is not essential for the SCR mechanism to generate iPSCs.

What factor is required for iPSC formation? Recent studies have revealed that there is a positive feedback regulation loop between miR-302 and Oct4-Sox2-Nanog-Lin28 expression (Fig. 8.5) [23, 25, 60]. Meanwhile, we found the miR-302-induced Oct4-Sox2-Nanog-Lin28 expression [23, 31]; Oct4 and Sox2, the two most essential reprogramming factors in both Yamanaka's and Thomson's methods, were also identified to be the upstream transcription factors for stimulating miR-302 expression [60, 61]. Yet, how does this positive feedback loop mechanism function to initiate global demethylation? Later, we further found that miR-302 targets several important epigenetic regulators, such as lysine-specific histone demethylases 1 and 2 (namely AOF1/2, KDM1B/1, or LSD2/1), MECP1/2 and DNMT1, to trigger global demethylation [25]. In particular, both AOF1 and AOF2 function to repress gene transcription by demethylating histone 3 on lysine 4 (H3K4) [62–64]. Inhibition of AOF2 by its antagonist tranylcypromine has been

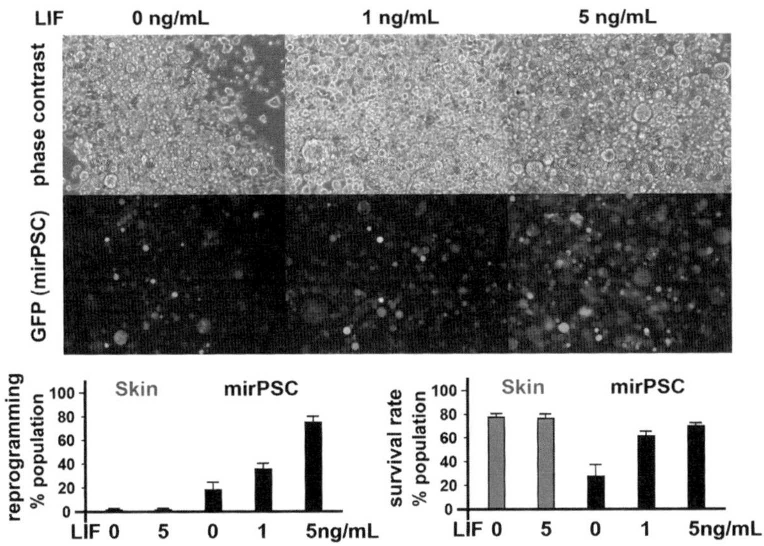

Fig. 8.4 Properties of human mirPSCs similar to mouse ESCs isolated from an earlier embryonic stage. Human mirPSCs form three-dimensional cell colonies (*red fluorescent*) and are sensitive to the growth stimulation of LIF. *Upper panels* show that LIF stimulation significantly increases both of the reprogramming efficiency and survival rates of mirPSCs in a dose-dependent matter, while *lower* charts indicate the relative levels of changes corresponding to the LIF concentration

shown to increase H3K4 methylation (H3K4me2/3), a chromosomal mark for ESCs, and to promote Oct4 overexpression in embryonal carcinoma cells [62, 63]. In transgenic knockout mice, loss of either AOF1 or AOF2 substantially elevates H3K4me2/3 [64, 65]. AOF1-knockout mice fail to set up de novo DNA methylation imprints during oogenesis [64], while AOF2 deficiency causes embryonic lethality due to a progressive loss of genomic DNA methylation and lack of global cell differentiation [65]. Therefore, silencing of AOF1/2 is sufficient to induce global demethylation. To this, our recent finding of miR-302-mediated AOF1/2 silencing has confirmed this mechanistic pathway in human iPSCs [25]. In addition, since AOF2 is required for stabilizing DNMT1 and preserving its activity [65], the silencing of AOF2 by miR-302 also inhibits DNMT1-mediated maintenance of global DNA methylation [25]. This result is further supported by the data of online miRNA-target prediction program provided by the European Bioinformatics Institute EMBL-EBI (http://www.ebi.ac.uk/enright-srv/microcosm/cgi-bin/targets/v5/detail_view.pl?transcript_id=ENST00000359526), showing that miR-302a-d can directly target DNMT1 for gene silencing. As a result, through the positive feedback loop mechanism shown in Fig. 8.5, both miR-302 and Oct4-Sox2-Nanog-Lin28 can induce global demethylation, but at different efficiency rates. Given that global demethylation is directly initiated by miR-302-mediated epigenetic gene silencing, the formation of mirPSCs usually takes much shorter time than

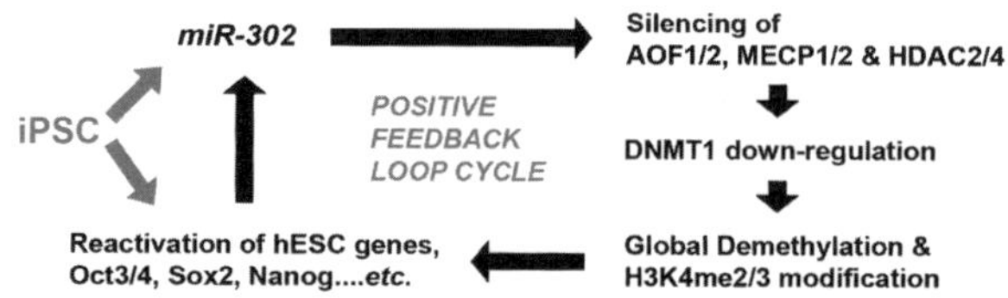

Fig. 8.5 Currently established SCR mechanism for iPSC generation. miR-302 silences AOF1/2 and DNMT1 activities and, in conjunction with the co-suppression of MECP1/2 and HDAC2, results in global DNA demethylation and chromosomal H3K4me2/3 modification. Subsequently, these epigenetic reprogramming events induce ESC-specific gene expression, in particular Oct4, Sox2, and Nanog, which in turn further stimulate miR-302 expression to form a positive feedback loop cycle crucial for the maintenance of SCR. Based on this mechanistic model, there are two methods for inducing iPSC formation: one is to force the miR-302 expression and the other is to introduce the co-expression of four defined factors (either Oct4-Sox2-Klf4-cMyc or Oct4/Sox2-Nanog-Lin28). Both methods can trigger the activation of this cycling SCR mechanism; however, miR-302 directly induces global demethylation to initiate SCR while the four factors indirectly function through miR-302 expression

does the four-factor-induced iPSC generation (approximately 1–2 weeks versus 2–3 weeks).

Induced global demethylation has been found to promote Oct4–Nanog overexpression in early mouse embryos and mouse-human fused heterokaryons [66, 67]. Our recent studies further demonstrated that elevated miR-302 expression over 1.1–1.3 folds of the normal hESC level (approximately 0.9–1.0 million copies per hESC) triggers both global demethylation and co-expression of Oct4, Sox2, and Nanog in human iPSCs [24, 25]. The expression of Lin28 and many other hESC marker genes was observed 1–3 days later than the presence of Oct4-Sox2-Nanog elevation. A similar miR-302 transfection approach was also shown to increase Oct4–Nanog expression by two folds in hESCs [68]. Further research in hESCs revealed that miR-302 can directly silence nuclear receptor subfamily 2, group F, number 2 (NR2F2), a transcriptional repressor against Oct4 expression, to activate Oct4 expression [69]. These findings suggest that global demethylation activates ESC-specific gene expression in particular, Oct4 and Nanog. Although we currently do not know the details of this event in iPSCs, the iPSCs generated by various methods from various somatic cell types all share a similar epigenetic and transcriptomic profile like the one found in ESCs, indicating the importance of global demethylation in erasing and resetting different somatic epigenomes and transcriptomes into an ESC-like state. In this state, we have performed microarray analyses to show over 91–93% similarity in the genome-wide gene expression patterns between mirPSCs/iPSCs and hESC H1/H9 cells [23, 25]. Global demethylation has also been evaluated in mirPSCs/iPSCs, highly resembling the hESC-specific pattern particularly in the promoter regions of hESC genes, *Oct4*, *Nanog*, *Sox2*, and *UTF1*. Taken together, all current studies have established a fact that miR-302 is able to directly induce global demethylation and Oct4-Nanog activation essential for the initiation of SCR to form iPSCs.

8.4 Demethylation is Induced by iPSC Cytoplasm Rather Than the Nucleus

As DNA demethylation occurs in the cell nucleus while miR-302 functions in the cytoplasm, SCNT is the best method to test this epigenetic reprogramming mechanism. SCNT is a well-established technology to generate ESC-like pluripotent stem cells by hybridizing a somatic cell nucleus into the oocyte cytoplasm [70, 71]. When following the same strategy to transfer a somatic cell nucleus into the mirPSC cytoplasm, we observed that most (93%) of the hybrid cells were successfully reprogrammed to mirPSC-like pluripotent cells possessing all ESC-like properties [25]. Global demethylation was quickly detected in a few days after SCNT. Conversely, transferring a mirPSC nucleus into the somatic cell cytoplasm failed to form any viable cell. Hence, the earliest reprogramming effector, the "initiator", resides in the cytoplasm rather than nucleus of a mirPSC. This finding is coincident with the previous SCNT results using the oocyte cytoplasm, which contains several miR-302 homologs that may have the same reprogramming function, such as miR-200c and miR-371–373. Given that miR-302 is a cytoplasmic effector whereas Oct4, Sox2, and Nanog are all nuclear transcription factors, it is clear that miR-302 is responsible for initiating SCR through global demethylation in the SCNT-induced iPSCs. On the other hand, the transfer of mirPSC nucleus fails to induce pluripotency, suggesting that somatic cytoplasm may lack an effector required for iPSC generation. Alternatively, somatic cytoplasm may contain an inhibitor directed against iPSC formation; however, this possibility has been ruled out by further induction of miR-302 elevation in the hybrid cells containing the mirPSC nucleus, showing that the efficiency of iPSC generation is increased proportional to miR-302 elevation [25].

8.5 Mechanism of Global Demethylation

A passive demethylation mechanism was first found in mirPSCs soon after the silencing of AOF1/2 and MECP1/2 by miR-302. When tracing the mirPSC formation induced by elevated miR-302 expression in transfected somatic cells, we found that the decreases of AOF1/2, MECP1/2, DNMT1, and HDAC2 were most prominent 3 days after the miR-302 elevation, while the expression of Oct4, Sox2, and Nanog was increased to a maximal level at the 5th day. During this period, almost no cell division was detected in the first three days and then only one or two at most were found in the 4th and 5th days. This cell division-dependent process is highly similar to the zygotic demethylation of a maternal genome, but now occurring in both of the parental genomes in mirPSCs. Due to the miR-302-targeted silencing of AOF1/2, MECP1/2, and especially DNMT1, DNA methylation cannot be replicated during the S-phase cell cycle, subsequently leading to global demethylation in following few cell divisions.

However, this passive mechanism will generate two hemimethylated cells in every mirPSC colony, an event that has not yet been identified. Whether these hemimethylated cells are quickly degraded by programmed cell death (apoptosis) during SCR or further demethylated by another active mechanism remains to be determined. To this, since we have recently observed the elevation of activation-induced cytidine deaminase (AID) expression in response to the silencing of AOF2 by miR-302 [25], it is possible that an active demethylation mechanism may be involved in the global demethylation of mirPSC genomes as well.

Previous studies have demonstrated that AOF2 stabilizes DNMT1 and preserves its activity on the maintenance of global DNA methylation [65], whereas AID is involved in global demethylation and Oct4-Nanog activation in early mouse embryos and mouse-human fused heterokaryons [66, 67]. It is conceivable that the deficiency of DNMT1 activity may cause mirPSC genomes to become more susceptible to the AID activity. In mammals, AID is expressed in B cells, PGCs, oocytes, and early stage embryos and is responsible for the deamination of 5mC that generates a cytosine (C)-thymine (T) conversion and hence results in T-guanine (G) mismatch base pairing [72, 73]. To explain the possible mechanism underlying the correction of this T-G mismatch pairing during DNA demethylation, a mammalian version of the base excision DNA repair (BER) pathway has been proposed to enzymatically replace the mismatched T with a C [74, 75]; however, the enzyme capable of carrying out this function has not yet been identified in mammals. On the other hand, another theory involving the excision repairing of 5hmC has also been postulated. In this model, Tet familial enzymes first convert 5mC–5hmC and then 5hmC can be further converted to C by spontaneous loss of its formaldehyde group [76] or by BER or a currently undefined DNA repair system. The formation of 5hmC has been identified in ESCs and iPSCs and may enhance passive demethylation in that DNMT1 does not recognize 5hmC as a substrate for replication [19, 77]. Nevertheless, in view of recent studies there are several inconsistencies with this theory [19, 78]. First, Tet expression is subject to Oct4 regulation, while global demethylation occurs before Oct4 activation [19]. Second, Tet depletion results in no effect on Oct4, Sox2, and Nanog expression in ESCs [19]. Last, Tet depletion in 2-cell-stage mouse embryos affects trophectoderm development and related developmental signaling, a stage much later than the zygotic demethylation event [77]. According to these unsolved questions, the involvement of an active mechanism in global demethylation remains elusive.

8.6 Conclusion

SCR demethylation is a forced epigenetic reprogramming mechanism comparable to zygotic demethylation. As shown in Fig. 8.1, this forced mechanism bypasses germline demethylation and directly induces ESC-like iPSC formation without imprint erasure and reestablishment. As a result, there are three major defects in the SCR-induced iPSCs. First, preserving parental imprints is beneficial for generating patient-friendly iPSCs or tissue clones, yet is not useful for creating universally compatible iPSCs for urgent treatments. It often takes several weeks or months to generate iPSCs from patients. Second, some parental germline elements required for normal zygotic development are not presented in iPSCs, such as paternal protamines and maternal oocyte transcripts. Loss of these parental elements may cause some developmental abnormalities in iPSCs. Last, due to lack of normal induction from fertilization, iPSCs often require a compensatory stimulation from oncogene activities (i.e., c-Myc and Klf4) to increase their survival rate, which may also lead to high tumorigenicity. Conceivably, it is not surprising that some researchers consider the four-factor-induced iPSCs as possible cancer stem cells; however, this problem can be corrected if we understand the mechanism how natural embryogenesis controls the ESC pluripotency without leading to tumorigenicity. For instance, we have recently found that miR-302 not only silences multiple epigenetic regulators to induce iPSC formation but also suppresses both cyclin E-CDK2 and cyclin D-CDK4/6 pathways to attenuate iPSC proliferation, a safeguard mechanism to prevent tumor formation during reprogramming [24, 25]. Additionally, our studies pinpoint that miR-302-induced SCR and cell cycle attenuation are two parallel events, acting together at almost the same miR-302 concentration to induce iPSC formation while preventing stem cell tumorigenicity [24]. Thus, by mimicking the functional role of miR-302, we may overcome the tumorigenicity problem of iPSCs, one of the major roadblocks hindering the development of stem cell therapy.

Global demethylation is the first sign of SCR required for the epigenetic reprogramming of somatic cell nuclei to form pluripotent stem cells [10]. Through deciphering the targeted epigenetic genes of miR-302, the most abundant miRNA family in hESCs, we revealed for the first time the mechanism underlying global demethylation in iPSCs, as shown in Fig. 8.5. Human somatic cells can be reprogrammed to iPSCs when the cellular miR-302 expression is beyond 1.1–1.3 folds of the level found in blastocyst-derived hESCs, as determined by northern blot analysis [23, 25]. This reprogramming mechanism functions through miR-302-mediated silencing of six important epigenetic regulators, AOF2 (KDM1, LSD1), AOF1, MECP1-p66, MECP2, DNMT1, and HDAC2, to cause global demethylation and the subsequent activation of Oct4-Sox2-Nanog-Lin28 expression for iPSC induction. Notably, silencing AOF2 also suppresses DNMT1 activity and further enhances global demethylation during SCR. Resupplementing AOF2 and/or DNMT1 into mirPSCs can partially disrupt global demethylation and induces cell differentiation, suggesting that concurrent silencing of these epigenetic targets is essential for the onset of global demethylation. Given that miR-302 is highly expressed in both hESCs and iPSCs, our findings have established a novel link between zygotic reprogramming and SCR, providing an epigenetic regulation mechanism responsible for global demethylation in both events.

Our recent studies have demonstrated a passive DNA demethylation pathway responsible for preventing the replication of somatic methylation during iPSC divisions [25]. On the other hand, the same findings also indicated that another active DNA demethylation pathway may be involved in the removal of the original somatic methylation sites leftover in the hemimethylated genomes of some iPSCs. AID was found to be a possible player in this active mechanism. The next questions are how miR-302-mediated gene silencing stimulates this active mechanism and what factors may be involved in addition to AID? Since there are no parental germline elements to support this active mechanism like the zygotic demethylation in early embryos, the demethylation of both parental genomes likely occurs simultaneously in iPSCs. Or, is it possible for the passive mechanism to carry out such global demethylation in both parental genomes? Alternatively, since we have frequently observed a small number of partially reprogrammed and apoptotic cells in the colonies of the first iPSC generation, it is possible that the iPSCs containing such a hemimethylated genome may be partially reprogrammed and eventually degraded by apoptosis. A thorough comparison between a single somatic cell and its iPSC methylation maps along the process of SCR may help to answer these questions. Based on the current understanding, global demethylation often occurs under a condition similar to early zygotic development. Due to this similarity, the epigenetic reprogramming mechanism of iPSCs may also serve as a comparable model for studying the zygotic demethylation occurring in early embryos.

Using the forced expression of miR-302 or four defined factors, we now can manually induce SCR to replace the natural two step germline-zygotic reprogramming mechanism for producing pluripotent stem cells. Although this artificial induction method is convenient, many parental elements are missing in the induced iPSCs. In oocyte cytoplasm, there are several miR-302 homologs that are abundantly expressed and may interact with each other to provide a similar or better reprogramming effect than miR-302, yet their functions remain unknown. As there is no transcription activity in mature oocytes and one-cell-stage zygotes, post-transcriptional mechanisms are essential for the natural formation of pluripotency in ESCs [79]. Our current finding of miR-302-mediated global demethylation may just reflect such a common functionality for this group of homologous miRNAs. Nowadays, overwhelming stem cell studies are focused on the effects of protein–protein interactions, while the regulatory mechanism of miRNAs on ESC-associated proteins is largely overlooked. Although some large-scale microarray and computer screening have been recently performed to predict the reprogramming-associated miRNAs, only a few of them can provide a thoroughly experimental study in individual miRNA function based on the dose-dependent feature of miRNA regulation. After our efforts in this direction, a step-wise protocol for studying the miRNA-mediated reprogramming mechanisms has been established in hope that more stem cell researchers may take the advantage of modern miRNA technologies. Given that SCR-associated global demethylation is modulated by a miRNA rather than protein, future studies using combined miRNA and iPSC technologies for improving stem cell applications is highly expected.

References

1. Hajkova P, Erhardt S, Lane N, Haaf T, El-Maarri O, Reik W, Walter J, Surani MA (2002) Epigenetic reprogramming in mouse primordial germ cells. Mech Dev 117:15–23
2. Szabó PE, Mann JR (1995) Biallelic expression of imprinted genes in the mouse germ line: implications for erasure, establishment, and mechanisms of genomic imprinting. Genes Dev 9:1857–1868
3. Mayer W, Niveleau A, Walter J, Fundele R, Haaf T (2000) Demethylation of the zygotic paternal genome. Nature 403:501–502
4. Santos F, Hendrich B, Reik W, Dean W (2002) Dynamic reprogramming of DNA methylation in the early mouse embryo. Dev Biol. 241:172–182
5. Stöger R, Kubicka P, Liu CG, Kafri T, Razin A, Cedar H, Barlow DP (1993) Maternal-specific methylation of the imprinted mouse Igf2r locus identifies the expressed locus as carrying the imprinting signal. Cell 73:61–71
6. Tremblay KD, Saam JR, Ingram RS, Tilghman SM, Bartolomei MS (1995) A paternal-specific methylation imprint marks the alleles of the mouse H19 gene. Nat Genet 9:407–413
7. Takahashi K, Yamanaka S (2006) Induction of pluripotent stem cells from mouse embryonic and adult fibroblast cultures by defined factors. Cell 126:663–676
8. Chamberlain SJ, Chen PF, Ng KY, Bourgois-Rocha F, Lemtiri-Chlieh F, Levine ES, Lalande M (2010) Induced pluripotent stem cell models of the genomic imprinting disorders Angelman and Prader-Willi syndromes. Proc Natl Acad Sci U S A 107:17668–17673
9. Kim K, Doi A, Wen B, Ng K, Zhao R, Cahan P, Kim J, Aryee MJ, Ji H, Ehrlich LI, Yabuuchi A, Takeuchi A, Cunniff KC, Hongguang H, McKinney-Freeman S, Naveiras O, Yoon TJ, Irizarry RA, Jung N, Seita J, Hanna J, Murakami P, Jaenisch R, Weissleder R, Orkin SH, Weissman IL, Feinberg AP, Daley GQ (2010) Epigenetic memory in induced pluripotent stem cells. Nature 467:285–290
10. Simonsson S, Gurdon J (2004) DNA demethylation is necessary for the epigenetic reprogramming of somatic cell nuclei. Nat Cell Biol 6:984–990
11. Carlson LL, Page AW, Bestor TH (1992) Properties and localization of DNA methyltransferase in preimplantation mouse embryos: implications for genomic imprinting. Genes Dev 6:2536–2541
12. Bestor TH, The DNA (2000) Methyltransferases of mammals. Hum Mol Genet 9:2395–2402
13. Vassena R, Dee Schramm R, Latham KE (2005) Species-dependent expression patterns of DNA methyltransferase genes in mammalian oocytes and preimplantation embryos. Mol Reprod Dev 72:430–436
14. Monk M, Boubelik M, Lehnert S (1987) Temporal and regional changes in DNA methylation in the embryonic, extraembryonic and germ cell lineages during mouse embryo development. Development 99:371–382
15. Hirasawa R, Chiba H, Kaneda M, Tajima S, Li E, Jaenisch R, Sasaki H (2008) Maternal and zygotic Dnmt1 are necessary and sufficient for the maintenance of DNA methylation imprints during preimplantation development. Genes Dev 22:1607–1616
16. Oswald J, Engemann S, Lane N, Mayer W, Olek A, Fundele R, Dean W, Reik W, Walter J (2000) Active demethylation of the paternal genome in the mouse zygote. Curr Biol 10:475–478
17. Weaver JR, Susiarjo M, Bartolomei MS (2009) Imprinting and epigenetic changes in the early embryo. Mamm Genome 20:532–543. doi:10.1007/s00335-009-9225-2
18. Feng S, Jacobsen SE, Reik W (2010) Epigenetic reprogramming in plant and animal development. Science 330:622–627
19. Koh KP, Yabuuchi A, Rao S, Huang Y, Cunniff K, Nardone J, Laiho A, Tahiliani M, Sommer CA, Mostoslavsky G, Lahesmaa R, Orkin SH, Rodig SJ, Daley GQ, Rao A (2011) Tet1 and tet2 regulate 5-hydroxymethylcytosine production and cell lineage specification in mouse embryonic stem cells. Cell Stem Cell 8:200–213
20. Meehan RR, Lewis JD, McKay S, Kleiner EL, Bird AP (1989) Identification of a mammalian protein that binds specifically to DNA containing methylated CpGs. Cell 58:499–507

21. Tate P, Skarnes W, Bird A (1996) The methyl-CpG binding protein MeCP2 is essential for embryonic development in the mouse. Nat Genet 12:205–208
22. Hendrich B, Bird A (1998) Identification and characterization of a family of mammalian methyl-CpG binding proteins. Mol Cell Biol 18:6538–6547
23. Lin SL, Chang D, Chang-Lin S, Lin CH, Wu DTS, Chen DT, Ying SY (2008) Mir-302 reprograms human skin cancer cells into a pluripotent ES-cell-like state. RNA 14:2115–2124
24. Lin SL, Chang D, Ying SY, Leu D, Wu DTS (2010) MicroRNA miR-302 inhibits the tumorigenecity of human pluripotent stem cells by coordinate suppression of CDK2 and CDK4/6 cell cycle pathways. Cancer Res 70:9473–9482
25. Lin SL, Chang D, Lin CH, Ying SY, Leu D, Wu DTS (2011) Regulation of somatic cell reprogramming through inducible mir-302 expression. Nucleic Acids Res 39:1054–1065
26. Suh MR, Lee Y, Kim JY, Kim SK, Moon SH, Lee JY, Cha KY, Chung HM, Yoon HS, Moon SY, Kim VN, Kim KS (2004) Human embryonic stem cells express a unique set of microRNAs. Dev Biol 270:488–498
27. Wilson KD, Venkatasubrahmanyam S, Jia F, Sun N, Butte AJ, Wu JC (2009) MicroRNA profiling of human-induced pluripotent stem cells. Stem Cells Dev 18:749–758
28. Li Z, Yang CS, Nakashima K, Rana TM (2011) Small RNA-mediated regulation of iPS cell generation. EMBO J 30:823–834
29. Puca AA, Daly MJ, Brewster SJ, Matise TC, Barrett J, Shea-Drinkwater M, Kang S, Joyce E, Nicoli J, Benson E, Kunkel LM, Perls T (2000) A genome-wide scan for linkage to human exceptional longevity identifies a locus on chromosome 4. Proc Natl Acad Sci USA 98:10505–10508
30. Lin SL, Chang D, Wu DY, Ying SY (2003) A novel RNA splicing-mediated gene silencing mechanism potential for genome evolution. Biochem Biophys Res Commun 310:754–760
31. Lin SL, Ying SY (2008) Role of mir-302 microRNA family in stem cell pluripotency and renewal. In: Ying SY (ed) Current perspectives in microRNAs. Springer Publishers press, New York, pp 167–185
32. Barroso-delJesus A, Romero-López C, Lucena-Aguilar G, Melen GJ, Sanchez L, Ligero G, Berzal-Herranz A, Menendez P (2008) Embryonic stem cell-specific miR302-367 cluster: human gene structure and functional characterization of its core promoter. Mol Cell Biol 28:6609–6619
33. Ying SY, Lin SL (2004) Intron-derived microRNAs—fine tuning of gene functions. Gene 342:25–28
34. Li Y, Pei J, Xia H, Ke H, Wang H, Tao W (2003) Lats2, a putative tumor suppressor, inhibits G1/S transition. Oncogene 22:4398–4405
35. Lin SL, Ying SY (2006) Gene silencing in vitro and in vivo using intronic microRNAs. In: Ying SY (ed) MicroRNA protocols. Humana press, Totowa, pp 295–312
36. Lin SL, Kim H, Ying SY (2008) Intron-mediated RNA interference and microRNA (miRNA). Front Biosci 13:2216–2230
37. Lin SL, Chang D, Ying SY (2005) Asymmetry of intronic pre-microRNA structures in functional RISC assembly. Gene 356:32–38
38. Danin-Kreiselman M, Lee CY, Chanfreau G (2003) RNAse III-mediated degradation of unspliced pre-mRNAs and lariat introns. Mol Cell 11:1279–1289
39. Yi R, Qin Y, Macara IG, Cullen BR (2003) Exportin-5 mediates the nuclear export of pre-microRNAs and short hairpin RNAs. Genes Dev 17:3011–3016
40. Lund E, Guttinger S, Calado A, Dahlberg JE, Kutay U (2004) Nuclear export of microRNA precursors. Science 303:95–98
41. Zhang G, Taneja KL, Singer RH, Green MR (1994) Localization of pre-mRNA splicing in mammalian nuclei. Nature 372:809–812
42. Lewis BP, Green RE, Brenner SE (2003) Evidence for the widespread coupling of alternative splicing and nonsense-mediated mRNA decay in humans. Proc Natl Acad Sci U S A 100:189–192

43. Grimm D, Streetz KL, Jopling CL, Storm TA, Pandey K, Davis CR, Marion P, Salazar F, Kay MA (2006) Fatality in mice due to oversaturation of cellular microRNA/short hairpin RNA pathways. Nature 441:537–541
44. Lin SL, Chang SJE, Ying SY (2006) Transgene-like animal model using intronic microRNAs. Methods Mol Biol 342:321–334
45. Lin SL, Chang SJE, Ying SY (2006) First in vivo evidence of microRNA-induced fragile X mental retardation syndrome. Mol Psychiatry 11:616–617
46. Lin SL, Ying SY (2009) Recent progress of the polymerase II-mediated intronic microRNA expression system. In: Gaur RK, Rossi J (eds) Regulation of gene expression by small RNAs. CRC press, Florida, pp 275–300
47. Zhou HSL, Xia XG, Xu Z (2005) An RNA polymerase II construct synthesizes short hairpin RNA with a quantitative indicator and mediates high efficient RNAi. Nucl Acid Res 33:e62
48. Chung KH, Hart CC, Al-Bassam S, Avery A, Taylor J, Patel PD, Vojtek AB, Turner DL (2006) Polycistronic RNA polymerase II expression vectors for RNA interference based on BIC/miR-155. Nucl Acid Res 34:e53
49. Yu J, Vodyanik MA, Smuga-Otto K, Antosiewicz-Bourget J, Frane JL, Tian S, Nie J, Jonsdottir GA, Ruotti V, Stewart R, Slukvin II, Thomson JA (2007) Induced pluripotent stem cell lines derived from human somatic cells. Science 318:1917–1920
50. Wernig M, Meissner A, Foreman R, Brambrink T, Ku M, Hochedlinger K, Bernstein BE, Jaenisch R (2007) In vitro reprogramming of fibroblasts into a pluripotent ES-cell-like state. Nature 448:318–324
51. Cowan CA, Klimanskaya I, McMahon J, Atienza J, Witmyer J, Zucker JP, Wang S, Morton CC, McMahon AP, Powers D, Melton DA (2004) Derivation of embryonic stem-cell lines from human blastocysts. N Engl J Med 350:1353–1356
52. Becker KA, Ghule PN, Therrien JA, Lian JB, Stein JL, van Wijnen AJ, Stein GS (2006) Self-renewal of human embryonic stem cells is supported by a shortened G1 cell cycle phase. J Cell Physiol 209:883–893
53. Tesar PJ, Chenoweth JG, Brook FA, Davies TJ, Evans EP, Mack DL, Gardner RL, McKay RD (2007) New cell lines from mouse epiblast share defining features with human embryonic stem cells. Nature 448:196–199
54. Brons IG, Smithers LE, Trotter MW, Rugg-Gunn P, Sun B, Chuva de Sousa Lopes SM, Howlett SK, Clarkson A, Ahrlund-Richter L, Pedersen RA, Vallier L (2007) Derivation of pluripotent epiblast stem cells from mammalian embryos. Nature 448:191–195
55. Hanna J, Saha K, Pando B, van Zon J, Lengner CJ, Creyghton MP, van Oudenaarden A, Jaenisch R (2009) Direct cell reprogramming is a stochastic process amenable to acceleration. Nature 462:595–601
56. Rowland BD, Bernards R, Peeper DS (2005) The KLF4 tumour suppressor is a transcriptional repressor of p53 that acts as a context-dependent oncogene. Nat Cell Biol 7:1074–1082
57. Jiang J, Chan YS, Loh YH, Cai J, Tong GQ, Lim CA, Robson P, Zhong S, Ng HH (2008) A core Klf circuitry regulates self-renewal of embryonic stem cells. Nat Cell Biol 10:353–360
58. Kim J, Chu J, Shen X, Wang J, Orkin SH (2008) An extended transcriptional network for pluripotency of embryonic stem cells. Cell 132:1049–1061
59. Nandan MO, Yang VW (2009) The role of Krüppel-like factors in the reprogramming of somatic cells to induced pluripotent stem cells. Histol Histopathol 24:1343–1355
60. Card DA, Hebbar PB, Li L, Trotter KW, Komatsu Y, Mishina Y, Archer TK (2008) Oct4/Sox2-regulated miR-302 targets cyclin D1 in human embryonic stem cells. Mol Cell Biol 28:6426–6438
61. Marson A, Levine SS, Cole MF, Frampton GM, Brambrink T, Johnstone S, Guenther MG, Johnston WK, Wernig M, Newman J, Calabrese JM, Dennis LM, Volkert TL, Gupta S, Love J, Hannett N, Sharp PA, Bartel DP, Jaenisch R, Young RA (2008) Connecting microRNA genes to the core transcriptional regulatory circuitry of embryonic stem cells. Cell 134:521–533
62. Lee MG, Wynder C, Cooch N, Shiekhattar R (2005) An essential role for CoREST in nucleosomal histone 3 lysine 4 demethylation. Nature 437:432–435. doi:10.1038/nature04021

63. Lee MG, Wynder C, Schmidt DM, McCafferty DG, Shiekhattar R (2006) Histone H3 lysine 4 demethylation is a target of nonselective antidepressive medications. Chem Biol 13:563–567
64. Ciccone DN, Su H, Hevi S, Gay F, Lei H, Bajko J, Xu G, Li E, Chen T (2009) KDM1B is a histone H3K4 demethylase required to establish maternal genomic imprints. Nature 461:415–418
65. Wang J, Hevi S, Kurash JK, Lei H, Gay F, Bajko J, Su H, Sun W, Chang H, Xu G, Gaudet F, Li E, Chen T (2009) The lysine demethylase LSD1 (KDM1) is required for maintenance of global DNA methylation. Nat Genet 41:125–129
66. Bhutani N, Brady JJ, Damian M, Sacco A, Corbel SY, Blau HM (2010) Reprogramming towards pluripotency requires AID-dependent DNA demethylation. Nature 463:1042–1047
67. Popp C, Dean W, Feng S, Cokus SJ, Andrews S, Pellegrini M, Jacobsen SE, Reik W (2010) Genome-wide erasure of DNA methylation in mouse primordial germ cells is affected by AID deficiency. Nature 463:1101–1105. doi:10.1038/nature08829
68. Rosa A, Spagnoli FM, Brivanlou AH (2009) The miR-430/427/302 family controls mesendodermal fate specification via species-specific target selection. Dev Cell 16:517–527
69. Rosa A, Brivanlou AH (2011) A regulatory circuitry comprised of miR-302 and the transcription factors OCT4 and NR2F2 regulates human embryonic stem cell differentiation. EMBO J 30:237–248
70. Wilmut I, Schnieke AE, McWhir J, Kind AJ, Campbell KH (1997) Viable offspring derived from fetal and adult mammalian cells. Nature 385:810–813
71. Wakayama T, Perry AC, Zuccotti M, Johnson KR, Yanagimachi R (1998) Full-term development of mice from enucleated oocytes injected with cumulus cell nuclei. Nature 394:369–374
72. Conticello SG, Langlois MA, Yang Z, Neuberger MS (2007) DNA deamination in immunity: AID in the context of its APOBEC relatives. Adv Immunol 94:37–73
73. Morgan HD, Dean W, Coker HA, Reik W, Petersen-Mahrt SK (2004) Activation-induced cytidine deaminase deaminates 5-methylcytosine in DNA and is expressed in pluripotent tissues: implications for epigenetic reprogramming. J Biol Chem 279:52353–52360
74. Morgan HD, Santos F, Green K, Dean W, Reik W (2005) Epigenetic reprogramming in mammals. Hum Mol Genet 1:R47–R58
75. Rai K, Huggins IJ, James SR, Karpf AR, Jones DA, Cairns BR (2008) DNA demethylation in zebrafish involves the coupling of a deaminase, a glycosylase, and gadd45. Cell 135:1201–1212
76. Privat E, Sowers LC (1996) Photochemical deamination and demethylation of 5-methylcytosine. Chem Res Toxicol 9:745–750
77. Valinluck V, Sowers LC (2007) Endogenous cytosine damage products alter the site selectivity of human DNA maintenance methyltransferase DNMT1. Cancer Res 67:946–950
78. Ito S, D'Alessio AC, Taranova OV, Hong K, Sowers LC, Zhang Y (2010) Role of Tet proteins in 5mC to 5hmC conversion, ES-cell self-renewal and inner cell mass specification. Nature 466:1129–1133
79. Tang F, Kaneda M, O'Carroll D, Hajkova P, Barton SC, Sun YA, Lee C, Tarakhovsky A, Lao K, Surani MA (2007) Maternal microRNAs are essential for mouse zygotic development. Genes Dev 21:644–648

Chapter 9
Microtechnological Approaches in Stem Cell Science

Faramarz Edalat, Jae Min Cha, Hojae Bae, Sam Manoucheri, Sang Bok Kim and Ali Khademhosseini

Abstract Stem cells are versatile materials in biomedical research with applications in regenerative medicine, drug discovery, biosensors, and developmental and disease models. Many of the current technologies used by biologists are restricted to a macroscale level of control, which has impeded precise manipulation of spatial and temporal features of the cellular microenvironment. Microtechnologies developed in engineering are powerful tools for addressing existing challenges in stem cell biology by permitting control of cellular behaviors at scales matching those of biological processes. Many of these approaches demonstrate feasibility in opening a new era in the therapeutic application of stem cells. In this chapter, we introduce biologically relevant microtechnologies and present their applications in stem cell science.

Abbreviations

μCP	Microcontact printing
AFM	Atomic force microscope
BMP	Bone morphogenetic protein

F. Edalat and J. M. Cha contributed equally to this work.

F. Edalat · J. M. Cha · H. Bae · S. Manoucheri · S. B. Kim · A. Khademhosseini (✉)
Department of Medicine, Brigham and Women's Hospital, Center for Biomedical Engineering, Harvard Medical School, Cambridge, MA, USA
e-mail: alik@rics.bwh.harvard.edu

F. Edalat · J. M. Cha · H. Bae · S. Manoucheri · S. B. Kim · A. Khademhosseini
Harvard-MIT Division of Health Sciences and Technology,
Massachusetts Institute of Technology, Cambridge, MA, USA

A. Khademhosseini
Wyss Institute for Biologically Inspired Engineering,
Harvard University, Boston, MA, USA

H. Baharvand and N. Aghdami (eds.), *Advances in Stem Cell Research*,
Stem Cell Biology and Regenerative Medicine, DOI: 10.1007/978-1-61779-940-2_9,

CAD	Computer-aided design
DM	Differentiation medium
DPN	Dip pen nanolithography
EB	Embryoid body
EBL	Electron beam lithography
ECM	Extracellular matrix
ESC	Embryonic stem cell
FGF	Fibroblast growth factor
GelMA	Gelatin methacrylate
GM	Growth medium
HA	Hyaluronic acid
HepG2	Human hepatocellular carcinoma cell line
HUVEC	Human umbilical vein endothelial cells
MIMIC	Micromolding in capillaries
MSC	Mesenchymal stem cell
nCP	Nanocontact printing
NIL	Nanoimprint lithography
PCL	Polycaprolactone
PCR	Polymerase chain reaction
PDMS	Poly(dimethylsiloxane)
PEG	Poly(ethylene glycol)
PLGA	Poly-L-lactic/glycolic acid
PLLA	Poly-L-lactic acid
PMMA	Poly(methyl methacrylate)
PPL	Polymer pen lithography
PS	Polystyrene
RT	Reverse transcriptase
SiO_2	Silicon dioxide
SMC	Smooth muscle cell
SPL	Scanning probe lithography
STM	Scanning tunneling microscope
UV	Ultraviolet
Wnt	Wingless-type MMTV integration site family

9.1 Introduction

Human pluripotent stem cells are capable of generating $\sim$200 different cell types existing in the body. In contrast, adult stem cells are multipotent, yet provide a more efficient differentiation pathway over embryonic and induced pluripotent stem cells. These capabilities have brought stem cell science to the forefront of regenerative medicine, drug discovery, biosensors, and studying disease [1, 2].

However, the fruition of the potential benefits of stem cells has been hampered by the difficulty in controlling their self-renewal and differentiation [3]. One approach to assert control over these cells is by investigating and understanding the input provided by their environment. Indeed, stem cells are situated in unique microenvironments, or niches, that convey dynamic biochemical and biophysical signals [4, 5]. These signals are provided by the extracellular matrix (ECM), cytokines and metabolites, and support cells [6]. Stem cells interact with these niche components at the nano- and microscale dimensions, leading to highly organized cellular processes, such as proliferation, apoptosis, migration, and differentiation. Currently, conventional cell culture systems lack the means to recapitulate the minuscule details of the niche and suffer from low reproducibility, low-throughput, and high cost. Microtechnologies offer an innovative route to emulating and manipulating biological structures to properly assess and direct stem cell fate in a reproducible, high-throughput, and cost-effective manner. In this chapter, we introduce microscale technologies relevant for studying biological phenomena and then discuss applications of these technologies in directing stem cell fate.

9.2 Microtechnology: A Primer for Stem Cell Biologists

Microtechnology refers to the fabrication and processing of structures on the micron and sub-micron scale. This technology was first developed by the microelectronics industry to increase the density of transistors in an integrated circuit as predicted by Moore's Law [7]. Since then, there has been an explosion of new microfabrication methods that have been applied to biomedical research. These methods have come to have an increasing influence in stem cell biology given that biological structures have nano- and micron-scale features [8, 9]. For instance, ECM structures are on the order of nanometers in diameter, whereas eukaryotic cells have lengths of 10–100 μm [10]. Thus, microfabrication has enabled bioengineers and stem cell biologists to have increasing control in their studies and design systems in a more physiologically relevant manner. Integrating these technologies with the realm of biology requires achieving several design criteria including accessibility, low cost, high resolution, reproducibility, and large surface area ($\sim 1\ \text{cm}^2$) exposure. In the following section, a set of microfabrication technologies known as lithography that has been applied to stem cell science, will be introduced and compared (Table 9.1). Emerging technologies that have not yet been utilized in biological sciences but have great potential to make an impact will also be mentioned.

Table 9.1 Comparison of various microfabrication methods

Method	Materials	Resolution (nm)	Cost	Access	Through-put	Surface area	Biological applications
Photo-lithography	SiO_2, PMMA, PS, PLLA, PDMS, PEG, HA	45 [159]	++	+	+	++	2D polymer/cell patterning [16] 3D cell-laden hydrogels [17, 18]
Electron beam lithography	SiO_2, PMMA, PLGA, PS, PCL	3–5 30–40 (large area) [23]	++	+	+	+	Topography [24, 25]
Focus ion beam lithography	PMMA	20 [27]	++	+	+	+	Protein/nucleic acid patterning [160]
Soft lithography	PDMS, PCL, PS, PLLA	Typically 1μm [30] 42 (nCP) [35]	+	++	++	++	Shear stress[38] Protein/cell patterning [46–48, 82, 83, 93] Topography [42–45, 99] GF concentration gradients [49, 50] Bioanalytics [51–53, 115, 123, 124] Cell-cell studies [103, 104] Microbioreactor [132–136]
Nanoimprint lithography	PMMA, PLLA, PCL, PS	2 [55]	+	++	++	++	Protein patterning [57] Topography [56, 58]
Dip pen lithography	Synthetic polymers, peptides, proteins, nucleic acids	15 [62]	++	+	+/++	+/++	Peptide, protein, nucleic acid patterning [69–72] Nanoarrays [73, 74]
Ejecting methods	Synthetic polymers, proteins, cells	Hundreds of nm to <10μm [12]	+	++	++	++	Protein/cell patterning [77, 78] 3D tissue construction [79, 80] Microarrays [85–87]

+: low; ++: high; *HA*, hyaluronic acid; *nCP*, nanocontact printing; *PCL*, polycaprolactone; *PDMS*, poly(dimethylsiloxane); *PEG*, poly(ethylene glycol); *PLGA*, poly-L-lactic/glycolic acid; *PLLA*, poly-L-lactic acid; *PMMA*, poly(methyl methacrylate); *PS*, polystyrene; SiO_2, silicon dioxide

9.2.1 Conventional Lithography

Lithography is the term used for the transfer of fine-scale patterns onto a substrate. Conventional lithography refers to a set of techniques that use high energy irradiation, which encompasses photo- and scanning beam lithography, two of the earliest microtechnologies used in microelectronics [11].

9.2.1.1 Photolithography

Photolithography is a technique that uses light to generate structures on a surface. This method is based on the selective ultraviolet (UV) irradiation of a photoreactive material (a monomer, oligomer, or polymer), called a photoresist, pre-coated on a substrate such as a silicon wafer or glass (Fig. 9.1a) [12]. This exposure causes cross-linking, polymerization, or degradation of irradiated material. Subsequently, if needed, dissolution of selected areas can be achieved by developing in an organic solvent [13]. The selective UV exposure is done by illuminating the light through a mask with features designed using a computer-aided design (CAD) software. Mask-free methods such as multi-beam interference [14] and two-photon absorption [15] techniques are recent techniques of exposing specified areas of the photoresist to generate 3D structures as opposed to the 2D patterns created using masks. The resulting patterned structure can then be used on its own or as a bas-relief master for molding other polymers. Numerous photoreactive synthetic and biological polymers have been used in photolithography to generate patterned structures, either as 2D cellular scaffolds or as cell-encapsulating materials to study a range of stem cell behaviors [16–18].

Conventional photolithography is advantageous given its reproducibility, patterning over large surface areas, and relatively high resolution. If higher resolutions are needed, however, a shorter wavelength light (extreme UV lithography), unconventional masks, or different photoresists must be used, all of which are associated with higher costs. Also, given that UV light, photoinitiators, and/or organic solvents may be present in the fabrication process, only biological entities compatible with these processing steps should be used [12]. Another disadvantage is the difficulty of access to microfabrication facilities. However, shared foundries have enabled increasing access to scientists without such facilities.

9.2.1.2 Scanning Beam Lithography

One of the restrictions of conventional photolithography is the diffraction constraints of UV light that results in lower resolutions than desired. To overcome this problem, lithographic techniques that use particles with shorter wavelengths are employed. Scanning beam lithography is a set of conventional techniques that use high energy particle beams with short wavelengths, such as electrons or ions, to impart patterns in

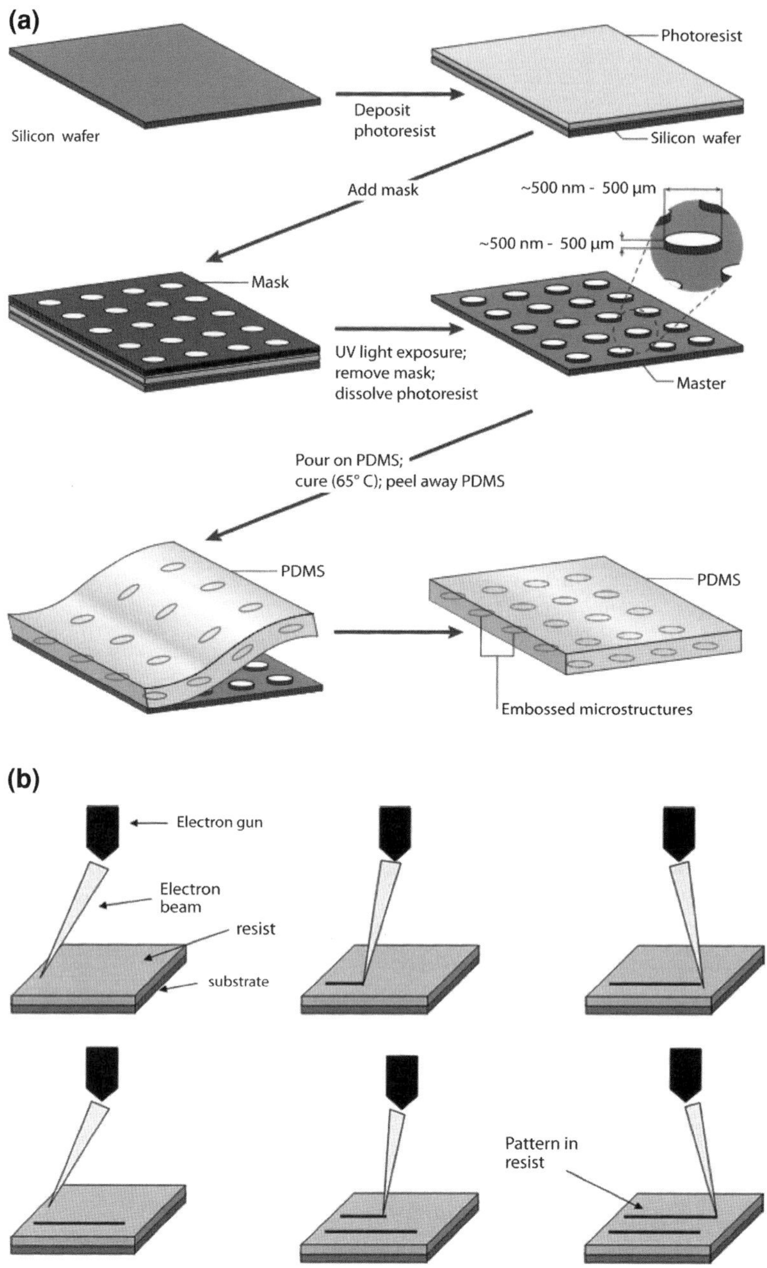

appropriate resists. In electron beam lithography (EBL), the selected areas of electron-sensitive resist are exposed to electron beams resulting in their differential solubility in solvents (Fig. 9.1b) [19]. Poly(methyl methacrylate) is the most widely

◀**Fig. 9.1** Conventional lithography methods. **a** Photolithography: silicon wafers are used as substrates for spin-coating a photoresist layer, followed by mask placement and UV light illumination. The unexposed areas of the photoresists dissolve in a solvent and wash away resulting in a master. The master can then be used as a mold by pouring and curing PDMS polymer resulting in embossed microstructures [30]. Adapted by permission from Macmillan Publishers Ltd: [Nature Reviews Microbiology], copyright (2007). **b** Electron beam lithography: An electron gun exposes selected areas of spin-coated electron-sensitive resist to electron beams [19]. Reprinted with permission from Springer Science + Business Media: [Annals of Biomedical Engineering], copyright (2006)

used resist; however, other synthetic polymers [20], proteins [21], inorganic materials [22], and self-assembled monolayers can be patterned using EBL. The patterns are programmable and resolutions of 3–5 nm can be obtained, although for larger surface areas, the resolution decreases to 30–40 nm [23]. These patterns can be used by themselves or as masters to study the effects of nanotopography on stem cells [24, 25].

Focused ion beam lithography is another method that uses heavy ions such as gallium to selectively construct a pattern by modification of a resist layer, sputtering, or deposition on a substrate [26]. This method yields topographies with a resolution of approximately 20 nm and a smallest lateral dimension of 5 nm [27]. Scanning beam lithography's shortcomings are its high cost, slow processing speed, and low-throughput nature that have limited its widespread use.

9.2.2 Soft Lithography

Whereas scanning beam lithography suffers from low-throughput and high cost, soft lithography has been used by many biologists given its ease of use, low cost, and high-throughput. Soft lithography is a set of methods that use a soft flexible material, called an elastomer, to pattern micron and sub-micron scale structures or molecules on a surface (Fig. 9.2) [28]. Poly(dimethylsiloxane) (PDMS) is the most common elastomer used, however, other polymers, such as polyimides, polyurethanes, and natural polymers can also be used. PDMS is especially advantageous given its flexibility, biocompatibility, and permeability to gases such as oxygen and carbon dioxide [29]. Also, its transparency to light makes it ideal for real-time optical imaging. In soft lithography, molds generated by other lithographic techniques (e.g. photolithography) are used to emboss structures in the elastomer in a single step process. The structured elastomer can then be used as a tool in molding, printing, or embossing [30].

The various techniques in soft lithography include replica molding, nano- and microcontact printing (μCP), micromolding in capillaries (MIMIC), microtransfer molding, and microfluidics [30]. In replica molding, patterned elastomers are used as molds to create bas-relief structures in other polymers (Fig 9.2a). This technique can be used to fabricate microscale stencils (Fig. 9.2fi). Stencils are a thin layer of material (commonly a polymer) containing specified holes of desired

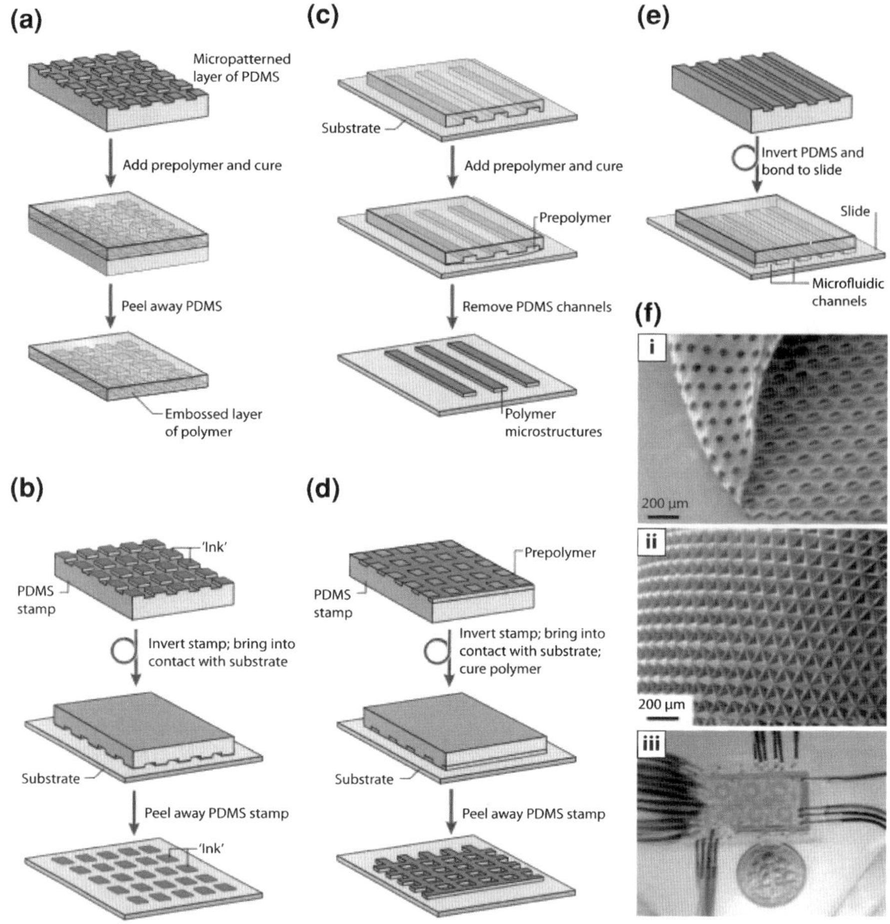

Fig. 9.2 Soft lithographic techniques. **a** Replica molding. **b** Microcontact printing. **c** Micromolding in capillaries. **d** Microtransfer molding. **e** Microfluidics. **f** Examples of soft lithographic products. **fi** Microscale stencils made from poly(dimethylsiloxane) via replica molding. **fii** An example of a distinct geometry that can be formed using replica molding. **fiii** A microfluidic chip used for microbial culture [30]. Adapted by permission from Macmillan Publishers Ltd: [Nature Reviews Microbiology], copyright (2007)

dimensions and shapes, that have been utilized in studying heterotypic cell–cell interactions [31]. μCP uses a topographically patterned elastomer as a stamp to transfer 'ink' to a surface (Fig 9.2b) [32]. The ink may consist of proteins, nucleic acids [33], or cell suspensions. μCP can routinely produce patterns with dimensions in the submicron range; however, higher resolutions are hampered by the lateral diffusion of ink molecules and loss of mechanical integrity of the soft elastomer at the sub-micrometer scale [34]. Nanocontact printing overcomes these obstacles by using stiffer elastomers and higher molecular weight inks to generate sub-100 nm scale features [35]. MIMIC and microtransfer molding techniques are similar in that gaps in the elastomeric structures are filled with prepolymer after and prior to conformal contact with a surface, respectively, followed by curing and peeling off the elastomer (Fig. 9.2c, d) [36, 37].

In microfluidics, devices are created by placing PDMS with embossed channels against glass to form closed channels (Fig. 9.2e, fiii). The manipulation of fluids inside these channels allows for studying the effect of shear force, generated by flowing fluids, on stem cells [38]. Given the micron-scale dimensions of microfluidic channels, flow is laminar and mixing occurs by diffusion [29]. This property has been utilized to generate precise biomolecular concentration gradients to mimic the spatial concentration gradients of growth factors and morphogens present in biological systems [39, 40]. Furthermore, these devices can be scaled-up by constructing multilayer microfluidic constructs [41].

Soft lithography has gained wide acceptance in stem cell biology given its reproducibility and ability to generate arrays of biomolecules. This microtechnology allows for construction of nanotopographies [42–45], precise control over cell adhesion and shape [46–48], and generation of biomolecular concentration gradients [49, 50] that has provided insight into stem cell behavior. Furthermore, microfluidics has spurred innovation of miniaturized bioanalytical tools, such as nanoliter-scale polymerase chain reaction (PCR) [51], single-molecule genomic sequencing [52], and picoliter-scale cell sorting [53] that will replace or come to act synergistically with current analytical methods.

9.2.3 Nanoimprint Lithography

While in soft lithography, molds are soft elastomers, in nanoimprint lithography (NIL) rigid molds are used to transfer nano- and microarchitectures to polymers. The polymers are cured either by cooling or UV-irradiation (Fig. 9.3a, b) [54]. In the former method, known as thermal-NIL, a silicon mold is pressed against a thermoplastic polymer at conditions above the glass-transition temperature followed by cooling below this limit to trap the pattern within the polymer. Alternatively in UV-NIL, also known as step-and-flash imprint lithography, a transparent rigid mold may be pressed against a photosensitive polymer precursor, followed by UV exposure to cross-link the prepolymer and transfer topographical features of the mold [12]. The remaining layer may be stripped by etching or another material can be added on the pattern by chemical vapor deposition. NIL can produce structures with features as low as 2 nm [55]. NIL has not made a considerable impact in the biological sciences; however, it has been used for nucleic acid manipulation [56], protein patterning [57], and cellular alignment [58].

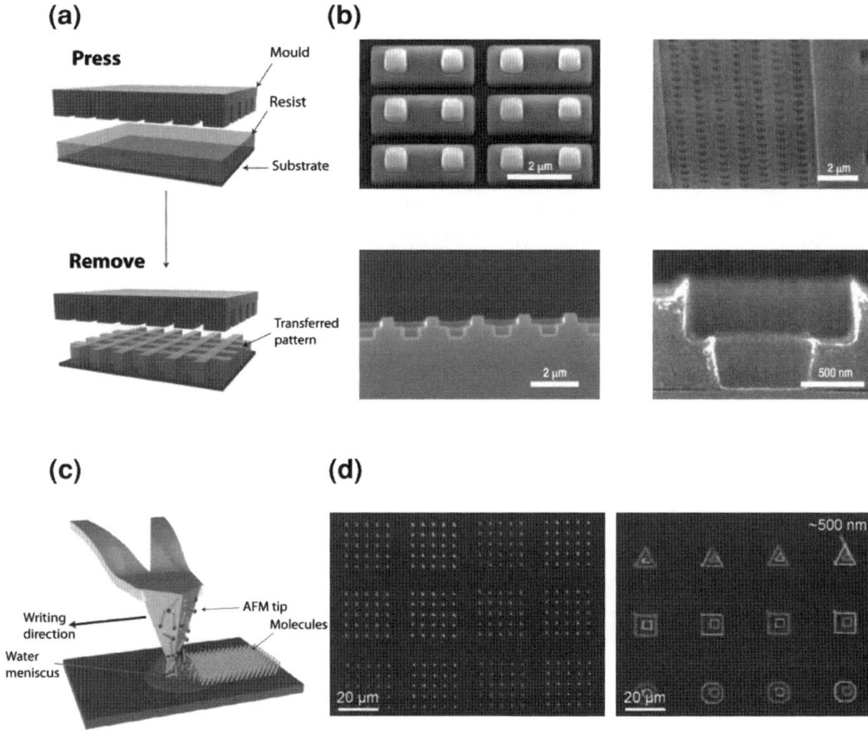

Fig. 9.3 Nanoimprint lithography (NIL) and dip pen nanolithography (DPN). **a** In NIL, a mold is pressed against the resist layer followed by thermal or UV curing [12]. **b** Scanning electron micrographs of structures obtained with this method are shown on the right [157]. Adapted with permission from American Institute of Physics: [Journal of Vacuum Science & Technology B], copyright (2006). **c** Dip pen nanolithography (DPN): an atomic force microscope probe is used to 'write' on a substrate using selected molecules as 'ink' [12]. Adapted by permission from Macmillan Publishers Ltd: [Nature Materials], copyright (2008). **d** Examples of parallel DPN showing patterns of two different fluorescently tagged proteins [64]. Adapted by permission from John Wiley and Sons: [Small], copyright (2008)

9.2.4 Direct-Write Methods

Direct-write methods involve the transfer of molecules from a probe or nozzle onto a substrate's surface. This can be accomplished by using scanning probe microscopes such as an atomic force microscope (AFM) or a scanning tunneling microscope (STM) to deposit various molecules, the process that is termed scanning probe lithography (SPL). Alternatively, the nozzles in inkjet printing or robotic deposition can be used to deposit materials on surfaces [12].

9.2.4.1 Dip Pen Lithography

Scanning probe lithography takes advantage of the nano-sized probe tip of AFM or STM to alter material structure down to the atomic or molecular level via modification, elimination, or deposition [59]. In traditional SPL such as nanoshaving or nanografting, the probe applies force on the material to remove a patterned layer [60]. However, these 'destructive' methods, similar to conventional lithographies, limit the number of usable substrates. Thus SPL has been expanded to be used as a direct-write tool in a 'constructive' manner by depositing materials onto surfaces, as in dip pen nanolithography (DPN) (Fig. 9.3c). In DPN, the probe acts as a 'pen' carrying molecules of interest or 'ink', and as it comes into contact with the substrate's surface, forms a meniscus. This meniscus acts as a conduit for the capillary transport of molecules onto the surface [61]. The movement of the probe is controlled via a computer and hence the pattern generated can be programmed [59]. Ultrahigh resolution can be obtained ranging from 15 nm to several microns [62]. DPN has surpassed the low-throughput nature of traditional SPL by parallelization and multiplexing using an array of probes that can simultaneously deposit different molecules on the same surface (Fig. 9.3d) [63, 64].

In recent years, other lithographic methods have been integrated with DPN to combine the strengths and reduce the limitations of each technique [65]. For instance, polymer pen lithography (PPL) combines microcontact printing with DPN, by using an array of pyramidal elastomeric tips that absorb ink and act as reservoirs to digitally print molecules onto a surface [66]. PPL can contain thousands to 10^8 tips that can produce patterns of 80 nm to 10 μm at a fraction of the cost of DPN [60]. In another method, termed beam pen lithography, an opaque metal coats the elastomeric tips except at its apex to direct light [67]. Since the tips are flexible, they can come in contact with the substrate surface and light is passed directly from the tip, acting as an aperture, onto the surface. This technique overcomes the diffraction constraints of traditional photolithography and is high-throughput.

DPN has been used to pattern a variety of molecules, such as synthetic polymers [68], peptides [69], proteins [70, 71], and nucleic acids [72]. In stem cell science, variations in stem cell populations and cell culture substrates can reduce a study's reproducibility. Hence, there is a dire need for the production of homogenous arrays of substrates. The introduction of parallel DPN (Fig. 9.3d) has allowed for the generation of reproducible micro- and nanoarrays of biomolecules to study cell-material interactions, cell adhesion, and stem cell differentiation [73, 74].

9.2.4.2 Ejecting Techniques

Ejecting methods employ a nozzle that extrudes ink onto a surface. Such methods include inkjet printing and robotic deposition. The patterns are digitally programmed and can be 2D or 3D. In 3D patterning, a layer-by-layer deposition of cells and matrices is accomplished for the formation of tissue-like structures [75].

The resolutions of these methods range from hundreds of nanometers for the robotic method to less than 10 μm for inkjet printing [12]. Advantages include preserving the structure and functionality of molecules transported, high-throughput, and compatibility over large surface areas.

The ejecting methods can be used to fabricate cellular scaffolds or form 2D patterns of synthetic polymers, proteins, and stem cells to study cell–matrix or cell-material interactions [76–78]. Also, with the advent of 3D deposition, cellular scaffolds [79] and complex biological structures can be fabricated to study cell–cell interactions and to engineer tissues and organs [80].

9.3 Applications of Microtechnology in Stem Cell Biology

As outlined above, microtechnological approaches can yield substrates with unique surface properties and features of differing geometries and sizes. These technologies have been applied to deconstruct the stem cell niche into its components. While simplistic, this approach allows a reductionist reasoning to untangle the complex molecular network that is at play in the niche. This makes it possible to examine the effects of stem cell niche's elements on an individual basis. In this next section, we present the use of microtechnology in studying cell-substrate and cell–cell interactions. Then, the roles of microfluidics in creating dynamic biochemical environments and analytical tools are highlighted. This section ends with a discussion on strategies for creating more complex tissue-like structures.

9.3.1 Controlling Cell-Substrate Interactions

Microfabrication can be used to generate substrates with desired 2D geometric patterns to induce a variety of cellular responses, such as proliferation [81], apoptosis [82], polarization [83], and differentiation [46–48]. In addition to controlling the 2D microenvironment, lithographic techniques can construct 3D nano- and microscale topographies on substrates to modulate cell behavior [84].

9.3.1.1 2D Biomolecular Patterns

Generation of spatial biochemical patterns can be achieved by deposition of cell adhesive molecules on a substrate's surface using a variety of microtechnologies. Soft lithography, in particular μCP, inkjet printing, and robotic deposition have emerged as versatile tools in creating 2D biomolecular patterns. Numerous cell adhesive molecules, both synthetic and natural, have been used to study cell-material or cell-ECM interactions. Given the large number of combinations of extracellular matrices and synthetic materials, and the problem of predicting their effects on stem cells, high-throughput combinatorial screening assays have emerged as a solution. Robotic deposition has been the technology of choice to create microarrays of biomaterials consisting of different combinations of ECM proteins or synthetic materials. For example, Anderson et al. [85] tested over 1,700 synthetic polymer-human embryonic stem cell (ESC) interactions to study their effects on cell attachment, proliferation, and differentiation (Fig. 9.4a). In another work, a microarray containing 32 combinations of different ECM molecules (laminin, fibronectin, collagen I, III and IV) was used to screen for mouse ESC differentiation [86]. The array demonstrated a ~140-fold difference in hepatic differentiation between the least and the most efficient ECM combinations. This platform was further extended to study the effects of ECM and growth factors on cardiac lineage differentiation. The chip was compartmentalized into several wells that contained mixtures of growth factors, such as wingless-type MMTV integration site family (Wnt) 3a, bone morphogenetic protein (BMP) 4, fibroblast growth factor (FGF) 4, and activin A [87]. 3D microarrays have also been developed for screening human ESCs by utilizing ECM components to make a variety of cell-encapsulated hydrogels [88]. Hydrogels are 3D cross-linked polymeric networks with high water content that have similar structural properties as ECM [89, 90]. While these combinatorial libraries are high-throughput and cost-effective, one of their caveats is the possibility of diffusion of molecules from one region of the chip to another, leading to cross-contamination [91]. To address this problem, each distinct microregion can be isolated by sandwiching cell-seeded microwells with a microarray of posts containing a combinatorial array of chemicals [92].

ECM interacts with cells by binding to transmembrane receptors through its integrin binding sites. Such a biochemical signal also imparts physical changes through the modulation of cell's cytoskeletal machinery, leading to cell morphogenesis. These bio-physicochemical cues are inherent in the microarchitecture of the niche through the particular arrangement of ECM molecules and neighboring cells. In vitro, cell shape can be controlled by fabricating geometric patterns of cell adhesive molecules, which have been shown to affect stem cell behavior [16, 46–48]. For instance, mesenchymal stem cell (MSC) differentiation was directed via controlling the size of fibronectin islands created by μCP [46, 47]. Small islands reduced cell spreading and promoted a round morphology, whereas larger islands increased cellular spreading and favored a flattened shape (Fig. 9.4b). Adipogenesis and osteogenesis were favored by small and large islands, respectively, in the presence of mixed osteogenic/adipogenic media. When transforming growth factor-β was used instead of mixed media, small islands promoted chondrogenesis and large islands directed myogenesis.

Micropatterning of substrates not only can be used to study cell-substrate interaction, but also to form homogenous embryoid bodies (EBs). For instance, circular patterns of ECM of varying diameters were used to form different human ESC colony sizes [93]. The colonies were then transferred to suspension culture where they formed EBs of uniform size. The size of the colony determined the differentiation trajectory with smaller colonies exhibiting an increased endodermal/neural gene expression ratio. Subsequently in EBs derived from these

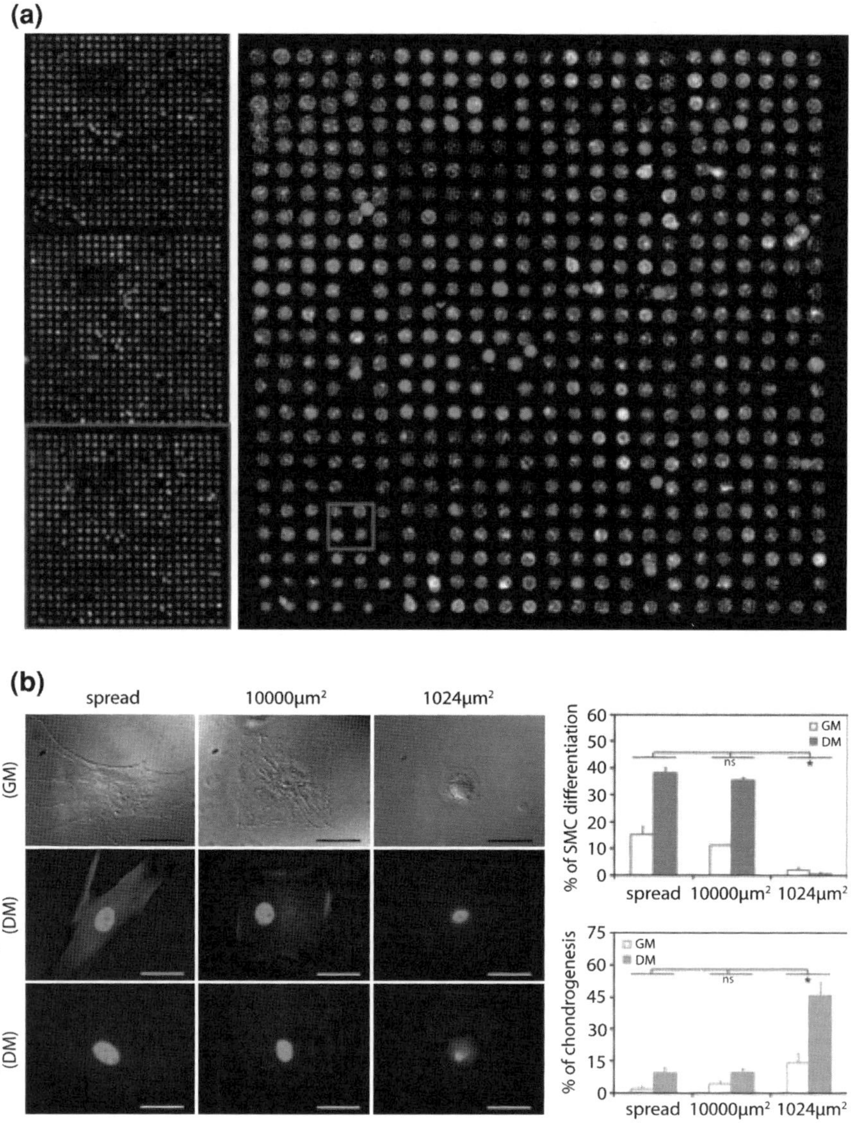

Fig. 9.4 2D Biomolecular patterns. **a** Microarray of synthetic materials seeded with human embryonic stem cells stained against cytokeratin 7 (*green*) and vimentin (*red*) [85]. Adapted by permission from Macmillan Publishers Ltd: [Nature Biotechnology], copyright (2004). **b** Effect of cell shape on differentiation: microcontact printing was used to create different cell adhesive geometries, leading to different cell shapes. This in turn resulted in the preferential differentiation of mesenchymal stem cells to chondrocytes on small islands (1,024 μm^2) and to smooth muscle cells (SMC) on large islands (10,000 μm^2) in the presence of transforming growth factor-β. Immunofluorescent staining was done against calponin (SMC marker) and collagen II (chondrocyte marker). GM, growth medium; DM, differentiation medium; scale bar, 50 μm [47]. Adapted by permission from John Wiley and Sons: [Stem Cells], copyright (2010)

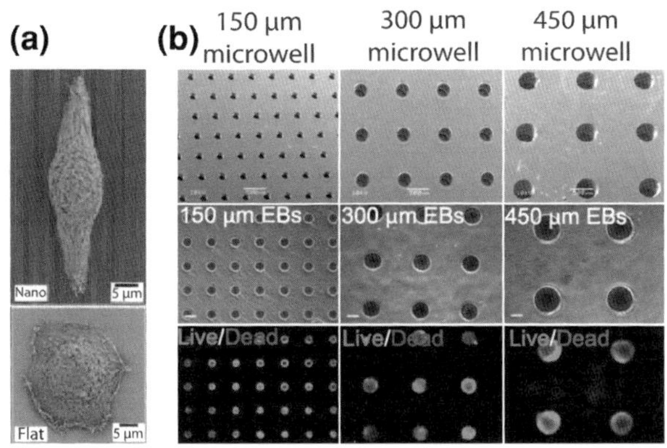

Fig. 9.5 Topography. **a** Response of corneal epithelial cell to nanograting. Scanning electron micrographs (SEM) of an epithelial cell on a nanograting topography (*top*) and unpatterned surface (*bottom*), illustrating the elongation and alignment of the cell along the grating axis [97]. Adapted by permission from Company of Biologists Ltd: [Journal of Cell Science], copyright (2003). **b** Microwell system for homogenous embryoid body (EB) generation. SEM (*top row*) and phase contrast (*middle row*) images of different sized microwells. Fluorescent images (calcein AM/ethidium homodimer assay) of EBs in microwells are shown in the *bottom row* [99]. Copyright (2009) National Academy of Sciences, USA

colonies, there was a greater induction of mesoderm and cardiac lineages with larger EB sizes.

9.3.1.2 Topographical Patterns

There is a great interest in the development of novel 3D culture systems given the lack of physiological relevance of 2D culture platforms [94, 95]. Scaffolds with micro- and nanoscale 3D geometrical features can be fabricated to investigate interaction of cells with these topographical cues. Cells sense and respond to these 3D geometric cues through contact guidance and cytoskeletal reorganization [96]. Numerous microtechnologies including photo- and soft lithography [44, 45], EBL [25], and nanoimprinting [58] have been used to more precisely mimic the micron and submicron topographies of the in vivo environment [19]. For instance, 70 nm wide ridges of silicon oxide fabricated by EBL resulted in cytoplasmic and actin microfilament alignment of human corneal epithelial cell (Fig. 9.5a) [97]. In another example, nanograting structures fabricated by nanoimprinting resulted in an increased expression of neuronal markers in cultured MSCs, compared to unpatterned and micropatterned structures [44]. Although the mechanism behind the morphological and genetic changes induced by nanotopography is unclear, clustering of cell adhesion molecules such as integrins, or preferential actin polymerization are the leading theories in explaining this phenomenon [84, 98].

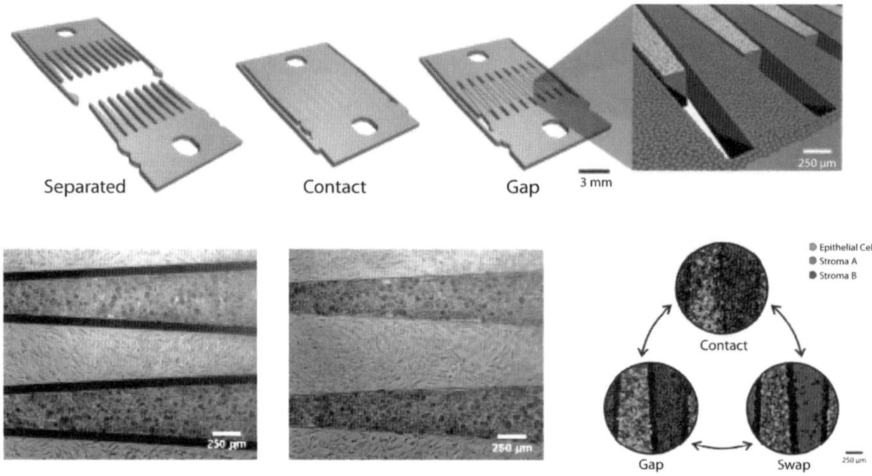

Fig. 9.6 Cell–cell interaction. A Micromechanical device used to create organized cocultures of hepatocytes and stromal cells and control the distance between these two cell types by movement of the interdigitating plates [102]. Copyright (2007) National Academy of Sciences, USA

One of the obstacles in stem cell biology has been cell population heterogeneity resulting in reduced reproducibility of experiments. Microwell arrays, fabricated via replica molding, have emerged as a powerful tool for generating homogenous size-based arrays of embryoid bodies (EBs) (Fig. 9.5b). The diameter of microwells determines the size of EBs produced within each well. It has been shown that differentiation trajectories can be altered by changing the size of EBs. In one study, differentiation to the cardiac phenotype was preferred in larger, compared to smaller EBs, whereas smaller EBs favored differentiation to the endothelial lineage, compared to larger EBs [99]. Furthermore, the surface of microwell arrays can be functionalized with ECM molecules to provide a biomimetic quasi-3D microenvironment [100, 101].

9.3.2 Controlling Cell–Cell Interactions

The dynamics of cell–cell communication is important for understanding a number of biological phenomena, such as the determination of cell fate. In vivo, stem cells are in contact, or in close proximity with neighboring cells and intercellular signals govern many stem cell biological processes. Hence, controlling cell–cell interactions provides an additional level of control over regulating stem cell differentiation. There have been many attempts to control the spatial organization of an individual cell or cellular colonies using a variety of microscale approaches. For example, a silicon platform consisting of two interdigitating pieces was fabricated by micromachining to dynamically manipulate heterotypic cell–cell interactions via adjusting the distance between the interdigitating plates (Fig. 9.6) [102]. Using this device, the dynamics of communication between hepatocytes and stromal cells were assessed, revealing that short distances from stromal cells (<400 μm) are required for the maintenance of hepatocytes.

Stencils are another tool commonly used to control cell–cell interactions. To this end, a dynamic environment involving coculture of murine ESCs with fibroblasts or hepatocytes can be created using microscale stencils [103]. Briefly, the stencil was placed against a flat fibronectin-coated PDMS substrate and murine ESCs were seeded. The ESCs adhered to the PDMS substrate through the holes of the stencil. After allowing time for cell attachment, the stencil was peeled off, removing cells that had not fallen through the stencil holes. Subsequently, fibroblasts or hepatocytes were seeded to create a coculture system.

Although the previous examples demonstrate heterotypic cell–cell interactions, homotypic stem cell communication can be used to direct differentiation. In work by Tang et al. [104], patterns of arginine-glycine-aspartate microislands were fabricated using photolithography. Each microisland was of such a size that allowed single cell attachment and could be next to a variable number of other microislands, thereby adjusting the number of cell–cell contacts. It was shown that cells on isolated microislands showed significantly less differentiation compared to those with a higher number of contacts to adjacent cells.

9.3.3 Microfluidics and Stem Cells

Microfluidics has emerged as an enabling apparatus for the dynamic perturbation of biophysical and biochemical factors similar to that experienced by stem cells in their niche. Furthermore, this technology can be used as a bioanalytical tool for applications such as cell sorting, and protein and nucleic acid analysis. In addition, studying the effects of shear stress, caused by flowing fluids in microchannels, on cellular functions is possible. Generation of cocultures [105, 106] to study heterotypic interactions and patterning of proteins and cells [107] to investigate cell-ECM cross talk can also be accomplished with these devices. Another advantage of microfluidics is the pico- to nanoliter volume requirement of cell culture reagents, which is the reason for their use as microbioreactors. In this section, we will discuss various microfluidic-based approaches relevant to stem cell studies.

9.3.3.1 Creating Dynamic Biochemical Cues

Cytokines play a major role in cellular signaling through binding to cell surface receptors that result in modulation of cell behaviors. Indeed, numerous cytokines have been used to direct differentiation of stem cells in static cell cultures [108–110]. Microfluidics has circumvented the lack of dynamic spatiotemporal control over cytokine presentation in standard culture methods. Numerous methods have been developed for the generation of concentration gradients using microfluidics (Fig. 9.7) [111, 112]. In one study, Park et al. differentiated neural progenitor cells

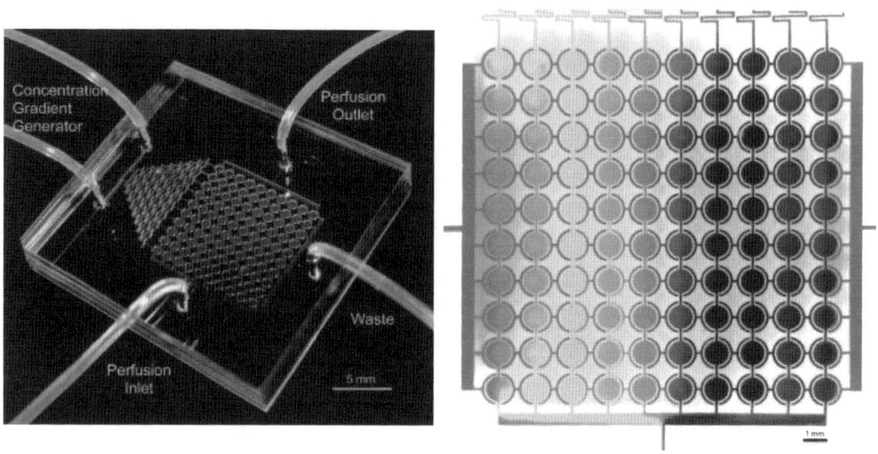

Fig. 9.7 Generation of a concentration gradient using microfluidics. A microfluidic device (*left*) fabricated from poly(dimethylsiloxane) showing its ports for media perfusion and cell/reagent loading. *Red*, *blue* and *yellow* dyes were used to create a concentration gradient (*right*) [158]. Adapted by permission from John Wiley and Sons: [Biotechnology and Bioengineering], copyright (2005)

into neurons using a concentration gradient of sonic hedgehog, BMP-4 and FGF-8 [49]. In another study, Chung et al. exposed human neural stem cells to a gradient of epidermal growth factor, FGF-2, and platelet-derived growth factor, and demonstrated that differentiation into astrocytes was concentration-dependent [50]. These gradients can be generated by attachment of molecules on a substrate, embedment within a 3D matrix, or under continuous flow.

9.3.3.2 Micro-Bioanalytical Systems

Microfluidic systems have been designed to address the challenges faced in typical RNA or DNA analysis platforms, including large consumption of expensive reagents and samples, and relatively slow analysis [113]. For instance, the Quake group developed a large-scale microfluidic chip with 1,000 independent picoliter-sized reaction chambers and over 3,500 microvalves for loading and compartmentalization for high-throughput bioassay applications [114]. Using multilayer microfluidic networks, 72 parallel reverse transcriptase-PCR (RT-PCR) reactions were done using picoliter volumes of samples. The cycling times for these reactions were much shorter given the high surface area to volume ratio of microfluidic channels [115]. In addition to nucleic acid analysis, microfluidics has been used to manipulate cells via electrical [53], optical [116], and magnetic [117] methods, which can be beneficial in cell sorting applications.

Microfluidics has enabled a new approach to the transport, mixing, sorting, and recovery of samples, thereby reducing the scale of experiments to that on a small chip [118]. These so-called 'lab-on-a-chip' devices or 'micro-total-analysis-systems' have numerous applications, including clinical diagnostics, and nucleic acid and protein manipulation [119]. Despite their potential, microfluidic devices suffer from complexity associated with their plumbing and numerous valves that has shunned mainstream biologists from accepting it. As in other industries, simplification of design parameters and versatility are needed to push these devices into widespread use [112].

9.3.3.3 Single-Cell Analytical Platforms

One of the difficulties of working with stem cells is the heterogeneity within each population, as well as the batch-to-batch variability. Conventional culture methods measure the average response of these cell populations, invariably skewing the results toward the characteristics of the majority-representing cell type and masking the effects of the rare or minor cell types present. Microfabrication methods such as replica molding [120] and microfluidics [121] have been used to generate microwells for single-cell analysis of hematopoietic stem cell line [122]. In work by Lecault et al., a microfluidic device containing 1,600 nL-sized chambers was used to capture a single hematopoietic stem cell per well [121]. This device was then used to introduce Steel factor in a spatial and time-dependent manner to study cell proliferation. The application of microfluidics has also been extended to single-cell gene expression profiling and genomic sequencing [123, 124].

9.3.3.4 Stem Cell Microbioreactors

Bioreactors are devices with designs based on bioprocess engineering principles to tightly control multiple cell culture parameters with the goal of driving a biological process in a robust, efficient, and reproducible manner [125]. In the case of stem cells, the desired biological processes include uncommitted cell expansion, directed differentiation, and formation of tissues. Unlike static cell cultures, bioreactors possess sensors to strictly control physiological parameters, such as oxygen tension, pH, and nutrient and waste concentrations [126]. The bioprocess control involved in constructing these devices is based on optimization of specific culture conditions and large-scale production and manufacturing. The optimization process is accomplished at reduced scales to decrease the consumption of reagents and their associated costs. The emergence of microfluidics in cell biology has allowed for a dramatic reduction in bioreactor scales, from milliliters to nano- and microliters. These so-called microbioreactors have grown substantially more complex by allowing automation, parallel multi-parametric experiments, and real-time imaging and analysis. Once optimization is accomplished by microbioreactors, large-scale production can ensue. Such manufacturing is needed to supply biological parts for cellular-based therapies and pharmaceutical research [127]. In this section we will focus only on microfluidic-based microbioreactors for stem cell culture; however, the reader is referred to several excellent articles for further reading on bioreactors [126, 128–131].

As mentioned previously, microfluidic devices are capable of manipulating fluids and controlling concentration of biomolecules in time and space. As opposed to stagnant culture methods where concentration gradients of molecules develop as cells take up nutrient and secrete waste, in microfluidics, a steady state is achieved by flowing fluid. This results in a constant supply of nutrients and removal of metabolic waste products, similar to the function performed by the microvasculature in vivo. Perfusion in microfluidic devices seeded with ESCs have demonstrated that cell proliferation is directly correlated with flow rate and pluripotency can be maintained for >7 days in a feeder-free culture system [132, 133]. One disadvantage with continuous perfusion systems is the elimination of autocrine and paracrine signals which play a crucial role in the stem cell niche. To reduce this limitation, microbioreactors with cyclic periods of flow and static incubation have been devised. In one instance, human ESCs cocultured with human foreskin fibroblasts were subjected to short periods of continuous flow and longer periods of stagnancy to demonstrate the feasibility of such a system [134].

In most bioreactors, convective mass transport is accomplished by controlled stirring to homogenize the soluble factors present in the cell microenvironment. However, achieving this at the micron-scale level is a challenge. One method recently developed uses convective flow as a stirring method [135]. This approach is based on creating minuscule temperature gradients that results in convection and mixing within a liquid medium and has been shown to support hematopoietic stem cells in an array of microliter-sized bioreactors. Another example of increased sophistication in these microbioreactor devices is the use of traps in a microfluidic system to capture ESCs and promote their proliferation into EBs [136].

9.3.4 Building Tissue Microarchitecture

The behavior of stem cells is critically dependent on their microenvironment. The niche provides a multitude of bio-physicochemical cues that are integrated by each stem cell resulting in perturbation of its molecular circuitry. Hence, valuable insight may be gained by mimicking the in vivo microarchitecture of tissues. Complex hierarchical structures can be built using a variety of microtechnologies, including inkjet printing and robotic deposition [75], electropatterning [137], and modular assembly [138]. In the next section, we discuss the role of hydrogels in constructing tissues as well as vascularization strategies for engineered tissue constructs.

9.3.4.1 Hydrogels and Tissue Assembly

The construction of biomimetic structures may aid in controlling the differentiation of stem cells to a target lineage and engineering of functionalized tissues. One of the promising avenues for engineering tissues is the use of hydrogels as cellular scaffolds. Hydrogels can be naturally-derived or synthetic. Naturally-derived hydrogels are often made from ECM components (collagen, hyaluronic acid, fibronectin) as well as other natural materials (chitosan, alginate, silk). These materials have the advantage of possessing biological cues [139]. In contrast, synthetic hydrogels are advantageous due to their low immunogenicity and control over physical and chemical properties [140]. The combination of hydrogels and microscale technologies such as photolithography and soft lithography has enabled the creation of cell-laden hydrogels with characteristic shapes and micron-scale sizes [141, 142]. For example, micropatterned gelatin methacrylate (GelMA) hydrogels were used to direct 3D cellular behavior [143, 144]. Cell confinement within GelMA hydrogel micro-constructs induced cellular alignment and elongation of numerous cell types, including cardiac stem cells.

Micron-sized hydrogels or microgels can be assembled in a step-wise fashion to form a higher-order macrostructure in an approach known as 'bottom-up' or modular assembly [138]. In this approach, each microgel is used as a building unit, composed of a particular polymer, containing a specific cell, and having a desired microarchitecture. This assembly technique has great potential in constructing complex tissue structures with high precision over its microarchitecture to better mimic native tissue structure and function [145–148]. The assembly of each unit occurs by harnessing the properties of surface tension with regard to hydrophilic hydrogels (Fig. 9.8). By introducing hydrophobic mineral oil to the media that the cell-laden hydrogels are in, hydrophilic microgels minimize surface energy via aggregation. Thereafter, the aggregated microgels are interconnected by exposure to UV light. In another similar technique, microgels were self-assembled at the air–liquid interface to centimeter-scale structures [147]. Briefly, polyethylene glycol (PEG) microgels of hundreds of microns in dimension were fabricated and randomly transferred onto the surface of a very dense, hydrophobic material, such as tetrachloride or perfluorodecalin. The microgels spontaneously assembled via hydrophilic interactions to minimize Gibb's free energy. Further, Yanagawa et al. [149] were able to build sheets of cell-encapsulated PEG microgels and assemble them in a 3D manner by stacking the sheets onto one another. Sodium alginate microgels were randomly interspersed within these PEG microgels and selectively dissolved after self-assembly to create a highly porous structure.

Potential drawbacks of these techniques include scalability and mechanical integrity. For instance, the ultimate size of the assembled tissues is limited to the order of millimeters, which is not applicable for clinical use. Moreover, mechanical properties of engineered tissues must be robust for transplantation. However, current bottom-up approaches mainly use hydrogels with low mechanical properties. Therefore, the following requirements need to be met before clinical applications are realized: (1) scale-up technologies to achieve clinically relevant length scale, (2) recapitulation of native mechanical properties, and (3) successful demonstration of functional improvement in diseased tissues in vivo.

9.3.4.2 Vascularization

Cells require a constant supply of oxygen and nutrients for survival and to function properly. In vitro, diffusion is the transport mechanism responsible for this supply in conventional culture systems. 3D tissue-engineered scaffolds of more than a few hundred micron thickness cannot rely on diffusion alone and need an alternative

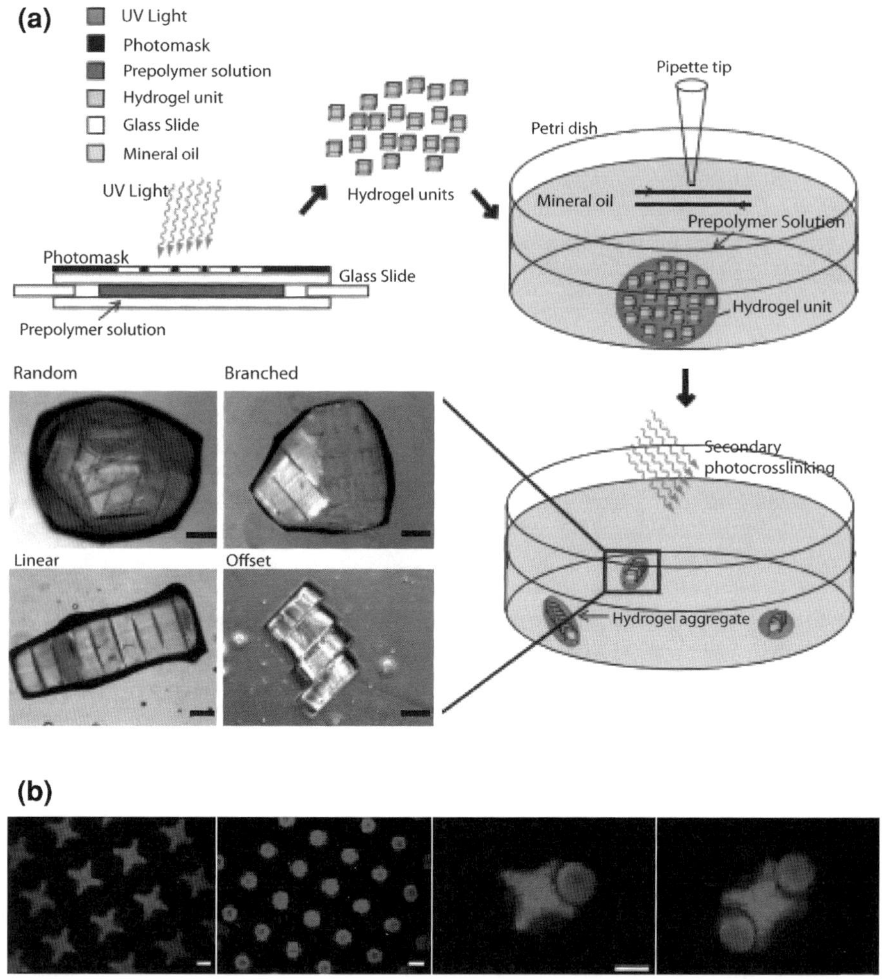

Fig. 9.8 Modular Assembly. **a** Schematic of modular assembly using hydrogels. Phase contrast images of assembled microgels are shown. Scale bar, 200 μm. **b** Microgels can be fabricated to have lock-and-key geometries. Shown here are fluorescent images of step-wise assembly of microgels. Green, FITC-dextran; red, Nile Red. Scale bar, 200 μm [148]. Copyright (2008) National Academy of Sciences, USA

mechanism of transport to ensure cell viability. Thus, similar to tissues in vivo, functional microvasculatures capable of molecular exchange need to be integrated into current engineered scaffolds [150, 151]. A hydrogel assembly technique has been developed to create a perfusable vascularized tissue in vitro (Fig. 9.9) [152]. Briefly, cylindrical human hepatocellular carcinoma line cell (HepG2)-laden collagen hydrogel building blocks were fabricated. Then, human umbilical vein endothelial cells (HUVEC) were seeded on these hydrogel blocks, forming a confluent endothelial layer on the surface. The HepG2-HUVEC modules were then

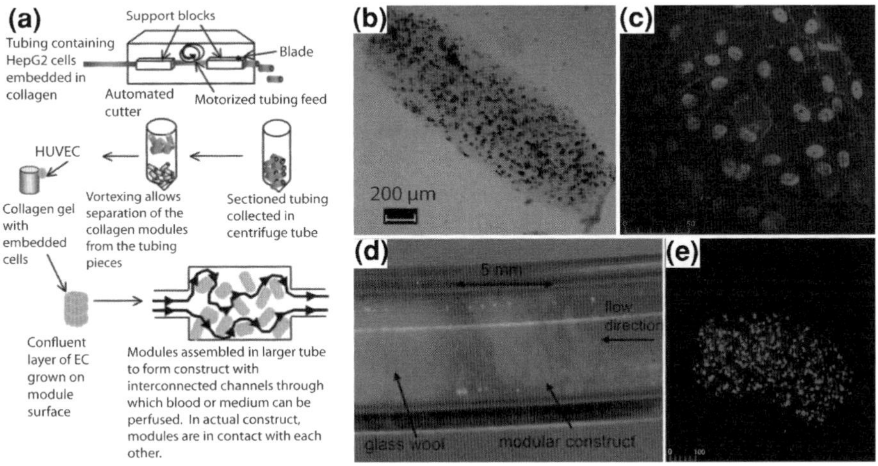

Fig. 9.9 Engineered vascularized tissue construct. **a** Fabrication schematic of the modular construct. **b** Phase contrast image of collagen hydrogel module encapsulated with hepatocytes (HepG2). **c** After seeding, endothelial cells (EC) formed a confluent layer on the collagen-hepatocyte module as demonstrated in this confocal micrograph. *Green*, VE-cadherin immunofluorescent staining. **d** Perfusion of the flow circuit containing modular constructs. **e** Collagen-hepatocyte-EC construct shown in this confocal microscope image after exposure to 7 days of perfusion. Hepatocytes are labeled with Vybrant CFDA SE [152]. Copyright (2006) National Academy of Sciences, USA

perfused in a larger tube that allowed the aggregation and compaction of the modules into a tube-like structure. The engineered devices were perfused with whole blood and demonstrated high cell viability and antithrombogenicity. This technique could potentially employ stem cells for creating perfused tissues. There are numerous other approaches toward engineering vascularized tissues that have been extensively reviewed in the literature [153–155].

9.4 Conclusion

The use of microtechnology has spawned a new era in the design of non-conventional culture systems that recapitulate the in vivo environment, fabrication of multifunctional integrated devices, and assembly of complex tissue-like constructs [156]. These devices allow for culturing of cells, perturbation of cellular microenvironment via biochemical (e.g., cytokines) and biophysical (e.g., topography) stimulation, cell sorting, and biochemical analysis (e.g., protein and nucleic analysis) in a more precise, reproducible, and cost-effective manner. Although microtechnologies have allowed an additional level of control over the self-renewal and differentiation of stem cells, further research at the interface of microtechnology and stem cell biology will be critical to aid in making the therapeutic potential of stem cells a reality.

References

1. Kim SB, Bae H, Cha JM, Moon SJ, Dokmeci MR, Cropek DM, Khademhosseini A (2011) A cell-based biosensor for real-time detection of cardiotoxicity using lensfree imaging. Lab Chip 11(10):1801–1807
2. Kiskinis E, Eggan K (2010) Progress toward the clinical application of patient-specific pluripotent stem cells. J Clin Invest 120(1):51–59
3. Daley GQ, Scadden DT (2008) Prospects for stem cell-based therapy. Cell 132(4):544–548
4. Voog J, Jones DL (2010) Stem cells and the niche: a dynamic duo. Cell Stem Cell 6(2):103–115
5. Morrison SJ, Spradling AC (2008) Stem cells and niches: mechanisms that promote stem cell maintenance throughout life. Cell 132(4):598–611
6. Discher DE, Mooney DJ, Zandstra PW (2009) Growth factors, matrices, and forces combine and control stem cells. Science 324(5935):1673–1677
7. Ito R, Okazaki S (2000) Pushing the limits of lithography. Nature 406(6799):1027–1031
8. Khademhosseini A, Langer R, Borenstein J, Vacanti JP (2006) Microscale technologies for tissue engineering and biology. Proc Natl Acad Sci U S A 103(8):2480–2487
9. Dolatshahi-Pirouz A, Nikkhah M, Kolind K, Dokmeci MR, Khademhosseini A (2011) Micro- and nanoengineering approaches to control stem cell-biomaterial interactions. J Funct Biomater 2(3):88–106
10. Stevens MM, George JH (2005) Exploring and engineering the cell surface interface. Science 310(5751):1135–1138
11. Gates BD, Xu Q, Stewart M, Ryan D, Willson CG, Whitesides GM (2005) New approaches to nanofabrication: molding, printing, and other techniques. Chem Rev 105(4):1171–1196
12. Nie Z, Kumacheva E (2008) Patterning surfaces with functional polymers. Nat Mater 7(4):277–290
13. del Campo A, Arzt E (2008) Fabrication approaches for generating complex micro-and nanopatterns on polymeric surfaces. Chem Rev 108(3):911–945
14. Moon JH, Ford J, Yang S (2006) Fabricating three dimensional polymeric photonic structures by multi beam interference lithography. Polym Adv Technol 17(2):83–93
15. Hahn MS, Miller JS, West JL (2006) Three dimensional biochemical and biomechanical patterning of hydrogels for guiding cell behavior. Adv Mater 18(20):2679–2684
16. Song W, Lu H, Kawazoe N, Chen G (2011) Adipogenic differentiation of individual mesenchymal stem cell on different geometric micropatterns. Langmuir 27(10):6155–6162
17. Bae H, Ahari AF, Shin H, Nichol JW, Hutson CB, Masaeli M, Kim SH, Aubin H, Yamanlar S, Khademhosseini A (2011) Cell-laden microengineered pullulan methacrylate hydrogels promote cell proliferation and 3D cluster formation. Soft Matter 7(5):1903–1911
18. Qi H, Du Y, Wang L, Kaji H, Bae H, Khademhosseini A (2010) Patterned differentiation of individual embryoid bodies in spatially organized 3D hybrid microgels. Adv Mater 22(46): 5276–5281
19. Norman JJ, Desai TA (2006) Methods for fabrication of nanoscale topography for tissue engineering scaffolds. Ann Biomed Eng 34(1):89–101
20. Idota N, Tsukahara T, Sato K, Okano T, Kitamori T (2009) The use of electron beam lithographic graft-polymerization on thermoresponsive polymers for regulating the directionality of cell attachment and detachment. Biomaterials 30(11):2095–2101
21. Christman KL, Schopf E, Broyer RM, Li RC, Chen Y, Maynard HD (2009) Positioning multiple proteins at the nanoscale with electron beam cross-linked functional polymers. J Am Chem Soc 131(2):521–527
22. Das G, Mecarini F, Gentile F, De Angelis F, Mohan Kumar H, Candeloro P, Liberale C, Cuda G, Di Fabrizio E (2009) Nano-patterned SERS substrate: application for protein analysis vs. temperature. Biosens Bioelectron 24(6):1693–1699
23. Vieu C, Carcenac F, Pepin A, Chen Y, Mejias M, Lebib A, Manin-Ferlazzo L, Couraud L, Launois H (2000) Electron beam lithography: resolution limits and applications. Appl Surf Sci 164(1–4):111–117

24. Biggs MJP, Richards RG, Gadegaard N, Wilkinson CDW, Oreffo ROC, Dalby MJ (2009) The use of nanoscale topography to modulate the dynamics of adhesion formation in primary osteoblasts and ERK/MAPK signalling in STRO-1+ enriched skeletal stem cells. Biomaterials 30(28):5094–5103
25. McMurray RJ, Gadegaard N, Tsimbouri PM, Burgess KV, McNamara LE, Tare R, Murawski K, Kingham E, Oreffo ROC, Dalby MJ (2011) Nanoscale surfaces for the long-term maintenance of mesenchymal stem cell phenotype and multipotency. Nat Mater 10(8):637–644
26. Watt F, Bettiol A, Van Kan J, Teo E, Breese M (2005) Ion beam lithography and nanofabrication: a review. Int J Nanosci 4(3):269–286
27. Kubena RL (1993) Resolution limits of focused-ion-beam resist patterning. Mater Res Soc Symp Proc 279:567–576
28. Xia Y, Whitesides GM (1998) Soft lithography. Annu Rev Mater Sci 28(1):153–184
29. Whitesides GM (2006) The origins and the future of microfluidics. Nature 442(7101): 368–373
30. Weibel DB, DiLuzio WR, Whitesides GM (2007) Microfabrication meets microbiology. Nat Rev Microbiol 5(3):209–218
31. Folch A, Jo BH, Hurtado O, Beebe DJ, Toner M (2000) Microfabricated elastomeric stencils for micropatterning cell cultures. J Biomed Mater Res 52(2):346–353
32. Kaufmann T, Ravoo BJ (2010) Stamps, inks and substrates: polymers in microcontact printing. Polym Chem 1(4):371–387
33. Xu C, Taylor P, Ersoz M, Fletcher PDI, Paunov VN (2003) Microcontact printing of DNA-surfactant arrays on solid substrates. J Mater Chem 13(12):3044–3048
34. Perl A, Reinhoudt DN, Huskens J (2009) Microcontact printing: limitations and achievements. Adv Mater 21(22):2257–2268
35. Li HW, Muir BVO, Fichet G, Huck WTS (2003) Nanocontact printing: A route to sub-50-nm-scale chemical and biological patterning. Langmuir 19(6):1963–1965
36. Kim E, Xia Y, Whitesides GM (1995) Polymer microstructures formed by moulding in capillaries. Nature 376(6541):581–584
37. Zhao XM, Xia Y, Whitesides GM (1996) Fabrication of three dimensional micro structures: microtransfer molding. Adv Mater 8(10):837–840
38. Toh YC, Voldman J (2011) Fluid shear stress primes mouse embryonic stem cells for differentiation in a self-renewing environment via heparan sulfate proteoglycans transduction. FASEB J 25(4):1208–1217
39. Burdick JA, Khademhosseini A, Langer R (2004) Fabrication of gradient hydrogels using a microfluidics/photopolymerization process. Langmuir 20(13):5153–5156
40. Sant S, Hancock MJ, Donnelly JP, Iyer D, Khademhosseini A (2010) Biomimetic gradient hydrogels for tissue engineering. Can J Chem Eng 88(6):899–911
41. Leclerc E, Sakai Y, Fujii T (2004) Microfluidic PDMS (polydimethylsiloxane) bioreactor for large scale culture of hepatocytes. Biotechnol Prog 20(3):750–755
42. Kim Y, Kim KS, Kounovsky KL, Chang R, Jung GY, de Pablo JJ, Jo K, Schwartz DC (2011) Nanochannel confinement: DNA stretch approaching full contour length. Lab Chip 11(10):1721–1729
43. Gerecht S, Bettinger CJ, Zhang Z, Borenstein JT, Vunjak-Novakovic G, Langer R (2007) The effect of actin disrupting agents on contact guidance of human embryonic stem cells. Biomaterials 28(28):4068–4077
44. Yim EKF, Pang SW, Leong KW (2007) Synthetic nanostructures inducing differentiation of human mesenchymal stem cells into neuronal lineage. Exp Cell Res 313(9):1820–1829
45. Fu J, Wang YK, Yang MT, Desai RA, Yu X, Liu Z, Chen CS (2010) Mechanical regulation of cell function with geometrically modulated elastomeric substrates. Nat Methods 7(9): 733–736
46. McBeath R, Pirone DM, Nelson CM, Bhadriraju K, Chen CS (2004) Cell shape, cytoskeletal tension, and RhoA regulate stem cell lineage commitment. Dev Cell 6(4): 483–495

47. Gao L, McBeath R, Chen CS (2010) Stem cell shape regulates a chondrogenic versus myogenic fate through Rac1 and N-Cadherin. Stem Cells 28(3):564–572
48. Kilian KA, Bugarija B, Lahn BT, Mrksich M (2010) Geometric cues for directing the differentiation of mesenchymal stem cells. Proc Natl Acad Sci U S A 107(11):4872–4877
49. Park JY, Kim SK, Woo DH, Lee EJ, Kim JH, Lee SH (2009) Differentiation of neural progenitor cells in a microfluidic chip generated cytokine gradient. Stem Cells 27(11): 2646–2654
50. Chung BG, Flanagan LA, Rhee SW, Schwartz PH, Lee AP, Monuki ES, Jeon NL (2005) Human neural stem cell growth and differentiation in a gradient-generating microfluidic device. Lab Chip 5(4):401–406
51. Liu J, Hansen C, Stephen SR (2003) Solving the "world-to-chip" interface problem with a microfluidic matrix. Anal Chem 75(18):4718–4723
52. Pushkarev D, Neff NF, Quake SR (2009) Single-molecule sequencing of an individual human genome. Nat Biotechnol 27(9):847–850
53. Fu AY, Chou HP, Spence C, Arnold FH, Stephen SR (2002) An integrated microfabricated cell sorter. Anal Chem 74(11):2451–2457
54. Guo LJ (2007) Nanoimprint lithography: methods and material requirements. Adv Mater 19(4):495–513
55. Hua F, Sun Y, Gaur A, Meitl MA, Bilhaut L, Rotkina L, Wang J, Geil P, Shim M, Rogers JA (2004) Polymer imprint lithography with molecular-scale resolution. Nano Lett 4(12): 2467–2471
56. Guo LJ, Cheng X, Chou CF (2004) Fabrication of size-controllable nanofluidic channels by nanoimprinting and its application for DNA stretching. Nano Lett 4(1):69–73
57. Hoff JD, Cheng LJ, Meyhöfer E, Guo LJ, Hunt AJ (2004) Nanoscale protein patterning by imprint lithography. Nano Lett 4(5):853–857
58. Subramani C, Cengiz N, Saha K, Gevrek TN, Yu X, Jeong Y, Bajaj A, Sanyal A, Rotello VM (2011) Direct fabrication of functional and biofunctional nanostructures through reactive imprinting. Adv Mater 23(28):3165–3169
59. Salaita K, Wang Y, Mirkin CA (2007) Applications of dip-pen nanolithography. Nat Nanotechnol 2(3):145–155
60. Braunschweig AB, Huo F, Mirkin CA (2009) Molecular printing. Nat Chem 1(5):353–358
61. Piner RD, Zhu J, Xu F, Hong S, Mirkin CA (1999) "Dip-pen" nanolithography. Science 283(5402):661–663
62. Hong S, Zhu J, Mirkin CA (1999) Multiple ink nanolithography: toward a multiple-pen nano-plotter. Science 286(5439):523–525
63. Mirkin CA (2007) The power of the pen: development of massively parallel dip-pen nanolithography. ACS Nano 1(2):79–83
64. Sekula S, Fuchs J, Weg Remers S, Nagel P, Schuppler S, Fragala J, Theilacker N, Franzreb M, Wingren C, Ellmark P, Borrebaeck CAK, Mirkin CA, Fuchs H, Lenhert S (2008) Multiplexed lipid dip pen nanolithography on subcellular scales for the templating of functional proteins and cell culture. Small 4(10):1785–1793
65. Giam LR, Mirkin CA (2011) Cantilever free scanning probe molecular printing. Angew Chem Int Edit 50(33):7482–7485
66. Huo F, Zheng Z, Zheng G, Giam LR, Zhang H, Mirkin CA (2008) Polymer pen lithography. Science 321(5896):1658–1660
67. Huo F, Zheng G, Liao X, Giam LR, Chai J, Chen X, Shim W, Mirkin CA (2010) Beam pen lithography. Nat Nanotechnol 5(9):637–640
68. Lim JH, Mirkin CA (2002) Electrostatically driven dip pen nanolithography of conducting polymers. Adv Mater 14(20):1474–1477
69. Jiang H, Stupp SI (2005) Dip-pen patterning and surface assembly of peptide amphiphiles. Langmuir 21(12):5242–5246
70. Lee KB, Park SJ, Mirkin CA, Smith JC, Mrksich M (2002) Protein nanoarrays generated by dip-pen nanolithography. Science 295(5560):1702–1705

71. Lee M, Kang DK, Yang HK, Park KH, Choe SY, Kang CS, Chang SI, Han MH, Kang IC (2006) Protein nanoarray on Prolinker^{TM} surface constructed by atomic force microscopy dip pen nanolithography for analysis of protein interaction. Proteomics 6(4):1094–1103
72. Demers LM, Ginger DS, Park SJ, Li Z, Chung SW, Mirkin CA (2002) Direct patterning of modified oligonucleotides on metals and insulators by dip-pen nanolithography. Science 296(5574):1836–1838
73. Wu CC, Reinhoudt DN, Otto C, Subramaniam V, Velders AH (2011) Strategies for patterning biomolecules with dip pen nanolithography. Small 7(8):989–1002
74. Curran JM, Stokes R, Irvine E, Graham D, Amro N, Sanedrin R, Jamil H, Hunt JA (2010) Introducing dip pen nanolithography as a tool for controlling stem cell behaviour: unlocking the potential of the next generation of smart materials in regenerative medicine. Lab Chip 10(13):1662–1670
75. Mironov V, Kasyanov V, Markwald RR (2011) Organ printing: from bioprinter to organ biofabrication line. Curr Opin Biotech 22:1–7
76. Kim JD, Choi JS, Kim BS, Chan YC, Cho YW (2010) Piezoelectric inkjet printing of polymers: stem cell patterning on polymer substrates. Polymer 51(10):2147–2154
77. Ker EDF, Chu B, Phillippi JA, Gharaibeh B, Huard J, Weiss LE, Campbell PG (2011) Engineering spatial control of multiple differentiation fates within a stem cell population. Biomaterials 32(13):3413–3422
78. Ilkhanizadeh S, Teixeira AI, Hermanson O (2007) Inkjet printing of macromolecules on hydrogels to steer neural stem cell differentiation. Biomaterials 28(27):3936–3943
79. Woodfield TBF, Malda J, De Wijn J, Peters F, Riesle J, Van Blitterswijk CA (2004) Design of porous scaffolds for cartilage tissue engineering using a three-dimensional fiber-deposition technique. Biomaterials 25(18):4149–4161
80. Calvert P (2007) Printing cells. Science 318(5848209):208–209
81. Folkman J, Moscona A (1978) Role of cell shape in growth control. Nature 273(5661): 345–349
82. Chen CS, Mrksich M, Huang S, Whitesides GM, Ingber DE (1997) Geometric control of cell life and death. Science 276(5317):1425–1428
83. Thery M, Racine V, Piel M, Pepin A, Dimitrov A, Chen Y, Sibarita JB, Bornens M (2006) Anisotropy of cell adhesive microenvironment governs cell internal organization and orientation of polarity. Proc Natl Acad Sci U S A 103(52):19771–19776
84. Bettinger CJ, Langer R, Borenstein JT (2009) Engineering substrate topography at the micro and nanoscale to control cell function. Angew Chem Int Edit 48(30):5406–5415
85. Anderson DG, Levenberg S, Langer R (2004) Nanoliter-scale synthesis of arrayed biomaterials and application to human embryonic stem cells. Nat Biotechnol 22(7):863–866
86. Flaim CJ, Chien S, Bhatia SN (2005) An extracellular matrix microarray for probing cellular differentiation. Nat Methods 2(2):119–125
87. Flaim CJ, Teng D, Chien S, Bhatia SN (2008) Combinatorial signaling microenvironments for studying stem cell fate. Stem Cells Dev 17(1):29–39
88. Yang F, Cho SW, Son SM, Hudson SP, Bogatyrev S, Keung L, Kohane DS, Langer R, Anderson DG (2010) Combinatorial extracellular matrices for human embryonic stem cell differentiation in 3D. Biomacromolecules 11(8):1909–1914
89. Peppas NA, Hilt JZ, Khademhosseini A, Langer R (2006) Hydrogels in biology and medicine: from molecular principles to bionanotechnology. Adv Mater 18(11):1345–1360
90. Guillame-Gentil O, Semenov O, Roca AS, Groth T, Zahn R, Vörös J, Zenobi Wong M (2010) Engineering the extracellular environment: strategies for building 2D and 3D cellular structures. Adv Mater 22(48):5443–5462
91. Fernandes TG, Diogo MM, Clark DS, Dordick JS, Cabral JM (2009) High-throughput cellular microarray platforms: applications in drug discovery, toxicology and stem cell research. Trends Biotechnol 27(6):342–349
92. Wu J, Wheeldon I, Guo Y, Lu T, Du Y, Wang B, He J, Hu Y, Khademhosseini A (2011) A sandwiched microarray platform for benchtop cell-based high throughput screening. Biomaterials 32(3):841–848

93. Bauwens CL, Peerani R, Niebruegge S, Woodhouse KA, Kumacheva E, Husain M, Zandstra PW (2008) Control of human embryonic stem cell colony and aggregate size heterogeneity influences differentiation trajectories. Stem Cells 26(9):2300–2310
94. Cukierman E, Pankov R, Stevens DR, Yamada KM (2001) Taking cell-matrix adhesions to the third dimension. Science 294(5547):1708–1712
95. Doyle AD, Wang FW, Matsumoto K, Yamada KM (2009) One-dimensional topography underlies three-dimensional fibrillar cell migration. J Cell Biol 184(4):481–490
96. Rorth P (2011) Whence directionality: guidance mechanisms in solitary and collective cell migration. Dev Cell 20(1):9–18
97. Teixeira AI, Abrams GA, Bertics PJ, Murphy CJ, Nealey PF (2003) Epithelial contact guidance on well-defined micro- and nanostructured substrates. J Cell Sci 116(10): 1881–1892
98. Guilak F, Cohen DM, Estes BT, Gimble JM, Liedtke W, Chen CS (2009) Control of stem cell fate by physical interactions with the extracellular matrix. Cell Stem Cell 5(1):17–26
99. Hwang YS, Chung BG, Ortmann D, Hattori N, Moeller HC, Khademhosseini A (2009) Microwell-mediated control of embryoid body size regulates embryonic stem cell fate via differential expression of WNT5a and WNT11. Proc Natl Acad Sci U S A 106(40): 16978–16983
100. Dusseiller MR, Schlaepfer D, Koch M, Kroschewski R, Textor M (2005) An inverted microcontact printing method on topographically structured polystyrene chips for arrayed micro-3-D culturing of single cells. Biomaterials 26(29):5917–5925
101. Ochsner M, Dusseiller MR, Grandin HM, Luna-Morris S, Textor M, Vogel V, Smith ML (2007) Micro-well arrays for 3D shape control and high resolution analysis of single cells. Lab Chip 7(8):1074–1077
102. Hui EE, Bhatia SN (2007) Micromechanical control of cell–cell interactions. Proc Natl Acad Sci U S A 104(14):5722–5726
103. Wright D, Rajalingam B, Selvarasah S, Dokmeci MR, Khademhosseini A (2007) Generation of static and dynamic patterned co-cultures using microfabricated parylene-C stencils. Lab Chip 7(10):1272–1279
104. Tang J, Peng R, Ding J (2010) The regulation of stem cell differentiation by cell-cell contact on micropatterned material surfaces. Biomaterials 31(9):2470–2476
105. Chung S, Sudo R, Mack PJ, Wan CR, Vickerman V, Kamm RD (2009) Cell migration into scaffolds under co-culture conditions in a microfluidic platform. Lab Chip 9(2):269–275
106. Chung S, Sudo R, Vickerman V, Zervantonakis IK, Kamm RD (2010) Microfluidic platforms for studies of angiogenesis, cell migration, and cell–cell interactions. Ann Biomed Eng 38(3):1164–1177
107. Chiu DT, Jeon NL, Huang S, Kane RS, Wargo CJ, Choi IS, Ingber DE, Whitesides GM (2000) Patterned deposition of cells and proteins onto surfaces by using three-dimensional microfluidic systems. Proc Natl Acad Sci U S A 97(6):2408–2413
108. Schuldiner M, Yanuka O, Itskovitz-Eldor J, Melton DA, Benvenisty N (2000) Effects of eight growth factors on the differentiation of cells derived from human embryonic stem cells. Proc Natl Acad Sci U S A 97(21):11307–11312
109. Bhardwaj G, Murdoch B, Wu D, Baker D, Williams K, Chadwick K, Ling L, Karanu F, Bhatia M (2001) Sonic hedgehog induces the proliferation of primitive human hematopoietic cells via BMP regulation. Nat Immunol 2(2):172–180
110. de Haan G, Weersing E, Dontje B, van Os R, Bystrykh LV, Vellenga E, Miller G (2003) In vitro generation of long-term repopulating hematopoietic stem cells by fibroblast growth factor-1. Dev Cell 4(2):241–251
111. Keenan TM, Folch A (2008) Biomolecular gradients in cell culture systems. Lab Chip 8(1):34–57
112. Dertinger S, Chiu D, Jeon N, Whitesides G (2001) Generation of gradients having complex shapes using microfluidic networks. Anal Chem 73(6):1240–1246
113. Hong JW, Quake SR (2003) Integrated nanoliter systems. Nat Biotechnol 21(10): 1179–1183

114. Thorsen T, Maerkl SJ, Quake SR (2002) Microfluidic large-scale integration. Science 298(5593):580–584
115. Marcus JS, Anderson WF, Stephen R (2006) Parallel picoliter RT-PCR assays using microfluidics. Anal Chem 78(3):956–958
116. Wang M, Tu E, Raymond D, Yang J, Zhang H, Hagen N, Dees B, Mercer E, Forster A, Kariv I, Marchand PJ, Butler WF (2004) Microfluidic sorting of mammalian cells by optical force switching. Nat Biotechnol 23(1):83–87
117. Lee H, Liu Y, Ham D, Westervelt RM (2007) Integrated cell manipulation system-CMOS/ microfluidic hybrid. Lab Chip 7(3):331–337
118. Arora A, Simone G, Salieb-Beugelaar GB, Kim JT, Manz A (2010) Latest developments in micro total analysis systems. Anal Chem 82(12):4830–4847
119. Salieb-Beugelaar GB, Simone G, Arora A, Philippi A, Manz A (2010) Latest developments in microfluidic cell biology and analysis systems. Anal Chem 82(12):4848–4864
120. Lutolf MP, Doyonnas R, Havenstrite K, Koleckar K, Blau HM (2009) Perturbation of single hematopoietic stem cell fates in artificial niches. Integr Biol 1(1):59–69
121. Lecault V, VanInsberghe M, Sekulovic S, Knapp DJHF, Wohrer S, Bowden W, Viel F, McLaughlin T, Jarandehei A, Miller MM, Falconnet D, White AK, Kent DG, Copley MR, Taghipour FJ, Eaves CJ, Humphries RK, Piret JM, Hansen CL (2011) High-throughput analysis of single hematopoietic stem cell proliferation in microfluidic cell culture arrays. Nat Methods 8(7):581–586
122. Dykstra B, Ramunas J, Kent D, McCaffrey L, Szumsky E, Kelly L, Farn K, Blaylock A, Eaves C, Jervis E (2006) High-resolution video monitoring of hematopoietic stem cells cultured in single-cell arrays identifies new features of self-renewal. Proc Natl Acad Sci USA 103(21):8185–8190
123. Zhong JF, Chen Y, Marcus JS, Scherer A, Quake SR, Taylor CR, Weiner LP (2008) A microfluidic processor for gene expression profiling of single human embryonic stem cells. Lab Chip 8(1):68–74
124. Kalisky T, Quake SR (2011) Single-cell genomics. Nat Methods 8(4):311–314
125. Kirouac DC, Zandstra PW (2008) The systematic production of cells for cell therapies. Cell Stem Cell 3(4):369–381
126. Burdick JA, Vunjak-Novakovic G (2009) Engineered microenvironments for controlled stem cell differentiation. Tissue Eng Part A 15(2):205–219
127. Vunjak-Novakovic G, Scadden DT (2011) Biomimetic platforms for human stem cell research. Cell Stem Cell 8(3):252–261
128. Krawetz R, Taiani JT, Liu S, Meng G, Li X, Kallos MS, Rancourt DE (2010) Large-scale expansion of pluripotent human embryonic stem cells in stirred-suspension bioreactors. Tissue Eng Part C Methods 16(4):573–582
129. King JA, Miller WM (2007) Bioreactor development for stem cell expansion and controlled differentiation. Curr Opin Chem Biol 11(4):394–398
130. Lock LT, Tzanakakis ES (2009) Expansion and differentiation of human embryonic stem cells to endoderm progeny in a microcarrier stirred-suspension culture. Tissue Eng Part A 15(8):2051–2063
131. Placzek MR, Chung I-M, Macedo HM, Ismail S, Blanco TM, Lim M, Cha JM, Fauzi I, Kang Y, Yeo DCL, Ma CYJ, Polak JM, Panoskaltsis N, Mantalaris A (2009) Stem cell bioprocessing: fundamentals and principles. J R Soc Interface 6(32):209–232
132. Kim L, Vahey MD, Lee HY, Voldman J (2006) Microfluidic arrays for logarithmically perfused embryonic stem cell culture. Lab Chip 6(3):394–406
133. Titmarsh D, Hidalgo A, Turner J, Wolvetang E, Cooper-White J (2011) Optimization of flowrate for expansion of human embryonic stem cells in perfusion microbioreactors. Biotechnol Bioeng. doi:10.1002/bit.23260
134. Korin N, Bransky A, Dinnar U, Levenberg S (2009) Periodic "flow-stop" perfusion microchannel bioreactors for mammalian and human embryonic stem cell long-term culture. Biomed Microdevices 11(1):87–94

135. Luni C, Feldman HC, Pozzobon M, De Coppi P, Meinhart CD, Elvassore N (2010) Microliter-bioreactor array with buoyancy-driven stirring for human hematopoietic stem cell culture. Biomicrofluidics 4(3):034105
136. Khoury M, Bransky A, Korin N, Konak LC, Enikolopov G, Tzchori I, Levenberg S (2010) A microfluidic traps system supporting prolonged culture of human embryonic stem cells aggregates. Biomed Microdevices 12(6):1001–1008
137. Albrecht DR, Underhill GH, Wassermann TB, Sah RL, Bhatia SN (2006) Probing the role of multicellular organization in three-dimensional microenvironments. Nat Methods 3(5): 369–375
138. Nichol JW, Khademhosseini A (2009) Modular tissue engineering: engineering biological tissues from the bottom up. Soft Matter 5(7):1312–1319
139. Kraehenbuehl TP, Langer R, Ferreira LS (2011) Three-dimensional biomaterials for the study of human pluripotent stem cells. Nat Methods 8(9):731–736
140. Liu SQ, Tay R, Khan M, Rachel Ee PL, Hedrick JL, Yang YY (2010) Synthetic hydrogels for controlled stem cell differentiation. Soft Matter 6(1):67–81
141. Koh WG, Revzin A, Pishko MV (2002) Poly(ethylene glycol) hydrogel microstructures encapsulating living cells. Langmuir 18(7):2459–2462
142. Karp JM, Yeo Y, Geng W, Cannizarro C, Yan K, Kohane DS, Vunjak-Novakovic G, Langer RS, Radisic M (2006) A photolithographic method to create cellular micropatterns. Biomaterials 27(27):4755–4764
143. Nichol JW, Koshy ST, Bae H, Hwang CM, Yamanlar S, Khademhosseini A (2010) Cell-laden microengineered gelatin methacrylate hydrogels. Biomaterials 31(21):5536–5544
144. Aubin H, Nichol JW, Hutson CB, Bae H, Sieminski AL, Cropek DM, Akhyari P, Khademhosseini A (2010) Directed 3D cell alignment and elongation in microengineered hydrogels. Biomaterials 31(27):6941–6951
145. Du Y, Lo E, Vidula MK, Khabiry M, Khademhosseini A (2008) Method of bottom-up directed assembly of cell-laden microgels. Cell Mol Bioeng 1(2):157–162
146. Mironov V, Boland T, Trusk T, Forgacs G, Markwald RR (2003) Organ printing: computer-aided jet-based 3D tissue engineering. Trends Biotechnol 21(4):157–161
147. Zamanian B, Masaeli M, Nichol JW, Khabiry M, Hancock MJ, Bae H, Khademhosseini A (2010) Interface-directed self-assembly of cell-laden microgels. Small 6(8):937–944
148. Du Y, Lo E, Ali S, Khademhosseini A (2008) Directed assembly of cell-laden microgels for fabrication of 3D tissue constructs. Proc Natl Acad Sci U S A 105(28):9522–9527
149. Yanagawa F, Kaji H, Jang Y-H, Bae H, Yanan D, Fukuda J, Qi H, Khademhosseini A (2011) Directed assembly of cell-laden microgels for building porous three-dimensional tissue constructs. J Biomed Mater Res A 97A(1):93–102
150. Jain RK, Au P, Tam J, Duda DG, Fukumura D (2005) Engineering vascularized tissue. Nat Biotechnol 23(7):821–823
151. Tulloch NL, Muskheli V, Razumova MV, Korte FS, Regnier M, Hauch KD, Pabon L, Reinecke H, Murry CE (2011) Growth of engineered human myocardium with mechanical loading and vascular coculture. Circ Res 109(1):47–59
152. McGuigan AP, Sefton MV (2006) Vascularized organoid engineered by modular assembly enables blood perfusion. Proc Natl Acad Sci U S A 103(31):11461–11466
153. Kaully T, Kaufman-Francis K, Lesman A, Levenberg S (2009) Vascularization—the conduit to viable engineered tissues. Tissue Eng Part B Rev 15(2):159–169
154. Lovett M, Lee K, Edwards A, Kaplan DL (2009) Vascularization strategies for tissue engineering. Tissue Eng Part B Rev 15(3):353–370
155. Novosel EC, Kleinhans C, Kluger PJ (2011) Vascularization is the key challenge in tissue engineering. Adv Drug Deliv Rev 63(4–5):300–311
156. El-Ali J, Sorger PK, Jensen KF (2006) Cells on chips. Nature 442(7101):403–411
157. Schmid GM, Stewart MD, Wetzel J, Palmieri F, Hao J, Nishimura Y, Jen K, Kim EK, Resnick DJ, Liddle JA, Willson CG (2006) Implementation of an imprint damascene process for interconnect fabrication. J Vac Sci Technol B Microelectron Nanometer Struct 24(3):1283–1291

158. Hung P, Lee P, Sabounchi P, Lin R, Lee L (2005) Continuous perfusion microfluidic cell culture array for high-throughput cell-based assays. Biotechnol Bioeng 89(1):1–8
159. Rothschild M (2005) Projection optical lithography. Mater Today 8(2):18–24
160. Jiang J, Li X, Mak WC, Trau D (2008) Integrated direct DNA/protein patterning and microfabrication by focused ion beam milling. Adv Mater 20(9):1636–1643

Chapter 10
Application of Micro/Nanotechnology to Stem Cell Research and Technology

Amir R. Aref

Abstract Emerging technology has the potential to provide solutions to the devastating complications of illnesses, for people of all ages and genders and all backgrounds. Nevertheless, there are difficulties. Perhaps the most challenging area is transplantation, and in particular using stem cells. Transplantation implies contact, hence surface interactions, between the stem cell and the host tissue. Attachment and spreading of a cell on a substratum are the first part of the process that leads to the ultimate assimilation of the new cell into the host tissue. Together with confocal microscopy, we have exploited a uniquely powerful non-invasive optical technique and a 3-D microfluidic system by integrating a hydrogel scaffold into a PDMS device for cell growth, with co-culture capability to quantity attachment and spreading, and determine how the cell environment (the substratum, which might be tissue or an artificial non-living implant (prosthesis); the complex liquid medium bathing the cell; and the possible presence of congeners) influence attachment and spreading. This novel microfluidic platform has proven to be a versatile and powerful tool to study cell migration for various biological applications. This chapter highlights an overview of the application of Micro/Nanotechnology to stem cell research and technology.

10.1 Overview of Stem Cells

The fact is that a great number of people now are suffering from diabetes, kidney failure, spinal cord injury, Parkinson's disease and many types of cancers. Most of us share at least a piece of this experience, our loved ones, in times of pain or need,

A. R. Aref (✉)
Department of Biological Engineering,
Massachusetts Institute of Technology, Cambridge, MA, USA
e-mail: aref@mit.edu

H. Baharvand and N. Aghdami (eds.), *Advances in Stem Cell Research*,
Stem Cell Biology and Regenerative Medicine, DOI: 10.1007/978-1-61779-940-2_10,

reaching out, looking to us for comfort, for a way to stop their suffering. This is the quest we join, together, when we contemplate the promise of stem cell research, debate its proper methods and work toward making our hopes a reality curing disease and disability through stem cell derived therapies.

However, stem cell research starts with understanding its potential for you and me, our parents and children, our friends and families. For people with Parkinson's disease, stem cell research holds the promise of replacing destroyed specialised brain cells. For spinal injury patients, it offers the potential for regeneration of neural tissue, which would reconnect the pathways of sensation and motor control and allow them to walk again, or talk again, or hug their child again.

It may also provide solutions to the devastating complications of diabetes: blindness, kidney failure, amputation and cardiovascular disease. Stem cell research offers hope for people with a great diversity of illnesses, for people of all age and genders and all backgrounds. It offers hope for each of us, and that hope is not measured by numbers, for a review, see e.g. [1].

10.1.1 Nanotechnology and Where is It Used Now

Nanotechnology has become one of the most important and exciting fields during the last few years uniting physics, chemistry, engineering and biology. It shows great promise for providing us with breakthroughs that may change the direction of technological advances in a wide range of applications (for a review, see e.g. [2]). Nanotechnology is thus defined and concerned with the design, synthesis, characterisation and application of materials and devices with control at least one dimension on the nanometre scale (see, e.g. [3]).

The real meaning of nanotechnology is therefore size and control. The consideration of nanotechnology is rather new, but it is by no means is based with atoms and molecules. In fact, 'the disciplines of physics, chemistry and biology have long dealt with atoms and molecules, their behaviour and their manipulation; and quantum mechanics is already firmly established as the science of the absolutely small' see Fig. 10.1 [4].

It was first described in 1959 by Feynman in his famous talk 'There's plenty of room in the bottom'. He introduced the idea of making things smaller, so small that ultimately the manipulation of single atoms would be possible. He proposed that such an improvement in physics would eventually find applications in other sciences such as biology, chemistry and electronics [5]. Despite Feynman's vision and enthusiasm, the progress of nanotechnology really took off in the 1980s when the development of apparatus such as the scanning tunnelling microscope [6] resulted in the opportunity to 'see' things at the nanometre scale.

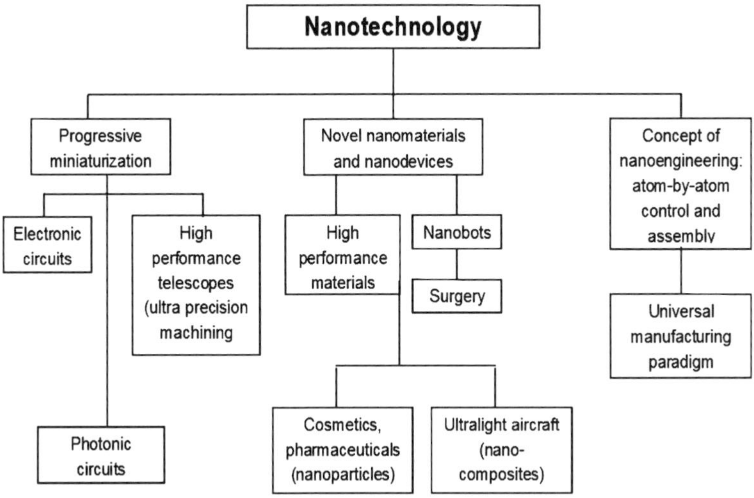

Fig. 10.1 From *left* to *right*, the indirect, direct and conceptual branches of nanotechnology, with examples [4]

10.1.2 Nanotechnology and Biology

Nanotechnology and biology are young and rapidly evolving fields of research at the junction of biotechnology and nanoscience, these two areas, each of which combines advances in science and engineering. Although the field of nanobiotechnology is still in a fairly embryonic state and the applications identification process are also used to generate related to health and environmental issues. Currently, nanobiotechnology is a field that concerns the utilisation of biological systems optimised which capable the evolution such as cells, cellular components, nucleic acids and proteins, to fabricate functional nanostructured comprised of organic and inorganic materials. Nanobiotechnology and Bionanotechnology are essentially synonyms [4] refer to the materials and processes at the nanometer scale that are based on biological, biomimetic or biologically molecules and also concern the modification and application of instruments that can precisely control the biological systems and initially designed to generate and manipulate nanostructured materials, to basic and applied studies of fundamental biological processes (for a review, see e.g. [7]).

Nanotechnology is beginning to help advance the equally pioneering field of stem cell research, with devices that can precisely control stem cells and provide nanoscaffolds and magnetic tracking systems. The characteristics of stem cells indicate that these cells, are used in developmental biology studies, have the potential to provide an unlimited supply of different cell types for tissue replacement, drug discovery and functional genomics applications [8].

10.1.2.1 Nanotechnology and the Cellular Environment

Most biological systems are highly complex systems which have evolved over time to respond to various changes in a cell's environment. In general, cellular behaviour is derived from intrinsic and extrinsic factors. The intrinsic factors are the internal genetic make-up of a cell defined by gene regulatory networks. The regulation of the gene networks is through stochastic and deterministic intracellular events. Although these events regulate the cell behaviour at times in an autonomous manner, many of the signals that regulate cellular behaviour are extrinsic and derived from the surrounding environment. The ability to control cell behaviour is important for a wide range of applications in biotechnology and tissue engineering. Developmental biology has shown that the environment plays an important role in controlling cellular differentiation, proliferation and of particular interest in controlling the cellular environment is to control the interactions at the nano length scales. In the body, cells interact with their surrounding tissues on a sub-cellular level as well as on a tissue level. For example, many extracellular matrix (ECM) fibres arc on the multinanometer length scales. Therefore, the ability to control the in vitro cellular environment on the nano-length scales by using engineering and materials sciences approaches is of great interest in controlling cell behaviour (for review, see e.g. [9]).

Nanotechnology has been increasingly used to control the cellular environment. For example, nanomaterials such as self-assembled peptides and electrospun nanofibers are emerging as powerful technologies to generate 3-D structures. In addition, nanoscale topography has been shown to be a powerful regulator of cell behaviour such as alignment, migration and differentiation (see e.g. [10]).

10.1.2.2 Nanotechnology and Stem Cells

During the last few years, stem cell biology has pioneered a rapid advancement, leading to rising hopes that stem cells, migrating towards a lesion target area, will contribute to functional improvement. Stem cell research area has recently generated more public and professional interest than almost any other topic in biology. One reason stem cells capture the imagination of so many is the promise that understanding their unique properties may provide deep insights into the biology of stem cells, as well as a path towards treatments for a variety of degenerative illnesses. And although the field of stem cell biology has grown rapidly, there exists considerable confusion and disagreement as to the nature of stem cells (for review, see e.g. [11]).

Therefore, there is a need to develop methods for visualising transplanted stem cells for a better understanding of their migrational dynamics and differentiation process and of their regeneration potential. Initial cell surface reactions may trigger multiple responses, which in turn result in either spreading or detachment of the cell. Cell adhesion and mobility depend strongly on the interactions between cells and ECM substrates [12].

10.1.2.3 Cell Adhesion

When cells are cultured outside the body in an artificial environment, they inevitably lose many of their in vivo characteristics. An essential principle is that cells are almost never in direct contact with an artificial surface: there is almost always an intermediate layer of protein, and if the protein is not provided beforehand, the cells will typically synthesise and excrete it themselves before bonding to the surface, and a second principle is that the behaviour of the cells is modified by their interaction with the surface.

In the human body, cells are in contact with a complex combination of proteins, proteoglycans and glycoproteins that make up the basement membrane and the ECM of a tissue. These acellular portions of a cell's natural environment vary from tissue to tissue in both composition and structure. When cultured on flat surfaces, such as the standard petri dish, cells will often lose their native morphology. Cells typically flatten and spread out on flat surfaces, and as a population (for reviews see e.g. [10]). Cells grown in a liquid culture medium will usually be spherical as expected from a simple application of surface energy minimisation. A fascinating and complex sequence of reactions results when a cell lands on a solid substratum. The shape change happens from sphere to segment and during this time the cell-substratum contact increases from zero to tens of square micrometers [13].

Spreading can take place longer to perform and many molecules are able to mediate cell adhesion, but few are able to support cell spreading. Initially, exploratory filopodia stick out from the cell, driven by localised actin polymerisation. Complicated topographical, mechanical and chemical information is conveyed from specialised receptors to the nucleus of the cell, and a suitable response is generated.

10.1.2.4 Cell Binding Mechanisms (Morphology)

Suspended cell morphology is that of a sphere and when attaching to a surface they first bind and then spread as they become more established. Figure 10.2a depicts rodent fibroblasts 10 min after attachment with the cells still mainly spherical in appearance. The same cells at 180 min are fully spread, morphologically distinct and well established (Fig. 10.2b).

The cells of multicellular organisms bind to the ECM and each other. The ECM consists of macromolecules secreted by cells such as collagen, proteoglycans and glycoproteins (e.g. fibronectin and laminin) that interact with the cell through integrin receptors at the cell surface. The glycoproteins give an ordered structure to the ECM and the integrins bind to these. Integrin receptors are attached to filaments within the cell cytoplasm, span the cell membrane and attach to arginine-glycine-aspartate (RGD) domains on the proteoglycan [14, 15].

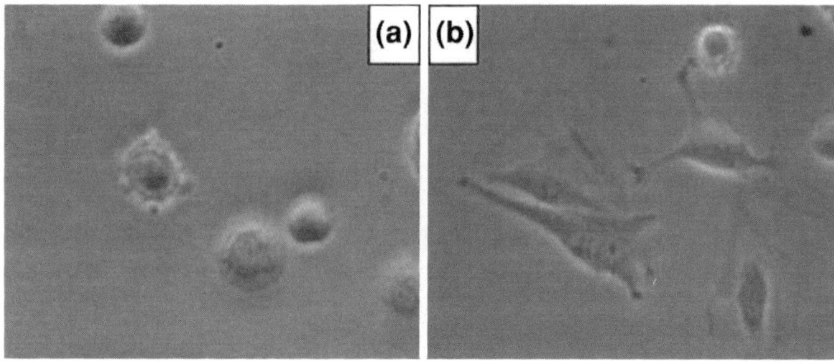

Fig. 10.2 Cells attachment on the glass surface after 10 min (**a**) and after 180 min (**b**)

10.1.2.5 Cell–Cell Signalling and Internal Cell Signalling

Cell to cell signalling involves the release of a signalling molecule from one cell which binds to a specific receptor on another cell. This can be paracrine, endocrine of autocrine and in addition cells can also communicate via gap junctions. Binding of the signalling molecule results in a receptor conformational change which initiates secondary signalling cascades that travel through the cell's internal machinery [16]. By changing the cell's environment, one also changes the cell's genetic expression. Genes that are normally expressed can be upregulated or downregulated, while genes that normally are not expressed at all may be turned on. This may lead to very different functionality in cultured cells compared to their in vivo counterparts [10].

Cells Behavioural on Bionanomaterials

Attachment and spreading is a key behavioural indicator of living cells. Nanoobjects materials offer many possibilities for the modification of various materials turning them into nanomaterials, such as nanofibers and nano carbon tubes for optimal attachment, proliferation and spreading of different kinds of cells (reviewed see e.g. [17]). Once cells plated onto artificial adhesive surfaces, cells first flatten and deform extensively as they spread. At the molecular level, the interaction of membrane-based integrins with the ECM has been exposed to initiate a complex flow of signalling events, which subsequently triggers cellular morphological changes and results in the generation of contractile forces [18].

Stem cells are a useful tool at this stage for investigation methods relating to the extraction of specific cell types from mixed cell population and to study the differentiation events. Studies of the interactions between substrate topography and stem cells have been encompassed a wide variety of cell types and substratum features including grooves, ridges, steps, pores, wells, nodes and adsorbed protein fibres.

Extracellular Matrix Enhancement

Tissues are complex and are typically organised into a well-defined, 3-D structure in our bodies. This architecture contributes significantly to the biological functions in the tissues, including the supply of oxygen and nutrient support, and a spatial environment for the cells to grow [19]. In this respect, there are three key factors to be considered for the success of tissue regeneration: cells, scaffold (i.e. the ECM) and cell–matrix (scaffold) interaction. With the ability to form nanolength structures, a drive to mimic the ECM and form scaffolds that are an artificial ECM suitable for tissue formation has begun. These scaffolds typically attempt to mimic collagen, a natural ECM component of nearly every tissue such as bone, skin, tendon, ligament and so forth [20].

When a cell comes into contact with biomaterial, it will perceive the chemistry of a surface using integrin transmembrane proteins to find suitable sites for adhesion, growth and maturation. In vitro, cells will produce distinct morphologies when motile and when adhered and entering the S-phase (for those capable of undergoing division) [21].

Tissue structure and function depend greatly on the arrangement of cellular and non-cellular components at the micro- and nanoscale level. The ECM provides a substrate with specific ligands for cell adhesion, movement, proliferation and function by providing a range of growth factors.

The nanoscaled dimensions of the physical structure of native ECM are known very well, for example, in a classic connective tissue, structural protein fibres such as collagen and elastin fibres have diameters that range from several tens to several hundred nanometers. It is reasonable to expect that an ECM-mimicking tissue-engineered scaffold will play a similar role to promote tissue regeneration in vitro as native ECM does in vivo and indicate that the chemistry of the ECM is of comparable importance to the morphology in influencing cell behaviours (see e.g. [22, 23]).

Understanding of how the microenvironment can influence the cell behaviour will aid the development of the next generation of scaffolds for tissue engineering and stem cell applications. Therefore, the design of nano-featured tissue scaffolds is exiting field and novel work, opening a new area in tissue engineering (for reviews see e.g. [24]). Extracellular matrix components have demonstrated that the presence of these molecules affects morphology, proliferation and morphogenesis of differentiated cells [25].

10.1.2.6 Nanomaterials

The role of the surface chemistry on cell behaviour has been explored extensively in the past decades [26]. Substratum topography has direct effects on the abilities of cells to orient themselves, migrate and produce organised cytoskeletal arrangements [27]. The nano revolution offers great promise for new creation of biomaterials and biomedical surfaces.

Nanostructured surfaces, such as polyamide nanofibers, nanoparticles and carbon nano tubes are considered for various applications in stem cell biology and technology. Most nanomaterials in various physical or chemical forms may be able to enhance the cell response selectively for biological application in tissue integration.

Bionanomaterials based on nanostructures define in the 1–100 nm range; have been used to create materials that have novel properties and functions. In particular, nanofibers, carbon nanotubes (CNTs) and magnetite nanoparticles are now used to target synthetic ECM and plasmids to specific cell types while protecting these macromolecules from enzymatic degradation [28, 29]. Additionally, 'nanoparticles have been proposed for the treatment of many diseases that need constant drug concentration in the blood or drug targeting to specific cells or organs' [30]. Nanomaterials can be classified as two main groups: natural or manmade. The former are produced naturally: such as proteins in the body, which control things like transfer of information or repairing cells.

A living cell is a gorgeous example of a highly sophisticated, controlled system of nature's nanomaterials, proteins and other biomolecules, which self-assemble into various supermolecular architectures and interact in a well-defined manner at the nanometer scale. Nanocomposites permit materials combining a greater variety of attributes than is possible with simpler materials [13] and also nanodevices permit the development of information processing and thus responsivity to be incorporated into coatings, combined with ultracompact artificial sensors. Flemming [31] have described the topography of the basement membrane and reviewed the effects and fabrication of synthetic nanostructured surfaces on cell behaviour. Rosenthal [32] have prepared serotonin-labelled nanocrystals and demonstrated that these nanocrystals inhibit serotonin transport activity in transfected cells.

Carbon Nanotubes

Another class of nanomaterials are carbon nanomaterials, which are a macromolecular form of carbon with high potential for biological applications due in part to their unique mechanical, physical and chemical properties. Carbon nanomaterials are strong, flexible, may conduct electrical current and can be functionalised with different molecules, properties that may be useful in basic and applied biological research. Carbon nanotubes have attracted a great deal of attention due to their unique structural, electrical and mechanical properties [33]. Recently, there has been an intense interest in exploring some of their novel properties, such as superior strength, flexibility, electrical conductivity and availability of chemical functionalisation [34] for biological applications both at molecular and cellular levels.

Studies of the interaction between CNTs and living cells are still limited. Mattson [35] reported the feasibility of using CNTs as a substrate for neuronal growth. Hu [36] have reported chemically functionalised CNTs as substrates for neuronal growth and studied fully interaction between CNTs and living mammalian cells and their results show that the chemical fictionalisation of multi-wall carbon nanotubes (MWCNTs) possess a diameter almost 100 nm and aspect ratios that are similar to those of small nerve fibres.

That idea of reconnecting the pathway of sensation signal after neuronal injury such as spinal cord injury is not new and during a last few years advances in nanotechnology have stimulated a renewed interest in the development of novel neural biomaterials that could potentially be used for reconnecting neurons. CNTs are a novel form of carbon made of a rolled layer of grapheme, which appears well suited to the design of novel neural biomaterials [37].

Nanofibers

Nanofibrous scaffolds structured by electrospinning certain polymers are highly porous have a high surface area to volume ratio, and morphological properties similar to collagen fibrils [38]. These physical characteristics promote favourable biological responses of seeded cells within these scaffolds including enhanced cell attachment, proliferation and maintenance of the chondrocytic phenotype [39]. Attempts to make nanofibrous structures began for tissue formation to make scaffolds that are an artificial ECM substitute. These scaffolds aim to mimic collagen, a natural ECM component in tissues such as ligament, tendon and bone and even skin and they reach scales between 50 and 500 nm [20].

Magnetite Nanoparticles

Iron oxide nanoparticles are the end products of a wide variety of physical, chemical and biological processes and have attracted a considerable attention during the last decade, and they have been of great interest in many important technological applications. The use of magnetite nanoparticles in clinical and medicine is important and has considerable promise for applications in the biomedical and diagnostic fields such as diagnostic medicine, targeted drug delivery, hyperthermic treatment for malignant cell, cell labelling, tissue repair and magnetic resonance imaging (MRI) [40, 41].

Another application of magnetic nanoparticles is in separation because the reactivity of the particles can be adapted by modifying the surface coatings on the nanoparticles. The nanoscale particles can be recovered for reuse and they have very high surface areas. The magnetic nanoparticles which are widely used for biomedical applications are Fe_3O_4 and related oxides, which are chemically stable, nontoxic, non-carcinogenic and have useful magnetic properties.

10.1.2.7 Labelling and Tracking of Stem Cells

During the last few years, stem cell biology has experienced a rapid advancement, leading to rising hopes that stem cells migrating towards a lesion target area, will contribute to functional improvement. There is a need to develop non-invasive methods to visualise transplanted stem cells [42, 43] for a better understanding of their migrational dynamics and differentiation processes and of their regeneration potential [44].

Stem cells can be labelled with antibody-bound magnetic nanoparticles and tracked [45]. Björklund [46] transplanted ESCs in a Parkinson rat model,[1] and for tracking they used iron-based nanoparticles. Examination of the literature thus far shows nanoparticles are feasible as imaging contrast agents for magnetic resonance and optical imaging.

MRI Contrast Agents

One of the main imaging technology and extraordinary 3-D capabilities is used by most research studies to track stem cells in vivo is MRI [43]. In general, nanoparticulates provide enormous surface area, which can be functionalised to serve as a scaffold for targeting ligands and magnetic labels and a wide variety [47] of iron oxide-based nanoparticles,[2] have been developed that differ in hydrodynamic particle size and surface coating material (dextran, starch, albumin, silicones).

Optical Labelling

Biological probes are indispensable tools for studying biological samples, cells in culture and animal models. Exogenous probes are frequently multifunctional, having one component that can detect a biological molecule or event, and another component that reports the presence of the probe. Fluorescent dyes are used regularly for determining the presence and location of biological molecules in cultured cells and tissue sections. The fluorescent moiety reports the presence and other parts of that molecule bind to the biological molecule. However, the technique is only semiquantitative, precision of is limited by optical resolution, and many dyes are toxic to cells.

Moreover, studies of biomolecules at the single molecule level often involve large teams of specialised biophysicists, sophisticated and finely tuned equipment, and long periods of time to record and analyse relatively few events. Moreover, these studies also suffer from other limitations of certain organic dyes, such as photobleaching, instability and low quantum efficiency [48, 49].

---

[1] The aim of the study was to examine of the potential of ESCs to differentiate into dopamine neurons and integration within the host brain.

[2] Iron oxide nanoparticles are used of a wide variety of physical, chemical and biological processes such as diagnostic medicine.

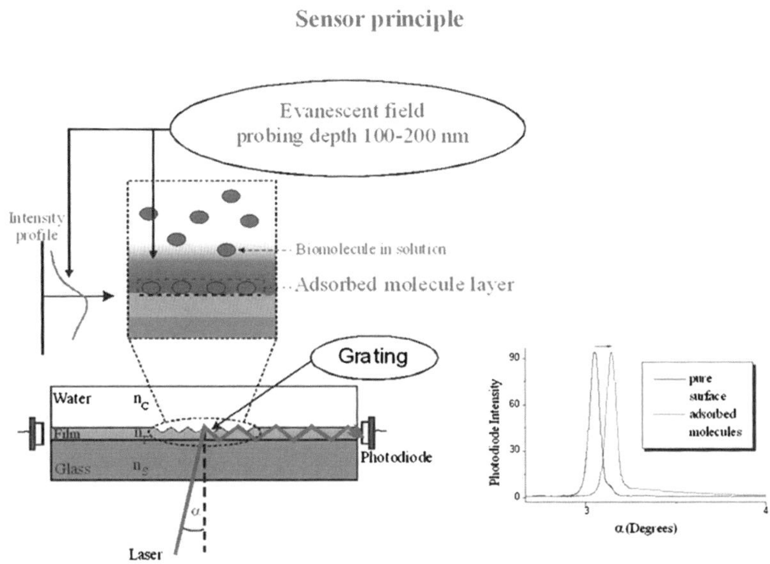

Fig. 10.3 Schematic illustration of the adsorption of proteins and cells on the waveguide [80]

10.1.2.8 Label-Free Optical Sensing

While techniques involving radioactive and fluorescent labelling have provided insight and many clues as to the nature of various cellular processes the fact remains that they require a modification to the biological molecule of interest. This will inevitably affect the function of the molecule [50]. Therefore, non-invasive label-free optical sensing and other novel techniques with undoubtedly play an important role in future investigations of cellular and subcellular systems.

Optical Waveguide Lightmode Spectroscopy

Optical waveguide lightmode spectroscopy (OWLS) is extremely powerful technique for the precise quantification of thin films material deposition on a solid surface and has mainly been used to study the adsorption of protein layers [50]. This technique is also very well suited for cell-based experiments which investigate cell binding and cell spreading parameters [49–51]. Quartz crystal microbalance (QCM) investigations [51, 52] also give information on attachment kinetics; however, with very little information on spreading [53]. OWLS was first used to measure protein adsorption kinetics. Later it was shown that the non-invasive and on-line measurements possible with OWLS are also attractive for applications with cells [54]. In particular, the adhesion and spreading of cells such as fibroblast strains was investigated, and have demonstrated the possibility of OWLS as a complete cell sensor by looking at cell adhesion and morphology changes of fibroblasts [55], Fig. 10.3.

Application of OWLS in Stem Cell Researches

One of the main current problems in the stem cell field is to determine the kinetics of differentiation, and how it is influenced by various environmental factors. At present, optical microscopy is the main observational tool in use. Although it is easy to implement experimentally, quantitative analysis of the results is rather difficult, and some degree of subjectivity in the interpretation is almost inevitable.

The spreading of a living cell in contact with the surface of a substratum is one of the fundamental responses of the cell to such contact [56]. There are myriads of cell types and environments—including here both the nature of the substratum and of the liquid medium bathing cell and substratum—and not all combinations thereof lead to spreading (see e.g. [54]). Moreover, the actual temporal course of the spreading, i.e., its kinetics, depends sensitively on cell type and environment [50, 56, 63]. It follows that these kinetics should be a sensitive indicator thereof, and, *a fortiori*, of differentiation, provided that an appropriately accurate way of measuring them can be found. Optical (e.g. [56]) or electron (e.g. [57]) microscopies are really unsuitable for this purpose, not only because of the difficulty of getting good time resolution (although modern techniques of digital photography have to some extent overcome that difficulty), but also because the information obtained is typically rather incomplete (for example, just a cross section, i.e., either plan or elevation).

Sometimes, more complete information can be obtained, such as using techniques of confocal microscopy, but these have the disadvantage of requiring the cell to be fed with fluorescent dyes, thereby altering its physiology in usually unwanted ways.

Nevertheless, a suitably informative technique capable of high temporal resolution has been developed, notably optical waveguide lightmode spectroscopy (OWLS), which is based on perturbation of the evanescent field generated at the surface of an optical waveguide by the presence of the cells [58]. The method relies on the refractive index of material of the cell being different from that of the medium bathing the cell. Since, the evanescent field decays exponentially from the waveguide surface (the cell substratum) into the medium, the perturbation depends on the number of cells per unit area, the refractive index of the cells and its distribution within them, as well as the size and shape of the cells [58].

One of the main goals in current stem cell research is the controlled differentiation of a pluripotent cell into a specific end form. This is typically achieved by the addition of certain molecules to the ambient medium. Although some patterns of differentiation are well-characterised empirically, the process as a whole depends on complex internal regulatory machinery inside the cell, and is not well understood. At present, the identity of a differentiated form has to be established by laborious morphological and other ways of phenotypical characterisation (including the expression pattern of selected genes) that may take days or even weeks to be concluded. It would, therefore, be a great advantage if an in situ technique were available with which the course of differentiation could be followed as it took place. In this work, I show that high resolution OWLS can be adapted to provide such a technique.

Stem Cell-Substratum Interactions

Up until now cell biologists investigating the attachment and spreading of *living* cells at solid–liquid interfaces, the quantification of which is a fundamental part of studies of the biocompatibility of medical devices, for example, have had little alternative but to use optical microscopy, but effective exploitation of this technique (e.g. using a confocal arrangement) requires the cells to be fed on relatively large quantities of toxic dye molecules, which to a greater or lesser degree affect the physiology and hence behaviour of the cells. A non-imaging alternative based on analysis of the perturbation of the evanescent field generated at the surface of an optical waveguide has been developed more recently, but this does not allow signals emanating from the cell to be separated from signals emanating from proteins, for example, microexudate from the cell, deposited on the waveguide. The significant advance reported in this work describes how a deeper and more extensive analysis of the waveguide signals enables protein deposition, and/or focal adhesions, to be measured separately from overall cell shape changes. The method is immediately applicable to high-throughput screening [59].

Modulation of the coupling of light into a waveguide via a grating is used to investigate in the first section the attachment of human embryonal carcinoma stem cells to three substrata, silica–titania (representative of artificial implants), poly-L-lysine (a commonly used laboratory cell culture substrate) and mucin (the coating of the mucosa), and in the second section the attachment at different seeding number of cells. By analysing selected optical parameters of the cell-waveguide system we distinguish the formation of filopodia, and the overall change in its shape (spreading). Moreover, the kinetics of these processes with excellent time resolution were obtained [59, 60].

Kinetics, Substrates Interactions (PLL, Mucin, Silica–Titania)

Within the (eucaryotic) cell, the nucleus, rich in nucleic acids, has a higher refractive index than the cytoplasm; and spreading typically implies movement of the nucleus nearer to the substratum, increasing the optical contrast between the covered and uncovered zones. The micro-morphology of the cell-substratum 'contact', already referred to, will also affect the optical contrast sensed by the exponentially decaying evanescent wave. Finally, the cell may secrete macromolecules as part of its adaptive response to the substratum.

These diverse effects are taking place simultaneously and are prima facie difficult to separate from one another. Here we show that by simultaneously comparing the evolution of the incoupling peak width with the evolution of the incoupling peak position (which depends on the propagation constant of the guided lightmode) it is possible to separate microexudate secretion (which does not affect incoupling peak width), filopodium formation, and cell body spreading, without perturbing the cell. Such observations are not possible with optical microscopy, let alone with high time resolution.

This work was investigated three distinctively different cell substrata [61]. Silica–titania is representative of artificial implants (titanium is the most common material used for prostheses and in the body is invariably oxidised; the salient surface energetic features, most notably the Lewis acids/base parameters, of titanium are almost the same as those of silicon [62], poly-L-lysine (a very commonly used cell culture substrate in the laboratory) and mucin, found ubiquitously coating the mucosae [18], of which it is the major macromolecular component; mucins comprise a family of heavily glycosylated, high molecular weight glycoproteins with carbohydrate side chains that make up some 50–80% of the molecule by weight [63].

Figure 10.4 shows the microscope images of the cells on the three different substrata. Even visually, it is very apparent that the cells have not spread on the mucin (a), have significantly spread in a circularly symmetrical fashion on the PLL (b), and have spread in an elongated (c) fashion on the $Si(Ti)O_2$. The corresponding kinetic data showing the evolution of the incoupling angle directly related to the waveguide propagation constant, eq: ($\alpha = \sin\, \alpha + \lambda\kappa/\Lambda$) and the incoupling peak width are shown in Fig. 10.5.

There are very striking differences between the cell responses to the three substrata. From Fig. 10.4, it is clear that the cells spread well on PLL, less well on $Si(Ti)O_2$, and perhaps not at all on mucin; this merely confirms the visual observations. Figure 10.4b, however, clearly implies that the cells do not spread at all on mucin, and the small and slow increase in $\Delta\alpha$ for the cells on mucin is likely due to microexudate, i.e., the ECM molecules secreted by the cell in an attempt to render an uncongenial substratum congenial. In fact, this is suggested by two facts: (i) the absence of any change in incoupling peak width; and (ii) the absence of the sigmoidal kinetics of increase of propagation constant, which appears invariably to be associated with cell spreading (e.g. [64]), whereas protein deposition kinetics usually give the hyperbolic behaviour seen here (e.g. [64–66]).

The evolution of Δw reaches a peak when 50% of the surface is covered by the cells. That cells on PLL reach this peak much faster than on $Si(Ti)O_2$ is not surprising, given the more rapid evolution of $\Delta\alpha$. What is revealing is that the value of $\Delta\alpha$ at the moment when Δw reaches its peak is much greater for PLL than for $Si(Ti)O_2$. We propose that this indicates that in cell-substratum contact is stronger (especially through more filopodia) for the former than for the latter. The fact that the Δw peak height is much greater for the PLL also corroborates this inference, for it implies that the optical contrast between the cell-covered and the uncovered regions of the waveguide is greater.

It is intriguing that what might be considered to be a cell that has not fully recovered from the storage in liquid nitrogen shows very different behaviour (see Fig. 10.5). Firstly, the temporal evolution of $\Delta\alpha$ is not sigmoidal, but hyperbolic. Secondly, there is a distinct lag before Δw starts to increase. Assuming that the initial increase of $\Delta\alpha$ is due to protein secretion (which cannot cause Δw to increase); note also the hyperbolic kinetics, characteristic of protein deposition,

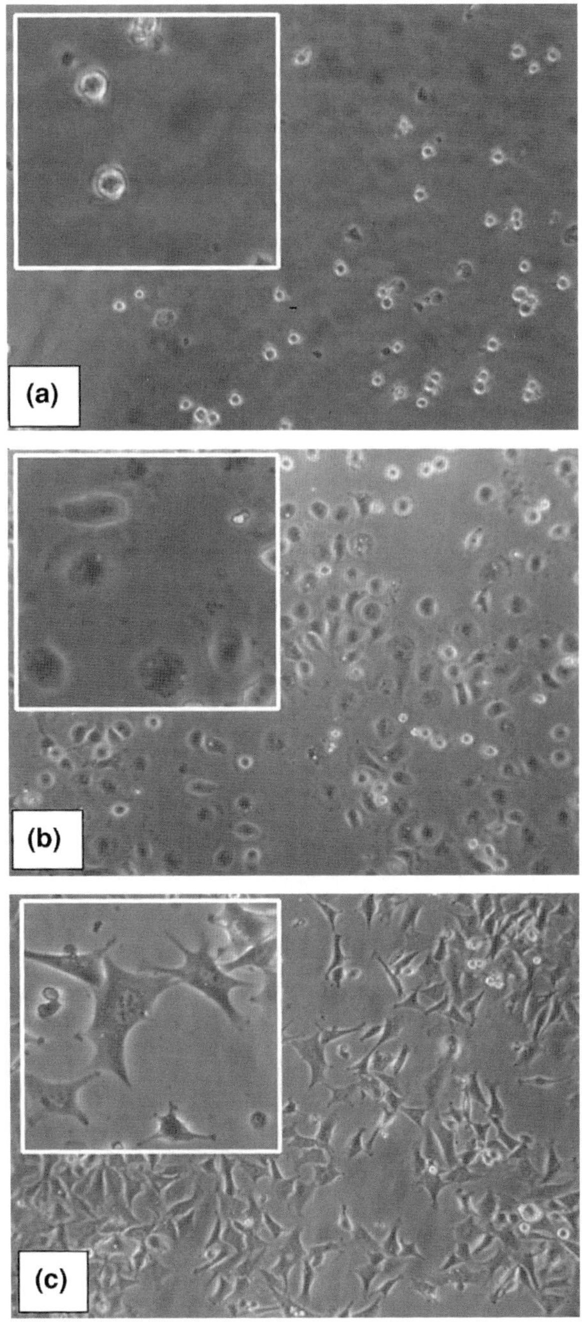

Fig. 10.4 Optical micrographs of TERA2 stem cells on **a** mucin, **b** PLL, **c** Si(Ti)O_2 after 120 min of culture. Each main image is 850 μm wide; the insets are 170 μm wide

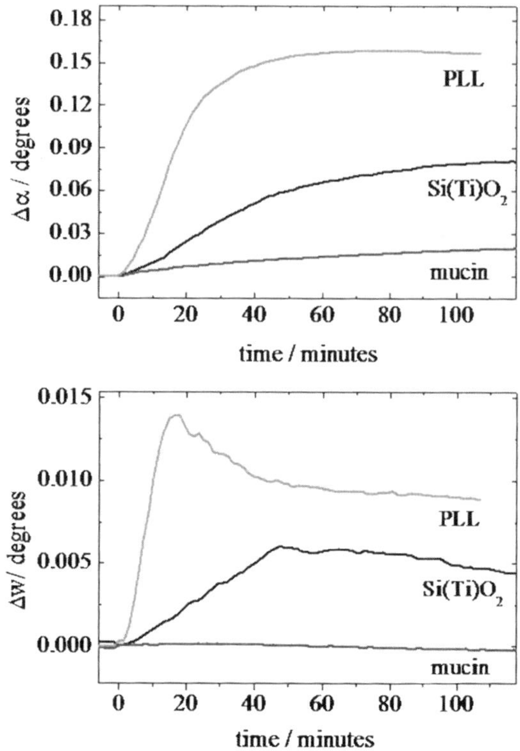

Fig. 10.5 *Upper panel*, **a** incoupling angle, given as the difference $\Delta\alpha$ between the actual value and the value immediately prior to seeding the waveguide with 5×10^4 TERA2.spl2 stem cells, grown for at least 2 days with at least one passage, in a total volume of 550 μL on (from *top* to *bottom*) PLL, $Si(Ti)O_2$, and mucin. *Lower panel*, **b** incoupling peak width, given as the difference Δw between the actual value and the value immediately prior to seeding the waveguide with the cells

e.g. [64], we estimate the amount of protein after 30 min to be about 1.6 ng/mm^2. From the rather low value of the Δw peak height, we also infer that the cell-substratum contact is weaker than for cells that have undergone more passages. This can be made quantitative: we propose that the value of $\Delta\alpha$ at the instant when Δw is at its peak is equated with the cell-substratum adhesivity (a composite quantity indicating the strength of the cell-substratum contact), [59].

10.1.3 Application of Stem Cells and Nanotechnology

Stem cells have generated a great deal of interest as a source of cells for tissue engineering or cell-based therapy. These include directing the differentiation of stem cells (using controlled microenviroments or genetic engineering) to ensure their safety and efficiency in vivo. We are getting close to the day [67] when stem cells can be manipulated in culture to produce fully differentiated cells that can be used to create and repair specific organs.

Currently, more than 74,000 patients in the United States are awaiting organ transplantation, and only 21,000 people receive transplants annually [67], at this time nanoengineering and biological limitations such as material biocompatibility, synthetic scaffolds that support tissue growth by serving as the ECM associated and immune system's reaction to shed antigens by the transplanted cells are some of the challenge that prevent these systems from widespread clinical applications. In tissue engineering, biological and engineering principles are combined to produce cell-based substitutes with or without the use of materials. A variety of studies support the concept that nanotechnology application of instruments that can precisely control stem cells, originally designed to generate and manipulate nanostructured materials, to basic and applied studies of essential biological processes.

10.1.4 Evaluation of Toxicity of Nanomaterials with Stem Cells

There is a serious lack of information concerning the effect of nanoparticles on human health and the environment, such widespread application of nanomaterials may cause significant potentials for human exposure. Therefore, appropriate risk assessment and management of nanomaterials should be performed to assess and regulate these nanomaterials to protect human health and the environment [68]. These nanomaterials are unique because their physical behaviour when measuring less than 100 nm changes from classical to becoming by quantum physics and are small enough to penetrate even very small capillaries throughout the body by one route but be widely disseminated to various organs and tissue [69]. Thus, they can affect the physiology of any cell and the cell exposed to foreign particles initiate an immunological response. Also, particles may have an effect on the cellular transcription process and change at the molecular level.

This consideration is importance for the development of stem cells research, where the effects of nanoparticles on their potential for self-renewal and differentiation are largely unknown. Braydich–Stolle [30] by using mouse spermatogonial stem cell line as a model to assess nanotoxicity in the male germ line in vitro demonstrated a concentration-dependent toxicity for all types of particles tested, whereas the soluble salts had no significant effect. Silver nanoparticles were the most toxic while molybdenum trioxide (MoO_3) nanoparticles were less toxic.

Single wall carbon nano tubes (SWNTs) in suspension in the culture medium were incorporated into the cell cytoplasm by macrophages and leukaemia cells without affecting the cell population growth [70]. But it has been shown that CNT substrates decreased keratinocyte [71], glial [72] and HEK293 cell survival significantly [73] raising important concerns about the biocompatibility of the nanomaterials.

Nanoparticles may also gain access to the body by ingestion or be generated within joints as wear particles from the bearing surfaces of a prosthetic replacement. The sites of deposition of these particles include the liver and spleen, in which organ there may be general consequences of exposure, for instance the stimulation of immune systems [69].

Despite there is a large body of evidence about the effects of nanoparticles on the living systems and specially cells from which they are constituted, these effects are largely unknown and remain to be investigated about the biological consequences of exposure to nanoparticles. In conclusion, nanoparticles represent a new challenge to those involved with toxicology and biocompatibility and little is known of the biological effects to nanoparticles and the increasing availability of sophisticated methods of evaluating biological phenomena, including developmental biology, drug discovery and molecular biology especially as it is applied in immunology and genetics, present opportunities for unfolding knowledge in this exciting and important area [69].

10.1.5 Microfluidic Technology and Cell Therapy

Microfluidic technology has shown promise in additional venues requiring a model ECM, chemotactic gradient and cell–cell interactions [74], but there are limited studies on stem cell research, cancer cell. This technology requires further development before it can be applied to the quantitative assessment of molecular and cellular level behaviour metastatic potential in a clinical setting. Here we describe the initial work that led to the present research and demonstrate the potential of the microfluidic technology (Fig. 10.6).

The microfluidic device contains two or three independent flow channels separated by a 3-D gel scaffold as illustrated in Fig. 10.6. In microfluidic systems, small volume of solvent, sample and reagents are moved through microchannels embedded in a chip and miniaturised biological assays into these chips include DNA sequencing, polymerase chain reaction, electrophoresis, DNA separation, enzymatic assays, immunoassays, cell counting, cell sorting and cell culture [75].

Poly (dimethylsiloxane) (PDMS) elastomer is a specially well–suited material for the fabrication of microfluidic devices for biological assays purposes since it is inexpensive, flexible, optically transparent (and compatible with most of the optical methods available for detection), impermeable, non–toxic to cells, permeable to gases and easily bonded to other surfaces [76].

10.1.5.1 Cancer Cell (Aggregate A549, Human Lung Cancer Cell) Interaction with an Endothelial (HUVEC) Cell Layer

Interactions between tumour (aggregate) cancer cells and an intact endothelial cell layer have been observed in microfluidic system. While recognising that A549 (lung cancer cells) are known to Epithelial Mesenchymal Transition process (EMT), they stimulate to migrate and interact with an endothelial cell layer in the model system as shown in Fig. 10.7. The human lung cancer cell line (A549) migrated through collagen gel toward the endothelial monolayer (GFP expressing

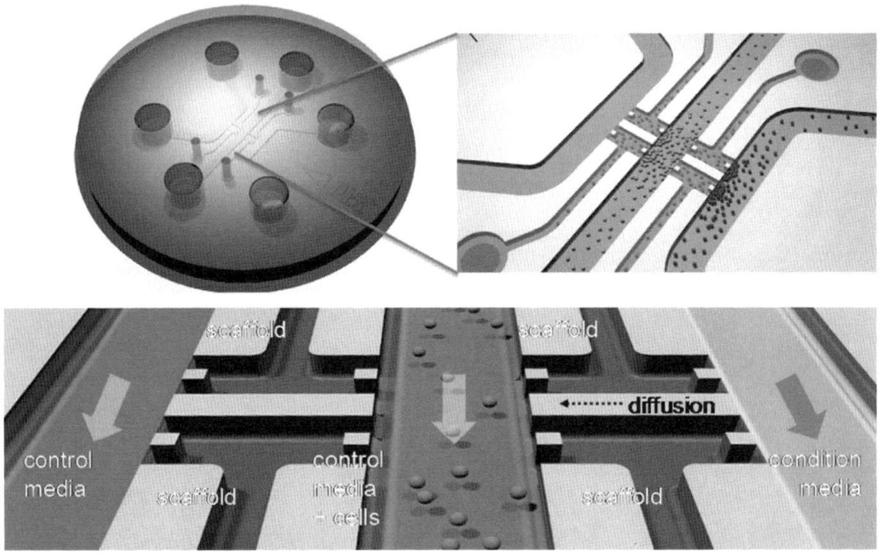

Fig. 10.6 Three channel device capabilities include simultaneous 2-D and 3-D culture, excellent imaging, multi-culture up to five different cell types, concentration gradients and flow [74]

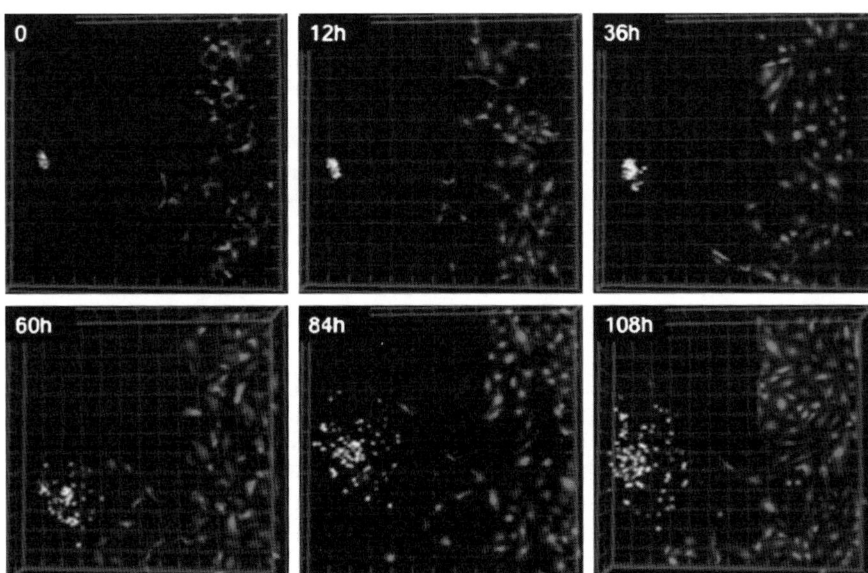

Fig. 10.7 Human lung cancer (A549) aggregate cells migrate towards a hUVEC [human umbilical cord blood cells, GFP (*green*)] layer in the 3-D gel model system. While cancer cells (*red*) are dispersed to intravasate, here they provide a proof of illustrating the ability of the 3-D system to initiate tumour-endothelial cell interaction. Each image is 1,500 μm wide

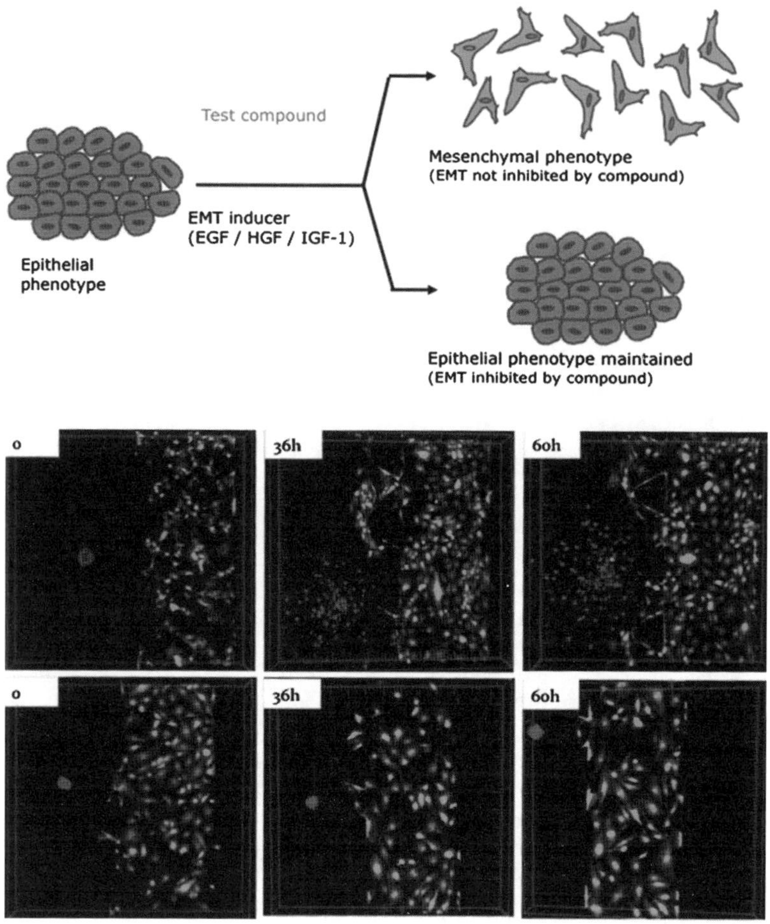

Fig. 10.8 An example of huge number of drugs, which were inhibited dispersion of cancer cells by transcriptional factors such as TGFb. Schematic of our drug cancer target to keep cell an epithelial phenotype and does not allow them to go to the mesenchymal stage (*top one*). A candidate of TGFb inhibitor that can inhibit the A549 at 500 nM (*low panel*) and the control device is shown on the *top panel*. Each image is 1,500 μm wide

HUVEC) making contact through a cellular extension. The system shows clear potential to explore migration and intravasation of cancer cells (Fig. 10.7).

10.1.5.2 Drug Screening in 3D Microfluidic System

Many signalling pathway trigger Epithelial Mesenchymal Transition in carcinogenesis. The signalling pathways include those triggered by different members of the TGFb superfamily, EGF, HGF and many others [77]. Small-molecule inhibitors directed against the pathways have been effective in pre-clinical and clinical trials (Fig. 10.8). Although originally developed as inhibitors of cell proliferation or angiogenesis, it is likely that these molecules interfere with EMT [78].

10.1.6 Future Possibilities

Tissue engineering at the nanoscale level is leading to the development of viable substitutes that can restore, maintain or improve the function of human tissue [79]. Regenerating tissue can be achieved by using biomaterials to convey signals to surrounding tissues to recruit cells that promote inherent regeneration or by using stem cells and a biomaterial scaffold to act as a framework for developing tissue.

Nanobiomaterials may serve as an important component to imparting novel properties to the biomatrix for directing stem cell proliferation, differentiation, functionalisation and transplantation.

Stem cells nanoengineering is beginning to help advance and labelling and tracking systems to create significant advances in vivo monitoring of engineered tissues. However, there will be some limitations to the nanomaterial since it is not biodegradable. In order for nanotechnology applications to develop to their fullest potential, it will be important for stem cell scientists and physicians and chemists to participate and contribute to the scientific process alongside physical and chemical science and engineering colleagues. While new uses of nanomaterials for biomedical applications are being developed concerns about cytotoxicity may be mitigated by chemical functionalisation.

References

1. Moore MT, Levine SR (2006) People help drive progress. In: Lanza R (ed) Essentials of stem cell biology. Elsevier, Burlington, p 513
2. Poole CP, Owens FJ (2003) Introduction to nanotechnology. Wiley, New York
3. Ozin GA, Arsenault AC (2005) Nanochemistry. RSC Publishing, Cambridge
4. Ramsden JJ (2005) What is nanotechnology? Nanotechnol Precept 1:3–17
5. Feynman RP (1959) There's plenty of room at the bottom. http://www.zyvex.com/nanotech/feynman.html. Accessed 10th Sept 2007
6. Binnih B, Rohrer H (1982) Scanning tunneling microscopy. Helv Phys Acta 55:38–47
7. Niemeyer C (2004) Nanobiotechnology: concepts, applications and perspectives. Wiley-VCH, Weinheim
8. Baharvand H, Zare N (2008) Nanotechnology application in stem cell biology and technology. In: Reisner D, Bronzino J (eds) Bionanotechnology. CRC Press, Boca Raton, p 1–43
9. Khademhosseini A, Langer R, Borenstein J, Vacanti J (2006) Microscale technologies for tissue engineering and biology. Proc Natl Acad Sci U S A 103:2480–2487
10. Khademhosseini A (2008) Micro- and nanoengineering of the cell microenvironment: technologies and applications. Artech House, Norwood
11. Lanza R, Gearhart J, Hogan B, Melton D, Pedersen R (2006) Essentials of stem cell biology. Academic Press, New York

12. Zamir E, Geiger B (2001) Components of cell-matrix adhesions. J Cell Sci 114:3577–3579
13. Ramsden JJ (2008) Biomedical Surfaces, 1st edn. Artech House, Norwood
14. Pritinder K, Li A (2000) Adhesive properties of human basal epidermal cells: an analysis of keratinocyte stem cells, transit amplifying cells and postmitotic differentiating cells. J Invest Dermatol 114:413–420
15. McColl J, Horvath R, Aref A, Larcombe L, Morgan S, Chianella I, Yakubov G, Ramsden J (2009) Polyphenol control of cell spreading on glycoprotein substrates. J Biomater Sci Polym Ed 20:841–851
16. Pennisi E (1998) How a growth control pathway takes a wrong turn to cancer. Science 281:1439–1441
17. Dinesh S (2006) Stem cell attachment to layer-by-layer assembled TiO_2 nanoparticle thin films. Biomaterials 27:4294–4303
18. Griffith LG (2002) Emerging design principles in biomaterials and scaffolds for tissue engineering. Ann N Y Acad Sci 961:83–95
19. Bissell MJ, Barcellos-Hoff MH (1987) The influence of extracellular matrix on gene expression: is structure the message? J Cell Sci Suppl 8:327–343
20. Smith LA, Ma PX (2004) Nano-fibrous scaffolds for tissue engineering. Coll Surf B 39:125–131
21. Dalby MJ, Riehle M, Sutherland S, Agheli H, Curtis A (2004) Changes in fibroblast morphology in response to nano-columns produced by colloidal lithography. Biomaterials 25:5415–5422
22. Langer R, Vacanti JP (1993) Tissue engineering. Science 260:920–926
23. Zuwei Ma, Masaya T (2005) Potential of nanofiber matrix as tissue engineering scaffolds. Tissue Eng 11:101–109
24. Cukierman E (2001) Taking cell-matrix adhesions to the third dimension. Science 294:1708–1712
25. Nur EKA (2005) Three dimensional nanofibrillar surfaces induce activation of Rac. Biochem Biophys Res Commun 331:428–434
26. Senesi GS, D'Aloia E , Gristina R, Favia P, d'Agostino R (2007) Surface characterization of plasma deposited nano-structured fluorocarbon coatings for promoting in vitro cell growth. Surf sci 601:1019–1025
27. Meng J, Li S, Jie M, Hua K, Guangjin Z, Wang C, Xu L, Xie S, Xu H (2006) Using single-walled carbon nanotubes nonwoven films as scaffolds to enhance long-term cell proliferation in vitro. J Biomed Mater Res Part A 79A:298–306
28. Chavany C, Behmoaras T, Puisieux F, Helene C (1994) Adsorption of oligonucleotides onto polyisohexycya nanocryslate nanoparticles protects them against nucleases and increases their cellular uptake. Pharm Res 11:1370–1378
29. Janes K, Calvo P, Alonso MJ (2001) Polysacharaide colloidal particles as delivery systems for macromolecules. Adv Drug Deliv Rev 47:83–97
30. Braydich-Stolle L, Hussain S, Schlager J, Hofmann M (2005) In vitro cytotoxicity of nanoparticles in mammalian germline stem cells. Toxicol Sci 88:412–419
31. Flemming RG, Murphy CJ, Abrams GA, Goodman SL, Nealey P (1998) Effect of synthetic micro-and nano-structured surfaces on cell behaviour. Biomaterials 20:573–588
32. Rosenthal S, Tolinson I, Adkins E (2001) Targeting cell surface receptor with ligand-conjugated nanocrystals. J Am Chem Soc 124:4586–4594
33. Paradise M, Goswami T (2007) Carbon nanotubes—Production and industrial applications. Mater Design 28:1477–1489
34. Chen J, Hamon MA, Hu H, Chen Y, Rao A, Eklund P, Haddon PC (1998) Solution properties of single-walled carbon nanotubes. Sceince 282:95–98
35. Mattson M, Haddon RC, Rao AM (2000) Molecular functionalization of carbon nanotubes and use as substrates for neuronal growth. J Mol Neurosci 123:3838
36. Hui H, Ni Y, Montana V, Haddon R, Parpura V (2004) Chemically functionalized carbon nanotubes as substate for neuronal growth. Nano Lett 4:507–511

37. Lovat V, Pantarotto D, Lagostena L (2005) Carbon nanotube substrates boost neuronal electrical signalling. Nano Lett 5:1107–1110
38. Li WJ (2002) Electrospun nanofibrous structure: a novel scaffold for tissue engineering. J Biomed Mater Res 60:613–621
39. Li WJ (2003) Biological response of chondrocytes cultured in three-dimensional nanofibrous poly (epsilon-caprolactone) scaffolds. J Biomed Mater Res 67A:1105–1114
40. Ansari F, Horvath R, Aref A, Ramsden JJ (2008) Bacterial adsorption onto a thin Fe_3O_4 magnetic nanofilms. In: 11th annual nanotechnology conference and trade show, Boston, MA, U S A, June 1–5, p 66
41. Berry CC, Curtis A (2003) Functionalisation of magnetic nanoparticles for applications in biomedicine. J Phys D Appl Phys 36:198–206
42. Riviere C (2005) Iron oxide nanoparticle-labeled rat smooth muscle cells: cardiac MR imaging for cell graft monitoring and quantitation. Radiology 235:959–967
43. Shapiro EM (2006) In vivo detection of single cells by MRI. Magn Reson Med 55:242–249
44. Hoehn M (2002) Monitoring of implanted stem cell migration in vivo: a highly resolved in vivo magnetic resonance imaging investigation of experimental stroke in rat. Proc Natl Acad Sci U S A 99:16267–16272
45. Payne AG (2004) Using immunomagnetic technology and other means to facilitate stem cell homing. Med Hypotheses 62:718–720
46. Bjorklund LM (2002) Embryonic stem cells develop into functional dopaminergic neurons after transplantation in a Parkinson rat model. Proc Natl Acad Sci 99:2344–2349
47. Bulte J, Kraitchman D (2004) Iron oxide MR contrast agents for molecular and cellular imaging. NMR Biomed 17:484–499
48. Guo P (2005) RNA nanotechnology: engineering, assembly and applications in detection, gene delivery and therapy. J Nanosci Nanotechnol 5:1964–1982
49. Vo-Dinh T (2006) Nanoprobes and nanobiosensors for monitoring and imaging individual living cells. Nanomed Nanotechnol Biol Med 2:22–30
50. Ramsden JJ (1995) Experimental methods for investigating protein adsorption kinetics at surfaces. Q Rev Biophys 27:41–105
51. Gryte DM, Ward MD, Hu W-S (1993) Real-time measurement of anchorage-dependent cell adhesion using a quartz crystal microbalance. Biotechnol Prog 9:105–108
52. Fredrikkson C, Kihlmann S, Rodahl M, Kasemo B (1998) The piezoelectric quartz crystal mass and dissipation sensor: a means of studying cell adhesion. Langmuir 14:248–251
53. Nimeri G, Fredrikkson C, Elwing H, Liu L, Rodahl M, Kasemo B (1998) Neutrophil interaction with protein–coated surfaces studied by an extended quartz crystal microbalance technique. Coll Surf B Biointerfaces 11:255–264
54. Li S-Y, Ramsden JJ, Prenosil JE, Heinzle E (1994) Measurement of adhesion and spreading kinetics of baby hamster kidney and hybridoma cells using an integrated optical method. Biotechnol Prog 10:520–524
55. Hug TS, Prenosil JE, Morbidelli M (2000) Optical waveguide lightmode spectroscopy as a new method to study adhesion of anchorage-dependent cells as an indicator of metabolic state. Biosens Bioelectron 16:865–874
56. Taylor AC (1961) Attachment and spreading of cells in culture. Exp Cell Res Suppl 8:154–173
57. Hanein D et al (1993) Selective interactions of cels with crystal surfaces. J Cell Sci 104:275–288
58. Ramsden JJ (1995) Optical method for measurement of number and shape of attached cells in real time. Cytometry 19:97–102
59. Aref A, Horvath R, McColl J, Ramsden, JJ (2009) Optical monitoring of stem cell-substratum interactions. J Biomed Opt 14:010501
60. Aref AR, Jeremy JR (2010) Nanotechnology applied to stem cell- substratum interactions: models and experiment. VDM Verlag, Saarbrücken
61. Aref A, Horvath R, Ramsden JJ (2010) Spreading kinetics for quantifying cell state during stem cell differentiation. J Biol Phys Chem 10:145–151

62. Cacace MG, Landau EM, Ramsden JJ (1997) The Hofmeister series: salt and solvent effects on interfacial phenomena. Q Rev Biophys 30:241–278
63. Shi L, Ardehali R, Caldwell KD, Valint P (2000) Mucin coating on polymeric material surfaces to suppress bacterial adhesion. Coll Surf B 17:229–239
64. Ramsden JJ, Bachmanova GI, Archakov AI (1994) Kinetic evidence for protein clustering at a surface. Phys Rev E 50:5072–5076
65. Ramsden JJ, Statist J (1993) Review of new experimental techniques for investigating random sequential adsorption. Physics 73:853–877
66. Ramsden JJ, Mate M (1998) Journal chemistry Society. Faraday Trans 94:783–788
67. Levenberg S, Khademhosseini A, Langer R (2006) Embryonic stem cell in tissue engineering. In: Lanza R (ed) Essentials of stem cell biology. Elsevier, San Diego
68. Gonsalves K, Craig RH, Cato TL, Lakshmi SN (2008) Biomedical nanostructures. Wiley, Hoboken
69. Revell PA (2006) The biological effects of nanoparticles. Nanotechnol Precept 2:283–298
70. Solter D, Skreb N, Damjanov I (1970) Extra uterine growth of mouse egg cylinders results in malignant teratoma. Nature 227:503–504
71. Shvedova AA (2003) Exposure to carbon nanotube material: assessment nanotube cytotoxicity using human keratinocyte cells. J Toxicol Environ Health 66:1909–1926
72. McKenzie JL (2004) Decreased functions of astrocytes on carbon nanofiber materials. Biomaterials 25:1309–1317
73. Cui D (2005) Effect of single wall carbon nanotubes on human HEK293 cells. Toxicol Lett 155:73–85
74. Vickerman V, Blundo J, Chung S, Kamm R (2008) Design, fabrication and implementation of a novel multi-parameter control microfluidic platform for three-dimensional cell culture and real-time imaging. Lap Chip 8(9):1468–1477
75. Whitesides G, Mc Donald (2002) Poly(dimethylsiloxane) as a material for fabricating microfluidic devices. Acc Chem Res 35(7):491–499
76. Sia S, Whitesides G (2003) Microfluidic devices fabricated in Poly (dimethylsiloxane) for biological studies. Electrophoresis 24(21):3563–3576
77. Thiery JP, Hervé Acloque R, Huang YJ, Angela M (2009) Epithelial-mesenchymal transitions in development and disease. Cell 25(139):871–890
78. Chua KN, Ma J, Thiery JP (2008) Targeted therapies in control of EMT in carcinoma and fibrosis. Drug Discov Today 4:261–267
79. Valentinuzzi MA (2004) Primer for bioengineering. World Scientific Publishing Co., Inc., Hackensack
80. Horvath R, Henrik CP, Nina S (2005) Monitoring of living cell attachment and spreading using reverse symmetry waveguide sensing. Appl Phys Lett 86:071101

Chapter 11
Spermatogonial Stem Cells

Hossein Azizi, Sabine Conrad, Thomas Skutella and Irma Virant-Klun

Abstract The process of spermatogenesis is initiated and maintained by a rare population of single spermatogonial stem cells (SSCs). The SSCs are attached to the basement membrane of the seminiferous tubules and are characterized by typical morphological criteria. SSCs are the important starting point as part of a robust stem cell system of the testis involved in spermatogenesis and reproduction. The isolation and cultivation of human SSCs would significantly contribute to the increasing knowledge of human germ and stem cell biology. Although still a difficult task, the newly established enrichment and in vitro propagation of spermatogonia that carry the male genome from generation to generation provides an important step for future transplantation and restoration of fertility in the clinic.

11.1 Introduction

The strategy of the isolation and short-term cultivation of spermatogonia in our hands is a prerequisite for the natural versatility of these adult stem cells. By our group the separation of human spermatogonial stem/progenitor cells has been achieved with magnetic-activated cell sorting (MACS), using the antibody to

H. Azizi · T. Skutella (✉)
Institute for Anatomy and Cell Biology, University of Heidelberg, Heidelberg, Germany
e-mail: skutella@ana.uni-heidelberg.de

S. Conrad
Insitute of Anatomy, University of Tübingen, Tübingen, Germany

I. Virant-Klun
Department of Obstetrics and Gynaecology, University Medical Centre Ljubljana, Ljubljana, Slovenia

H. Baharvand and N. Aghdami (eds.), *Advances in Stem Cell Research*,
Stem Cell Biology and Regenerative Medicine, DOI: 10.1007/978-1-61779-940-2_11,

CD49f (integrin alpha-6), followed by matrix seletion (collagen non-binding, laminin binding) to purify/enrich the spermatogonial stem cells (SSCs) from the somatic cells of the human testis. The CD49f/matrix-selected SSCs are easily distinguishable from other germ and somatic cells by cell size and morphological criteria. We molecularly characterized these cells during short-term culture by collecting them with the micromanipulation system and by performing germ cell gene expression profiling with Fluidigm real-time PCR technology. It became evident that the fingerprint of short-term cultured human SSCs (<2 weeks) included the high expression of a number of germ cell markers and the missing or low expression of somatic markers. Additionally, several other groups successfully established in parallel similar techniques and improved approaches to enrich and culture spermatogonia even for longer time periods, including the in vitro propagation of human SSCs from frozen-thawed testicular biopsies of infertile men for up to 15 weeks. Since it is now possible to isolate and efficiently culture spermatogonia, there is major interest to understand the self-renewal and germ-associated networks of human adult SSCs, which have not been identified so far. This new knowledge would be important to better understand the molecular mechanisms of fertility/infertility and testicular cancer manifestation.

In recent years, modern biomedicine in the field of male infertility treatment has significantly progressed, and has become increasingly advanced. Despite, there is a certain proportion of patients who still remain infertile, especially men with non-obstructive azoospermia or with no testicular sperm available for in vitro fertilization. This might be a natural condition, or a consequence of chemo- or radiotherapy in young cancer patients. In certain groups of patients autotransplantation of frozen-thawed spermatogonia to restore fertility seems to be a realistic approach. The most important might be cryopreservation of spermatogonia in prepubertal boys with cancer for later autologous use and fertility restoration. This could also be a method of choice in patients with difficult inflammation disease, treated by non-steroidal anti-inflammatory or immunosuppressive drugs, which may induce the reversible or irreversible infertility. Moreover, the results of studies in humans and animal models show that it is possible to in vitro culture cells expressing markers of pluripotent stem cells—germinal stem cells—from spermatogonia, thus bringing a new possibility to in vitro culture sperm and treat severe male infertility in the future. These new principles and methodologies might lead to the achievement of a healthy baby to couples with severe male infertility and save the life of men with testicular cancer one day.

11.2 Anatomical and Histological Basics on Mammalian Testis

The testes of mammals are paired organs, which are mainly responsible for the production of spermatozoa and synthesis of steroids. The human testis is divided into 400–500 lobes separated by septula, which are connected and encased by a connective tissue capsule, the tunica albuginea. The lobuli constitute the parenchyma of the testis with 1–4 tubuli seminiferi in each lobe. In the testes of humans and other mammals, the cellular architecture of spermatogenesis is situated inside the epithelium of the seminiferous tubules, coiled structures forming hairpin loops that empty at both ends into the tubuli recti and rete testis. Each of the convoluted tubuli is 30–70 cm long and has a diameter of approximately 150–200 μm. Spermatogenesis takes place uniformly over the inner surface of the seminiferous epithelium. This epithelium is composed of different stages of germ cells and supporting Sertoli cells, which are in contrast to the germ cells mitotically inactive in adults. Sertoli cells are believed to be the key players controlling the germ cells. Sertoli cells build two separated environments inside the tubules with tight junctions, which constitute the anatomical basis of the blood-testis barrier. Thereby, the junctional complexes separate a basal and adluminal compartment of the epithelium. The basal is the stem cell compartment. The space between the basement membrane and the tight junction is occupied by all stages of the mitotically active spermatogenic cells. During the onset the meiotic prophase, the germ cells translocate to the adluminal compartment across the tight junctions of the Sertoli cells. This is subsequently followed by movement of the spermatocytes toward the lumen. While the Sertoli cells occupy the complete epithelium, the ordering of the germ cells in the epithelium is stratified. On the outer side the seminiferous tubules are enveloped by a limiting membrane, which is composed of contractile peritubular myoid cells interposed between connective tissue layers of collagen and elastic fibers. The intervening interstitial space (12% of testis volume) between the tubules is occupied by connective tissue, immune cells, blood vessels, and groups of testosterone producing Leydig cells. Yoshida et al. [1] elegantly demonstrated a vascular and interstitial tissue-associated niche for undifferentiated spermatogonia in the mouse.

11.3 Spermatogonial Stem Cells

The process of spermatogenesis is initiated and maintained by a rare population of single SSCs. In the adult testis of rodents these cells have been estimated to comprise only approximately 0.02–0.03% of the total number of germ cells [2]. SSCs are attached to the basement membrane of the seminiferous tubules and are characterized by morphological criteria, such as basal localization, missing intercellular connections to other spermatogonia, clear cytoplasm, and high nucleus-to-cytoplasm ratio. The nucleus of the stem cells contains a mottled appearance with dark speckles of heterochromatin, moderately dense heterochromatin, and rarefied areas [3].

According to a well-established general scheme of spermatogonial self-renewal in rodents, A-single (A_s) spermatogonia are the separated unique stem cells of spermatogenesis, which have the capacity to divide into new A_s stem cells or into differentiated A-paired (A_{pr}) spermatogonia that remain interconnected by an intercellular bridge. The A_{pr} spermatogonia themselves divide into chains of

A-aligned (A_{al}) spermatogonia that can divide further into chains of cells. The A_{al} form A_1 to A_4 spermatogonia and then intermediate and type B spermatogonia and the primary spermatocytes [4]. Human SSCs have been first identified by Clemont [5], who suggested that both types of A spermatogonia, dark and pale were stem cells. According to this model the dark spermatogonia were the presumptive reserve cells, while the pale spermatogonia will continuously recycle and will give rise to type B spermatogonia, which further differentiate to spermatocytes.

In humans the process of spermatogenesis is completed and developed from few self-renewing single SSCs quite late at puberty, at about 10–13 years after birth. The first type B spermatogonia and subsequent more differentiated germ cell stages are first visible much earlier, from 4 to 5 years onwards [6]. The production of haploid spermatozoa is a complex cell differentiation process that involves supporting cells, myoid cells, Sertoli and Leydig cells, and blood vessels in close proximity to the tubules. The myoid cells are localized directly under the basement membrane and are among the candidate niche somatic cells and produce colony-stimulating factor (CSF1) [7, 8]. Csf1 is not only expressed in the peritubular myoid cells, but also in interstitial cells and increases the stem cell activity of the SSCs that are maintained in culture in the presence of glial-derived nerve growth factor (GDNF) [8]. It has been observed that Csf1 does not directly influence the growth rate of SSCs but enhances the colony-forming activity of these cells. The endocrine Leydig cells are located outside the tubules and produce steroid hormones (testosterone), which in conjunction with hypothalamic FSH regulate the Sertoli cells in the germinal epithelium of the testicular tubules [9, 10]. The GDNF and SCF (stem cell factor) producing Sertoli cells are the supporting cells inside the testicular tubules, which accompany germ cells during spermatogenesis in a synthytium-like form. The Sertoli cells are themselves connected with junctional complexes ("blood-testis barrier") constructing the basal "GDNF" stem cell compartment and separating the SSCs from the apical "SCF" compartment of terminal differentiating cells. A key regulator of the SSC niche is GDNF, which acts through Ret receptor tyrosine kinase and GFRa1 co-receptor, which forms a receptor complex on the surface of A_s, A_{pr}, and A_{al} [11]. The downstream signaling pathways activated by GDNF in SSCs are the PI3 K/Akt pathway, members of the Src kinase family, and the Ras/Erk1/2 pathway [12–14].

SSCs are surrounded on their lateral and apical surface with the cell membranes of Sertoli cells and only the basal cell surface oriented toward the basement membrane is not covered by them. Recently, blood vessels have been discussed as being part of the stem cell niche [15].

During in vitro SSC culturing potential effects of a number of soluble factors have been described. A potential role for basic fibroblast growth factor (bFGF) has been postulated [16, 17]. Important questions to be answered are how to identify the composition of the key regulators of SSC function and how all the secreted and non-secreted factors involved in the niche function could be used to optimize human SSC short-term and long-term culturing. This would also involve a detailed understanding of the downstream events that affect the intracellular machinery to control the stem cell status of SSCs and to hinder differentiation.

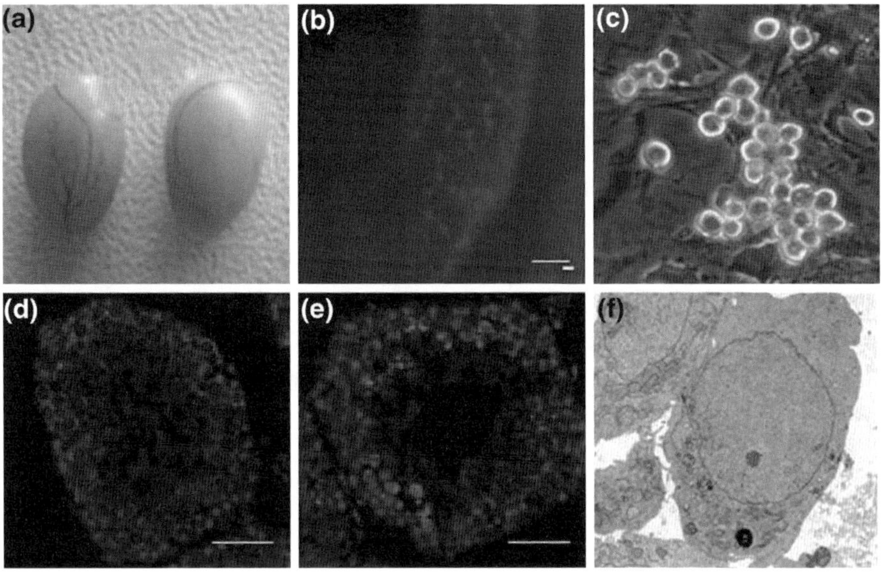

Fig. 11.1 **a** Normal adult mouse testes; **b** expression of Oct4 in the seminiferous tubule from Oct4 promoter GFP mice; **c** mSSC colonies cultured on CF1 feeder cells; **d** expression of VASA in the mouse seminiferous tubule; **e** expression of DAZL in the mouse seminiferous tubule; **f** electron microscopy of mouse SSCs in culture

11.4 Isolation of SSCs

While it is not really necessary to use cell sorting techniques when generating SSC cultures with neonatal mice, in older mice useful protocols for the isolation of the few SSCs from the mass of testis cells have been established [18, 19]. These include fluorescence-activated cell sorting (FACS), magnetic-activated cell sorting (MACS), matrix selection or morphology-based selection. For sorting cell surface markers for positive selection CD49f, CD29, GPR125, CD9, CD90 (Thy1), GFR-α1, and for negative selection CD117 (c-kit) can be used [20–27]. Furthermore, in a very elegant way reporter mice are available, in which a GFP construct is under control of the promoter of germ cell-enriched genes such as STRA8, GPR125, or OCT4 (see Fig. 11.1) [18, 19, 28]. The GFP-positive SSCs can then be FACS-sorted and cultured. When mimicking parts of the molecular environment of the SSC niche, the enriched cells can be long-term cultured without loss of their properties in the presence of growth factors and hormones [17]. The cells retain their diploid karyotype and are characterized by the expression of germ cell-specific markers and a complete androgenic imprinting pattern [29]. In an important functional assay, after transplantation of the cultured cells into the seminiferous tubules of infertile recipient mice, spermatogenesis can be re-initiated [20, 21, 30, 31].

11.5 Culturing of SSCs

SSCs are the important starting point as part of a robust stem cell system of the testis, involved in spermatogenesis and reproduction. In mice SSC cultures are usually established from pup testes (5–12 days postpartum), because SSCs are enriched at this stage of development, but these cultures can also be generated from neonate [17] or adult mouse testis cells [32]. Cultures from neonatal animals are not directly derived from SSCs but from gonoytes, the founder cells of SSCs (for detailed review on gonocytes see Culty 2009) [33]. In postnatal mice, gonocytes are the only germ cell type in the tubules and are located up to the center of the germinal epithelium. Within about 5 days after birth gonocytes migrate to the seminiferous tubule basement membrane to become the first NGN3-positive SSCs, which also retain the expression of OCT4, PLZF, and RET [15]. These NGN3–positive SSCs contribute to the leading edge of pubertal spermatogenesis, which gives rise to the first spermatozoa that are released into the lumen of the testicular tubules around postnatal day 35, five weeks after birth [34]. In our experience mouse SSC cultures are easily obtained from postnatal animals >5 days postpartum and show all characteristic of spermatogonia (see Figs. 11.1, 11.2, 11.3, 11.4). The propagation of SSC cultures from late pubertal animals (4–6 weeks of age) is more difficult and time consuming [18]. Ko et al. [28] derived the adult SCC culture from pubertal mice 35 days postpartum. Only Seandel [19] used older mice. Up to now, no real adult mouse SSC cell culture research projects have been published, which would only most remotely fulfill the criteria of culturing human cells from adult patients. To provide a real picture for human SCC research it might be important to derive and culture SSCs from adult animals systematically with increasing age. For the culture of rodent SSCs, GDNF is essential to maintain and expand the cells [16, 17, 35]. The proliferative effect of GDNF in both mice and rats can be supported and enhanced by the addition of soluble GFRa1 (the receptor for GDNF) and FGF2 to the SSC culture medium [16, 36]. The substitution of leukemia inhibitory factor (LIF) and fetal bovine serum (FBS) to SSC cultures is unnecessary and even detrimental [16]. Normally SIM mouse embryo-derived thioguanine and ouabain resistant (STO) feeder or mouse embryonic fibroblast (MEF) feeder cells are usually used by different groups [16, 37]. Shinohara et al. [38] demonstrated that mouse SSCs can also be cultured and maintained in feeder-free conditions.

11.6 Culturing of Human SSCs

The isolation and cultivation of human SSCs would significantly contribute to the increasing knowledge of human germ and stem cell biology. Although a difficult task, the newly established enrichment and in vitro culture of spermatogonia provided an important step for future transplantation and restoration of fertility in

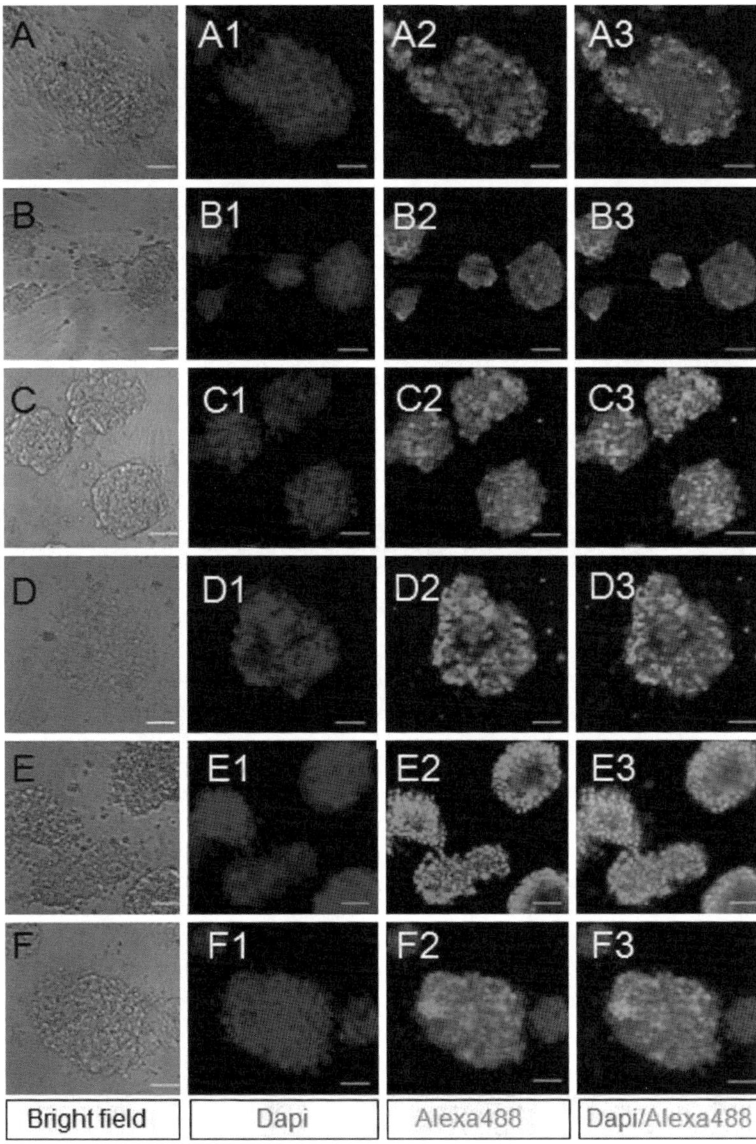

Fig. 11.2 Expression of germ cell marker by mouse SSCs; **a** DAZL, **b** VASA; **c** PLZF, **d** GFRα1 and of pluripotency markers in cultured SSCs; **e** Oct-4 and **f** KLF

the clinic (reviewed by Hwang and Lamb [39]). Sadri-Ardekani et al. [40] have provided first evidence for a potential clinical application by the in vitro propagation of human prepubertal and adult SSCs. Furthermore, understanding the molecular mechanisms of germ stem cells in relation to germ cell tumor development is of massive clinical importance [41].

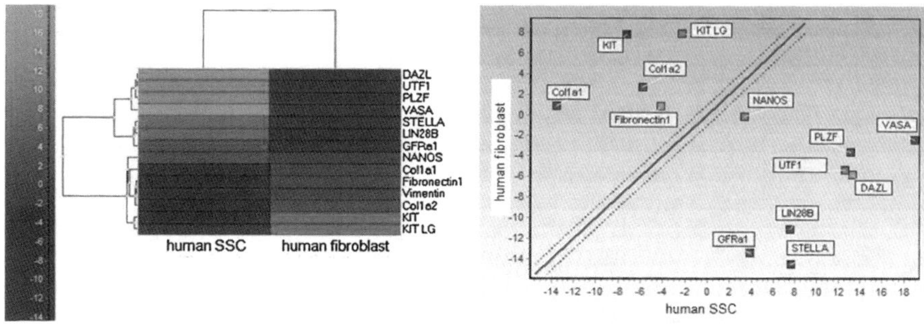

Fig. 11.3 *Left* Expression of germ- and somatic-specific genes in human SSCs and testicular fibroblasts. *Right* Heatmap clustering. Scatter plot: *red*-upregulated, *green*-downregulated genes

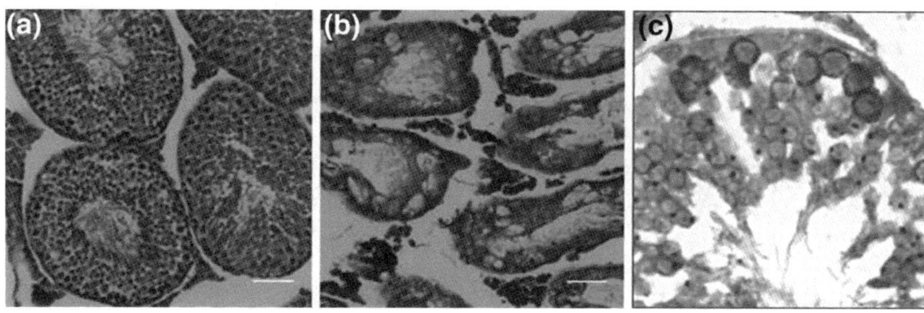

Fig. 11.4 **a** Hematoxylin-Eosin (HE) staining of the normal seminiferous tubules in mice; **b** HE staining of the busulfan-treated seminiferous tubules in mice; **c** transplanted VASA-positive SSCs in the busulfan-treated seminiferous tubule of mouse

The strategy of the isolation and short-term cultivation of spermatogonia in our hands is a prerequisite for the natural reprogramming of these adult stem cells [42]. By our group, the separation of human spermatogonial stem/progenitor cells has been achieved with MACS, using the antibody to CD49f (integrin alpha-6), followed by matrix seletion (collagen non-binding, laminin binding) to purify/enrich the SSCs from the somatic cells of the human adult testis. The CD49f/matrix-selected spermatogonia stem cells are easily distinguishable from other germ and somatic cells by cell size and morphological criteria. SSCs are single round cells with a diameter of 6–12 μm and a high nucleus/cytoplasm ratio.

We molecularly characterized these cells during short-term culture (<2 weeks) by collecting them with the micromanipulation system and by performing germ cell gene expression profiling with Fluidigm real-time PCR technology. This multiplex nanofluid PCR technology allows the simultaneous quantification of multiple germ cell and somatic markers in parallel. From Fig. 11.3 it is evident that the fingerprint of short-term cultured human SSCs included the high expression of a number of germ cell markers (DAZL, UTF1, PLZF, VASA, STELLA, LIN28B, GFRA1, NANOS) and the missing or low expression of somatic markers (COL1A1, COL1A2, FIBRONECTIN, VIMENTIN, KIT, KITLIG).

In the last years several groups successfully established in parallel similar techniques and improved approaches to enrich and culture spermatogonia even for longer time periods [43–46]. Sadri-Ardekani et al. [43] established the in vitro propagation of human SSCs from frozen-thawed testicular biopsies (100–200 mg tissue pieces) for up to 15 weeks. The cultured cells proliferated, revealed characteristics of germ cells, and repopulated testicular tubules of mice after transplantation. In 2011, this group described the propagation of PLZF, ITGA6, ITGB1-positive germ cell clusters from adult spermatogonia for up to 28 weeks. According to this study long-term cultured SSCs repopulated testicular tubules of bulsulfan-treated mouse. Also Chen et al. [44] established large-scale human spermatogonia cultures supported by human embryonic stem cell-derived fibroblast-like cells for at least 2 months. These cells retained high levels of alkaline phosphatase activity and were positively stained for antistage-specific embryonic antigen SSEA-1, POU5F1, and CD49f and also expressed the genes *POU5F1* (*OCT4*), *SOX3*, and *STRA8*. In parallel, He et al. [45] isolated human SSCs by MACS with GPR125 antibody. These cells co-expressed CD49f, CD90 (thy-1), and the GDNF receptor GFRA1 and were cultured for 2 weeks in a proliferative undifferentiated state. Furthermore, Lim et al. [46] improved the efficiency of isolation and optimized the proliferative potential of human SSCs. After cell sorting using collagen in combination with MACS with CD9 antibody and a simple long-term proliferation system large numbers of high-purity SSCs proliferated. These authors reported that the cells, which were positive for SSC-specific markers GFRα-1 and CD49f proliferated and maintained their characteristics for more than 12 passages (>6 months) in vitro. Lim et al. [46] also provided some evidence for in vitro spermatogenesis by differentiating long-term cultured SSCs by the calcium alginate encapsulation method for 2 weeks. Meiotic or haploid germ cells were analyzed by SCP3 immunostaining and flow cytometry. Lim et al. found meiotic and haploid germ cells in the in vitro-differentiated group. Furthermore, acrosome granules of putative round spermatids with diameters of 7–10 μm provided positive signals after intra-acrosomal protein and TRITC-peanut agglutinin staining. Izadyar et al. [47] reported the MACS isolation and enrichment of SSEA-4$^+$ cells from human testis biopsies. These cells had a SSEA-4$^+$, CD49f$^+$, GPR125$^+$, and c-kit-phenotype and the short-term cultured cells repopulated tubules of busulphan-treated mouse testis. Recently, Mirzapour et al. [48] successfully co-cultured isolated human SSCs on Sertoli cells. Cultured cells were obtained from patients with maturation arrest of spermatogenesis. The largest number of colonies was observed in SSCs co-cultured with Sertoli cells and the largest clusters by the addition of bFGF and LIF to the culture medium. The isolated SSCs were positive for spermatogonial cell markers, such as OCT4, STRA8, PIWIL2, and VASA, but negative for NANOG. Also here, the transplantation technique indicated that hSSCs have good efficiency for colonization of mouse seminiferous tubules after proliferation in the culture system.

Since it is now possible to isolate and efficiently culture spermatogonia [42, 43, 45–48], there is major interest to understand the self-renewal and germ-associated networks of human adult SSCs, which have not been identified so far.

At present, the functional characterization of human SSCs is still limited. While it is possible to functionally differentiate mouse spermatogonia in vivo in a well-established transplantation model [20, 21, 30, 31], this approach is still not suitable for human cells. We could also provide some evidence for the presence of human SSCs after short-term culture by using xenotransplantation to seminiferous tubules of busulphan-treated mice (unpublished data). Although migration and colonization of SSCs to the basement membrane of the seminiferous tubule could be observed, this human-mouse xenotransplantation model does not provide any evidence and therefore is not a functional assay to prove the presence of SSCs in a testicular cell population. In contrast to the transplantation with mouse SSCs, after the colonization of mouse seminiferous tubules with human SSCs no spermatogenesis has been observed by us and other groups in this model system [40, 48]. An alternative approach for the functional characterization of human SSCs might be the organotypic slice culture with human testis slices, as previously experienced in the mouse testis model [49]. In addition, SSCs have recently been shown to reprogram spontaneously to pluripotent embryonic stem (ES)-like cells that can differentiate into a number of cell lineages comprising the three embryonic germ layers [42, 46, 50, 51].

Up to now our knowledge regarding the molecular mechanisms that define human SSCs is significantly limited. The rarity of testicular tissue available for a research purpose (e.g., donated by organ donors), the relatively low number of adult stem cells in the testis, the unspecificity of unique surface markers, and the absence of a robust proliferative culture system to support their self-renewal have prevented so far the efficient isolation and culture of SSCs with high purity for further study.

11.7 Severe Male Infertility and Stem Cells

In recent years, modern biomedicine in the field of male infertility treatment has significantly progressed, and has become increasingly advanced. Despite, there is a certain proportion of patients who still remain infertile, especially men with non-obstructive azoospermia or with no testicular sperm available for in vitro fertilization. This might be a natural condition, or a consequence of chemo- or radiotherapy in young cancer patients. In all these patients diagnostic biopsy is performed and no sperm can be found in their testicular tissue due to severe infertility (e.g., Sertoli cell-only syndrome, maturation arrest). At present, the infertility of these patients can be treated only by in vitro fertilization of female partner's oocytes by donated sperm. Therefore, there are several studies trying to help these patients, and enable them to get their own child in the future.

11.8 Spermatogonia and Infertility

A better understanding of biological mechanisms involved in the differentiation of spermatogonia into male gametes—sperm—would enable new approaches to the treatment of male infertility in the future. It is known that interphase prespermatogonial stem cells, present in the immature testes of prepubertal boys are reactivated during puberty and enter into the process of intense mitotic divisions in adult testicles, in the basal compartment of seminiferous tubules. They represent (pre)spermatogonial stem cells, which are unipotent and can differentiate only into sperm in natural conditions. The mechanisms which sustain (pre)spermatogonial stem cells in the non-differentiated state or trigger the formation of tumor or further differentiation to sperm are still poorly understood. It seems that Sertoli cells have an important role through their epigenome-chromatin modifier Swi-independent3a (Sin3a). When Sin3a of Sertoli cells is destroyed in the fetal testicles, mature animals show progressive loss of differentiating spermatogonia, maturation arrest of spermatids, and intense degeneration of germ cells, as shown in the animal model [52]. Moreover, Sertoli cells may attract spermatogonia by releasing the specific substances—chemokines—and sustain them at the non-differentiated state. The release of Sertoli cell chemokines seems to be regulated by the expression of gene *Etv 5*, as evidenced in the mouse model [53]. It becomes more and more evident that genes *PLZF*, *STAT 3*, and some other genes may play an important role in differentiation of SSCs [54]. Any abnormality in these important mechanisms may lead to the lack of sperm production, and to male infertility.

11.9 Transplantation of Spermatogonia in the Animal Model

The complex procedure of spermatogonia transplantation includes the removal of spermatogonia from the testis of donor and their transfer into the testis of the recipient, where they enter into the process of spermatogenesis and develop into sperm. Most studies performed until now have been performed in the animal model, e.g., mouse or rat. The procedure of spermatogonia transplantation has been pioneering by the group of Dr. Ralph Brinster and includes microinjection of the suspension of different types of testicular cells from the donor testis into the recipient testis, through the capsule into the *rete testis* [31]. Microinjection can be performed manually or by the micromanipulation system. Regardless of the position of transplantation, the success of the transplantation procedure depends on the retrograde flow of testicular cell suspension from the *rete testis* into the seminiferous tubules.

Before the microinjection, donated testicular cells are enzymatically treated and the resulting suspension of separated cells includes spermatogonia. The concentration of approximately 5×10^7 testicular cells per milliliter of suspension is needed to fill from 30 to 100% of testicular tubules in the recipient testicle. The presence of spermatogonia in the recipient testicular tubules is estimated by trypan blue staining. In the recipient testes of experimental animals the natural spermatogenesis and the presence of sperm must be avoided, therefore the infertile, genetically modified animals are usually used, or infertility is induced by the cytotoxic agent busulfan. After microinjection, the transplanted spermatogonia need several months to settle in the recipient testis. After that, the histological evaluation of the recipient testis is performed to confirm the presence of de novo formed germ cells using immunohistochemical staining on beta galactosidase. Sperm developed de novo can be checked and aspirated from the epididymis, and used for in vitro fertilization in the animal model. In general, only small proportions of spermatogonia survive, go through the process of meiosis, and develop to sperm after transplantation.

11.10 Transplantation of Human Spermatogonia into the Mouse Testicles

There have been several unsuccessful attempts of human spermatogonia transplantation into the mouse testicles. However, in 2010 a relatively successful transplantation of human testicular tissue into the mouse testes was published [55]. A small piece of testicular tissue of a 3-month-old boy with testicular cancer was transplanted into the testes of immunodefficient (SCID) mouse without spermatogenesis and sperm to preserve the boy's spermatogonia. At the time of transplantation only type A spermatogonia were present in the boy's testicular tissue. One year after transplantation spermatogonia of type B and spermatids were also found present in the transplanted testicular tissue. Spermatocytes in the testicular tissue underwent the process of meiosis, as revealed by positive immunocytostaining on the expression of the genes *BOULE* in *CDC25A*. Ultrastructural and immunohistochemical analyses also revealed the maturation of Sertoli and Leydig cells in the transplanted testicular tissue. The maturation status of the transplanted piece of boy's testicular tissue was better than in the in vivo condition. The authors concluded that this type of methodology could be used in the treatment of male infertility as well.

Izadyar et al. purified spermatogonia from the biopsies of infertile men with obstructive azoospermia, and transplanted them into the busulfan-treated mouse testes [47]. Extensive colonization of human cells in the mouse testes indicated the presence of highly enriched populations of SSCs in the SSEA-4 (+) sorted cells. The subpopulations of human spermatogonia that colonized mouse testes were positively stained for CD49f, GPR-125, Nanog, and Oct4 indicating the existence of a population of cells among human spermatogonia expressing pluripotent characteristics.

Mitchell et al. have proposed that xenografting of human fetal testicular tissue into the mouse testicles might be a new possible approach to study the fetal testis development and germ cell differentiation, and its susceptibility to disruption by exogenous factors, e.g., environmental pollutants [56]. They xenografted testicular tissue of 10 human fetuses at 9–18 weeks of gestation into the testicles of nude mice for 6 weeks. Human fetal testicular tissue xenografts demonstrated a normal structure, function, and development after xenografting, including normal germ cell differentiation.

These kinds of studies bring new scientific solutions on the one hand and open certain ethical questions and dilemmas, which still need to be resolved, on the other.

11.11 Potential Clinical Use of Spermatogonia Transplantation in Human Medicine

In spite of many studies, it is still not clear, whether transplantation of human spermatogonia will really be an important method for fertility preservation in the future. It is becoming more evident, however, that it would be possible to regulate male fertility by transplantation of spermatogonia. Theoretically, it would be possible to restore male infertility by transplantation of donated spermatogonia or by transplantation of own spermatogonia into the recipient testicle. This might bring a problem of histoincompatibility, because spermatogonia would be released from the protective barrier of Sertoli cells and the recipient's body might reject them. Nowroozi et al. have found that nice cell cultures of spermatogonia can be established by testicular biopsies of infertile men with maturation arrest at the premeiotic spermatogonia stage, regardless of their demographic and clinical parameters [57]. However, in certain groups of patients autotransplantation of spermatogonia to restore fertility seems to be a more realistic approach.

The most important is cryopreservation of spermatogonia in prepubertal boys with cancer for later autologous use and fertility restoration [58]. Although chemo- and radiotherapy are successful methods of cancer treatment, they might impair the spermatogenesis and result in azoospermia. During oncotherapy spermatogonia can be damaged, resulting in the lack of sperm production and infertility. Cryopreservation of spermatogonia before the treatment may provide the possibility of autotransplantation of non-damaged spermatogonia and fertility restoration. Ginsberg et al. from the Division of Oncology, Department of Pediatrics, The Children's Hospital of Philadelphia, Pennsylvania, USA, have already reported the cryopreservation of testicular tissue before chemo- or radiotherapy in 16 prepubertal cancer patients, aged from 3 months to 14 years [59]. In these young boys the testicular tissue was cryopreserved for a long-term storage and potential autologous use to restore their fertility in the future. The authors have found that testicular biopsy in these young cancer patients was a completely safe procedure and had no negative consequences. Hermann et al. have found that spermatogonia can be successfully separated from the contaminating malignant cells by fluorescence-activated cell sorting prior to spermatogonial stem cell transplantation, and that post-sorting purity checks are required to confirm elimination of malignant cells [60].

The cryopreservation of spermatogonia for later autotransplantation to restore the fertility could be a method of choice also in patients with difficult inflammation disease (e.g., rheumatic disease), treated by non-steroidal anti-inflammatory or immunosuppressive drugs [61]. These drugs may induce the reversible infertility, but alkylating agents may also induce irreversible infertility. In these patients, the fertility could be preserved by autotransplantation of their spermatogonia, frozen before the treatment.

11.12 Difficulties Related to Transplantation of Spermatogonia in Humans

It has already become possible to establish long-term (up to 15 weeks) cell cultures of human spermatogonia, as revealed in men who underwent orchiectomy as part of the prostate cancer treatment [40]. But, the cryopreservation of spermatogonia still seems to be a problem. Spermatogonia are very sensitive cells. The slow, step-wise freezing procedure by using a cryoprotectant dimethyl sulfoxide is a suboptimal procedure in the animal model. Spermatogonia survive the freeze-thawing procedure and they proliferate when transplanted into the recipient testicle, but only a low proportion of them undergo the process of meiosis and differentiate into sperm. However, it has been reported that it is possible to establish long-term cell cultures by frozen-thawed spermatogonia, persisting in vitro for over 14 months and 65 passages [62]. These spermatogonia were successfully transplanted into the testicles of mice, previously sterilized by busulfan. In these mice, transplanted spermatogonia normally matured and they had a normal offspring after mating [62]. Some authors propose the cryopreservation and transplantation of pieces of testicular tissue rather than spermatogonia, because the cryopreservation procedure is more successful in testicular tissue than in spermatogonia [63]. Curaba et al. found that also vitrification of testicular tissue is a suitible method of choice to preserve spermatogonia [64].

Using whole genome approach, highly significant differences in gene expression levels were correlated with the appearance of spermatogonia, including 239 best candidates of human spermatogonia [65]. The authors concluded that these biomarkers, especially the surface markers FGFR3 and DSG2, may facilitate the isolation and enrichment of human stem and/or progenitor spermatogonia, and therefore provide efficient studies of long-term maintenance of human spermatogonia, spermatogonial self-renewal, clonal expansion, and differentiation.

Most studies of spermatogonia transplantation have been performed in animal models, especially rodents. However, the structure of the human testicle is much more complex than that of rodents. The human testicle consists of hundreds of seminiferous tubules, therefore the transplantation of spermatogonia into the human testicles has not yet reached the phase to be clinically used. In spite of anatomical limitations of human testicle due to its complex structure, only a small number of mature sperm would be needed to treat infertility. The method of intracytoplasmic sperm injection (ICSI), one of in vitro fertilization procedures, requires the microinjection of only one sperm into a female partner's oocyte; fertilization and pregnancy are therefore possible even at a very low number of sperm in the male partner.

Since spermatogonial stem cell transplantation and testicular tissue grafting may have important clinical implications, the safety of these promising techniques has still to be proven. Goossens et al. have found that the expression levels of genes *DNMT1* and *DNMT3A*, the DNA methylation status, and most of the stage-specific histone modifications after spermatogonial stem cell transplantation and

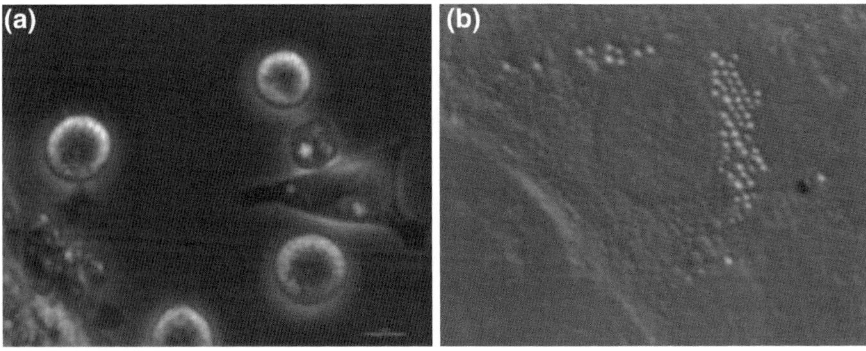

Fig. 11.5 Human testicular stem cell culture. **a** four spermatogonia; **b** very small embryonic-like stem cells in adult testicular tissue. Scale Bar: a,b = 10 μm

testicular tissue grafting are not different from those in fertile adult controls [66]. However, after spermatogonial stem cell transplantation the stage-dependent expression of *H4K5ac* and *H4K8ac* was altered in elongated spermatids. The authors have concluded that intratesticular tissue grafting might be the better choice for fertility restoration, because disruption of the stem cell niche might influence the epigenetic patterns of germ cells. Since the function of *H4K5ac* and *H4K8ac* in spermatogonia and spermatocytes is still to be elucidated, in-depth epigenetic analyses are proposed to test the safety of the procedure.

11.13 Pluripotent Stem Cells in Adult Mammalian Testicles

In the mature testicles, spermatogonia are unipotent stem cells, which are able to develop only in one type of cells—sperm—in natural conditions. The researchers have already successfully established long-persisting cell populations of spermatogonia in patients with obstructive or non-obstructive azoospermia in the programme of assisted reproduction [46].

The results of studies in humans [42, 67] and in the animal model [68, 69] show that it is possible to culture in vitro cells expressing markers of pluripotent stem cells—germinal stem cells—from spermatogonia (Fig. 11.5a). In appropriate culture conditions spermatogonia might dedifferentiate in the process of reprogramming. Germinal stem cells express different markers of pluripotent stem cells and are able to differentiate in vitro into cells of all three germ layers (endoderm, mesoderm, and ectoderm) and to form teratoma in vivo after injection into SCID mice. Additionally, in these cells a similar global gene expression profile can be confirmed than in embryonic stem cells. Moreover, recently there has been some new evidence [70] that in the adult human testicles there are present small putative stem cells (Fig. 11.5b) comparable to very small embyonic-like stem cells (VSELs), found in other tissues and organs, such as ovaries [71, 72], bone marrow, and umbilical cord [73].

These small cells expressed some markers of pluripotency and are proposed to persist in the adult human tissues and organs from the embryonic period of life [73]. Because expressing markers of pluripotency, both types of testicular stem cells theoretically might be in vitro differentiated to the sperm.

11.14 Spermatogenesis In Vitro

In spite of the complexity and relatively scarce knowledge of the subject, mouse embryonic stem cells have already been successfully differentiated into the functional sperm. Nayernia et al. have successfully differentiated mouse embryonic stem cells into sperm and fertilized mouse oocytes with them using the method of ICSI [74]. The embryos developed after ICSI were transferred into the oviducts of female mouse, resulting in the birth of healthy offspring. Based on the experience with mouse embryonic stem cells, Nolte et al. have differentiated mouse germinal stem cells retrieved from spermatogonia in a *Stra8*-EGFP transgene mouse into haploid sperm by double cell selection and addition of retinoic acid to the culture medium [75]. ICSI-derived embryos were transferred into the oviducts of the female mouse, resulting in the birth of healthy offspring. Based on the experience with mouse embryonic stem cells, Nolte et al. have differentiated mouse germinal stem cells retrieved from spermatogonia in a *Stra8*-EGFP transgene mouse into haploid sperm by double cell selection and addition of retinoic acid to the culture medium [76]. They found that approximately 5% of human iPSCs differentiated to primordial germ cells (PGCs) following induction with bone morphogenetic proteins. Surprisingly, human iPSCs formed meiotic cells with extensive synaptonemal complexes and also post-meiotic haploid germ cells with a similar pattern of ACROSIN staining as usually observed in the human spermatids. Human testicular stem cells have not been differentiated to sperm up-to-date.

11.15 Conclusions

The creation of new knowledge on the subject of spermatogonia, stem cells, and spermatogenesis in vitro provides new insights into the physiology of adult human testicles, the mechanisms involved in male fertility/infertility, and the manifestation of testicular cancer. In spite of the rapid development of new biotechnological principles and methods, a potential introduction of these new approaches into the reproductive medicine still seems to be distant. The main scientific tasks for the present are to further explore SSCs and very small embryonic-like stem cells in the adult human testicle and to gain additional knowledge about the real genetic/epigenetic status of the in vitro cultured stem cells and sperm produced in vitro. In the future, these new principles and methodologies might lead to the achievement of a healthy baby to couples with severe male infertility and save the life to men with testicular cancer.

References

1. Yoshida S, Sukeno M, Nabeshima Y (2007) A vasculature-associated niche for undifferentiated spermatogonia in the mouse testis. Science 317(5845):1722–1726
2. Tegelenbosch RA, de Rooij DG (1993) A quantitative study of spermatogonial multiplication and stem cell renewal in the C3H/101 F1 hybrid mouse. Mutat Res 290(2):193–200
3. Chiarini-Garcia H, Russell LD (2002) Characterization of mouse spermatogonia by transmission electron microscopy. Reproduction 123(4):567–577
4. de Rooij DG (2001) Proliferation and differentiation of spermatogonial stem cells. Reproduction 121(3):347–354
5. Clermont Y (1966) Renewal of spermatogonia in man. Am J Anat 118(2):509–524
6. Paniagua R, Nistal M (1984) Morphological and histometric study of human spermatogonia from birth to the onset of puberty. J Anat 139(Pt 3):535–552
7. Kokkinaki M, Lee TL, He Z, Jiang J, Golestaneh N, Hofmann MC, Chan WY, Dym M (2009) The molecular signature of spermatogonial stem/progenitor cells in the 6-day-old mouse testis. Biol Reprod 80(4):707–717
8. Oatley JM, Oatley MJ, Avarbock MR, Tobias JW, Brinster RL (2009) Colony stimulating factor 1 is an extrinsic stimulator of mouse spermatogonial stem cell self-renewal. Development 136(7):1191–1199
9. Gnessi L, Fabbri A, Spera G (1997) Gonadal peptides as mediators of development and functional control of the testis: an integrated system with hormones and local environment. Endocr Rev 18(4):541–609
10. Meroni SB, Riera MF, Pellizzari EH, Cigorraga SB (2002) Regulation of rat Sertoli cell function by FSH: possible role of phosphatidylinositol 3-kinase/protein kinase B pathway. J Endocrinol 174(2):195–204
11. Meng X, Lindahl M, Hyvonen ME, Parvinen M, de Rooij DG, Hess MW, Raatikainen-Ahokas A, Sainio K, Rauvala H, Lakso M, Pichel JG, Westphal H, Saarma M, Sariola H (2000) Regulation of cell fate decision of undifferentiated spermatogonia by GDNF. Science 287(5457):1489–1493
12. Braydich-Stolle L, Kostereva N, Dym M, Hofmann MC (2007) Role of Src family kinases and N-Myc in spermatogonial stem cell proliferation. Dev Biol 304(1):34–45
13. Oatley JM, Avarbock MR, Brinster RL (2007) Glial cell line-derived neurotrophic factor regulation of genes essential for self-renewal of mouse spermatogonial stem cells is dependent on Src family kinase signaling. J Biol Chem 282(35):25842–25851
14. He Z, Jiang J, Kokkinaki M, Golestaneh N, Hofmann MC, Dym M (2008) Gdnf upregulates c-Fos transcription via the Ras/Erk1/2 pathway to promote mouse spermatogonial stem cell proliferation. Stem Cells 26(1):266–278
15. Yoshida S, Sukeno M, Nakagawa T, Ohbo K, Nagamatsu G, Suda T, Nabeshima Y (2006) The first round of mouse spermatogenesis is a distinctive program that lacks the self-renewing spermatogonia stage. Development 133(8):1495–1505
16. Kubota H, Avarbock MR, Brinster RL (2004) Growth factors essential for self-renewal and expansion of mouse spermatogonial stem cells. Proc Natl Acad Sci U S A 101(47):16489–16494
17. Kanatsu-Shinohara M, Ogonuki N, Inoue K, Miki H, Ogura A, Toyokuni S, Shinohara T (2003) Long-term proliferation in culture and germline transmission of mouse male germline stem cells. Biol Reprod 69(2):612–616
18. Guan K, Nayernia K, Maier LS, Wagner S, Dressel R, Lee JH, Nolte J, Wolf F, Li M, Engel W, Hasenfuss G (2006) Pluripotency of spermatogonial stem cells from adult mouse testis. Nature 440(7088):1199–1203
19. Seandel M, James D, Shmelkov SV, Falciatori I, Kim J, Chavala S, Scherr DS, Zhang F, Torres R, Gale NW, Yancopoulos GD, Murphy A, Valenzuela DM, Hobbs RM, Pandolfi PP, Rafii S (2007) Generation of functional multipotent adult stem cells from GPR125 + germline progenitors. Nature 449(7160):346–350

20. Shinohara T, Avarbock MR, Brinster RL (1999) beta1- and alpha6-integrin are surface markers on mouse spermatogonial stem cells. Proc Natl Acad Sci U S A 96(10):5504–5509
21. Shinohara T, Avarbock MR, Brinster RL (2000) Functional analysis of spermatogonial stem cells in Steel and cryptorchid infertile mouse models. Dev Biol 220(2):401–411
22. Kubota H, Avarbock MR, Brinster RL (2003) Spermatogonial stem cells share some, but not all, phenotypic and functional characteristics with other stem cells. Proc Natl Acad Sci U S A 100(11):6487–6492
23. Kanatsu-Shinohara M, Toyokuni S, Shinohara T (2004) CD9 is a surface marker on mouse and rat male germline stem cells. Biol Reprod 70(1):70–75
24. Buageaw A, Sukhwani M, Ben-Yehudah A, Ehmcke J, Rawe VY, Pholpramool C, Orwig KE, Schlatt S (2005) GDNF family receptor alpha1 phenotype of spermatogonial stem cells in immature mouse testes. Biol Reprod 73(5):1011–1016
25. Fujita K, Ohta H, Tsujimura A, Takao T, Miyagawa Y, Takada S, Matsumiya K, Wakayama T, Okuyama A (2005) Transplantation of spermatogonial stem cells isolated from leukemic mice restores fertility without inducing leukemia. J Clin Invest 115(7):1855–1861
26. Hofmann MC, Braydich-Stolle L, Dettin L, Johnson E, Dym M (2005) Immortalization of mouse germ line stem cells. Stem Cells 23(2):200–210
27. Tokuda M, Kadokawa Y, Kurahashi H, Marunouchi T (2007) CDH1 is a specific marker for undifferentiated spermatogonia in mouse testes. Biol Reprod 76(1):130–141
28. Ko K, Tapia N, Wu G, Kim JB, Bravo MJ, Sasse P, Glaser T, Ruau D, Han DW, Greber B, Hausdorfer K, Sebastiano V, Stehling M, Fleischmann BK, Brustle O, Zenke M, Scholer HR (2009) Induction of pluripotency in adult unipotent germline stem cells. Cell Stem Cell 5(1):87–96
29. Kanatsu-Shinohara M, Inoue K, Lee J, Yoshimoto M, Ogonuki N, Miki H, Baba S, Kato T, Kazuki Y, Toyokuni S, Toyoshima M, Niwa O, Oshimura M, Heike T, Nakahata T, Ishino F, Ogura A, Shinohara T (2004) Generation of pluripotent stem cells from neonatal mouse testis. Cell 119(7):1001–1012
30. Brinster RL, Avarbock MR (1994) Germline transmission of donor haplotype following spermatogonial transplantation. Proc Natl Acad Sci U S A 91(24):11303–11307
31. Brinster CJ, Ryu BY, Avarbock MR, Karagenc L, Brinster RL, Orwig KE (2003) Restoration of fertility by germ cell transplantation requires effective recipient preparation. Biol Reprod 69(2):412–420
32. Kubota H, Avarbock MR, Brinster RL (2004) Culture conditions and single growth factors affect fate determination of mouse spermatogonial stem cells. Biol Reprod 71(3):722–731
33. Culty M (2009) Gonocytes, the forgotten cells of the germ cell lineage. Birth Defects Res C Embryo Today 87(1):1–26
34. Kluin PM, Kramer MF, de Rooij DG (1982) Spermatogenesis in the immature mouse proceeds faster than in the adult. Int J Androl 5(3):282–294
35. Koruji M, Movahedin M, Mowla SJ, Gourabi H, Arfaee AJ (2009) Efficiency of adult mouse spermatogonial stem cell colony formation under several culture conditions. In Vitro Cell Dev Biol Anim 45(5–6):281–289
36. Ryu BY, Kubota H, Avarbock MR, Brinster RL (2005) Conservation of spermatogonial stem cell self-renewal signaling between mouse and rat. Proc Natl Acad Sci U S A 102(40):14302–14307
37. Mohamadi SM, Movahedin M, Koruji SM, Jafarabadi MA, Makoolati Z (2011) Comparison of colony formation in adult mouse spermatogonial stem cells developed in Sertoli and STO coculture systems. Andrologia. doi: 10.1111/j.1439-0272.2011.01201.x. [Epub ahead of print]
38. Kanatsu-Shinohara M, Miki H, Inoue K, Ogonuki N, Toyokuni S, Ogura A, Shinohara T (2005) Long-term culture of mouse male germline stem cells under serum-or feeder-free conditions. Biol Reprod 72(4):985–991
39. Hwang K, Lamb DJ (2011) Diagnostic testicular biopsy before varicocele repair plays a realistic and important role. Fertil Steril 95(2):488
40. Sadri-Ardekani H, Akhondi MA, van der Veen F, Repping S, van Pelt AM (2011) In vitro propagation of human prepubertal spermatogonial stem cells. JAMA 305(23):2416–2418

41. Lee J, Shinohara T (2011) Epigenetic modifications and self-renewal regulation of mouse germline stem cells. Cell Res 21(8):1164–1171
42. Conrad S, Renninger M, Hennenlotter J, Wiesner T, Just L, Bonin M, Aicher W, Buhring HJ, Mattheus U, Mack A, Wagner HJ, Minger S, Matzkies M, Reppel M, Hescheler J, Sievert KD, Stenzl A, Skutella T (2008) Generation of pluripotent stem cells from adult human testis. Nature 456(7220):344–349
43. Sadri-Ardekani H, Mizrak SC, van Daalen SK, Korver CM, Roepers-Gajadien HL, Koruji M, Hovingh S, de Reijke TM, de la Rosette JJ, van der Veen F, de Rooij DG, Repping S, van Pelt AM (2009) Propagation of human spermatogonial stem cells in vitro. JAMA 302(19):2127–2134
44. Chen B, Wang YB, Zhang ZL, Xia WL, Wang HX, Xiang ZQ, Hu K, Han YF, Wang YX, Huang YR, Wang Z (2009) Xeno-free culture of human spermatogonial stem cells supported by human embryonic stem cell-derived fibroblast-like cells. Asian J Androl 11(5):557–565
45. He Z, Kokkinaki M, Jiang J, Dobrinski I, Dym M (2010) Isolation, characterization, and culture of human spermatogonia. Biol Reprod 82(2):363–372
46. Lim JJ, Sung SY, Kim HJ, Song SH, Hong JY, Yoon TK, Kim JK, Kim KS, Lee DR (2010) Long-term proliferation and characterization of human spermatogonial stem cells obtained from obstructive and non-obstructive azoospermia under exogenous feeder-free culture conditions. Cell Prolif 43(4):405–417
47. Izadyar F, Wong J, Maki C, Pacchiarotti J, Ramos T, Howerton K, Yuen C, Greilach S, Zhao HH, Chow M, Chow YC, Rao J, Barritt J, Bar-Chama N, Copperman A (2011) Identification and characterization of repopulating spermatogonial stem cells from the adult human testis. Hum Reprod 26(6):1296–1306
48. Mirzapour T, Movahedin M, Tengku Ibrahim TA, Haron AW, Nowroozi MR, Rafieian SH (2011) Effects of basic fibroblast growth factor and leukaemia inhibitory factor on proliferation and short-term culture of human spermatogonial stem cells. Andrologia. doi: 10.1111/j.1439-0272.2010.01135.x. [Epub ahead of print]
49. Sato T, Katagiri K, Gohbara A, Inoue K, Ogonuki N, Ogura A, Kubota Y, Ogawa T (2011) In vitro production of functional sperm in cultured neonatal mouse testes. Nature 471(7339):504–507
50. Golestaneh N, Beauchamp E, Fallen S, Kokkinaki M, Uren A, Dym M (2009) Wnt signaling promotes proliferation and stemness regulation of spermatogonial stem/progenitor cells. Reproduction 138(1):151–162
51. Kossack N, Meneses J, Shefi S, Nguyen HN, Chavez S, Nicholas C, Gromoll J, Turek PJ, Reijo-Pera RA (2009) Isolation and characterization of pluripotent human spermatogonial stem cell-derived cells. Stem Cells 27(1):138–149
52. Payne CJ, Gallagher SJ, Foreman O, Dannenberg JH, Depinho RA, Braun RE (2010) Sin3a is required by sertoli cells to establish a niche for undifferentiated spermatogonia, germ cell tumors, and spermatid elongation. Stem Cells 28(8):1424–1434
53. Simon L, Ekman GC, Garcia T, Carnes K, Zhang Z, Murphy T, Murphy KM, Hess RA, Cooke PS, Hofmann MC (2010) ETV5 regulates sertoli cell chemokines involved in mouse stem/progenitor spermatogonia maintenance. Stem Cells 28(10):1882–1892
54. Hobbs RM, Seandel M, Falciatori I, Rafii S, Pandolfi PP (2010) Plzf regulates germline progenitor self-renewal by opposing mTORC1. Cell 142(3):468–479
55. Sato Y, Nozawa S, Yoshiike M, Arai M, Sasaki C, Iwamoto T (2010) Xenografting of testicular tissue from an infant human donor results in accelerated testicular maturation. Hum Reprod 25(5):1113–1122
56. Mitchell RT, Saunders PT, Childs AJ, Cassidy-Kojima C, Anderson RA, Wallace WH, Kelnar CJ, Sharpe RM (2010) Xenografting of human fetal testis tissue: a new approach to study fetal testis development and germ cell differentiation. Hum Reprod 25(10):2405–2414
57. Nowroozi MR, Ahmadi H, Rafiian S, Mirzapour T, Movahedin M (2011) In vitro colonization of human spermatogonia stem cells: effect of patient's clinical characteristics and testicular histologic findings. Urology 78(5):1075–1078

58. Wyns C, Curaba M, Vanabelle B, Van Langendonckt A, Donnez J (2010) Options for fertility preservation in prepubertal boys. Hum Reprod Update 16(3):312–328
59. Ginsberg JP, Carlson CA, Lin K, Hobbie WL, Wigo E, Wu X, Brinster RL, Kolon TF (2010) An experimental protocol for fertility preservation in prepubertal boys recently diagnosed with cancer: a report of acceptability and safety. Hum Reprod 25(1):37–41
60. Hermann BP, Sukhwani M, Salati J, Sheng Y, Chu T, Orwig KE (2011) Separating spermatogonia from cancer cells in contaminated prepubertal primate testis cell suspensions. Hum Reprod 26(12):3222–3231
61. Silva CA, Bonfa E, Ostensen M (2010) Maintenance of fertility in patients with rheumatic diseases needing antiinflammatory and immunosuppressive drugs. Arthritis Care Res (Hoboken) 62(12):1682–1690
62. Yuan Z, Hou R, Wu J (2009) Generation of mice by transplantation of an adult spermatogonial cell line after cryopreservation. Cell Prolif 42(2):123–131
63. Schlatt S, Ehmcke J, Jahnukainen K (2009) Testicular stem cells for fertility preservation: preclinical studies on male germ cell transplantation and testicular grafting. Pediatr Blood Cancer 53(2):274–280
64. Curaba M, Poels J, van Langendonckt A, Donnez J, Wyns C (2011) Can prepubertal human testicular tissue be cryopreserved by vitrification? Fertil Steril 95(6), 2123.e9–12
65. von Kopylow K, Kirchhoff C, Jezek D, Schulze W, Feig C, Primig M, Steinkraus V, Spiess AN (2010) Screening for biomarkers of spermatogonia within the human testis: a whole genome approach. Hum Reprod 25(5):1104–1112
66. Goossens E, Bilgec T, Van Saen D, Tournaye H (2011) Mouse germ cells go through typical epigenetic modifications after intratesticular tissue grafting. Hum Reprod 26(12):3388–3400
67. Skutella T (2009) Induced pluripotent stem cells from adult testis: a new source of stem cells? Regen Med 4(1):3–5
68. Izadyar F, Pau F, Marh J, Slepko N, Wang T, Gonzalez R, Ramos T, Howerton K, Sayre C, Silva F (2008) Generation of multipotent cell lines from a distinct population of male germ line stem cells. Reproduction 135(6):771–784
69. Ko K, Arauzo-Bravo MJ, Kim J, Stehling M, Scholer HR (2010) Conversion of adult mouse unipotent germline stem cells into pluripotent stem cells. Nat Protoc 5(5):921–928
70. Bhartiya D, Kasiviswanathan S, Unni SK, Pethe P, Dhabalia JV, Patwardhan S, Tongaonkar HB (2010) Newer insights into premeiotic development of germ cells in adult human testis using Oct-4 as a stem cell marker. J Histochem Cytochem 58(12):1093–1106
71. Virant-Klun I, Zech N, Rozman P, Vogler A, Cvjeticanin B, Klemenc P, Malicev E, Meden-Vrtovec H (2008) Putative stem cells with an embryonic character isolated from the ovarian surface epithelium of women with no naturally present follicles and oocytes. Differentiation 76(8):843–856
72. Parte S, Bhartiya D, Telang J, Daithankar V, Salvi V, Zaveri K, Hinduja I (2011) Detection, characterization, and spontaneous differentiation in vitro of very small embryonic-like putative stem cells in adult mammalian ovary. Stem Cells Dev 20(8):1451–1464
73. Ratajczak MZ, Shin DM, Liu R, Marlicz W, Tarnowski M, Ratajczak J, Kucia M (2010) Epiblast/germ line hypothesis of cancer development revisited: lesson from the presence of Oct-4 + cells in adult tissues. Stem Cell Rev 6(2):307–316
74. Nayernia K, Nolte J, Michelmann HW, Lee JH, Rathsack K, Drusenheimer N, Dev A, Wulf G, Ehrmann IE, Elliott DJ, Okpanyi V, Zechner U, Haaf T, Meinhardt A, Engel W (2006) In vitro-differentiated embryonic stem cells give rise to male gametes that can generate offspring mice. Dev Cell 11(1):125–132
75. Nolte J, Michelmann HW, Wolf M, Wulf G, Nayernia K, Meinhardt A, Zechner U, Engel W (2010) PSCDGs of mouse multipotent adult germline stem cells can enter and progress through meiosis to form haploid male germ cells in vitro. Differentiation 80(4–5):184–194
76. Panula S, Medrano JV, Kee K, Bergstrom R, Nguyen HN, Byers B, Wilson KD, Wu JC, Simon C, Hovatta O, Reijo Pera RA (2011) Human germ cell differentiation from fetal- and adult-derived induced pluripotent stem cells. Hum Mol Genet 20(4):752–762

Chapter 12
Cellular Reprogramming and Fate Conversion

Masaki Ieda

Abstract Cellular reprogramming and fate conversion have long been recognized in somatic nuclear transfer and cell fusion experiments. The potency of specific transcription factors as cell fate determinants was first demonstrated by the discovery of MyoD, a master gene for skeletal muscle differentiation, and by the subsequent identification of several genes as lineage-converting factors within the blood cell lineage. These pioneer works led to the landmark study by Dr. Yamanaka and colleagues in 2006 that is reprogramming of somatic cells into a pluripotent state by transduction of the four stem cell-specific transcription factors, Oct4, Sox2, Klf4, and c-Myc. This study fundamentally altered the approach to regenerative medicine and also inspired a broad strategy to generate desired cell types by introducing combinations of lineage-specific transcription factors. In fact, it has been demonstrated that a diverse range of cell types, such as pancreatic β cells, neurons, cardiomyocytes, and hepatocytes, could be directly induced from heterologous cells by combinations of lineage-specific reprogramming factors. This chapter reviews the pioneer and recent works of cellular reprogramming and fate conversion, and also discusses the future perspective and challenges of using this technology in regenerative medicine.

M. Ieda
Department of Clinical and Molecular Cardiovascular Research,
Keio University School of Medicine, Shinjuku-ku, Tokyo, Japan

M. Ieda (✉)
Department of Cardiology, Keio University School of Medicine,
Shinjuku-ku, Tokyo, Japan
e-mail: mieda@z8.keio.jp

H. Baharvand and N. Aghdami (eds.), *Advances in Stem Cell Research*,
Stem Cell Biology and Regenerative Medicine, DOI: 10.1007/978-1-61779-940-2_12,

12.1 Introduction

Developmental biologists have long recognized that differentiated cells maintain their state for years and rarely switch to a new differentiation state. Indeed, once cells reach a stable position by a progressive or sequential differentiation from the pluripotent state, some become terminally differentiated and undergo no further cell division [1, 2]. However, experiments performed several decades ago showed that dormant gene expression programs could be induced in differentiated cells by cell fusion or nuclear transfer to produce a different cell type [3–6]. Moreover, lineage conversions have been accomplished simply through the introduction of defined transcription factors [7–13]. While cell type conversion phenomena are scientifically intriguing, the usefulness of this technology for regenerative medicine remained largely neglected until the generation of induced pluripotent stem cells (iPSCs) in mouse by Takahashi and Yamanaka in 2006 and in human in 2007 [14–16].

Pluripotent reprogramming of cells by defined factors has revolutionized the field of regenerative medicine, with projected applications in cell therapy, disease modeling, and personalized medicine [17–27]. The discovery of iPSCs further delivered the direct generation of specific cell types in an approach called lineage reprogramming. Such a strategy involves reprogramming mature cell types directly into another cell type without passing through a stem cell by the combination of several lineage-specific factors. Recent studies demonstrated that direct lineage reprogramming could yield a diverse range of medically relevant cell types including pancreatic β cells, neurons, cardiomyocytes, and hepatocytes [28–32]. Here we review the background and recent studies of cellular reprogramming and fate conversion with their potential future applications in regenerative medicine.

12.2 MyoD, a Master Regulator for Skeletal Muscle Differentiation

MyoD is a classic example of a master gene for cell differentiation in the sense that transduction of this gene is sufficient to activate the genetic program for muscle differentiation in non-muscle cells [33, 34]. In 1979, Taylor and Jones [35, 36] found that treatment of the 10T1/2 mouse embryonic fibroblast cell line with 5-azacytidine (5-azaC), an inhibitor of DNA methylation, induced differentiation of these cells into skeletal muscle cells. This observation led to the hypothesis that DNA demethylation and the resulting activation of unidentified genes were responsible for the conversion to muscle cells [37]. A subtractive hybridization experiment comparing untreated and 5-azaC-treated fibroblasts led to the identification of a cDNA encoding MyoD [7]. When overexpressed in primary fibroblasts or in a wide variety of other cell types, including pigment, nerve, fat, and liver cells, MyoD induced reprogramming into skeletal muscle cells [10, 11].

These findings provided the first evidence that a single gene can initiate a complex program of differentiation, thereby acting as a master switch.

The ability of MyoD to convert fibroblasts and other cell types into skeletal muscle strongly implicated a central role in myogenesis, and subsequent studies sought to determine the biological roles and molecular mechanisms of MyoD function in muscle development [38]. The MyoD protein contains a basic helix-loop-helix (bHLH) motif that is common to a large family of transcription factors [38–40]. Indeed, the strongly related bHLH proteins, Myf5, Mrf4, and myogenin, are also expressed in skeletal muscle, with each factor playing a sequential and synergistically crucial role in muscle cell specification and differentiation in vivo [41–43]. It was therefore proposed that lineage-associated transcription factors, which help to establish and maintain cellular identity during development, could also change cell fate when ectopically expressed in certain heterologous cells.

12.3 Lineage Specification and Conversion in Hematopoietic Cells

Following the discovery of MyoD, conversion of one cell type into another was predominantly demonstrated among hematopoietic cells, which are an ideal cell type for lineage conversion experiments because the cellular lineages are well defined. Ectopic expression of GATA1, the erythroid-megakaryocyte- affiliated transcription factor, produced erythroid-megakaryocytic-eosinophil-basophils from granulocyte–macrophage progenitors freshly isolated from normal bone marrow [44–46]. C/EBPβ, a transcription factor required for the formation of granulocyte–macrophage precursors, can convert committed B and T cell progenitors into functional macrophages at frequencies approaching 100% [8, 9, 47]. Mature immunoglobulin-producing B cells could also be switched into macrophages by C/EBPβ, although at lower frequencies.

Inactivation of key regulators may also lead to the activation of earlier genetic programs in committed cells, resulting in their dedifferentiation and activation of multi-lineage potential. For instance, deletion of the Pax5 transcription factor induced dedifferentiation of B cells into common hematopoietic progenitor cells [48–50]. Under appropriate culture conditions, these cells may then differentiate into granulocyte/macrophage, T cell, dendritic, natural killer, and osteoclast cell lineages.

12.4 Induction of Pluripotency by Defined Factors

In the 1960s, John Gurdon and colleagues [5, 51] demonstrated that transferring the nucleus of a differentiated frog cell into an enucleated egg could reprogram the cell back to the totipotency of a zygote, and thus give rise to a whole new frog. Following this striking demonstration using frog cells, reprogramming into pluripotent states by somatic nuclear transfer in mammals was first reported in 1997 with the birth of Dolly the sheep, and subsequently accomplished in many other mammalian species [4, 6, 52–58]. These impressive demonstrations of the inherent plasticity and potency of cells led to a search for reprogramming factors able to convert differentiated cells back to a pluripotent stage.

Takahashi and Yamanaka [14] achieved a breakthrough in this field by overexpressing four embryonic stem cell (ESC)-specific transcription factors in somatic fibroblasts to induce a pluripotent state, resulting in the so-called iPSC. Using retroviral vectors, they expressed 24 candidate genes and selected for reprogramming cells by incorporating neomycin resistance and β-galactosidase reporter genes into Fbx15, a gene specifically expressed in pluripotent stem cells [59]. The combination of 24 factors activated Fbx15 and induced the formation of drug-resistant colonies with characteristic ESC morphology. Successive selection rounds to eliminate individual dispensable factors led to the minimally required core set of four genes, comprising Oct4, Sox2, Klf4, and c-Myc and referred to as the Yamanaka factors. Many laboratories have since improved iPSC generation techniques to show that murine iPSCs share all defining features with naive mouse ESCs, including expression of pluripotency markers, reactivation of both X chromosomes, and the ability to generate chimeric mice [60–67].

iPSCs can be derived from a number of different species, including humans, rats, and rhesus monkeys, by expression of the four Yamanaka factors [15, 16, 68–70]. Similarly, iPSCs have been derived from other somatic cell populations, such as keratinocytes, neural cells, stomach and liver cells, and melanocytes, as well as from terminally differentiated lymphocytes, demonstrating the universality of the induced pluripotency [71–75]. Importantly, reprogramming can be induced not only by Oct4, Sox2, Klf4, and c-Myc, but also by alternative combinations that employ Nanog, Lin28, ESRRB, NR5A2, and other genes that help to establish the core transcriptional circuitry of stem cells [16, 76, 77].

12.5 In Vivo Reprogramming of Pancreatic Exocrine Cells to Endocrine β Cells

The generation of iPSCs sparked the new and related idea of converting mature cell types directly into another cell type without becoming a stem cell using a combination of several lineage-specific factors [28, 30, 32, 78, 79]. Zhou et al. [32] showed that adenoviral gene transfer of transcription factors could efficiently reprogram pancreatic exocrine cells into functional β-cells in mouse [32], using Neurogenin 3, Pdx1, and Mafa, which are known to be important in the embryonic development of pancreas and β-cells [80]. The induced β -cells were indistinguishable from endogenous islet β-cells in structure and expressed genes essential for β-cell function. Importantly, the induced β-cells ameliorated hyperglycemia in type 1 diabetic mice by remodeling local vasculature and secreting insulin.

Thus, although the new β-cells were not reorganized into complete islet structures and this may limit their effectiveness, this study provided the first evidence of cellular reprogramming in vivo by defined factors.

12.6 Conversion of Mesodermal Fibroblast Cells into Ectodermal Neural Cells

It is then conceivable that while direct cell type conversion within the same lineage may be easy, converting to a completely different lineage could be challenging. To this end, Vierbuchen et al. [30] converted mouse dermal fibroblasts into functional neurons in vitro using neuronal lineage-specific transcription factors, Ascl1, Brn2, and Myt1l. These induced neuronal (iN) cells expressed multiple neuron-specific proteins, generated action potentials, and formed functional synapses. The same group recently reported that when combined with the bHLH transcription factor NeuroD1, these three factors could also convert human fibroblasts into iN cells [81] that also showed typical neuronal morphologies and expressed multiple neuronal markers, even after downregulation of the exogenous transcription factors. However, the iN cells in this study represented a heterogeneous population of excitatory neurons and it was unclear whether direct lineage reprogramming could produce a specific neuronal subtype. Caiazzo et al. [82] reported that a combination of three transcription factors, Ascl1, Nurr1, and Lmx1a, could generate functional dopaminergic neurons from mouse and human fibroblasts without reverting to a progenitor cell stage, as well as from Parkinson's disease patients. These studies thus revealed that lineage conversions are not restricted to within the same lineage or germ layer, since fibroblasts are mesodermal in origin and neurons are derived from ectoderm.

12.7 Direct Reprogramming of Fibroblasts into Cardiomyocytes by Defined Factors

While embryonic mesoderm can be induced to generate cardiomyocytes, no master regulator of cardiac differentiation has been identified that parallels MyoD [7, 82, 83]. We thus hypothesized that a combination of key developmental cardiac genes is required to directly convert fibroblasts into cardiomyocytes. To test this, we selected 14 genes as candidates for cardiac reprogramming; all are specifically expressed in embryonic cardiomyocytes and are critical for cardiac cell fate specification as demonstrated by knockout mouse studies [84]. Cardiac fibroblasts were isolated from heart explants of transgenic mice expressing EGFP under a cardiac-specific alpha myosin heavy chain (αMHC) promoter and the fibroblast cells not expressing EGFP were used for screening [85, 86]. Transduction of all 14 factors into fibroblasts induced 1.7% of GFP^{+} cells, and serial reduction of individual factors demonstrated that a combination of three factors, Gata4, Mef2c, and Tbx5, was sufficient for GFP$^+$ cell induction (around 15%). We designated these GFP$^+$ cardiomyocyte-like cells induced cardiomyocytes (iCMs) [31]. Gata4, Mef2c, and Tbx5 are core cardiac transcription factors in early heart development [87–89] that interact with each other to coactivate cardiac gene expression and promote cardiomyocyte differentiation [90–92].

Despite the described progress, the genetic and epigenetic status of directly induced cells and the exact mechanism of lineage reprogramming remains unclear. We addressed these questions by microarray and epigenetic analyses of iCMs and by cell fate mapping studies. The iCMs are similar, although not identical, to cardiomyocytes in genetics and epigenetics status. Their global gene expression profile resembles that of neonatal cardiomyocytes, but is different from that of the original cardiac fibroblasts. The histone modification and DNA methylation patterns of iCMs were also similar to those of cardiomyocytes [93], and a subset of iCMs contracted spontaneously after 4 weeks of culture. Indeed, lineage-mapping experiments with Mesp1-Cre/R26R-YFP and Isl1-Cre/R26R-YFP reporter mice suggested direct reprogramming of the fibroblasts to the iCM fate without reversion to a cardiac mesoderm/progenitor stage [83, 94–96]. We therefore took the fibroblasts at an early stage of reprogramming and injected them into the mouse hearts. These cells differentiated in situ and formed small isolated cardiomyocyte-like cell clusters that were positive for cardiac markers. This is the first study representing the global gene expression profiles and epigenetic status of directly induced cells, and demonstrating direct evidence of cell fate in lineage conversion using several reporter mice.

12.8 Induced Hepatocyte-like Cells Repair Damaged Liver After Transplantation

Overexpression of lineage-specific transcription factors directly converts terminally differentiated cells into some other lineages; however, it remained unclear whether these lineage-converted cells could reconstitute damaged tissues in vivo. In 2011, two groups clearly demonstrated that induced hepatocyte-like (iHep) cells directly generated from fibroblasts reconstituted damaged hepatic tissues after cell transplantation [28, 29]. Huang et al. [28] demonstrated the direct induction of functional iHep cells from mouse tail-tip fibroblasts by transduction of Gata4, Hnf1a, and Foxa3, and inactivation of p19Arf. These iHep cells showed typical epithelial morphology, expressed hepatic genes, and acquired hepatocyte functions. Notably, transplanting the iHep cells into fumarylacetoacetate-hydrolase-deficient mice repopulated the damage liver and rescued almost half of the recipients from death by restoring liver function. The authors proposed that Gata4 and Foxa3 probably act as 'pioneer factors' to trigger a global chromatin modification during the hepatic conversion [97–99]. Similarly, Sekiya et al. [29] identified three specific combinations of two transcription factors, comprising

Hnf4a plus Foxa1, Foxa2, or Foxa3, that could convert mouse embryonic and adult fibroblasts into iHep cells, which were also functional and rescued damaged hepatic tissues after transplantation. These studies provide novel strategies to generate functional hepatocyte-like cells for the purpose of liver engineering and regenerative medicine. However, because the iHep cell transplantation experiments produced only partial rescue and that cells are still not identical to the bona fide hepatocytes, further modifications for generating iHep cells are needed to apply this approach in clinical settings. Importantly, both studies showed no tumor formation using the iHep cells.

12.9 Transient Overexpression of Pluripotency-Inducing Factors and Subsequent Extrinsic Signals Instruct Lineage Determination

Direct reprogramming of somatic cells to a pluripotent state using the Yamanaka factors entails extensive genetic and epigenetic resetting. Thus, it is possible that the reprogramming factors function to erase cell identity by epigenetic mechanisms and that subsequent exposure to external lineage-specific signals directs lineage specification and induction of terminal differentiation. Efe et al. [100] showed that transient overexpression of Oct4, Sox2, Klf4, and c-Myc and subsequent exposure to cardiogenic media converted mouse fibroblasts directly into spontaneously contracting cardiomyocytes, via a cardiac progenitor cell state with no pluripotent intermediate. In contrast to our cardiac reprogramming protocol, they showed that spontaneous contractions began after 11 days and achieved a sixfold better yield efficiency of generating cTnT + cardiomyocytes. The latter results reflect that mitotically active cardiac precursor cells were generated first rather than terminally differentiated cardiac cells.

Efe et al. [101] further demonstrated similar induction of neural progenitor cells from mouse fibroblasts by transient expression of the pluripotent factors and an exposure to the cell culture conditions favorable for neural precursor cells. For future clinical applications, these authors now need to confirm that their fibroblast-derived cardiomyocytes and neural progenitor cells are functionally integrated into the host tissue and do not form teratoma.

12.10 Comparison of Pluripotent Reprogramming and Lineage Reprogramming

As discussed above, reprogramming technologies have undergone extensive advances that may change the field of regenerative medicine in the future (Fig. 12.1) [23, 31, 100, 102]. The pluripotent reprogramming approach with directed differentiation is theoretically able to produce all desired cell types as

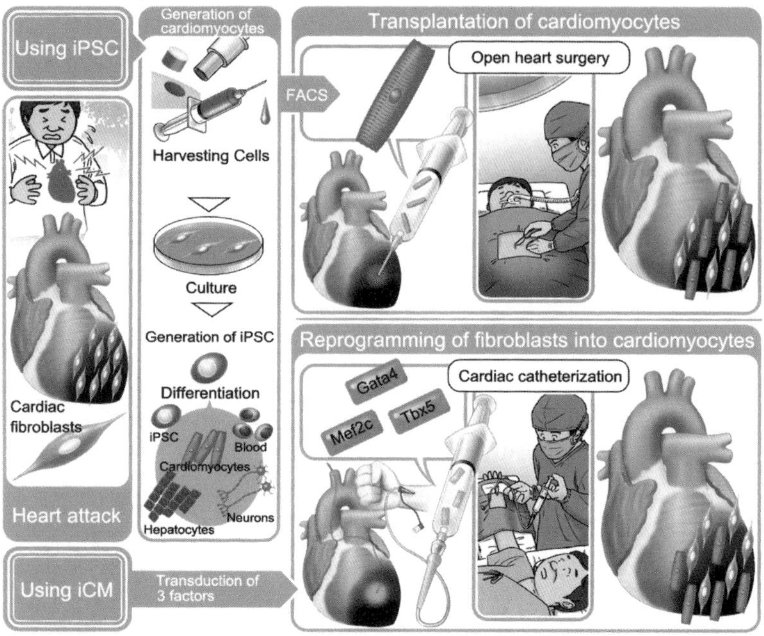

Fig. 12.1 Future cardiac regenerative therapy. Cardiomyocytes cannot regenerate after myocardial infarction and the lost myocardium is replaced by scar tissue which consists of abundant cardiac fibroblasts. In future cardiac regenerative therapy, induced pluripotent stem cell (iPSC)-based cardiac regeneration may improve cardiac function (*upper panel*). Direct cardiac reprogramming technology using induced cardiomyocytes (iCMs) may convert endogenous fibroblasts into cardiomyocytes by defined factors and skip several steps to generate cardiomyocytes (*lower panel*)

shown by recent ESC studies [101, 103]. A potential advantage of iPSCs is that pluripotent cells have nearly unlimited capacity to proliferate in culture and may provide abundant starting material to produce desired cell types. In contrast, the lineage reprogramming approach does not induce all cell types at present and directly induced cells do not proliferate infinitely in general. However, a possible advantage of lineage reprogramming is that trans differentiation among lineage-related cells may require only a limited amount of epigenetic resetting, resulting in faster and more efficient reprogramming compared with pluripotent induction. For example, iPSCs can be used in cardiac regeneration to generate graft cardiomyocytes, and after cardiac cell selection the cells could potentially be transplanted into the damaged hearts [104, 105]. Thus, it needs multiple steps for cardiac repair. In contrast, iCMs directly generated with Gata4/Mef2c/Tbx5 transduction could be transplanted into mouse hearts without cell sorting [31]. However, both cases rely on the graft cells being organized and successfully integrated into the host tissues to improve cardiac function, thus further investigations are needed.

Another advantage of lineage reprogramming is the possibility of converting cells directly in vivo by in situ regeneration and restoring function, as shown with pancreatic β-cell induction [32]. The potential of this being possible consistently raises several important advantages: first, the process is simple and short; second, avoiding reprogramming to pluripotent cells before lineage differentiation would greatly lower the risk of tumor formation; and third, direct injection of defined factors can avoid cell transplantation in which long-term cell survival might be challenging in some cell types such as cardiomyocytes [104, 106–108].

Studies over the past 5 years have induced a wide variety of cells from pluripotent cells using the Yamanaka factors, and iPSC properties have been characterized extensively. In contrast, lineage reprogramming using varied specific factors has just emerged and induced cell characters should be carefully examined in the future. Studies in human cells and understanding of molecular mechanisms of direct lineage reprogramming are necessary to advance this technology for clinical applications.

12.11 Conclusions

Cellular reprogramming and fate conversion have long been recognized as a possibility, although the impact of cell type conversion by defined genes was most prominently exemplified only recently by the discovery of iPSCs. This landmark finding fundamentally altered approaches and ideas about regenerative medicine and also provided a broad strategy to induce desired cell types by introducing lineage-specific factors. The numerous reprogramming studies have taught us that cells have more plasticity and cell fate can be molded more easily than previously thought once the right combination of transcription factors is identified. Notably, the issues of appropriate cellular differentiation, risk of tumor formation due to contaminating immature cells and insertional mutagenesis, and proper integration into the recipient organ require further and careful investigation [109–111] for future clinical applications to be realized.

Acknowledgments M. I. is supported in part by research grants from the Japan Science and Technology Agency's Core Research of Evolutional Science and Technology program. I apologize to authors whose work has not been cited due to space limitations.

References

1. Weintraub H, Groudine M (1976) Chromosomal subunits in active genes have an altered conformation. Science 193(4256):848–856
2. Goldberg AD, Allis CD, Bernstein E (2007) Epigenetics: a landscape takes shape. Cell 128(4):635–638

3. Blau HM (1989) How fixed is the differentiated state? Lessons from heterokaryons. Trends Genet 5(8):268–272
4. Wilmut I, Schnieke AE, McWhir J, Kind AJ, Campbell KH (1997) Viable offspring derived from fetal and adult mammalian cells. Nature 385(6619):810–813
5. Gurdon JB (1962) The developmental capacity of nuclei taken from intestinal epithelium cells of feeding tadpoles. J Embryol Exp Morphol 10:622–640
6. Hochedlinger K, Jaenisch R (2002) Monoclonal mice generated by nuclear transfer from mature B and T donor cells. Nature 415(6875):1035–1038
7. Davis RL, Weintraub H, Lassar AB (1987) Expression of a single transfected cDNA converts fibroblasts to myoblasts. Cell 51(6):987–1000
8. Xie H, Ye M, Feng R, Graf T (2004) Stepwise reprogramming of B cells into macrophages. Cell 117(5):663–676
9. Laiosa CV, Stadtfeld M, Xie H, de Andres-Aguayo L, Graf T (2006) Reprogramming of committed T cell progenitors to macrophages and dendritic cells by C/EBP alpha and PU.1 transcription factors. Immunity 25(5):731–744
10. Choi J, Costa ML, Mermelstein CS, Chagas C, Holtzer S, Holtzer H (1990) MyoD converts primary dermal fibroblasts, chondroblasts, smooth muscle, and retinal pigmented epithelial cells into striated mononucleated myoblasts and multinucleated myotubes. Proc Natl Acad Sci U S A 87(20):7988–7992
11. Weintraub H, Tapscott SJ, Davis RL, Thayer MJ, Adam MA, Lassar AB, Miller AD (1989) Activation of muscle-specific genes in pigment, nerve, fat, liver, and fibroblast cell lines by forced expression of MyoD. Proc Natl Acad Sci U S A 86(14):5434–5438
12. Nerlov C, Graf T (1998) PU.1 induces myeloid lineage commitment in multipotent hematopoietic progenitors. Genes Dev 12(15):2403–2412
13. Nerlov C, McNagny KM, Doderlein G, Kowenz-Leutz E, Graf T (1998) Distinct C/EBP functions are required for eosinophil lineage commitment and maturation. Genes Dev 12(15):2413–2423
14. Takahashi K, Yamanaka S (2006) Induction of pluripotent stem cells from mouse embryonic and adult fibroblast cultures by defined factors. Cell 126(4):663–676
15. Takahashi K, Tanabe K, Ohnuki M, Narita M, Ichisaka T, Tomoda K, Yamanaka S (2007) Induction of pluripotent stem cells from adult human fibroblasts by defined factors. Cell 131(5):861–872
16. Yu J, Vodyanik MA, Smuga-Otto K, Antosiewicz-Bourget J, Frane JL, Tian S, Nie J, Jonsdottir GA, Ruotti V, Stewart R, Slukvin II, Thomson JA (2007) Induced pluripotent stem cell lines derived from human somatic cells. Science 318(5858):1917–1920
17. Wu SM, Hochedlinger K (2011) Harnessing the potential of induced pluripotent stem cells for regenerative medicine. Nat Cell Biol 13(6):734
18. Park IH, Arora N, Huo H, Maherali N, Ahfeldt T, Shimamura A, Lensch MW, Cowan C, Hochedlinger K, Daley GQ (2008) Disease-specific induced pluripotent stem cells. Cell 134(5):877–886
19. Amabile G, Meissner A (2009) Induced pluripotent stem cells: current progress and potential for regenerative medicine. Trends Mol Med 15(2):59–68
20. Kamp TJ, Lyons GE (2009) On the road to iPS cell cardiovascular applications. Circ Res 105(7):617–619
21. Carvajal-Vergara X, Sevilla A, D'Souza SL, Ang YS, Schaniel C, Lee DF, Yang L, Kaplan AD, Adler ED, Rozov R, Ge Y, Cohen N, Edelmann LJ, Chang B, Waghray A, Su J, Pardo S, Lichtenbelt KD, Tartaglia M, Gelb BD, Lemischka IR (2010) Patient-specific induced pluripotent stem-cell-derived models of LEOPARD syndrome. Nature 465(7299):808–812
22. Moretti A, Bellin M, Welling A, Jung CB, Lam JT, Bott-Flugel L, Dorn T, Goedel A, Hohnke C, Hofmann F, Seyfarth M, Sinnecker D, Schomig A, Laugwitz KL (2010) Patient-specific induced pluripotent stem-cell models for long-QT syndrome. N Engl J Med 363(15):1397–1409

23. Yoshida Y, Yamanaka S (2010) Recent stem cell advances: induced pluripotent stem cells for disease modeling and stem cell-based regeneration. Circulation 122(1):80–87
24. Itzhaki I, Maizels L, Huber I, Zwi-Dantsis L, Caspi O, Winterstern A, Feldman O, Gepstein A, Arbel G, Hammerman H, Boulos M, Gepstein L (2011) Modelling the long QT syndrome with induced pluripotent stem cells. Nature 471(7337):225–229
25. Nelson TJ, Terzic A (2011) Induced pluripotent stem cells: an emerging theranostics platform. Clin Pharmacol Ther 89(5):648–650
26. Tiscornia G, Monserrat N, Belmonte JC (2011) Modelling long QT syndrome with iPS cells: be still, my beating heart. Circ Res 108(6):648–649
27. Yazawa M, Hsueh B, Jia X, Pasca AM, Bernstein JA, Hallmayer J, Dolmetsch RE (2011) Using induced pluripotent stem cells to investigate cardiac phenotypes in Timothy syndrome. Nature 471(7337):230–234
28. Huang P, He Z, Ji S, Sun H, Xiang D, Liu C, Hu Y, Wang X, Hui L (2011) Induction of functional hepatocyte-like cells from mouse fibroblasts by defined factors. Nature 475(7356):386–389
29. Sekiya S, Suzuki A (2011) Direct conversion of mouse fibroblasts to hepatocyte-like cells by defined factors. Nature 475(7356):390–393
30. Vierbuchen T, Ostermeier A, Pang ZP, Kokubu Y, Sudhof TC, Wernig M (2010) Direct conversion of fibroblasts to functional neurons by defined factors. Nature 463(7284):1035–1041
31. Ieda M, Fu JD, Delgado-Olguin P, Vedantham V, Hayashi Y, Bruneau BG, Srivastava D (2010) Direct reprogramming of fibroblasts into functional cardiomyocytes by defined factors. Cell 142(3):375–386
32. Zhou Q, Brown J, Kanarek A, Rajagopal J, Melton DA (2008) In vivo reprogramming of adult pancreatic exocrine cells to beta-cells. Nature 455(7213):627–632
33. Berkes CA, Tapscott SJ (2005) MyoD and the transcriptional control of myogenesis. Semin Cell Dev Biol 16(4–5):585–595
34. Tapscott SJ (2005) The circuitry of a master switch: myod and the regulation of skeletal muscle gene transcription. Development 132(12):2685–2695
35. Constantinides PG, Jones PA, Gevers W (1977) Functional striated muscle cells from non-myoblast precursors following 5-azacytidine treatment. Nature 267(5609):364–366
36. Taylor SM, Jones PA (1979) Multiple new phenotypes induced in 10T1/2 and 3T3 cells treated with 5-azacytidine. Cell 17(4):771–779
37. Lassar AB, Paterson BM, Weintraub H (1986) Transfection of a DNA locus that mediates the conversion of 10T1/2 fibroblasts to myoblasts. Cell 47(5):649–656
38. Pownall ME, Gustafsson MK, Emerson CP Jr (2002) Myogenic regulatory factors and the specification of muscle progenitors in vertebrate embryos. Annu Rev Cell Dev Biol 18:747–783
39. Ledent V, Vervoort M (2001) The basic helix-loop-helix protein family: comparative genomics and phylogenetic analysis. Genome Res 11(5):754–770
40. Ledent V, Paquet O, Vervoort M (2002) Phylogenetic analysis of the human basic helix-loop-helix proteins. Genome Biol **3**(6), RESEARCH0030
41. Buckingham M, Bajard L, Chang T, Daubas P, Hadchouel J, Meilhac S, Montarras D, Rocancourt D, Relaix F (2003) The formation of skeletal muscle: from somite to limb. J Anat 202(1):59–68
42. Molkentin JD, Olson EN (1996) Defining the regulatory networks for muscle development. Curr Opin Genet Dev 6(4):445–453
43. Perry RL, Parker MH, Rudnicki MA (2001) Activated MEK1 binds the nuclear MyoD transcriptional complex to repress transactivation. Mol Cell 8(2):291–301
44. Kulessa H, Frampton J, Graf T (1995) GATA-1 reprograms avian myelomonocytic cell lines into eosinophils, thromboblasts, and erythroblasts. Genes Dev 9(10):1250–1262
45. Visvader JE, Elefanty AG, Strasser A, Adams JM (1992) GATA-1 but not SCL induces megakaryocytic differentiation in an early myeloid line. EMBO J 11(12):4557–4564

46. Heyworth C, Pearson S, May G, Enver T (2002) Transcription factor-mediated lineage switching reveals plasticity in primary committed progenitor cells. EMBO J 21(14):3770–3781
47. Zhang P, Iwasaki-Arai J, Iwasaki H, Fenyus ML, Dayaram T, Owens BM, Shigematsu H, Levantini E, Huettner CS, Lekstrom-Himes JA, Akashi K, Tenen DG (2004) Enhancement of hematopoietic stem cell repopulating capacity and self-renewal in the absence of the transcription factor C/EBP alpha. Immunity 21(6):853–863
48. Nutt SL, Heavey B, Rolink AG, Busslinger M (1999) Commitment to the B-lymphoid lineage depends on the transcription factor Pax5. Nature 401(6753):556–562. doi:10.1038/44076
49. Rolink AG, Nutt SL, Melchers F, Busslinger M (1999) Long-term in vivo reconstitution of T-cell development by Pax5-deficient B-cell progenitors. Nature 401(6753):603–606
50. Cobaleda C, Jochum W, Busslinger M (2007) Conversion of mature B cells into T cells by dedifferentiation to uncommitted progenitors. Nature 449(7161):473–477
51. Gurdon JB (2006) From nuclear transfer to nuclear reprogramming: the reversal of cell differentiation. Annu Rev Cell Dev Biol 22:1–22
52. Eggan K, Baldwin K, Tackett M, Osborne J, Gogos J, Chess A, Axel R, Jaenisch R (2004) Mice cloned from olfactory sensory neurons. Nature 428(6978):44–49
53. Eggan K, Akutsu H, Loring J, Jackson-Grusby L, Klemm M, Rideout WM 3rd, Yanagimachi R, Jaenisch R (2001) Hybrid vigor, fetal overgrowth, and viability of mice derived by nuclear cloning and tetraploid embryo complementation. Proc Natl Acad Sci U S A 98(11):6209–6214
54. Inoue K, Wakao H, Ogonuki N, Miki H, Seino K, Nambu-Wakao R, Noda S, Miyoshi H, Koseki H, Taniguchi M, Ogura A (2005) Generation of cloned mice by direct nuclear transfer from natural killer T cells. Curr Biol 15(12):1114–1118
55. Li J, Ishii T, Feinstein P, Mombaerts P (2004) Odorant receptor gene choice is reset by nuclear transfer from mouse olfactory sensory neurons. Nature 428(6981):393–399
56. Gurdon JB, Byrne JA, Simonsson S (2003) Nuclear reprogramming and stem cell creation. Proc Natl Acad Sci U S A 100(Suppl 1):11819–11822
57. Gurdon JB, Byrne JA (2003) The first half-century of nuclear transplantation. Proc Natl Acad Sci U S A 100(14):8048–8052
58. Tamashiro KL, Wakayama T, Akutsu H, Yamazaki Y, Lachey JL, Wortman MD, Seeley RJ, D'Alessio DA, Woods SC, Yanagimachi R, Sakai RR (2002) Cloned mice have an obese phenotype not transmitted to their offspring. Nat Med 8(3):262–267
59. Tokuzawa Y, Kaiho E, Maruyama M, Takahashi K, Mitsui K, Maeda M, Niwa H, Yamanaka S (2003) Fbx15 is a novel target of Oct3/4 but is dispensable for embryonic stem cell self-renewal and mouse development. Mol Cell Biol 23(8):2699–2708
60. Boland MJ, Hazen JL, Nazor KL, Rodriguez AR, Gifford W, Martin G, Kupriyanov S, Baldwin KK (2009) Adult mice generated from induced pluripotent stem cells. Nature 461(7260):91–94
61. Kang L, Wang J, Zhang Y, Kou Z, Gao S (2009) iPS cells can support full-term development of tetraploid blastocyst-complemented embryos. Cell Stem Cell 5(2):135–138
62. Zhao XY, Li W, Lv Z, Liu L, Tong M, Hai T, Hao J, Guo CL, Ma QW, Wang L, Zeng F, Zhou Q (2009) iPS cells produce viable mice through tetraploid complementation. Nature 461(7260):86–90
63. Okita K, Ichisaka T, Yamanaka S (2007) Generation of germline-competent induced pluripotent stem cells. Nature 448(7151):313–317
64. Maherali N, Sridharan R, Xie W, Utikal J, Eminli S, Arnold K, Stadtfeld M, Yachechko R, Tchieu J, Jaenisch R, Plath K, Hochedlinger K (2007) Directly reprogrammed fibroblasts show global epigenetic remodeling and widespread tissue contribution. Cell Stem Cell 1(1):55–70
65. Wernig M, Meissner A, Foreman R, Brambrink T, Ku M, Hochedlinger K, Bernstein BE, Jaenisch R (2007) In vitro reprogramming of fibroblasts into a pluripotent ES-cell-like state. Nature 448(7151):318–324

66. Stadtfeld M, Maherali N, Breault DT, Hochedlinger K (2008) Defining molecular cornerstones during fibroblast to iPS cell reprogramming in mouse. Cell Stem Cell 2(3):230–240
67. Maherali N, Hochedlinger K (2008) Guidelines and techniques for the generation of induced pluripotent stem cells. Cell Stem Cell 3(6):595–605
68. Park IH, Lerou PH, Zhao R, Huo H, Daley GQ (2008) Generation of human-induced pluripotent stem cells. Nat Protoc 3(7):1180–1186
69. Li W, Wei W, Zhu S, Zhu J, Shi Y, Lin T, Hao E, Hayek A, Deng H, Ding S (2009) Generation of rat and human induced pluripotent stem cells by combining genetic reprogramming and chemical inhibitors. Cell Stem Cell 4(1):16–19
70. Liu H, Zhu F, Yong J, Zhang P, Hou P, Li H, Jiang W, Cai J, Liu M, Cui K, Qu X, Xiang T, Lu D, Chi X, Gao G, Ji W, Ding M, Deng H (2008) Generation of induced pluripotent stem cells from adult rhesus monkey fibroblasts. Cell Stem Cell 3(6):587–590
71. Aoi T, Yae K, Nakagawa M, Ichisaka T, Okita K, Takahashi K, Chiba T, Yamanaka S (2008) Generation of pluripotent stem cells from adult mouse liver and stomach cells. Science 321(5889):699–702
72. Aasen T, Raya A, Barrero MJ, Garreta E, Consiglio A, Gonzalez F, Vassena R, Bilic J, Pekarik V, Tiscornia G, Edel M, Boue S, Izpisua Belmonte JC (2008) Efficient and rapid generation of induced pluripotent stem cells from human keratinocytes. Nat Biotechnol 26(11):1276–1284
73. Seki T, Yuasa S, Oda M, Egashira T, Yae K, Kusumoto D, Nakata H, Tohyama S, Hashimoto H, Kodaira M, Okada Y, Seimiya H, Fusaki N, Hasegawa M, Fukuda K (2010) Generation of induced pluripotent stem cells from human terminally differentiated circulating T cells. Cell Stem Cell 7(1):11–14
74. Kim JB, Zaehres H, Wu G, Gentile L, Ko K, Sebastiano V, Arauzo-Bravo MJ, Ruau D, Han DW, Zenke M, Scholer HR (2008) Pluripotent stem cells induced from adult neural stem cells by reprogramming with two factors. Nature 454(7204):646–650
75. Hanna J, Markoulaki S, Schorderet P, Carey BW, Beard C, Wernig M, Creyghton MP, Steine EJ, Cassady JP, Foreman R, Lengner CJ, Dausman JA, Jaenisch R (2008) Direct reprogramming of terminally differentiated mature B lymphocytes to pluripotency. Cell 133(2):250–264
76. Mitsui K, Tokuzawa Y, Itoh H, Segawa K, Murakami M, Takahashi K, Maruyama M, Maeda M, Yamanaka S (2003) The homeoprotein Nanog is required for maintenance of pluripotency in mouse epiblast and ES cells. Cell 113(5):631–642
77. Ichida JK, Blanchard J, Lam K, Son EY, Chung JE, Egli D, Loh KM, Carter AC, Di Giorgio FP, Koszka K, Huangfu D, Akutsu H, Liu DR, Rubin LL, Eggan K (2009) A small-molecule inhibitor of tgf-Beta signaling replaces sox2 in reprogramming by inducing nanog. Cell Stem Cell 5(5):491–503
78. Szabo E, Rampalli S, Risueno RM, Schnerch A, Mitchell R, Fiebig-Comyn A, Levadoux-Martin M, Bhatia M (2010) Direct conversion of human fibroblasts to multilineage blood progenitors. Nature 468(7323):521–526
79. Kanazawa H, Ieda M, Kimura K, Arai T, Kawaguchi-Manabe H, Matsuhashi T, Endo J, Sano M, Kawakami T, Kimura T, Monkawa T, Hayashi M, Iwanami A, Okano H, Okada Y, Ishibashi-Ueda H, Ogawa S, Fukuda K (2010) Heart failure causes cholinergic transdifferentiation of cardiac sympathetic nerves via gp130-signaling cytokines in rodents. J Clin Invest 120(2):408–421
80. Zhou Q, Law AC, Rajagopal J, Anderson WJ, Gray PA, Melton DA (2007) A multipotent progenitor domain guides pancreatic organogenesis. Dev Cell 13(1):103–114
81. Pang ZP, Yang N, Vierbuchen T, Ostermeier A, Fuentes DR, Yang TQ, Citri A, Sebastiano V, Marro S, Sudhof TC, Wernig M (2011) Induction of human neuronal cells by defined transcription factors. Nature 476(7359):220–223
82. Takeuchi JK, Bruneau BG (2009) Directed transdifferentiation of mouse mesoderm to heart tissue by defined factors. Nature 459(7247):708–711

83. Bondue A, Lapouge G, Paulissen C, Semeraro C, Iacovino M, Kyba M, Blanpain C (2008) Mesp1 acts as a master regulator of multipotent cardiovascular progenitor specification. Cell Stem Cell 3(1):69–84
84. Ieda M, Tsuchihashi T, Ivey KN, Ross RS, Hong TT, Shaw RM, Srivastava D (2009) Cardiac fibroblasts regulate myocardial proliferation through beta1 integrin signaling. Dev Cell 16(2):233–244
85. Ieda M, Kanazawa H, Kimura K, Hattori F, Ieda Y, Taniguchi M, Lee JK, Matsumura K, Tomita Y, Miyoshi S, Shimoda K, Makino S, Sano M, Kodama I, Ogawa S, Fukuda K (2007) Sema3a maintains normal heart rhythm through sympathetic innervation patterning. Nat Med 13(5):604–612
86. Gulick J, Subramaniam A, Neumann J, Robbins J (1991) Isolation and characterization of the mouse cardiac myosin heavy chain genes. J Biol Chem 266(14):9180–9185
87. Zhao R, Watt AJ, Battle MA, Li J, Bondow BJ, Duncan SA (2008) Loss of both GATA4 and GATA6 blocks cardiac myocyte differentiation and results in acardia in mice. Dev Biol 317(2):614–619
88. Srivastava D (2006) Making or breaking the heart: from lineage determination to morphogenesis. Cell 126(6):1037–1048
89. Olson EN (2006) Gene regulatory networks in the evolution and development of the heart. Science 313(5795):1922–1927
90. Ghosh TK, Song FF, Packham EA, Buxton S, Robinson TE, Ronksley J, Self T, Bonser AJ, Brook JD (2009) Physical interaction between TBX5 and MEF2C is required for early heart development. Mol Cell Biol 29(8):2205–2218
91. Garg V, Kathiriya IS, Barnes R, Schluterman MK, King IN, Butler CA, Rothrock CR, Eapen RS, Hirayama-Yamada K, Joo K, Matsuoka R, Cohen JC, Srivastava D (2003) GATA4 mutations cause human congenital heart defects and reveal an interaction with TBX5. Nature 424(6947):443–447
92. Morin S, Charron F, Robitaille L, Nemer M (2000) GATA-dependent recruitment of MEF2 proteins to target promoters. EMBO J 19(9):2046–2055
93. Li B, Carey M, Workman JL (2007) The role of chromatin during transcription. Cell 128(4):707–719
94. Wu SM, Fujiwara Y, Cibulsky SM, Clapham DE, Lien CL, Schultheiss TM, Orkin SH (2006) Developmental origin of a bipotential myocardial and smooth muscle cell precursor in the mammalian heart. Cell 127(6):1137–1150
95. Laugwitz KL, Moretti A, Lam J, Gruber P, Chen Y, Woodard S, Lin LZ, Cai CL, Lu MM, Reth M, Platoshyn O, Yuan JX, Evans S, Chien KR (2005) Postnatal isl1+ cardioblasts enter fully differentiated cardiomyocyte lineages. Nature 433(7026):647–653
96. Saga Y, Miyagawa-Tomita S, Takagi A, Kitajima S, Miyazaki J, Inoue T (1999) MesP1 is expressed in the heart precursor cells and required for the formation of a single heart tube. Development 126(15):3437–3447
97. Zaret KS (2008) Genetic programming of liver and pancreas progenitors: lessons for stem-cell differentiation. Nat Rev Genet 9(5):329–340
98. Cirillo LA, Lin FR, Cuesta I, Friedman D, Jarnik M, Zaret KS (2002) Opening of compacted chromatin by early developmental transcription factors HNF3 (FoxA) and GATA-4. Mol Cell 9(2):279–289
99. Cirillo LA, Zaret KS (1999) An early developmental transcription factor complex that is more stable on nucleosome core particles than on free DNA. Mol Cell 4(6):961–969
100. Hochedlinger K (2010) From MYOD1 to iPS cells. Nat Rev Mol Cell Biol 11(12):817
101. Yamashita JK (2010) ES and iPS cell research for cardiovascular regeneration. Exp Cell Res 316(16):2555–2559
102. Efe JA, Hilcove S, Kim J, Zhou H, Ouyang K, Wang G, Chen J, Ding S (2011) Conversion of mouse fibroblasts into cardiomyocytes using a direct reprogramming strategy. Nat Cell Biol 13(3):215–222

103. Bu L, Jiang X, Martin-Puig S, Caron L, Zhu S, Shao Y, Roberts DJ, Huang PL, Domian IJ, Chien KR (2009) Human ISL1 heart progenitors generate diverse multipotent cardiovascular cell lineages. Nature 460(7251):113–117
104. Huber I, Itzhaki I, Caspi O, Arbel G, Tzukerman M, Gepstein A, Habib M, Yankelson L, Kehat I, Gepstein L (2007) Identification and selection of cardiomyocytes during human embryonic stem cell differentiation. Faseb J 21(10):2551–2563
105. Hattori F, Chen H, Yamashita H, Tohyama S, Satoh YS, Yuasa S, Li W, Yamakawa H, Tanaka T, Onitsuka T, Shimoji K, Ohno Y, Egashira T, Kaneda R, Murata M, Hidaka K, Morisaki T, Sasaki E, Suzuki T, Sano M, Makino S, Oikawa S, Fukuda K (2010) Nongenetic method for purifying stem cell-derived cardiomyocytes. Nat Methods 7(1):61–66
106. Laflamme MA, Chen KY, Naumova AV, Muskheli V, Fugate JA, Dupras SK, Reinecke H, Xu C, Hassanipour M, Police S, O'Sullivan C, Collins L, Chen Y, Minami E, Gill EA, Ueno S, Yuan C, Gold J, Murry CE (2007) Cardiomyocytes derived from human embryonic stem cells in pro-survival factors enhance function of infarcted rat hearts. Nat Biotechnol 25(9):1015–1024
107. Zhang M, Methot D, Poppa V, Fujio Y, Walsh K, Murry CE (2001) Cardiomyocyte grafting for cardiac repair: graft cell death and anti-death strategies. J Mol Cell Cardiol 33(5):907–921
108. Blum B, Benvenisty N (2008) The tumorigenicity of human embryonic stem cells. Adv Cancer Res 100:133–158
109. Yamashita T, Kawai H, Tian F, Ohta Y, Abe K (2010) Tumorigenic development of induced pluripotent stem cells in ischemic mouse brain. Cell Transplant 20(6):883–891
110. Miura K, Okada Y, Aoi T, Okada A, Takahashi K, Okita K, Nakagawa M, Koyanagi M, Tanabe K, Ohnuki M, Ogawa D, Ikeda E, Okano H, Yamanaka S (2009) Variation in the safety of induced pluripotent stem cell lines. Nat Biotechnol 27(8):743–745
111. Dai W, Field LJ, Rubart M, Reuter S, Hale SL, Zweigerdt R, Graichen RE, Kay GL, Jyrala AJ, Colman A, Davidson BP, Pera M, Kloner RA (2007) Survival and maturation of human embryonic stem cell-derived cardiomyocytes in rat hearts. J Mol Cell Cardiol 43(4):504–516

Chapter 13
Omics in Stem Cell Therapy: The Road Ahead

Kyunghee Byun, Goo-Bo Jeong, Trevor N. Collingwood and Bonghee Lee

Abstract Understanding self-renewal and differentiation of stem cells at molecular levels is essential for both basic research and clinical applications of stem cells. The complexity of stem cell fate is controlled by the sophisticated mechanisms of temporal and spatial gene expression and protein function. Thus, the capacity to control a gene's expression, build cellular reporters that reflect the activity state of lineage-specific endogenous genes, and either create or repair disease alleles is essential for realizing the potential of stem cell technology. In this chapter, we introduce zinc finger nucleases (ZFNs) as a tool to efficiently and selectively manipulate a given gene at the endogenous locus for advancing both our understanding as well as therapeutic application of stem cell biology. Proteins too are key players in the cell and often serve as biomarkers. They have diverse features that are not predictable from gene sequences or from the level of transcripts. Major breakthroughs in stem cell research were made by the identification of protein biomarkers such as colony-stimulating factors (CSFs) and cell-surface CD molecules. Recently, the Human Proteome Organization (HUPO) and Asia Oceanic HUPO (AOHUPO) launched initiatives for systematic MS-based proteomics as well as a gene coded proteomic approach to understanding stem cell differentiation. These studies will shed new light on stem cell biology and accelerate clinical applications of stem cells.

K. Byun · B. Lee (✉)
Center for Genomics and Proteomics and Stem Cell Core Facility,
Lee Gil Ya Cancer and Diabetes Institute,
Gachon University of Medicine and Science, Incheon, Korea
e-mail: bhlee@gachon.ac.kr

K. Byun · G.-B. Jeong · B. Lee
Department of Anatomy and Cell Biology, Graduate School of Medicine,
Gachon University of Medicine and Science, Incheon, Korea

T. N. Collingwood
Sigma-Aldrich, Inc., St. Louis, MO, USA

H. Baharvand and N. Aghdami (eds.), *Advances in Stem Cell Research*,
Stem Cell Biology and Regenerative Medicine, DOI: 10.1007/978-1-61779-940-2_13,

13.1 Introduction

Stem cells are widely and intensively studied in the field of life sciences due to their enormous potential to treat many incurable diseases. At the same time, there has been a wide range of social and ethical concerns on using human embryonic stem (ES) cells for research or therapeutic purposes because of their derivation source.

Stem cells are functionally defined as undifferentiated, primitive cells that retain the ability to reproduce themselves indefinitely ("self-renewal") and also have the ability to generate multiple types of cells in response to internal signals and external cues. Stem cells are largely categorized into two groups, ES cells and adult stem cells. Although these two types of stem cells have different merits and disadvantages, both may be valuable resources for future cell therapy. Recently, mouse and human somatic cells have been successfully reprogrammed into induced pluripotent stem (iPS) cells that exhibit similar characteristics to human ES cells [1–5]. iPS cells also have been spotlighted as an alternate source of pluripotent cells to the clinical patient's own stem cells.

These three types of stem cells have many advantages given their potential for regenerative therapy. Under the appropriate environmental conditions, they can be differentiated into most cell types in the body and can provide an unlimited number of cells for cell therapy. However, their applications for human treatment require special safety precautions. The biggest concern is the potential risk of genomic abnormality [6, 7] and tumor formation [8]. Other concerns are immune rejection of transplanted tissue, difficulty in achieving homogeneous cell populations after differentiation, and ethical controversy resulting from the destruction of embryos to generate new human ES cells. In contrast, adult stem cells have been used for cell therapy for a long time. More than 1,300 clinical trials with adult stem cells have been approved by the U.S. Food and Drug Administration (FDA). There are several reasons that account for the wide use of adult stem cells in clinical applications. These include very little chance of tumor formation, no ethical problems, the possibility of autologous transplantation (which causes no immune rejection), and easy differentiation into specific lineages of cells. However, the prevailing belief that adult stem cells do not form tumors needs thorough reevaluation. There has been a recent report that adipose tissue-derived human mesenchymal stromal cell (MSC) populations can immortalize and transform spontaneously. This MSC-TMC (transformed mesenchymal cells) transition happened after in vitro culture and expansion [9]. Furthermore, several other reports argue that some cancer cells are derived from adult stem cells in the body [10–13]. Currently, safety issues and techniques for in vitro expansion of the cells may be the two most important topics to develop additional successful cell therapies using adult stem cells.

Future studies should be focused on overcoming major obstacles that hinder successful stem cell therapy. First of all, obtaining autologous cell sources is required to avoid immune rejection. iPS cells have many advantages in this regard although more work is needed to bring iPS cells to clinical use. Patient-specific iPS cells thus have enormous promise for personalized cell therapy, as well as for research on human disease [4, 14]. Use of human iPS cells will require the development of efficient methods for performing gene targeting, a sequence-specific and permanent genome modification that exploits the cell's ability to perform homologous recombination (HR). Gene targeting by HR has played a critical role in genetic studies of various systems, including knock out/ knock in mouse models using mouse ES cells. This issue will be introduced in next section.

The second issue is preventing tumor formation. This is especially serious for clinical use of ES cells. Since tumor formation is thought to be caused by residual undifferentiated ES cells, the development of methods that remove these remnant ES cells is of great importance.

The third challenge is establishing standard protocols to efficiently differentiate stem cells into the cell types of interest. This will increase the efficacy of cell therapies, reduce side effects caused by the presence of unwanted cell types, and minimize tumor formation resulting from remnant undifferentiated ES or iPS cells.

All these needs are tightly associated with the activity of genes or proteins. Here we will discuss the current status and future perspective of genomics and proteomics-based stem cell research for successful stem cell therapy.

13.2 Why Stem Cell Genomics?

The complexity of stem cell biology is the product of myriad sophisticated mechanisms for temporal and spatial control of gene expression. The pursuit of safe, reproducible, and efficient strategies for reprogramming primary cells to the state of induced pluripotency, or driving differentiation of stem cells toward defined lineages is challenged by our need to better understand the role of the individual gene in these processes. Moreover, accurate real-time reporting on the differentiation state of the cell in response to manipulation is essential to elucidating and optimizing strategies for reliably generating the desired cellular characteristics. To this end, the capacity to (i) switch off/on or otherwise control a gene's expression, (ii) build cellular reporters that reflect the activity state of a lineage-specific endogenous gene, and (iii) either create or repair disease alleles is essential to realizing the potential of stem cell technology. For all the exquisite transcriptional, translational, and epigenetic governance the cellular environment exerts over a gene's activity, the product of its expression is still ultimately reliant on genomic structure and primary gene sequence. Thus, the development of tools to efficiently and selectively manipulate a given gene at the endogenous locus represents a critical need for advancing both our understanding as well as therapeutic application of stem cell biology.

13.3 Gene Correction Methods for Personalized Medicine

The causality of human disease is often linked to defined mutations in individual genes. Precise correction of these genetic lesions in patient-derived stem cells and iPS cells prior to in vitro differentiation and re-engraftment back into the donor is a critical barrier to the broad application of personalized autologous cell-based therapy. Classical gene targeting in mammalian cells combines positive and negative selection to isolate the rare targeted events [15]. This has been an effective strategy in mouse ES cells, but reports of gene targeting in other mammalian primary cell types are limited [16–21]. Recent evolution of recombinant adeno-associated virus-mediated approaches has however boosted the efficiency of classical gene targeting in human pluripotent stem cells [22–24].

Zinc finger nuclease (ZFN) technology has emerged as a highly efficient new tool for precise eukaryotic gene editing directly at the endogenous genomic locus (reviewed by [25]). ZFNs comprise a pair of engineered zinc finger DNA-binding domains each linked to a modified catalytic nuclease domain of the restriction enzyme FokI (Fig. 13.1; [26–29]. The DNA binding specificity can be engineered to direct the ZFN pair to the desired genomic locus to induce a double-strand DNA break (DSB) with high fidelity. The highly conserved natural cellular DNA repair pathways of nonhomologous end joining (NHEJ) or homology directed repair (HDR) function to resolve the ZFN-induced DSB (Fig. 13.2). NHEJ acts to efficiently rejoin the cleaved DNA ends but occasionally unfaithful repair can lead to variable small deletions or insertions at the site of DSB. When this occurs within coding sequence the consequence can be a shift in reading frame, deletion of critical residues or disruption of exon splicing—each of which can lead to a nonsense transcript and effective knock out of the target gene expression. Several studies have utilized this approach to generate single endogenous gene [30, 31] and multiple gene [32, 33] knock out in transformed cell lines as well as in human primary cells and hematopoietic stem cells [34–36]. An alternative DNA repair pathway is HDR. By this mechanism the cell normally uses a DNA template that is homologous to the damaged locus—typically a sister chromatid—to direct faithful repair of the lesion. However, early studies discovered that the DNA ends generated at a DNA break in mitotically dividing mammalian cells were recombinogenic [37] and that the efficiency of homologous recombination during repair of a DSB is several orders of magnitude greater than in the absence of the DSB [38]. Thus, when ZFNs are co-delivered to a cell with an exogenous homologous donor DNA carrying an investigator-defined genetic payload, the HDR pathway can instead use this donor as the repair template to insert the specified genetic modification. These alterations at the genomic locus can range from a single base change to site-specific insertion of at least 8 kb in size—without the need for antibiotic selection [39–42]. ZFN technology therefore represents a facile solution to a multitude of genomic editing needs, as summarized in Fig. 13.3.

Recent application of ZFNs to genome editing in human pluripotent stem cells has signaled an important advance toward the goal of patient-derived cell-based

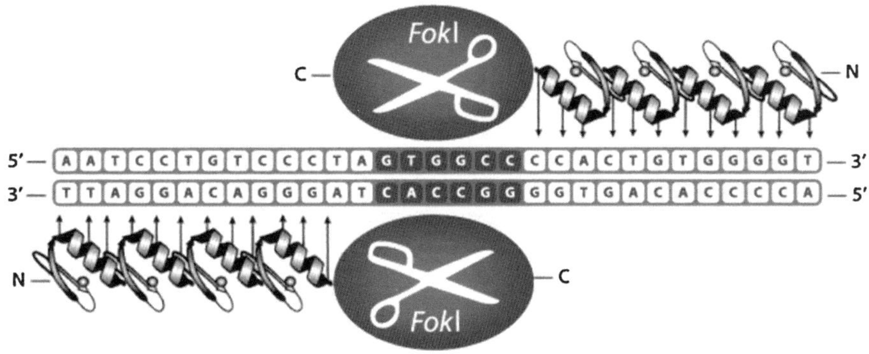

Fig. 13.1 Schematic of zinc finger nucleases binding a target sequence. Each zinc finger nuclease (ZFN) consists of two functional domains: A DNA-binding domain and a catalytic nuclease domain. The DNA-binding moiety comprise a chain of four to six zinc finger modules, each recognizing a unique triplet (3 bp) sequence of DNA, stitched together to form a zinc finger protein (ZFP) with specificity for 12–18 bp of DNA. The nuclease domain of the restriction enzyme FokI is attached to the ZFPs to generate the zinc finger nuclease. The FokI domain must dimerize to become catalytically active. ZFNs must therefore bind as a pair of proteins at the target site which typically comprise 24–36 bases of recognition sequence. The ZFN dimer then functions as a highly specific pair of 'genomic scissors' and cleaves the DNA

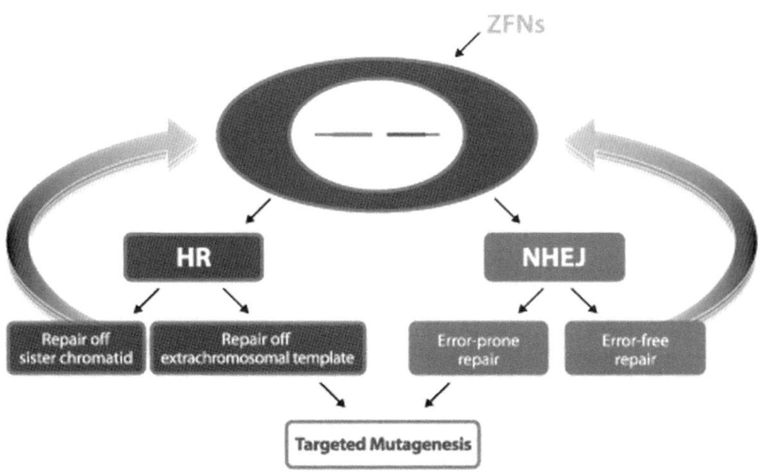

Fig. 13.2 The two natural DNA repair pathways invoked by a DNA double-strand break. The addition of zinc finger nucleases to the cell results in creation of a double-strand break at the target locus. This double-strand break is repaired by one of two endogenous repair pathways, either the non-homologous end joining (NHEJ) or the homology-dependent repair (HR) pathway. NHEJ is used to create gene knock outs while HR is typically utilized for targeted integration, gene tagging or codon conversion

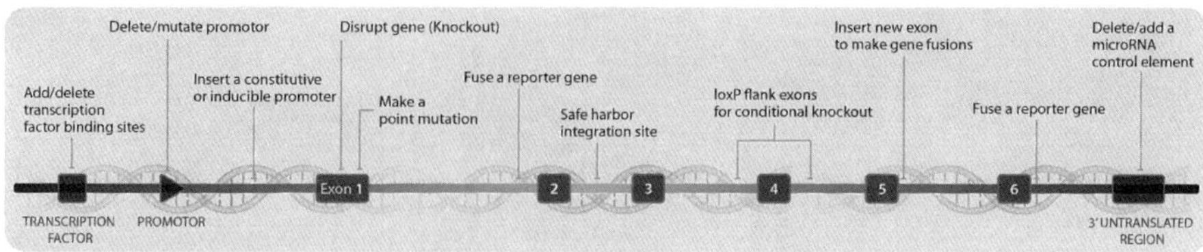

Fig. 13.3 Types of genomic modifications that may be mediated using ZFNs

therapy [43]. Initial studies used integration-defective lentivirus (IDLV) to deliver both ZFNs and the donor repair template into human ES cells to achieve targeted transgene insertion into the endogenous CCR5 locus—in the absence of antibiotic selection [44]. Targeting of the same locus in human mesenchymal stem cells using adenoviral-mediated ZFN delivery coupled with IDLV-mediated donor delivery was also successful—again without selection [45]. While viral strategies can overcome the challenges of low delivery efficiency to recalcitrant cell types, optimized electroporation mediated nucleic acid delivery coupled with positive selection strategies have led to successful inactivation of endogenous genes, such as PIG-A, by insertion of a drug resistance expression cassette into the gene's coding sequence in both human ES and iPS cells [36].

Oct4 gene expression is a hallmark of pluripotency. Hockemeyer and colleagues used ZFNs to insert a promoterless GFP cDNA with a 5′ splice acceptor into intron 1 of the Oct4 locus in ES cells. By this approach GFP expression was driven directly off the endogenous Oct4 promoter and so expressed only when the Oct4 promoter was active (in the pluripotent state), thereby creating an endogenous gene reporter of pluripotency [46]. Moreover, in an elegant strategy, inclusion of a promoterless puromycin resistance gene 3′ to the GFP and linked in frame by a 2A peptide sequence enabled the active Oct4 promoter to drive expression of both GFP and the resistance marker. Exposure to puromycin therefore selected for predominantly correctly targeted clones (up to 94%) rather than random integrants. Both the PIG-A and Oct4 genes are actively expressed in ES cells and are therefore presumed to reside in regions of open chromatin structure and, as such, be more accessible to ZFNs and the DNA repair machinery. However, Hockemeyer and colleagues also demonstrated GFP tagging at the transcriptionally silent PITX3 locus—both in ES and in iPS cells—thereby proving that endogenous gene expression is not a prerequisite for ZFN-mediated targeting [46]. Studies using an emerging DNA cleavage strategy—TAL nucleases—further validate the notion that the creation of targeted DSBs greatly facilitates genomic editing in these cell types [47].

Individual patient-derived iPS cells are providing new opportunities in modeling human disease "in a dish" [48]. Careful reprogramming of both wild type and diseased cells to become models for the tissues that are affected by the disease offers investigators a novel system for comparing the biology, and importantly, the drug sensitivity of affected and unaffected cells. However, notwithstanding the technical challenges of accurate and reproducible reprogramming, the molecular basis of a disease may differ between patients - even for apparently monogenic disorders. If two iPS lines from the same disease indication behave differently it is difficult to determine whether the cause is other disease-relevant genetic differences, or a variable consequence of reprogramming. Generation of isogenic pairs of wild type and mutant iPS cells which differ only at the disease-linked gene offers a solution resolving and eliminating variable genetic background as a confounding factor. Yet to achieve this goal requires a strategy for gene correction in patient-derived iPS cells (or mutation of wild type cells) that leaves only the disease linked mutation and no other genetic modification, such as selectable markers used for isolating the genomic modification events, or even the genetic "scar" left by recombinase-mediated removal of the marker. To this end, Soldner and colleagues have used ZFN-based genomic editing to generate isogenic sets of human disease and control pluripotent stem cells that differ solely by either of two Parkinson's disease susceptibility alleles in the α-synuclein gene [49]. Moreover, this was achieved using plasmid-based ZFN and donor delivery without use of antibiotic selection. A recent study has shown that single strand DNA oligonucleotides (ssODNs) can be effectively used as donor templates for creating ZFN-mediated point mutations, providing a simpler path to donor construction as well as eliminating the potential for a random donor integration [50]. Substitution of a plasmid donor with a 114 bp ssODN donor proved equally successful in generating genetically edited iPS cells [49]. This work represents an important milestone in creating patient-derived isogenic pluripotent stem cell models.

The studies above describe the control of gene expression and function by permanent targeted modification of the gene itself. This approach has the advantage of creating a defined and reproducible outcome. However, in some instances it may be desirable to regulate gene expression in a dose-dependant manner. Such strategies might include the insertion of inducible expression cassettes for transgenes (e.g., key regulators of pluripotency), or shRNA/miRNA that in turn regulate the expression of target genes. Recent reports have shown that ZFNs can be used to efficiently insert both inducible and constitutively active constructs specifically into a safe harbor locus (AAVS1) in pluripotent stem cells and so render their expression controllable by a small molecule drug [36, 46, 51]. Importantly, site-specific insertion into a safe harbor avoids the genomic and phenotypic uncertainty that surrounds random integration (e.g., epigenetic silencing of the transgene or disrupted regulation of other genes near the site of insertion) thereby providing the investigator with greater experimental control.

13.4 ZFN Advantages

The researcher's toolbox for control of stem cell biology must continue to expand if we are to realize the full potential of this exciting technology. Optimizing methods to direct cell fate are critical components for generating the array of pluripotent and differentiated states that will be required. The tagging of endogenous markers of pluripotency as well as those that define a differentiated lineage is an important tool to unraveling the complexities of stem cell biology. The challenge of unambiguous identification of lineage may be met through the creation of accurate genomic reporters of lineage. These strategies will likely include inserting into a safe harbor locus exogenous cassettes comprised of a reporter tag (e.g., GFP) under the control of critical regulatory elements from a lineage-specific gene. In a more physiologically representative approach, the endogenous gene itself may be tagged directly with the fluorescent marker, such as the Oct4 study described earlier. Finally, the ability to alter just a single codon enables better modeling of the functional consequences of disease alleles. With the genomic editing tools now available each approach appears feasible.

13.5 Stem Cell Proteomics: Why Stem Cell Proteomics?

Major breakthroughs in human stem cell research were made by the identification of proteins, such as colony-stimulating factors (CSFs) and cell-surface CD molecules. Proteins, key players in the cell, have diverse features that are not predictable from gene sequences or transcript levels. For example, posttranslational modifications (PTMs), protein–protein interactions, and subcellular locations affect the function and activity of proteins, but these are not predictable using genomics or transcriptomics technology.

Both basic and clinically oriented stem cell research are confronted with many open questions that can be most efficiently answered by proteomics. For instance, the cell surface proteins and signaling cascades of stem cells and their differentiated progenies are largely unknown, as are the differentiation-specific proteins that can be used as biomarkers of the intermediate or terminal steps of cell differentiation, or to discriminate tumorigenic cells from the pool [52]. A major challenge researchers now face is the systematic and optimized use of proteomics technology to decipher stem cell biology.

13.6 Stem Cell Proteomics: The Road Ahead

Antibody proteomics. Antibody proteomics, one of the major initiatives of the Human Proteome Organization (HUPO), is defined as the systematic generation and use of protein-specific antibodies to functionally explore the proteome [53]. Agaton et al. [54] showed that protein epitope signatures tags (PrESTs), unique epitopes present in native proteins, can be used as antigens and affinity ligands for cost-effective generation of highly selective, mono-specific antibodies or affinity-purified polyclonal antibodies. Currently, this approach has been extensively used to construct a comprehensive, antibody-based protein atlas of expression and

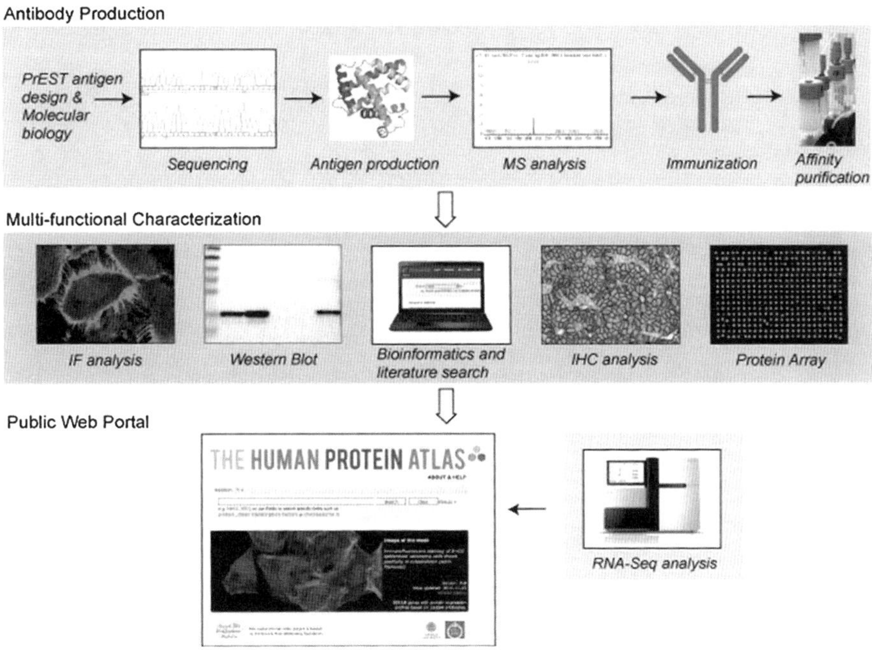

Fig. 13.4 HUPO Antibody initiative project: Workflow of "Antibody Production, Multi-functional Characterization, and Public Web Portal" (After Atlas Antibody: http://www.atlasantibody.com)

localization profiles in human normal and cancer tissues [55] (Fig. 13.4). As monoclonal antibodies have been systematically validated, the resources developed by the antibody proteomics initiative can be systematically used for stem cell research. For that purpose, well-defined panels of stem cells and their derivatives are required. Once prepared, these panels of cells can be efficiently screened using high-throughput cell microarray (tissue microarray) methods. This approach allows for detection of novel CD molecules involved in the differentiation of a variety of stem cells, and can be easily applicable to cell therapy by removing tumorigenic cells.

MS-based membrane proteomics. As previously mentioned, the antibody-based approach has both success and promise in identifying novel proteins, especially membrane proteins. However, this method has intrinsic limitations because it is an indirect method for characterizing proteins, and there can be cross-reactivity in antibody-antigen interactions [56]. Also, the antibody-based approach cannot provide much information about PTMs that may be important for protein function.

In contrast, a mass spectrometry (MS)-based method allows direct sequencing of peptides and proteins, and the technology is evolving as the cutting edge in MS-based proteomics [57] (Fig. 13.5) and can be used either as a method

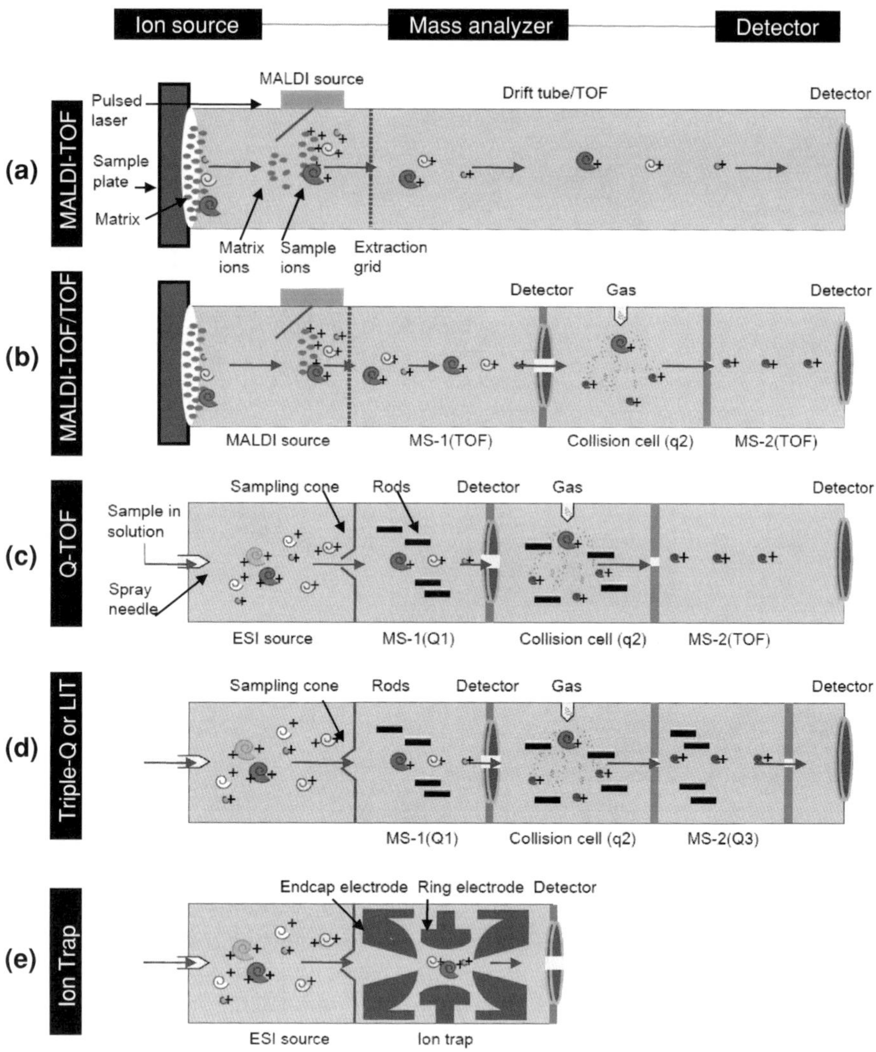

complementing the antibody-based approach or for direct identification and characterization of CD molecules. To that end, MS-based proteomics can first be used to identify molecular targets of antibodies or epitope mapping [58]. Secondly, a MS-based approach, coupled with membrane protein preparation and separation technologies, can be used to directly identify and characterize membrane proteins from cells and tissues [59, 60]. In 2005, Asia Oceania Human Proteome Organization (AOHUPO) launched the Membrane Proteomics Initiative (MPI), which aims to develop methods for characterizing membrane proteomes, and for characterizing the proteins of specific membrane systems. Since MPI has almost finished phase I (technology development) and phase II (large-scale analysis

◀**Fig. 13.5** Schematic overview of peptide sequencing by tandem MS (After Baharvand et al. [57]). The main components of a mass spectrometer are an ion source, one or more mass analyzers that measure the mass-to-charge ratio (m/z) of the ionized analytes, and a detector that registers the number of ions at each m/z value. The gas-phase ions are produced in the ion source, after which they enter the mass analyzer and are separated according to their mass/charge (m/z) ratios in the mass analyzer. MALDI (**a**, **b**) and ESI (**c**–**e**) are two techniques most commonly used to volatize and ionize the variety of biomolecules, including peptides, proteins, metabolites, and oligonucleotides to enable their measurement by MS. The mass analyzer is the central component behind the technology. Examples of mass analyzers currently used in proteomics research are: TOF, Q, triple Q or LIT, ion trap Fourier transform ion cyclotron resonance (FTICR), and Orbitrap. The TOF analyzers separate ions based on the differences in transit time from the ion source to the detector in flight tubes under vacuum (**a**–**c**). The Q analyzer transmits only ions within a narrow m/z range and uses the stability of the trajectory to separate these ions according to their m/z ratio on four parallel cylindrical metal rods (**c**). In triple Q or LIT, ions of a particular m/z are selected in a first section (Q1) and fragmented in a collision cell (q2), after which the fragments are separated in Q3. In the LIT, ions are captured in Q3. They are then excited via resonant electric field and the fragments are scanned out, creating the tandem mass spectrum (**d**). The ion trap analyzer captures or traps the ions, which are then subjected to MS or MS/MS analysis (**e**)

of membrane proteomes) successfully, the methods and experiences from MPI can readily be utilized for more complex areas, such as stem cell proteomics. Because human iPSCs can mimic cellular phenotypes of various intractable diseases, the MS-based proteomics can contribute to target discovery in the drug development process for these disease.

PTMs. PTMs play important roles in stem cell biology. Elliott et al. [61] showed that ~21% of the proteins identified in murine R1 ES cells had PTMs, and several of them, including Ras-GTPase activating protein binding protein 1 and phosphoglycerate kinase, had not been previously reported as associated with PTMs. Unwin et al. [62], revealed the importance of posttranslational control as a regulatory factor in primary hematopoietic stem cells. From the proteomics perspective, assaying changes in PTMs during stem cell differentiation in a high-throughput fashion is a key to understanding the underlying mechanisms of stemness and differentiation.

Proteomics also allows for quantitative analysis of subproteomes with a specific PTM (e.g. the phosphoproteome [63] and the methylproteome [64]). For example, Kratchmarova et al. [65] used MS-based approaches to quantitatively monitor tyrosine-phosphorylated proteins upon the differentiation of mesenchymal stem cells into bone-forming cells in response to growth factors. Recently, Van Hoof et al. [66] successfully reported phospho-protein profiling during early differentiation of human ES cells, which enables the identification of key protein activations in specific signaling pathways of the early differentiation process. They provide strong evidence of how PTM-based proteomics can contribute to understanding the mechanism of adaptor signaling at the stem cell level. This technique will likely be widely applied to mechanism-based research focusing on self-renewal and differentiation processes of stem cells.

Finding stem cell growth factors: Secretome. Understanding differentiation is important both for basic research and clinical applications. In stem cell research, finding growth factors in the secretome is a critical step toward understanding and controlling stem cell differentiation at the molecular level. Recently, a proteomics approach revealed the endodermal secretome network guiding stem cell cardiopoiesis from ES cells [67]. Evidence from normal and tumor cells also illustrate the linkage between signal transduction pathways and the secretomes [68, 69].

Most colony-stimulating factors (CSFs) were purified from media conditioned by various human tumor cells, and it is noteworthy that tumors secreting CSFs include not only hematopoietic cancers, such as a T-lymphoblast cell line [70], but also a human squamous cell line from a lower oral cavity [71] and a gastric cancer cell line [72], which are not related to hematopoiesis. Therefore, systematic screening of stem and cancer cell media, combined with extensive proteomic purification and characterization, may be required for high-throughput characterization of the growth factor repertoire of each stem cell line. Recently, we reported proteomic differences of various sized exosomes from human neural stem cells and their differentiated oligodendrocyte progeny [73]. In the study, we separated exosomes with Field-Flow Fractionation by 10 nm size differences, and showed that this step was useful for reproducibly analyzing the proteomics changes. These promising results suggest that well-designed sampling strategy and analysis technologies should be further developed to achieve better yields in screening the tumorigenic cells for safe stem cell therapy.

13.7 Future Perspectives

Our understanding of the functional consequence of disease-linked mutations even in the presence of diverse genetic backgrounds is expected to advance significantly with the advent of paired isogenic pluripotent cell-based models. While the current focus may be on using the isogenic cell line approach to enhance our understanding the biology of disease, the future of personalized stem cell therapy will likely lie in our ability to take a patient's own cells, correct the disease allele, and then return those cells to the patient in a genetically and developmentally correct format. ZFN-mediated genomic editing has demonstrated the creation or repair of disease alleles in stem cells without leaving behind any other detectable genetic change. While patient-specific stem cell therapy is still in its infancy, the advent of facile methods for precise genomic engineering illuminates a path to this goal. In 2007, the HUPO and International Society for Stem Cell Research (ISSCR) joint initiative, Proteome Biology of Stem Cells, was established as a collaborative effort bringing stem cell biologists and researchers together in proteomics [52, 74]. The aim of the initiative is to effect the implementation of cutting edge proteomic technology in stem cell research to further our understanding of stem cell biology. This has been prompted primarily by major breakthroughs in stem cell research, the potential of stem cells for biomedical application, and the awareness that proteomics may be able to accelerate this progress further and possibly open yet unexplored areas of research. Over the last 2 years, acting groups of the AOHUPO initiative have also started workshops to optimize protocols for human ES cells sampling, MS analysis, and bioinformatics analyses for preparing the AOHUPO stem cell initiative. The group has chosen human ES cells for initial study. Analysis of human ES cell membrane proteins will be used to standardize biomarker discovery, and this will help in providing standard guidelines for stem cell proteomics. The initiative will also be choosing the iPS cells and differentiated cells for membrane protein analysis of various intractable diseases. We believe that the initiative will produce many deliverables for use in drug discovery and cell therapy. The voyage to stem cell genomics and proteomics with cutting edge technology has already begun. We believe genomics and proteomics will have great impact on many aspects of the path to successful stem cell therapy.

Acknowledgments The authors thank Mrs. Jee Hyun Oh, Mr. Daesik Kim, Ms. Jin-A Kang, and Dr. Enkhjaigal Bayarsaikhan for their efforts. This research was supported by a grant from the National Research Foundation to KB (2011-0003677). Funding in part was also provided by a National Research Foundation (2010-0020573; BL) and the 21ST frontier science program SC-2110 of the Korea stem cell research center (BL).

References

1. Lowry WE, Richter L, Yachechko R, Pyle AD, Tchieu J, Sridharan R, Clark AT, Plath K (2008) Generation of human induced pluripotent stem cells from dermal fibroblasts. Proc Natl Acad Sci U S A 105(8):2883–2888
2. Mali P, Ye Z, Hommond HH, Yu X, Lin J, Chen G, Zou J, Cheng L (2008) Improved efficiency and pace of generating induced pluripotent stem cells from human adult and fetal fibroblasts. Stem Cells 26(8):1998–2005
3. Park IH, Zhao R, West JA, Yabuuchi A, Huo H, Ince TA, Lerou PH, Lensch MW, Daley GQ (2008) Reprograming of human somatic cells to pluripotency with defined factors. Nature 451(7175):141–146
4. Takahashi K, Tanabe K, Ohnuki M, Narita M, Ichisaka T, Tomoda K, Yamanaka S (2007) Induction of pluripotent stem cells from adult human fibroblasts by defined factors. Cell 131(5):861–872
5. Yu J, Vodyanik MA, Smuga-Otto K, Antosiewicz-Bourget J, Frane JL, Tian S, Nie J, Jonsdottir GA, Ruotti V, Stewart R, Slukvin II, Thomson JA (2007) Induced pluripotent stem cell lines derived from human somatic cells. Science 318(5858):1917–1920
6. Maitra A, Arking DE, Shivapurkar N, Ikeda M, Stastny V, Kassauei K, Sui G, Cutler DJ, Liu Y, Brimble SN, Noaksson K, Hyllner J, Schulz TC, Zeng X, Freed WJ, Crook J, Abraham S, Colman A, Sartipy P, Matsui S, Carpenter M, Gazdar AF, Rao M, Chakravarti A (2005) Genomic alterations in cultured human embryonic stem cells. Nat Genet 37(10):1099–1103
7. Draper JS, Smith K, Gokhale P, Moore HD, Maltby E, Johnson J, Meisner L, Zwaka TP, Thomson JA, Andrews PW (2004) Recurrent gain of chromosomes 17q and 12 in cultured human embryonic stem cells. Nat Biotechnol 22(1):53–54
8. Vogel G (2005) Cell biology. Ready or not? Human ES cells head toward the clinic. Science 308(5728):1534–1538
9. Rubio D, Garcia-Castro J, Martin MC, de la Fuente R, Cigudosa JC, Lloyd AC, Bernad A (2005) Spontaneous human adult stem cell transformation. Cancer Res 65(8):3035–3039

10. Pardal R, Clarke MF, Morrison SJ (2003) Applying the principles of stem-cell biology to cancer. Nat Rev Cancer 3(12):895–902
11. Marx J (2003) Cancer research. Mutant stem cells may seed cancer. Science 301(5638): 1308–1310
12. Clarke MF, Fuller M (2006) Stem cells and cancer: two faces of eve. Cell 124(6):1111–1115
13. Rajasekhar VK, Dalerba P, Passegue E, Lagasse E, Najbauer J (2007) Stem cells, cancer, and context dependence. Stem Cells 26:292–298
14. Yu J, Vodyanik MA, Smuga-Otto K, Antosiewicz-Bourget J, Frane JL, Tian S, Nie J, Jonsdottir GA, Ruotti V, Stewart R, Slukvin II, Thomson JA (2007) Induced pluripotent stem cell lines derived from human somatic cells. Science 318(5858):1917–1920
15. Thomas KR, Folger KR, Capecchi MR (1986) High frequency targeting of genes to specific sites in the mammalian genome. Cell 44(3):419–428
16. Davis RP, Ng ES, Costa M, Mossman AK, Sourris K, Elefanty AG, Stanley EG (2008) Targeting a GFP reporter gene to the MIXL1 locus of human embryonic stem cells identifies human primitive streak-like cells and enables isolation of primitive hematopoietic precursors. Blood 111(4):1876–1884
17. Di Domenico AI, Christodoulou I, Pells SC, McWhir J, Thomson AJ (2008) Sequential genetic modification of the hprt locus in human ESCs combining gene targeting and recombinase-mediated cassette exchange. Cloning Stem Cells 10(2):217–230
18. Irion S, Luche H, Gadue P, Fehling HJ, Kennedy M, Keller G (2007) Identification and targeting of the ROSA26 locus in human embryonic stem cells. Nat Biotechnol 25(12): 1477–1482
19. Ruby KM, Zheng B (2009) Gene targeting in a HUES line of human embryonic stem cells via electroporation. Stem Cells 27(7):1496–1506
20. Urbach A, Schuldiner M, Benvenisty N (2004) Modeling for Lesch-Nyhan disease by gene targeting in human embryonic stem cells. Stem Cells 22(4):635–641
21. Zwaka TP, Thomson JA (2003) Homologous recombination in human embryonic stem cells. Nat Biotechnol 21(3):319–321
22. Jang JH, Koerber JT, Kim JS, Asuri P, Vazin T, Bartel M, Keung A, Kwon I, Park KI, Schaffer DV (2011) An evolved adeno-associated viral variant enhances gene delivery and gene targeting in neural stem cells. Mol Ther 19(4):667–675
23. Khan IF, Hirata RK, Russell DW (2011) AAV-mediated gene targeting methods for human cells. Nat Protoc 6(4):482–501
24. Khan IF, Hirata RK, Wang PR, Li Y, Kho J, Nelson A, Huo Y, Zavaljevski M, Ware C, Russell DW (2010) Engineering of human pluripotent stem cells by AAV-mediated gene targeting. Mol Ther 18(6):1192–1199
25. Urnov FD, Rebar EJ, Holmes MC, Zhang HS, Gregory PD (2010) Genome editing with engineered zinc finger nucleases. Nat Rev Genet 11(9):636–646
26. Doyon Y, Vo TD, Mendel MC, Greenberg SG, Wang J, Xia DF, Miller JC, Urnov FD, Gregory PD, Holmes MC (2011) Enhancing zinc-finger-nuclease activity with improved obligate heterodimeric architectures. Nat Methods 8(1):74–79
27. Kim YG, Cha J, Chandrasegaran S (1996) Hybrid restriction enzymes: zinc finger fusions to Fok I cleavage domain. Proc Natl Acad Sci U S A 93(3):1156–1160
28. Miller JC, Holmes MC, Wang J, Guschin DY, Lee YL, Rupniewski I, Beausejour CM, Waite AJ, Wang NS, Kim KA, Gregory PD, Pabo CO, Rebar EJ (2007) An improved zinc-finger nuclease architecture for highly specific genome editing. Nat Biotechnol 25(7): 778–785
29. Szczepek M, Brondani V, Buchel J, Serrano L, Segal DJ, Cathomen T (2007) Structure-based redesign of the dimerization interface reduces the toxicity of zinc-finger nucleases. Nat Biotechnol 25(7):786–793
30. Malphettes L, Freyvert Y, Chang J, Liu PQ, Chan E, Miller JC, Zhou Z, Nguyen T, Tsai C, Snowden AW, Collingwood TN, Gregory PD, Cost GJ (2010) Highly efficient deletion of FUT8 in CHO cell lines using zinc-finger nucleases yields cells that produce completely nonfucosylated antibodies. Biotechnol Bioeng 106(5):774–783

31. Santiago Y, Chan E, Liu PQ, Orlando S, Zhang L, Urnov FD, Holmes MC, Guschin D, Waite A, Miller JC, Rebar EJ, Gregory PD, Klug A, Collingwood TN (2008) Targeted gene knockout in mammalian cells by using engineered zinc-finger nucleases. Proc Natl Acad Sci U S A 105(15):5809–5814
32. Cost GJ, Freyvert Y, Vafiadis A, Santiago Y, Miller JC, Rebar E, Collingwood TN, Snowden A, Gregory PD (2010) BAK and BAX deletion using zinc-finger nucleases yields apoptosis-resistant CHO cells. Biotechnol Bioeng 105(2):330–340
33. Liu PQ, Chan EM, Cost GJ, Zhang L, Wang J, Miller JC, Guschin DY, Reik A, Holmes MC, Mott JE, Collingwood TN, Gregory PD (2010) Generation of a triple-gene knockout mammalian cell line using engineered zinc-finger nucleases. Biotechnol Bioeng 106(1): 97–105
34. Holt N, Wang J, Kim K, Friedman G, Wang X, Taupin V, Crooks GM, Kohn DB, Gregory PD, Holmes MC, Cannon PM (2010) Human hematopoietic stem/progenitor cells modified by zinc-finger nucleases targeted to CCR5 control HIV-1 in vivo. Nat Biotechnol 28(8): 839–847
35. Perez EE, Wang J, Miller JC, Jouvenot Y, Kim KA, Liu O, Wang N, Lee G, Bartsevich VV, Lee YL, Guschin DY, Rupniewski I, Waite AJ, Carpenito C, Carroll RG, Orange JS, Urnov FD, Rebar EJ, Ando D, Gregory PD, Riley JL, Holmes MC, June CH (2008) Establishment of HIV-1 resistance in CD4+ T cells by genome editing using zinc-finger nucleases. Nat Biotechnol 26(7):808–816
36. Zou J, Maeder ML, Mali P, Pruett-Miller SM, Thibodeau-Beganny S, Chou BK, Chen G, Ye Z, Park IH, Daley GQ, Porteus MH, Joung JK, Cheng L (2009) Gene targeting of a disease-related gene in human induced pluripotent stem and embryonic stem cells. Cell Stem Cell 5(1):97–110
37. Rouet P, Smih F, Jasin M (1994) Expression of a site-specific endonuclease stimulates homologous recombination in mammalian cells. Proc Natl Acad Sci U S A 91(13): 6064–6068
38. Jasin M (1996) Genetic manipulation of genomes with rare-cutting endonucleases. Trends Genet 12(6):224–228
39. Moehle EA, Rock JM, Lee YL, Jouvenot Y, DeKelver RC, Gregory PD, Urnov FD, Holmes MC (2007) Targeted gene addition into a specified location in the human genome using designed zinc finger nucleases. Proc Natl Acad Sci U S A 104(9):3055–3060
40. Porteus MH, Baltimore D (2003) Chimeric nucleases stimulate gene targeting in human cells. Science 300(5620):763
41. Porteus MH, Carroll D (2005) Gene targeting using zinc finger nucleases. Nat Biotechnol 23(8):967–973
42. Urnov FD, Miller JC, Lee YL, Beausejour CM, Rock JM, Augustus S, Jamieson AC, Porteus MH, Gregory PD, Holmes MC (2005) Highly efficient endogenous human gene correction using designed zinc-finger nucleases. Nature 435(7042):646–651
43. Collin J, Lako M (2011) Concise review: putting a finger on stem cell biology: zinc finger nuclease-driven targeted genetic editing in human pluripotent stem cells. Stem Cells 29(7): 1021–1033
44. Lombardo A, Genovese P, Beausejour CM, Colleoni S, Lee YL, Kim KA, Ando D, Urnov FD, Galli C, Gregory PD, Holmes MC, Naldini L (2007) Gene editing in human stem cells using zinc finger nucleases and integrase-defective lentiviral vector delivery. Nat Biotechnol 25(11):1298–1306
45. Benabdallah BF, Allard E, Yao S, Friedman G, Gregory PD, Eliopoulos N, Fradette J, Spees JL, Haddad E, Holmes MC, Beausejour CM (2010) Targeted gene addition to human mesenchymal stromal cells as a cell-based plasma-soluble protein delivery platform. Cytotherapy 12(3):394–399
46. Hockemeyer D, Soldner F, Beard C, Gao Q, Mitalipova M, DeKelver RC, Katibah GE, Amora R, Boydston EA, Zeitler B, Meng X, Miller JC, Zhang L, Rebar EJ, Gregory PD, Urnov FD, Jaenisch R (2009) Efficient targeting of expressed and silent genes in human ESCs and iPSCs using zinc-finger nucleases. Nat Biotechnol 27(9):851–857

47. Hockemeyer D, Wang H, Kiani S, Lai CS, Gao Q, Cassady JP, Cost GJ, Zhang L, Santiago Y, Miller JC, Zeitler B, Cherone JM, Meng X, Hinkley SJ, Rebar EJ, Gregory PD, Urnov FD, Jaenisch R (2011) Genetic engineering of human pluripotent cells using TALE nucleases. Nat Biotechnol 29(8):731–734
48. Vogel G (2010) Stem cells. Diseases in a dish take off. Science 330(6008):1172–1173
49. Soldner F, Laganiere J, Cheng AW, Hockemeyer D, Gao Q, Alagappan R, Khurana V, Golbe LI, Myers RH, Lindquist S, Zhang L, Guschin D, Fong LK, Vu BJ, Meng X, Urnov FD, Rebar EJ, Gregory PD, Zhang HS, Jaenisch R (2011) Generation of isogenic pluripotent stem cells differing exclusively at two early onset Parkinson point mutations. Cell 146(2):318–331
50. Chen F, Pruett-Miller SM, Huang Y, Gjoka M, Duda K, Taunton J, Collingwood TN, Frodin M, Davis GD (2011) High-frequency genome editing using ssDNA oligonucleotides with zinc-finger nucleases. Nat Methods 8(9):753–755
51. DeKelver RC, Choi VM, Moehle EA, Paschon DE, Hockemeyer D, Meijsing SH, Sancak Y, Cui X, Steine EJ, Miller JC, Tam P, Bartsevich VV, Meng X, Rupniewski I, Gopalan SM, Sun HC, Pitz KJ, Rock JM, Zhang L, Davis GD, Rebar EJ, Cheeseman IM, Yamamoto KR, Sabatini DM, Jaenisch R, Gregory PD, Urnov FD (2010) Functional genomics, proteomics, and regulatory DNA analysis in isogenic settings using zinc finger nuclease-driven transgenesis into a safe harbor locus in the human genome. Genome Res 20(8):1133–1142
52. Krijgsveld J, Whetton AD, Lee B, Lemischka I, Oh S, Pera M, Mummery C, Heck AJ (2008) Proteome biology of stem cells: a new joint HUPO and ISSCR initiative. Mol Cell Proteomics 7(1):204–205
53. Uhlen M, Ponten F (2005) Antibody-based proteomics for human tissue profiling. Mol Cell Proteomics 4(4):384–393
54. Agaton C, Galli J, Hoiden Guthenberg I, Janzon L, Hansson M, Asplund A, Brundell E, Lindberg S, Ruthberg I, Wester K, Wurtz D, Hoog C, Lundeberg J, Stahl S, Ponten F, Uhlen M (2003) Affinity proteomics for systematic protein profiling of chromosome 21 gene products in human tissues. Mol Cell Proteomics 2(6):405–414
55. Uhlen M, Bjorling E, Agaton C, Szigyarto CA, Amini B, Andersen E, Andersson AC, Angelidou P, Asplund A, Asplund C, Berglund L, Bergstrom K, Brumer H, Cerjan D, Ekstrom M, Elobeid A, Eriksson C, Fagerberg L, Falk R, Fall J, Forsberg M, Bjorklund MG, Gumbel K, Halimi A, Hallin I, Hamsten C, Hansson M, Hedhammar M, Hercules G, Kampf C, Larsson K, Lindskog M, Lodewyckx W, Lund J, Lundeberg J, Magnusson K, Malm E, Nilsson P, Odling J, Oksvold P, Olsson I, Oster E, Ottosson J, Paavilainen L, Persson A, Rimini R, Rockberg J, Runeson M, Sivertsson A, Skollermo A, Steen J, Stenvall M, Sterky F, Stromberg S, Sundberg M, Tegel H, Tourle S, Wahlund E, Walden A, Wan J, Wernerus H, Westberg J, Wester K, Wrethagen U, Xu LL, Hober S, Ponten F (2005) A human protein atlas for normal and cancer tissues based on antibody proteomics. Mol Cell Proteomics 4(12):1920–1932
56. Bentley GA, Boulot G, Chitarra V (1994) Cross-reactivity in antibody-antigen interactions. Res Immunol 145(1):45–48
57. Baharvand H, Fathi A, van Hoof D, Salekdeh GH (2007) Concise review: trends in stem cell proteomics. Stem Cells 25(8):1888–1903
58. Yu L, Gaskell SJ, Brookman JL (1998) Epitope mapping of monoclonal antibodies by mass spectrometry: identification of protein antigens in complex biological systems. J Am Soc Mass Spectrom 9(3):208–215
59. Wu CC, MacCoss MJ, Howell KE, Yates JR 3rd (2003) A method for the comprehensive proteomic analysis of membrane proteins. Nat Biotechnol 21(5):532–538
60. Wu CC, Yates JR 3rd (2003) The application of mass spectrometry to membrane proteomics. Nat Biotechnol 21(3):262–267
61. Elliott ST, Crider DG, Garnham CP, Boheler KR, Van Eyk JE (2004) Two-dimensional gel electrophoresis database of murine R1 embryonic stem cells. Proteomics 4(12):3813–3832
62. Unwin RD, Smith DL, Blinco D, Wilson CL, Miller CJ, Evans CA, Jaworska E, Baldwin SA, Barnes K, Pierce A, Spooncer E, Whetton AD (2006) Quantitative proteomics reveals

posttranslational control as a regulatory factor in primary hematopoietic stem cells. Blood 107(12):4687–4694

63. Reinders J, Sickmann A (2005) State-of-the-art in phosphoproteomics. Proteomics 5(16): 4052–4061
64. Ong SE, Mittler G, Mann M (2004) Identifying and quantifying in vivo methylation sites by heavy methyl SILAC. Nat Methods 1(2):119–126
65. Kratchmarova I, Blagoev B, Haack-Sorensen M, Kassem M, Mann M (2005) Mechanism of divergent growth factor effects in mesenchymal stem cell differentiation. Science 308(5727): 1472–1477
66. Van Hoof D, Muñoz J, Braam SR, Pinkse MW, Linding R, Heck AJ, Mummery CL,Krijgsveld J (2009) Phosphorylation dynamics during early differentiation of human embryonic stem cells. Cell Stem Cell 5(2):214-226
67. Arrell DK, Niederlander NJ, Faustino RS, Behfar A, Terzic A (2008) Cardioinductive Network Guiding Stem Cell Differentiation Revealed by Proteomic Cartography of Tumor Necrosis Factor {alpha}-Primed Endodermal Secretome. Stem Cells 26(2):387–400
68. Jacobs JM, Waters KM, Kathmann LE (2008) Camp Ii, D.G., Wiley, H.S., Smith, R.D., Thrall, B.D.: The Mammary Epithelial Cell Secretome and Its Regulation by Signal Transduction Pathways. J Proteome Res 7(2):558–569
69. Khwaja FW, Svoboda P, Reed M, Pohl J, Pyrzynska B, Van Meir EG (2006) Proteomic identification of the wt-p53-regulated tumor cell secretome. Oncogene 25(58):7650–7661
70. Gasson JC, Weisbart RH, Kaufman SE, Clark SC, Hewick RM, Wong GG, Golde DW (1984) Purified human granulocyte-macrophage colony-stimulating factor: direct action on neutrophils. Science N Y 226(4680):1339–1342
71. Nomura H, Imazeki I, Oheda M, Kubota N, Tamura M, Ono M, Ueyama Y, Asano S (1986) Purification and characterization of human granulocyte colony-stimulating factor (G-CSF). EMBO J 5(5):871–876
72. Baba M, Hasegawa H, Nakayabu M, Shimizu N, Suzuki S, Kamada N, Tani K (1995) Establishment and characteristics of a gastric cancer cell line (HuGC-OOHIRA) producing high levels of G-CSF, GM-CSF, and IL-6: the presence of autocrine growth control by G-CSF. Am J Hematol 49(3):207–215
73. Kang D, Oh S, Ahn SM, Lee BH, Moon MH (2008) Proteomic analysis of exosomes from human neural stem cells by flow field-flow fractionation and nanoflow liquid chromatography-tandem mass spectrometry. J Proteome Res 7(8):3475–3480
74. Heck AJ, Mummery C, Whetton AD, Oh S, Lee B, Pera M, Lemischka I, Krijgsveld J (2007) Proteome biology of stem cells. Stem Cell Res 1:7–8

Chapter 14
Microglia: The Bodyguard and the Hunter of the Adult Neurogenic Niche

Jorge Valero, Maria Francisca Eiriz, Tiago Santos, Ismael Neiva, Raquel Ferreira and João O. Malva

Abstract The addition of new neurons in the mature central nervous system (CNS), which relies on millions of fine-tuned connections, is coordinated by a complex cascade of events. A minimum error could disturb the whole system. All this becomes even more complicated considering that neuroblasts can also move from their place of birth and cover long distances. In this scenario, the most insignificant process should be under the strictest control and any insurgency must be punished with death. For such purposes, microglia seem to be devoted to maintain things in correct order in the brain and to "use the force" when needed. However, on the special environment established in the adult neurogenic niches microglia should be more permissive but also be ready to react when needed. In this chapter, we analyze the particularities of microglial cells on charge of the surveillance of adult neurogenic niches and the tools they have to maintain

J. O. Malva (✉)
Laboratory of Biochemistry and Cell Biology, Faculty of Medicine,
Center for Neuroscience and Cell Biology, University of Coimbra, Coimbra, Portugal
e-mail: jomalva@fmed.uc.pt

J. Valero · T. Santos · I. Neiva · R. Ferreira
Center for Neuroscience and Cell Biology, University of Coimbra, Coimbra, Portugal
e-mail: jorge.valero@cnc.uc.pt

T. Santos
e-mail: tiagodesousasantos@gmail.com

I. Neiva
e-mail: ismael.neiva@cnc.uc.pt

R. Ferreira
e-mail: raquelmargarida@gmail.com

M. F. Eiriz
MIT-Portugal PhD Program, Bioengineering,
Center for Neuroscience and Cell Biology, University of Coimbra, Coimbra, Portugal
e-mail: francisca.eiriz@gmail.com

H. Baharvand and N. Aghdami (eds.), *Advances in Stem Cell Research*,
Stem Cell Biology and Regenerative Medicine, DOI: 10.1007/978-1-61779-940-2_14,

homeostasis. We comment on the current knowledge about the relationship between microglial cells and other elements of the neurogenic niche during healthy but also inflammatory situations. The activity of senescent and overreactive microglia seems to underlay most of the problems observed in neurodegenerative disorders. Therefore, we consider that an adequate understanding of microglial interaction with the neurogenic niche is essential for the proper development of brain repair therapies.

14.1 A Brief Introduction to Microglial Cells

Brain parenchyma is continuously surveyed by microglial cells which ensure the protection of the central nervous system (CNS). Microglia are small glial cells derived from myeloid precursors which enter the CNS during development and acquire a dendritic shape with branched, fine, and long processes which characterize the so-called "resting" microglia [1]. Microglial cells are often regarded as the resident macrophages of the CNS due to their ability to become phagocytes when needed. Apart from parenchymal microglia, other macrophage-like cells reside in the CNS: perivascular macrophages and pericytes. Microglial cells are difficult to distinguish from other macrophage-like cells [reviewed in 2]; thus, we will mainly use the terms microglia or microglial cells to refer to microglia/CNS invading macrophages residing in the brain. In the normal CNS, microglial cells play an active role in the clearance of apoptotic material, a process that may occur without inflammation. Accordingly, microglia display a wide range of receptors that mount a concerted response leading to phagocytosis and removal of cell debris [3–6]. Moreover, dying cells in the CNS also release signaling molecules that promote the recruitment of microglia and other phagocytes [7]. Although microglial cells were previously considered as quiescent cells in the absence of an activating/stimulating challenge, "resting" microglia are highly dynamic. Processes of nonactivated microglia continuously explore the surrounding environment forming a network which senses the state of the brain parenchyma [8, 9]. Furthermore, neurons express on their plasma membrane an array of molecules, some of them releasable, that control the activation state of microglial cells. These molecules constitute the "on/off" system. "Off" signals keep microglia on their nonactivated state, whereas their absence and/or presence of "on" molecules lead to their activation [10]. Once activated, microglial cells lose their ramified morphology, become amoeboid, and migrate to the site of injury, where they release several proinflammatory and trophic factors [11–14]. Microglia migrate by chemotaxis in response to various stimuli such as blood-derived factors from damaged blood vessels, extracellular ATP and ADP released after ischemic and traumatic CNS injuries, microbial signals, adhesion molecules, cytokines, and complement molecules, among others [8, 9, 15, 16]. Indeed, microglia evolved to express multiple membrane receptors able to identify a wide array of pathogen-

and damage-associated molecular determinants. Toll-like receptors (TLRs) are one of the most studied receptor families and, as such, are activated by bacterial lipopolysaccharides (LPS), hypomethylated DNA, flagellin, double-stranded RNAs, and other molecules [17]. When a pathological infection or injury occurs in the CNS, the interaction between TLRs and their ligands results in the activation of intracellular signaling cascades leading to the release of several factors in order to control and minimize the impact of CNS injuries. Proinflammatory molecules tend to eliminate any possible source of damage by restricting the spread of the injury into other areas of the brain parenchyma. However, an exacerbated proinflammatory reaction could affect healthy tissue and become detrimental. On the other hand, microglial anti-inflammatory response promotes cell survival and tissue recovery after injury [18]. Therefore, activated microglia release various molecules, which contribute to neural cell death or survival, such as interleukins (IL), growth factors, tumor necrosis factor-alpha (TNF-α), and monocyte chemotactic protein-1 (MCP-1), to mention a few [11]. Some of the molecules released by microglia induce blood–brain barrier leakage making it more permissive for leukocyte invasion [19]. In the special case of brain injury, T-lymphocytes may invade the brain and accumulate at inflammatory sites where, together with infiltrating microglial cells, they reinforce the immune response [20]. In conclusion, microglial cells are critical in the protection of CNS against infection, trauma, ischemia, brain tumors, and neurodegenerative processes.

Regional differences have been described in density, molecular profile, and responsiveness of microglia in the CNS. Such differences could be associated with functionality [reviewed in 21]. "Classical" functions of microglia have been related to inflammation, immune response, and debris clearance. However, nowadays, new aspects about the role of microglia in the CNS have emerged. In their review, Pont-Lezica et al. describe the association of microglia with four main elements: apoptotic neurons, developing vascular networks, developing axon tracks, or radial glia [22]. This distribution pattern suggests new microglial roles related to developmental cell death induction [23, reviewed in 24], angiogenesis [25, 26], axonal growth and guidance, and neuro- and gliogenesis [22]. Recent studies have also suggested the involvement of resting microglia on synaptogenesis and synaptic plasticity [24, 27]. Moreover, a role of microglia in the regulation and maintenance of adult neurogenesis has been proposed. The purpose of this chapter is to summarize the current knowledge referring to the interaction between microglia and adult neurogenesis and its possible implications on CNS disorders and pathologies.

14.2 Adult Neurogenic Niches

In the adult mammalian brain, new cells are generated through a continuous process called adult neurogenesis. This process occurs mainly in two mitotically restricted areas: the subventricular zone (SVZ) lining the lateral ventricles

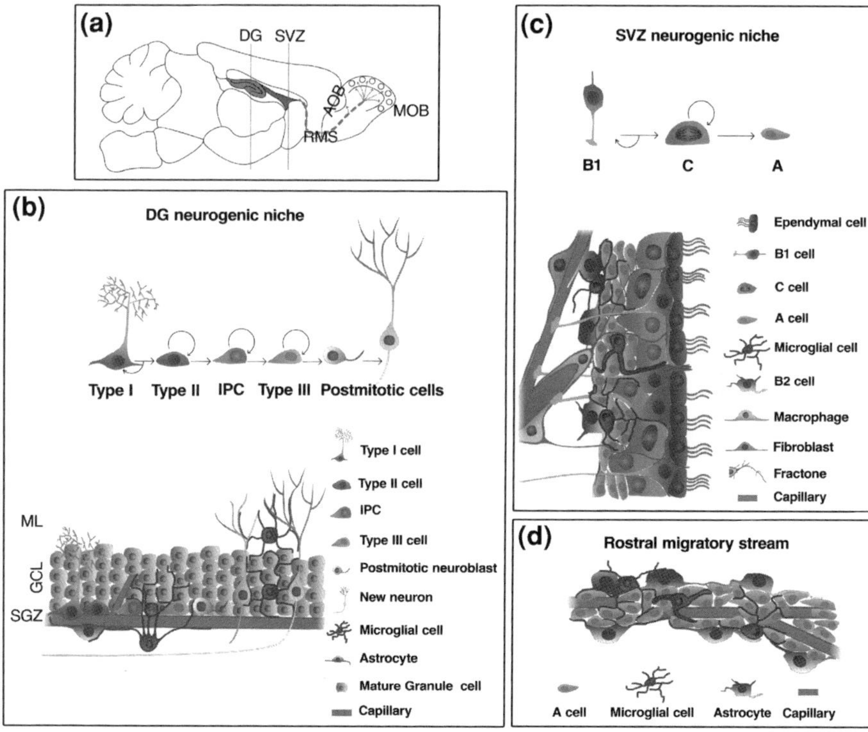

Fig. 14.1 Schematic representation of adult neurogenic niches. **a** Localization of neurogenic niches in the CNS. **b** *Top panel* represents cell lineages in the adult DG. *Bottom panel* shows a schematic representation of the organization of DG neurogenic niche. **c** *Top panel* represents cell lineages in the adult SVZ. *Bottom panel* shows a schematic representation of the organization of SVZ neurogenic niche. **d** Schematic representation of the organization of the rostral migratory stream. *AOB* Anterior Olfactory Bulb, *DG* Dentate Gyrus, *GCL* Granule Cell Layer, *IPC* Intermediate Progenitor Cell, *ML* Molecular Layer, *MOB* Main Olfactory Bulb, *RMS* Rostral Migratory Stream, *SGZ* Subgranular Zone, *SVZ* Subventricular Zone

[28–31]; and the subgranular zone (SGZ) of the hippocampal dentate gyrus (DG, Fig. 14.1a) [32–34]. These regions of the adult brain where constitutive neurogenesis takes place are called "neurogenic niches". Some of the cells present in the neurogenic niches have self-renewal and multipotency capacities, cardinal properties of stem cells. Accordingly, they are named neural stem cells [35–39]. However, the distinction between these stem cells and their derived precursor cells, which have a more restricted potential, is a scientific and technological challenge. Therefore, we will use the denomination neural progenitor cells (NPCs) to refer to both of them. New neurons have been implicated in the formation of new memories, olfactory functions, and potential recover of brain injury [40–42]. Thus, full knowledge of adult neurogenesis mechanisms is fundamental to better understand memory processing, olfaction, and also a scientific resource to develop new brain repair strategies.

Adult SVZ is a thin layer of cells that separates the striatum from the lateral ventricles (Figs. 14.1c and 14.2a$_2$). Here, the so-called B-cells originate C-cells which are actively proliferating precursors (Fig. 14.1c) [37]. C-cells auto-replicate and also give rise to immature neuroblasts (A-cells) that migrate tangentially to the accessory and main olfactory bulb (OB) through the rostral migratory stream (RMS, Fig. 14.1a, c) [43, 44]. Once in the OB, neuroblasts migrate radially and differentiate into new interneurons that acquire fully competent functions related to olfactory discrimination and memory [41, 45, 46]. B-cells express nestin, sex-determining region Y-box 2 (Sox2), glial fibrillary acidic protein (GFAP) and other astroglial markers [30, 47, 48], being considered as SVZ astrocytes. B1-cells, a particular subtype of B-cells, have been proposed to be the resident stem cells of the SVZ. They have some characteristics of radial glial cells such as specialized end feet on blood vessels and an apical ending in contact with ventricle fluid (Fig. 14.1c) [49, 50]. Endothelial cells from blood vessels are also active constituents of SVZ neurogenic niche as they foster stem cell proliferation [50] and also release important factors for NPC maintenance and differentiation. SVZ proliferating cells usually associate with blood vessels [51] or, alternatively, with a particular extracellular matrix named fractones, extending from capillaries (Fig. 14.1c) [52].

A-cells leave the SVZ integrating the chains of the RMS and they migrate tangentially to the main OB or accessory OB along specialized astrocytic tubular structures and blood vessels (Fig. 14.1d) [44, 53–55]. In the OB, neuroblasts detach from the chains and shift their tangential migration to a radial migration. Once they reach their final location in the OB, new neurons are selected to survive or to die by apoptosis. Finally, only few cells integrate into the existing circuitry mainly as mature granule cells or periglomerular neurons [56].

In the hippocampus, adult neurogenesis is responsible for the $\sim$10,000 new neurons generated daily at the DG [57]. Three main cellular players have been described in this neurogenic niche (Fig. 14.1b). Type-1 cells are considered the neural stem cells of DG, have a radial-glia GFAP-positive process and originate nestin-positive and proliferative type-2 cells [36, 58]. Type 1 cells have some similarities with B1 cells of SVZ: they express GFAP, Sox2, and present primary cilia [59, 60]. Type-3 cells, generated from an intermediate progenitor cell (a highly proliferative type-2 cell), are less proliferative, they do not express nestin but show typical markers of neuroblasts like doublecortin (Dcx) [61, 62]. These cells finally convert into postmitotic-cells which migrate a small distance into the granule cell layer (GCL, Fig. 14.1b). Most mature young neurons expand their axons until the *stratum lucidum* of the CA3 layer and develop a dendritic tree which reaches the molecular layer of the DG [63–66]. These new neurons go through an early survival period and many are eliminated by apoptosis [65, 67]. The surviving ones integrate into existing circuits having a role in memory function [40, 68].

14.2.1 Modulation of Adult Neurogenesis

Extrinsic and intrinsic factors regulate distinct aspects of adult neurogenesis: proliferation, migration, differentiation, survival, and integration [69, 70]. Cell cycle modulators, transcription factors, and epigenetic regulators are intrinsic cues regulating adult neurogenesis [69–72]. In turn, intrinsic factors are affected by a huge number of extrinsic factors: morphogens, adhesion molecules, growth factors, neurotrophins, cytokines, hormones, neurotransmitters, and neuropeptides [69, 70]. Physical exercise [73], environmental enrichment (EE) [74], diet [75], social interactions [76] and stress [77] are the main "environmental" factors that regulate adult neurogenesis. Surprisingly, most of them have almost no effects on SVZ neurogenesis. Only olfactory enrichment seems to affect OB-directed neurogenesis [78, 79]. Voluntary running, social interactions, EE, and dietary restriction increase adult neurogenesis by affecting cell proliferation or survival of new neurons in the DG [73–76]. Stress diminishes the production of new neurons in the SGZ, an effect mainly related with glucocorticoid levels [77]. Aging also negatively regulates adult neurogenesis in both DG and SVZ, affecting extrinsic and intrinsic factors [80]. Finally, adult neurogenesis is modified by brain damage or pathology, which has been described to either increase or reduce the production of new neurons and even to redirect them to affected regions. Microglial cells are able to produce, release and regulate, levels of the extrinsic factors affecting adult neurogenesis in healthy conditions as well as during aging, brain injury, and pathology [81–83]. Thus, the study of the role of microglia in the neurogenic niche could provide new therapeutic tools to modulate neurogenesis.

14.3 Microglial Cells in Neurogenic Niches

Microglial properties vary between regions in the CNS; their density, morphology, responsiveness, and pattern of receptor expression differ between distinct brain areas [21]. Considering the special characteristics of adult neurogenic niches, resident microglial cells likely play a role in support of neurogenesis. Taking into account the classical phagocytic role of microglia and the elevated incidence of apoptotic events occurring in adult neurogenic regions, a higher proportion of DG and OB microglia should be in a phagocytic-like state compared with other regions of the CNS [84]. Furthermore, microglial cells have been recently involved in other functions during development that may persist until adult life: axonal growth and guidance, angiogenesis and cell death induction [22]. Finally, the role of "resting" microglial cells on synaptogenesis and synaptic plasticity could be crucial during the integration of newly generated neurons into the preexistent CNS circuitry [24, 27]. Despite the extensive research that has been done on microglial cell physiology and function, little has been reported on the particularities of microglial modulation of the healthy adult neurogenic niche.

14.4 Microglial Cells in the Healthy Adult Subventricular Zone

Blakemore described in 1969 the existence of microglial cells subjacent to the ependyma in the SVZ of 3-month-old rats [85]. However, initial detailed studies of the adult SVZ cytoarchitecture just mentioned the presence of microglia [30, 86]. Mercier et al. observed microglial cells and perivascular macrophages when analyzing SVZ fractones in rodents and humans [52, 87]. A fractone is a continuous basal laminae with a base at the tip of a perivascular macrophage, a stem branching into the SVZ and bulbar-like structures underneath the ependyma (Fig. 14.1c). Perivascular macrophages establish contacts between them and with fibroblasts forming a network along the capillaries which enter the SVZ. Interestingly, fractones contact with SVZ ependymocytes, A, B, C-cells, and microglia. Considering that the meningeal fibroblast/macrophage network has continuity into the SVZ capillary network which gives rise to fractones, these structures may serve as communication channels between SVZ cells and other brain regions [52, 87]. Macrophages/microglia could provide the extracellular matrix constituents of the fractones and also produce growth factors, cytokines and chemokines diffusing and accumulating in this network, thereby influencing SVZ neurogenesis [52].

SVZ microglia have some special characteristics: they show less activation after cortical lesion [88] and they can be isolated and expanded more easily than microglia from non-neurogenic areas [89]. Goings et al. described that microglia from SVZ are normally ramified, express medium to high levels of the microglia/macrophage markers CD45 and CD11b (Fig. 14.2a–b), and proliferate more than microglia from adjacent areas. Cells highly stained for CD45 have shorter processes than medium stained ones, sometimes form clusters and frequently localize adjacent to blood vessels. These data indicate that SVZ microglial cells are in an intermediate state of constitutive activation [88]. In agreement, other authors showed that, in sham-operated rats, more than 20% of SVZ microglia express the phagocytic activity and microglial activation marker ectodermal dysplasia 1 (ED1), and that about 15, 2 and 1% of the total microglia population display an intermediate, amoeboid and round morphology, respectively [90]. Furthermore, some SVZ microglial cells lining dorsal and ventral edges of adult SVZ express the dendritic cell marker CD11c [91]. Interestingly, SVZ microglia have been found associated to B- and C-cells suggesting that microglial cells may be directly regulating NPCs function [92]. Indeed, some SVZ resting microglia containing the prostaglandin synthase cyclooxygenase 2 (Cox2) appear close to dividing cells. Since Cox2 inhibition induces a 90% decrease in the number of SVZ proliferating cells, prostaglandin synthesis by microglia could be related to the regulation of cell proliferation in normal conditions [93]. Therefore, microglial activity affects proliferation in the SVZ. Moreover, high doses of minocycline, an inhibitor of microglial activity, increase cell proliferation in the adult SVZ of mice [92]. Importantly, Walton et al. demonstrated that the presence of microglia on NPCs monolayer cultures is necessary for neurogenesis [94]. This effect is mediated by

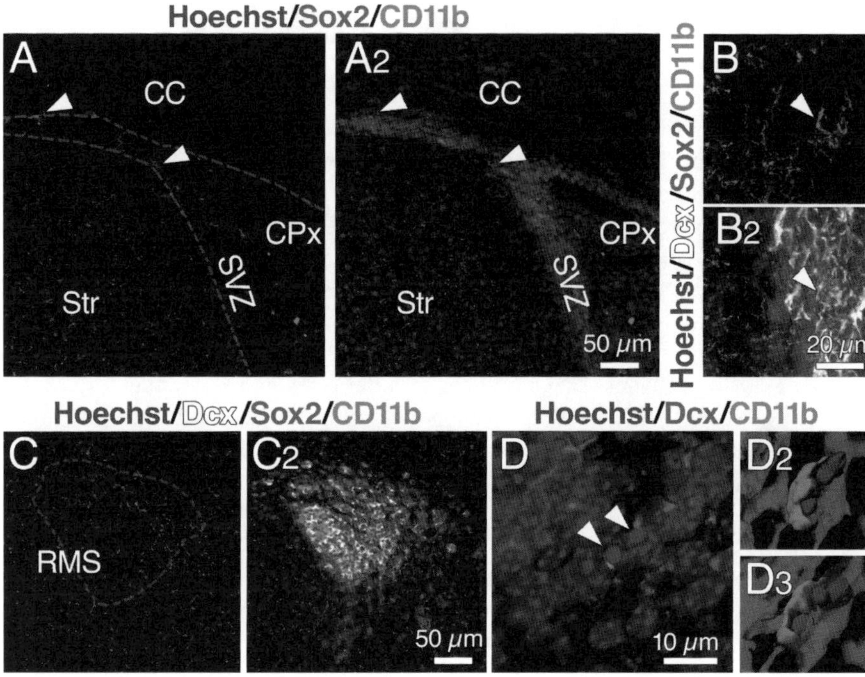

Fig. 14.2 Microglial cells in the Subventricular Zone and Rostral Migratory Stream. **a–d** Coronal section images from the brain of 2 month-old mice. **a–a_2:** Confocal microscopy image showing SVZ NPCs (Sox2+, *red*) and microglial cells (CD11b+, *green*) in the SVZ neurogenic niche and adjacent regions. Some microglia present high/medium intensity staining for CD11b and an intermediate activated-like morphology (*arrowheads*). **b–b_2** Confocal microscopy image of the SVZ neurogenic niche showing a CD11b+ (*green*) microglial cell (*arrowhead*) adjacent to NPCs (Sox2+, *red*) and neuroblasts (Dcx+, *white*). Note the different staining intensity between SVZ microglia and microglial cells from the adjacent region. **c–c_2** Confocal microscopy image showing microglial cells (CD11b+, *green*) in the RMS. Some microglial cells show higher staining for CD11b and are close to Sox2+ (*red*) and Dcx+ (*white*) cells. **d** Apoptotic-like (pyknotic) cells in the RMS (*arrowheads*). **d_2–d_3** Volume rendering images of confocal z-stacks showing the apoptotic-like cells pointed with *arrowheads* in (**d**). Note that one of the apoptotic-like cells is engulfed by a CD11b+ process suggesting ongoing phagocytosis by a microglial cell. *CC* Corpus Callosum, *CPx* Coroid Plexus, *Dcx* Doublecortin, *NPCs* Neural Progenitor Cells, *RMS* Rostral Migratory Stream, *Sox2* sex-determining region Y-box 2, *Str* Striatum, *SVZ* Subventricular Zone

soluble factors released by microglia since conditioned media from primary microglial cells restore inducible neurogenesis in microglia-depleted SVZ cultures and increase neuroblast production in normal NPCs cultures [92, 94]. Therefore, SVZ microglia modulate neurogenesis and may be considered a component of the neurogenic niche.

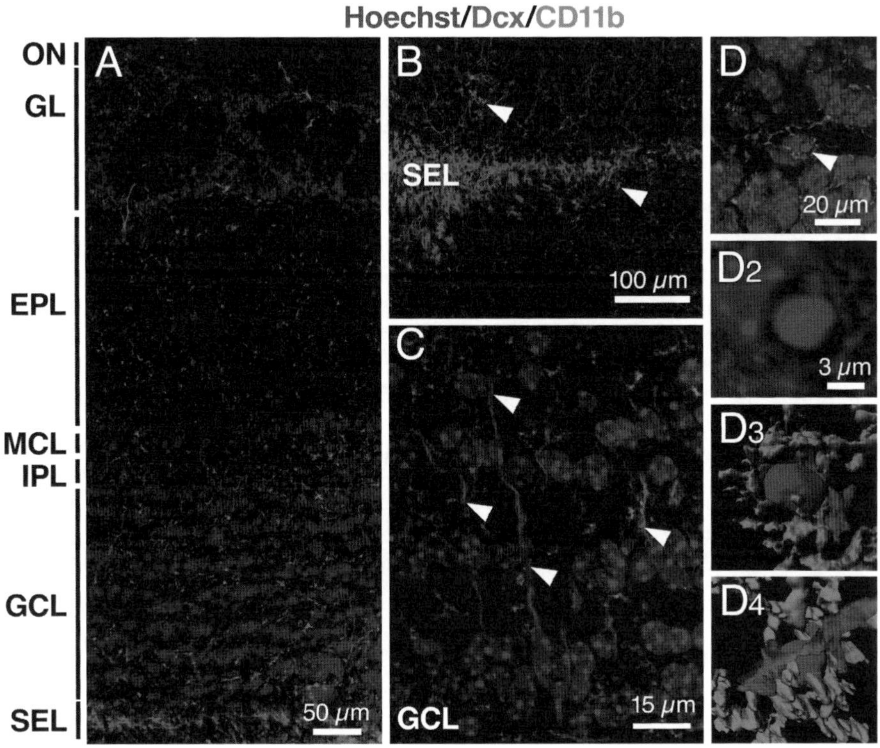

Fig. 14.3 Microglial cells in the main olfactory bulb. **a–d** Coronal section images from the olfactory bulb of 2 month-old mice. **a** Confocal microscopy images showing microglial cells (CD11b+, *green*), neuroblasts and young neurons (Dcx+, *red*), and cell nuclei (Hoechst staining, *blue*) in the main olfactory bulb. **b** Confocal microscopy image showing neuroblasts (Dcx+, *red*) in the SEL of the olfactory bulb. Note the presence of highly stained CD11b+ microglial cells in the inner part of the GCL, close to the SEL (*arrowheads*). **c** Confocal microscopy image of the GCL. *Arrowheads* point to regions of close apposition between processes of microglial cells (CD11b+, *green*) and new neurons (Dcx+, *red*) which suggests the existence of interactions between microglia and young neurons in the main olfactory bulb. **d** Apoptotic-like (pyknotic) cell in the junction between the GCL and SEL (*arrowhead*). $\mathbf{d_2}$ Magnification of the apoptotic-like cell pointed in (**d**). $\mathbf{d_3}$–$\mathbf{d_4}$ Volume rendering images of confocal z-stacks showing the same apoptotic-like cell of $\mathbf{d_2}$ panel. Note the CD11b+ process around the pyknotic cell suggesting the existence of phagocytosis. *Dcx* Doublecortin, *GL* Glomerular Layer, *GCL* Granule Cell Layer, *EPL* External Plexiform Layer, *IPL* Inner Plexiform Layer, *MCL* Mitral Cell Layer, *ON* Olfactory Nerve, *SEL* Subependymal Layer

14.4.1 Microglial Cells in the Rostral Migratory Stream and Main Olfactory Bulb

The maintenance of the special characteristics of SVZ microglia along the RMS is an interesting issue that has not been properly investigated (Fig. 14.2c–d). However, stellate microglial cells expressing CD11c exist in the rostral end of the RMS: the subepedymal layer (SEL) of the OB [91]. In the main OB of rodents most microglial cells show a highly branched and ramified morphology (Fig. 14.3a). Curiously, microglia adjacent to the SEL of the rat OB present a higher CD11b staining compared to microglial cells in the immediately adjacent GCL. These highly CD11b-stained cells emit short and hypertrophied processes resembling activated microglia (Fig. 14.3b). CD45 and Major Histocompatibility Complex-II (MHC-II) expressing microglial cells with amoeboid morphology also appear in the junction between the GCL and SEL. Interestingly, this region of the OB presents the highest concentration of pyknotic cells in the rat CNS. Some CD45 and MHC-II-positive cells also exist in the olfactory nerve and periglomerular zone, two regions which suffer continuous remodeling due to the renewal of olfactory axons and the arrival of new periglomerular cells [95]. Finally ED-1-positive (activated) microglia have been observed in the nerve layer, periglomerular zone and GCL of the rat OB [96]. This pattern of staining indicates the presence of active and phagocytic-like microglia in the normal OB, which are mainly concentrated close to the region of neuroblasts entrance (Fig. 14.3d).

14.4.2 T-Cells and Microglia/SVZ Neurogenic Niche Interaction

The influence of T-cells on the SVZ is suggested by different observations. First, severe combined immune deficiency (SCID) mice, which lack T and B-cells, show a 61% reduction of SVZ proliferating cells [97]. Second, in sham-operated rats, 1% of SVZ microglia express MHC-II, indicating that a direct communication between SVZ microglia and T-cells may exist [90]. Finally, SVZ manifests special chemoattractant properties for T-cells [98]. Although the presence of T-cells in the brain parenchyma and also in the SVZ is scarce [90] there are T-lymphocytes in the choroid plexus and meninges, where they contact antigen presenting cells and release cytokines to the cerebrospinal fluid (CSF) [reviewed in 99]. Importantly, CSF is in direct contact with SVZ B1-cells [49] and also with the meningeal-fractone network (Fig. 14.1b), thereby providing a vehicle for T-cell cytokine signaling [52, 87]. Furthermore, low levels of the T-cell proinflammatory cytokine IFN-γ or anti-inflammatory cytokine IL-4 promote neurogenesis in cocultures of microglia and SVZ NPCs from adult mice, whereas these cytokines have no effects on NPCs cultured alone. On the contrary, higher levels of IFN-γ are detrimental for neurogenesis [100]. Altogether, these data suggest a role for T-cells on the regulation of SVZ neurogenesis through interaction with microglial cells.

14.4.3 Microglial-Secreted Factors in the Normal SVZ Neurogenic Niche

IFN-γ which is expressed by T-cells, but also by microglia, has important effects on SVZ adult neurogenesis and has been detected in neurosphere cultures and SVZ tissue [98, 100, 101]. IFN-γ null mice show elevated production of neurospheres, increased proliferation in the SVZ and higher number of newly generated cells in the OB-GCL, suggesting a detrimental effect of IFN-γ on NPCs. [101]. Furthermore, IFN-γ and TNF-α may have an indirect effect on SVZ recruitment of microglia/macrophages by promoting production of the chemokine CXCL10 in this region [98]. Moreover, high levels of IFN-γ can lead to an increase of TNF-α production by microglial cells [100]. TNF-α and its receptors are expressed in the SVZ [102], indicating a direct action of TNF-α on the SVZ neurogenic niche. [98]. Indeed, this factor either decreases or increases proliferation of cultured SVZ NPCs [102–106]. Bernardino et al. demonstrated that treatment of NPCs derived from SVZ neurospheres with TNF-α at 1 ng/ml increases cell proliferation while higher concentrations (10 and 100 ng/ml) induce apoptosis. Therefore, low and high concentrations of this cytokine have opposite effects. Furthermore, the same authors showed that TNF-α (1 and 10 ng/ml) increases neuronal differentiation of SVZ NPCs via specific activation of TNF-α receptor I (TNF-RI) [104]. In agreement, mitotic activity induced by human TNF-α on SVZ neurospheres has been related to activation of the nuclear factor κB via TNF-RI [106]. In vivo, intracranial administration of TNF-α increases cell proliferation in the adult SVZ of rats [107]. However, deletion of TNF-RI or TNF-α does not affect cell proliferation in healthy conditions in the mouse SVZ, suggesting that a physiological low concentration of TNF-α has no effect on SVZ [102]. In conclusion, pro-neurogenic and anti-neurogenic effects of TNF-α in NPCs from the SVZ could be related with differential effects elicited by different concentrations of TNF-α and its binding to TNF-RI or TNF-RII, eventually triggering opposite effects [108]. Finally, recent data indicate that IL-6, which is present in the healthy SVZ [88], also plays an important role sustaining normal levels of proliferation and survival of NPCs in the SVZ [109].

Dcx-positive cells of SVZ express the MCP-1 receptor CCR2 and MCP-1 is able to induce migration and attract NPCs from SVZ in cell culture [110]. However, MCP-1 is produced by SVZ GFAP- and nestin- positive cells [111]. Therefore, its expression in this region may not be only attributable to microglia.

A summary of the data exposed here can be found in Table 14.1.

14.5 Microglial Cells in the Healthy Adult Dentate Gyrus

Density of microglial cells in mouse SGZ of DG is lower than in adjacent regions (Fig. 14.4a) and, contrary to the SVZ, Goings et al. described that most of them are in a resting state showing low levels of CD45 [88]. Nonetheless, others described that 25.65% of GCL microglia from 2-month-old mice are activated [112]. Similar to SVZ, DG microglia just experience a slight increase in CD45 staining intensity after cortical aspiration [88]. Some DG microglial cells also express Cox2 and its inhibition causes a 40% reduction in cell proliferation. However, whether this effect is mediated by the activity of Cox2 present in microglia is not clear as neurons and astrocytes also express Cox2 in the adult DG [93, 113].

Table 14.1 Effects of microglial-related factors in the healthy SVZ neurogenic niche

Factor	In vitro effect	In vivo effect	Refs.
Prostaglandin		Cox2 inhibition ↓ proliferation	[93]
IL-4	↑ neurogenenesis (microglia/ NPCs cocultures)		[100]
IFN-γ	*Low*	*Overexpression*	[98, 100, 101]
	↑ neurogenenesis (microglia/ NPCs cocultures)	↑ CXCL10 → microglia attraction	
	High	*Null mice*	
	↑TNF-α and ↓ neurogenesis	↑ proliferation and neurogenesis	
TNF-α	*Low*	*Iintracranial admin*	[98, 102–106]
	↑ proliferation and neuronal differentiation (via TNF-RI)	↑ proliferation	
		TNF-RI deletion	
		No effect	
	High	*Overexpression*	
	↑ cell death	↑ CXCL10 → microglial attraction	
	↓ or no effect in proliferation		
IL-6		*KO mice*	[109]
		↓ proliferation and survival	
MCP-1	Induce migration and attract NPCs		[110]

CD11c-positive microglial cells appear in the DG of adult mice extending their processes from one border to the other of the GCL. These CD11c-positive cells present a unique asymmetrical and dendriform morphology characteristic of the hippocampus (Fig. 14.4b) [91]. Furthermore, 37% of the NPCs and 30% of neuroblasts contact with microglial processes in the SGZ, suggesting that microglia play a role in surveillance of the functional status of the dentate gyrus neurogenic niche (Fig. 14.4b$_5$) [84, 91, 114, 115]. Importantly, nonactivated phagocytic microglia exert an important role removing apoptotic bodies from newly generated cells, that do not survive in the healthy SGZ (Fig. 14.4c). The majority of the apoptotic cells phagocyted by microglia in the SGZ correspond to 1–4 days-old cells in an intermediate state between late amplifying neuroprogenitor cells (type 2 cells) and early neuroblasts (type 3 cells), or ~8-days-old postmitotic differentiating newborn neurons. Indeed, most apoptotic cells of the SGZ contain nestin and/or polysialylated neural cell adhesion molecule (a marker of migrating neuroblasts) but not Dcx or NeuN. Most (91%) pyknotic cells of the SGZ are normally engulfed by phagocytic microglia processes. Furthermore, data indicate that SGZ microglia are highly efficient in the elimination of apoptotic debris [84]. However, it remains to be elucidated the existence of a similar phagocytic process in the OB, the RMS, and SVZ (Figs. 14.2d and 14.3d). Finally, whether these microglial cells actively participate in apoptotic induction of newly generated cells in the neurogenic niches is still unknown.

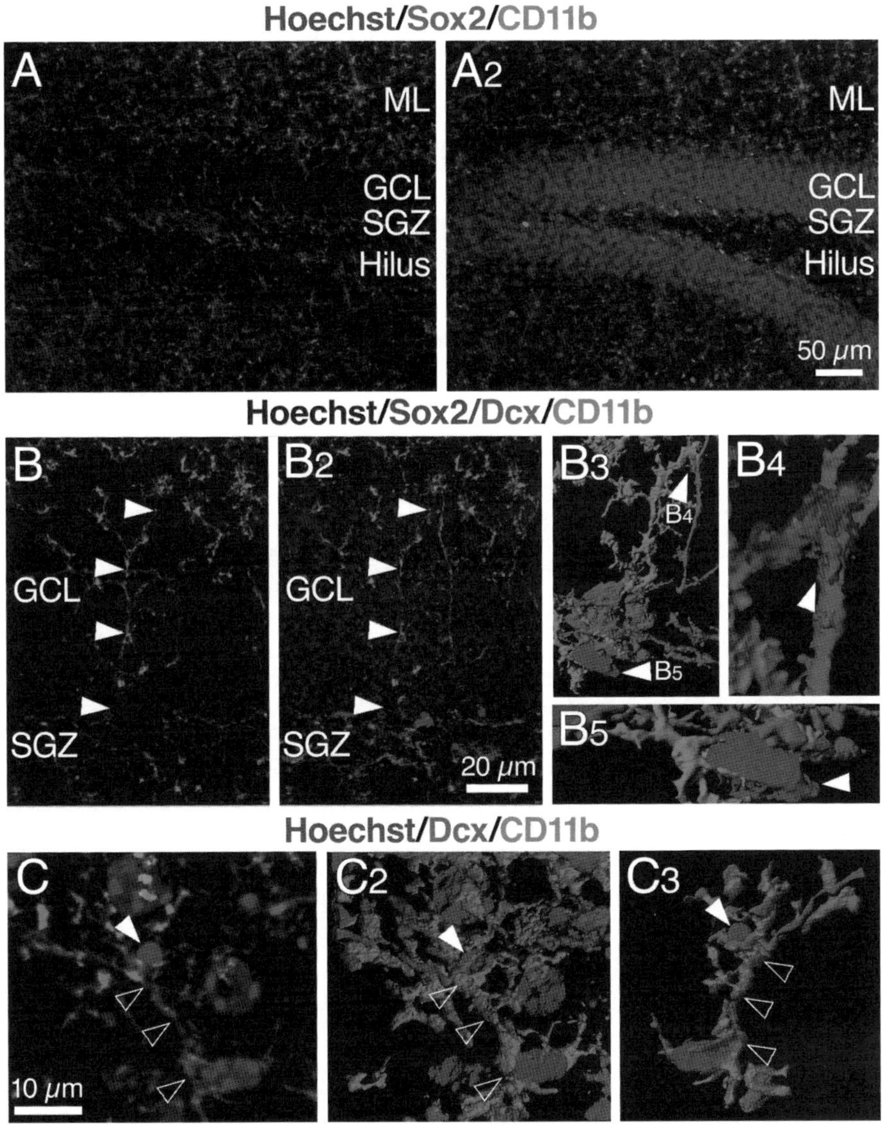

14.5.1 T-Cells and Microglia/DG Neurogenic Niche Interaction

Ziv et al. demonstrated a 60–68% decrease in the production of young neurons and 40% reduction in newly generated mature neurons in the DG of mice deficient in T-lymphocytes. This effect was mainly attributed to a decrease in cell proliferation. T-cells deficiency also decreases neuronal differentiation in the DG. Furthermore, injection of a source of T-cells partially restored normal levels of

◂**Fig. 14.4** Microglial cells in the dentate gyrus. **a–d** Coronal section images from the hippocampus of 2 month-old mice. **a–a_2** Confocal microscopy image showing microglial cells (CD11b+, *green*), NPCs (Sox2+, *red*), and cell nuclei (Hoechst staining, *blue*) in the dentate gyrus. Highly stained Sox2+ cells appear in the SGZ. Note the low density of microglial cells in the GCL compared with adjacent regions and how these cells extend their processes perpendicularly to the SGZ. **b–b_2** Confocal microscopy image showing the processes of a microglial cell extended from one to the other border of the GCL (*arrowheads*). In **b_2** NPCs cell bodies (Sox2+, *blue*) and neuroblasts (Dcx+, *red*) appear in the SGZ, while dendrites of new neurons extend through the GCL (Dcx+, *red*). **b_3–b_5** Volume rendering image of confocal z-stacks. **b_3** shows the same microglial cell (*green*) pointed with arrowheads in **b_2**. *Arrowheads* point to the areas amplified in **b_4** and **b_5**. **b_4** Close apposition between a microglial process (*green*) and a protuberance (*arrowhead*) in the dendrite of a new neuron (*red*). **b_5** Possible contact between microglial processes (CD11b+, *green*), neuroblasts (Dcx+, *red*), and NPCs (Sox2+, *blue*, *arrowhead*). These images suggest the existence of direct interactions between microglia and NPCs, neuroblasts, and newly formed neurons. **c** Confocal microscopy image of the SGZ showing an apoptotic-like cell (filled *arrowhead*) close to the process of a microglial cell (CD11b + , green, open arrowheads). **c_2–c_3** Volume rendering image of confocal z-stacks. **c_2** shows the same stained cells which appear in **c** while **c_3** shows a rotated scene of just few elements. Note that the apoptotic-like cell (filled *arrowhead*) is engulfed by the CD11b+ microglial process indicated by the open arrowheads, suggesting ongoing phagocytosis. *Dcx* doublecortin, *GCL* Granule Cell Layer, *ML* Molecular Layer, *NPCs* Neural Progenitor Cells, *SGZ* Subgranular Zone, *Sox2* sex-determining region Y-box 2

neurogenesis on the DG of mice lacking T and B-lymphocytes [97, 116]. The effect of T-cells on DG adult neurogenesis could be mediated by their influence on the proneurogenic state of microglial cells residing in, or close to, the SGZ [reviewed in 99]. Interestingly, microglial regulation of adult neurogenesis seems to be modulated only by T-cells directed against CNS self-antigens. In addition, mice with an excess of T-cells directed against a CNS auto-antigen showed improved spatial learning while depletion of CD4-cells impaired it [97, 116]. Thus, T-cell regulation of microglial activity has an important effect on DG adult neurogenesis and may influence cognition.

14.5.2 Microglial-Secreted Factors in the Normal DG Neurogenic Niche

Different IL modulate adult neurogenesis. Anti-inflammatory IL-10 increases proliferation without affecting differentiation of DG NPCs cocultured with microglial cells. This effect is mediated by regulatory properties of IL-10 on microglial cells and it is not observed when NPCs are cultured alone [117]. IL-2 knock-out mice exhibit augmented levels of IL-15, IL-12 and CXCL10 and also show increased neurogenesis [118]. Microglial cells are the most important producers of IL-1β in the brain [reviewed in 119], exerting a deleterious effect on adult neurogenesis after inflammation or during aging [115, 120, 121]. IL-1β affects DG adult neurogenesis in vivo by blocking cell proliferation, while no effects after intracerebroventricular administration of

IL-1β have been observed in the SVZ [122]. Proliferating cells, Sox2-positive and Dcx-positive cells express IL-1 receptor (IL-1R) suggesting that they could be directly affected by IL-1β [122]. Furthermore, increased levels of this cytokine reduce neuronal differentiation in the adult DG [123]. Interestingly, the absence of fractalkine "off" signaling reduces adult neurogenesis mainly due to the activation of microglia and increased production of IL-1β. Indeed, ramified microglial cells expressing fractalkine receptor (CX3CR1) are widely distributed into mice DG [115, 124]. Despite the constitutive expression of IL-1β in the hippocampus [125, 126] and the presence of IL-1R in NPCs, there are contradictory data about the effects of IL-1β in a healthy environment. Rats acutely treated with the IL-1RI antagonist (IL-1Ra) and IL-1RI null mice do not show changes in cell proliferation in the DG [122]. On the contrary, another study showed that chronic overexpression of IL-1Ra induces a decrease on cell proliferation without affecting neuronal differentiation in young mice [127]. Considering these data, further research should be done to clarify the effects of normal/low concentrations of IL-1-β in the adult DG. Something similar occurs with IL-6 and contradictory results have been obtained about the role of IL-6 in the DG neurogenic niche. IL-6 mRNA has not been detected under basal conditions in the hippocampus, but the mRNA of IL-6 receptor (IL-6R) is expressed in the DG [128]. Transgenic mice overexpressing IL-6 present decreased proliferation and survival, as well as a lower number of new neurons in the DG [129]. On the contrary, repeated but not single intra-hippocampal administration of IL-6 increases microglial reactivity and the number of proliferating cells in the DG of adult mice [130]. In agreement with this, IL-6 knock-out mice show decreased proliferation and survival of DG NPCs [109]. Moreover, mice overexpressing IFN-γ in the brain, at a rate that does not induce microglial activation, show increased neurogenesis in the DG and higher mRNA levels of TNF-α, IL-6 and the proneurogenic insulin-like growth factor (IGF-1) [131]. It would be particularly interesting to analyze whether this mouse model shows increased neurogenesis in the SVZ, eventually in contradiction with results obtained in the IFN-γ-null mice (commented above).

TNF-α is present in the DG [132] and its receptors, TNF-RI and TNF-RII, are expressed by hippocampal NPCs [108, 114, 133]. Mice lacking TNF-RI show increased cell proliferation and more newly generated neurons in the DG. On the contrary, TNF-RII-null mice show decreased cell proliferation but no effect on the final number of newly generated neurons [108]. Thus, in a nonchallenged environment, TFN-RI seems to play an important role down-regulating adult DG neurogenesis while TNF-RII has almost no effect. Contrary to what happens in SVZ, basal levels of TNF-α could reduce DG adult neurogenesis.

Transforming growth factor-β1 (TGF-β1) is expressed in the DG and increases upon aging [134, 135]. Some reports highlight a negative effect of TGF-β1 in DG neurogenesis, after its intracerebroventricular infusion or overexpression, mainly due to reduced cell proliferation of NPCs [136, 137]. However, other studies showed that, after induced neuroinflammatory conditions, high levels of TGF-β1 are correlated with increased neurogenesis, while its reduction, or the blockade of its type II receptor, decreases the number of newly generated cells in the DG. Nevertheless, the same authors showed that chronic overexpression of TGF-β1, or the blockade of TGF-β1 type-II receptor on control animals, did not impact neurogenesis [138, 139]. In fact, Battista et al. observed that incubation of DG NPCs with TGF-β1 decreases proliferation but stimulates neuronal differentiation [139]. Therefore, it is possible that TGF-β1 has a dual effect on neurogenesis, decreasing proliferation of NPCs and in parallel increasing their neuronal differentiation.

A summary of the data exposed here can be found in Table 14.2.

14.6 Microglia and Adult Neurogenesis Following Exercise and Environmental Enrichment

As previously mentioned, exposure of rodents to a stimulating environment promotes the incorporation of new neurons into the SGZ [140, 141]. Physical activity also increases neurogenesis in the DG mainly by inducing cell proliferation [142, 143]. Nonetheless, effects mediated by EE or exercise on adult neurogenesis are restricted to the DG neurogenic niche [144].

Elevated number of newly generated and total DG microglia have been observed after EE [97, 145, 146]. Effects of T-cells on microglia have been linked to EE-mediated alteration of adult neurogenesis. Indeed, T-lymphocytes expressing CNS self-antigens are necessary for this stimulation of neurogenesis [97]. Furthermore, EE also induces a 5-fold increase in the number of microglial cells competent to present antigens to T-cells (MHC-II-positive), in the adult rat hilus and GCL [97]. By contrast, effect of exercise on adult neurogenesis does not seem to depend on microglia and T-lymphocytes interaction mediated by MHC-II [112]. In fact, the relationship between expression of MHC-II and neurogenesis is not as clear since the quantity of MHC-II-expressing cells has been negatively correlated with the number of proliferating and Dcx-positive cells in the DG of aged rats [115, 120, 147, 148]. Enhancement on adult neurogenesis mediated by exercise is also associated with elevated levels of IGF-1 in the brain. Despite IGF-1 being secreted by microglia, the effect of voluntary exercise on DG neurogenesis has been related with increase of CNS uptake of circulating IGF-1 [143, 149]. However, an increase in microglial cells expressing IGF-1 was observed in adult mice with free access to a running wheel which also showed increased neurogenesis [150]. In fact, IGF-1 expression is linked to enhanced production of neuroblasts in the rat SVZ after stroke and with elevated cell proliferation in the mouse DG after seizures [90, 151]. Thus, microglial IGF-1 could have a role, although limited, on the effects of voluntary exercise on adult neurogenesis.

Table 14.2 Effects of microglial-related factors in the healthy DG neurogenic niche

Factor	In vitro effect	In vivo effect	Refs.
Prostaglandin		Cox2 inhibition ↓ proliferation	[93]
IL-1-β		↓ proliferation, survival, and neuronal differentiation	[115, 122, 123, 127]
		IL-1R null mice or IL-1Ra treated rats	
		No changes in proliferation	
		IL-1Ra chronic overexpression	
		↓ proliferation	
		No effects on neuronal differentiation	
IL-10	↑ proliferation (microglia/NPCs cocultures)		[117]
IL-2		*IL-2 KO mice*	[118]
		↑ neurogenesis and IL-15, IL-12 and CXCL10	
IL-6		*Overexpression*	[109, 129, 130]
		↓ proliferation, survival, and number of new neurons	
		Repeated intrahippocampal admin	
		↑ proliferation and microglial reactivity	
		KO mice	
		↓ proliferation, survival	
IFN-γ		*Overexpression which do not activate microglia*	[131]
		↑ neurogenesis, TNF-α, IL-6 and IGF-1	
TNF-α		*Lack of TNF-RI*	[108]
		↑ proliferation and number of new neurons	
		Lack of TNF-RII	
		↓ proliferation	
TGF-β1	↓ proliferation ↑ differentiation	*Intracerebroventricular infusion or overexpression*	[136–139]
		↓ proliferation	
		Chronic overexpression or Blockade of type II receptor	
		No effect on neurogenesis in normal conditions	

14.7 Microglia, Neuroinflammation, and Adult Neurogenesis

Neuroinflammation induces changes in the profile of microglia and this has been directly related with effects on adult neurogenesis in both DG and SVZ. Thus, activation of microglia and release of inflammatory molecules after administration of LPS, brain trauma, neurotoxic injury, irradiation, stroke, excitotoxicity, and status epilepticus (SE) have been related to changes in neurogenesis (Table 14.3).

14.7.1 LPS Effects on Adult Neurogenesis Mediated by Microglia

In general, LPS administration induces an inflammatory response in the brain [82, 152] and concomitant decrease in the survival of newly generated cells, without affecting proliferation [84, 153]. Indeed, early in vitro studies showed that reduction in NPCs survival is mediated by soluble factors released by microglial cells [154, 155]. Acute administration of LPS to microglial cell cultures increases microglial production of IL-1β, IL-6, TNF-α, prostaglandin E2 (PGE2), nitric oxide (NO), IL-1α, and IL-10 [155]. IL-1β negatively affects adult neurogenesis, mainly in the DG. Therefore, IL-1β increase after inflammatory response induced by LPS may contribute to the inhibition of adult neurogenesis in the hippocampus [122, 123, 127]. IL-6 is another important mediator of LPS-induced decrease on adult neurogenesis [154]. Interestingly, LPS provokes an increase on microglial secretion of IL-6 in vitro which is lower in microglial cells lacking the PGE2 receptor EP1. This receptor is also involved in the reduction of DG intermediate precursor cells that occurs after intracerebral administration of LPS in adult mice [156]. These evidences suggest a role for PGE2 in the upregulation of IL-6 production and reduction of DG adult neurogenesis induced by LPS. TNF-α has been involved in the induction of cell death in the SVZ-RMS after i.p. administration of LPS [153]. However, we should not forget that the effects of TNF-α could be pro- or anti-neurogenic depending on its concentration and interaction with TNFR-I or TNFR-II [104, 108]. NO, produced by nitric oxide synthases (NOS), is another modulator of neurogenesis [157–159]. Inducible NOS is not normally expressed by microglia in the healthy brain, but its production after inflammation has been related to decreased cell proliferation in the adult DG [160].

Acute and chronic treatment of microglia with LPS has different effects on the expression of pro- and anti-inflammatory molecules. IL-1β protein and TNF-α mRNA levels are normal after chronic treatment with LPS. Chronic exposure to LPS induces a smaller increase in IL-1α, IL-6 and NO concentrations, as compared with acute LPS treatment, and a robust increase of IL-10 and PGE2 [155]. IL-10 has been described as a neuroprotective cytokine associated with increase of NPCs proliferation in vitro [117]. Thus, chronically activated microglial cells could be more proneurogenic compared with acutely activated ones. Indeed, coculture of microglial cells, chronically exposed to LPS, with adult NPCs induces an increase in the proportion of GFAP-positive cells, without affecting the proportion of cells stained with a neuronal marker [155].

14.7.2 Stroke

The most common experimental model of stroke, middle cerebral artery occlusion (MCAO), increases NPCs proliferation on the DG and SVZ [reviewed in 161]. In this context, microglial activation after stroke in the DG neurogenic niche has been proposed to be either detrimental or beneficial [162, 163]. Effects of stroke in the SVZ have been highly analyzed. Stroke induces migration of SVZ newly formed neuroblasts toward the injured striatum, providing the affected region with neuronal replacement. However, just about 20% of new striatal neurons survive during the first 2 weeks [161]. Microglial state changes over time after stroke suggesting different effects at different time points. Early after insult, administration of an anti-inflammatory drug increases accumulation of young neurons, oligodendrocyte precursors, astrocytes, and NPCs in the affected striatum, while, on the opposite, decreases the proportion of activated microglia [164]. On the other hand, inactivation of microglial cells using minocycline decreases the production of SVZ neuroblasts after ischemia, suggesting a role of activated microglia on the upregulation of SVZ neurogenesis [165]. In fact, stroke increases the number of SVZ microglia [102]. In their study, Deierborg et al. showed that the higher number of microglial cells present in NPCs cultures from SVZ collected after stroke or excitotoxic lesion is related to elevated cell proliferation and increased number and size of neurospheres [102]. This effect on proliferation is mediated by microglial soluble factors [166]. Indeed, IGF-1 mediates the increase on proliferation induced by stroke in the DG and SVZ. NPCs of both DG and SVZ express IGF-1 receptor and its blockade prevents ischemia-induced proliferation [90, 167]. Galectin-3 overexpressed by striatal astrocytes and stroke-activated microglia showed a similar effect [168]. Mice lacking TNF-RI have increased levels of cell proliferation and SVZ neuroblasts 1 week after MCAO, suggesting a negative effect of TNF-α after stroke [102]. Osteopontin, a phosphorylated acidic glycoprotein, and MCP-1 promote the migration of neuroblasts to the injured striatum after MCAO [169, 170]. Thus, activated microglial cells seem to be directly involved in promoting production of new cells and the migration of neuroblasts from SVZ towards the ischemic region.

14.7.3 Status Epilepticus

Similar to stroke, after damage caused by *status epilepticus* (SE), aberrant migration of neuroblasts and granule cell dispersion occurs. SE also triggers an increase in cell proliferation and neurogenesis in the DG [151, 171, 172] and enhances cell proliferation in the SVZ [102]. In this case, SE-induced microglial activation has been negatively correlated with survival of new neurons in the hippocampus [173]. Indeed, SE induces formation of ectopic hilar basal dendrites and aberrant neuroblasts, an effect associated with microglial activation and Cox2 activity [174, 175]. Cox2 function has been also related to increased proliferation and production of new neurons and astrocytes after SE in the DG and SVZ [175, 176]. Similar to stroke, IGF-1 also has a role in the upregulation of cell proliferation after SE, an effect attributed to a CREB-mediated induction of IGF-1 transcription in DG microglia [151]. Finally, signaling through the two TNF-α receptors differentially affects adult neurogenesis after SE. TNF-RI interaction with TNF-α reduces NPCs proliferation and production of new neurons, whereas binding to TNF-RII increases the survival of new neurons in the DG [108]. However, in the SVZ, TNF-RI has no role on the increased production of neuroblasts after SE [102]. Thus, SE induction of proliferation in the DG and SVZ may be mediated by different mechanisms.

14.7.4 Irradiation

Intracranial ionizing radiation is used to treat brain tumors, but also impairs cognition and exerts negative effects on hippocampal-dependent memory and adult neurogenesis [177, 178]. In spite of the main effect of irradiation on neurogenic niches being mediated by direct DNA damage to NPCs and subsequent induction of apoptosis, recovery after irradiation has been related with microglial activity. Indeed, anti-inflammatory drugs increase the number of newborn neurons in the hippocampus after irradiation [154, 179]. Curiously, SVZ neurogenic niche recovers better than DG after irradiation, an effect that has been attributed mainly to differences on microglial response between these two regions [180]. Microglial-secreted factors seem to be involved on this regional difference, since irradiated SVZ microglia express higher mRNA levels of fibrillar growth factor 2 (FGF-2), leukemia inhibitory factor and platelet derived growth factor, as compared with irradiated hippocampal microglia (which showed down-regulated levels of FGF-2 transcripts). Furthermore, treating irradiated SVZ neurospheres with a cocktail containing the previous mentioned factors decreases apoptosis and increases proliferation [180]. Thus, microglial cells have a decisive role in the recovery of adult neurogenesis after irradiation and could be considered as a therapeutic target.

14.8 Aged Microglia and Aged Neurogenesis

Aging is one of the classical and stronger factors that negatively affect adult neurogenesis in DG and SVZ [reviewed in 80]. The decrease on DG and OB neurogenesis in old rodents has been associated with memory and fine olfactory discrimination deficits, respectively [181, 182]. Interestingly, a study showed that although the proportion of dividing NPCs is reduced in aged mice, the total amount of NPCs is not altered with aging [183]. In fact, the recovery of normal production of new neurons in the aged DG and SVZ can be achieved [142, 184–186], suggesting that the aged DG still possesses the potential to generate new neurons at the levels reached during young adulthood. However, a recent study showed depletion of NPCs in the aged DG [187].

Changes observed in "aged" microglia reduce their protective capacities and potentiate their proinflammatory response. Interestingly, voluntary exercise improves survival of newborn cells in the aged DG and induces a concomitant increase in the number of IGF-1-positive microglia, suggesting that aged microglial cells can be modulated to acquire a proneurogenic fate [150]. Alterations in the microenvironment of the aged brain have been hypothesized to be responsible for the reduced production of new neurons [18, 188–190]. Accordingly, upon aging, microglial cells acquire an activated-like phenotype characterized by morphological changes and increased mRNA expression of both proinflammatory (TNF-α, IL-1β, IL-6) and anti-inflammatory (IL-10, TGF-β1) cytokines that affect adult neurogenesis (Table 14.3) [191]. Aging also affects fractalkine/CX3CR1 "off" system. Protein levels of fractalkine are reduced while CX3CR1 levels are increased in the hippocampus of aged rats. Furthermore, exogenous administration of fractalkine reduces microglial activation in the aged hippocampus [115, 147]. Indeed, fractalkine/CX3CR1 system could mediate the aged-induced decrease on adult neurogenesis [further reviewed on 188]. Interestingly, increased levels of IL-1β mediate the downregulation of adult neurogenesis triggered by the reduction of fractalkine signaling to microglia [115]. Moreover, IL-1β is increased and affects neurogenesis in the aged hippocampus [120]. Recently, epigenetic modifications mediated by elevated levels of IL-1β in the aged hippocampus have been shown to underlay aging related decrease in adult neurogenesis [121].

14.9 Neurodegenerative Diseases, Microglia, and Adult Neurogenesis: The Triad

Microglial activation is a key hallmark of the major neurodegenerative diseases and, as discussed, may impact the regulation of neurogenesis (Table 14.4).

Multiple sclerosis (MS) is a generalized autoimmune disease of the CNS which develops concomitantly with infiltration of monocytes, T and B-cells into the brain

Table 14.3 Effects of activated microglia in neurogenic niches

Inflammatory agent	In vitro effect	In vivo effect	Microglial factors involved	Refs.
LPS	*Acute LPS-activated microglia* ↓ survival ↑ IL-10, PGE2, NO, IL-1α, IL-1β, IL-6, TNF-α *Chronic LPS-activated microglia* ↑ proportion of GFAP + cells ↑ IL-10, PGE2, NO, IL-1α, IL-6	↓ survival	iNOS → NO PGE2 → IL-6 TNF-α (SVZ)	[82, 152–156, 160]
Stroke	↑ proliferation ↑microglia	*SVZ* Migration to affected region ↑ proliferation and microglia *DG* ↑ proliferation *Minocycline treatmen*: ↑ new-neurons	*Proliferation* IGF-1, Galectin-3 TNF-α: negative effect *Migration* Osteopontin, MCP-1	[90, 102, 162–170]
Status epilecticus		Migration to affected region *SVZ* ↑ proliferation *DG* ↑ proliferation and neuronal production	Cox2 → PGE2 IGF-1 *TNF-α (DG)* *TNF-RI* ↓ proliferation and neuronal production *TNF-RII* ↑ survival	[102, 108, 151, 171–176]
Irradiation	↑ apoptosis	↓ neurogenesis *After irradiation* Microglia induce better recovery in SVZ than in DG	*SVZ* ↑ FGF-2, Leukemia inhibitory factor and platelet derived growth factor *DG* ↓ FGF-2	[154, 179, 180]
Aging		↓ proliferation	IL-1β	[120, 121, 183]

Table 14.4 Neurodegenerative diseases, microglia, and adult neurogenesis

Disease	In vitro effect	In vivo effect	References
Multiple sclerosis		*SVZ* Altered morphology ↓ proliferation	[92, 98]
Huntington's disease		*SVZ* Both ↑and no changed neurogenesis *DG* ↓ neurogenesis	[193, reviewed in 194, 195]
Parkinson's disease		↑ neurogenesis (SN, striatum, DG and SVZ)	[reviewed in 83]
Alzheimer's disease	*Microglial conditioned media from PS mutants* ↓ proliferation and neuronal differentiation *Microglia exposed to agregated amyloid β* ↓ survival	↓ neurogenesis	[reviewed in 199, 200]

parenchyma, and with concomitant activation of microglia [192]. In experimental autoimmune encephalomyelitis (EAE), the most used animal model of MS, the morphological organization of SVZ is altered during acute and chronic inflammation [98]. Specific infiltration of T-cells and macrophages takes place in the SVZ after EAE. These cells elevate the levels of cytokines like TNF-α and IFN-γ, which in turn upregulate the production of the NPCs chemokine CXCL10, a chemokine involved in the attraction of microglial/macrophage cells to the SVZ [98]. Most EAE animal models spontaneously recover from neuroinflammatory-induced demyelination and neurodegeneration, a feature that does not completely occur in human MS [192]. Interestingly, this remyelination is induced by oligodendrocytes arising from NPCs. Reduction of microglial activation increases NPCs proliferation and enhances the production of oligodendrocyte precursors in the SVZ [92]. Thus, endogenous neural precursors could be an ideal target to promote total remyelination on MS.

Huntington's disease (HD) is a hereditary disorder caused by a highly polymorphic CAG trinucleotide repeat expansion in the exon-1 of the gene encoding for huntingtin protein and characterized by striatal neurodegeneration, leading to typical motor and cognitive impairments. HD patients showed increased neurogenesis in the SVZ which also correlates with the severity of the illness [193]. However, decreased neurogenesis in the DG, and no changes in the SVZ, have been observed in mouse models of HD [reviewed in 194]. There is a strong inflammatory component on this disease, but a relationship between microglial activation, its effects on neurogenesis, and HD pathology has not been yet properly addressed [revised in 195].

Parkinson's disease (PD) is characterized by progressive degeneration of dopaminergic neurons of the *substantia nigra* (SN). Neuroinflammation and microglial cell activation is also a main signature of the disease. In animal models of PD, neurogenesis is increased in the SN, striatum, DG, and SVZ [reviewed in 83]. However, interestingly, the relationship between the neuroinflammatory response in PD animal models and increased adult neurogenesis has not been carefully addressed. Moreover, in animal models of PD, microglial cells of the SN enter in a "primed" state in which IL-1α and IL-1β transcription is enhanced, but their translation is blocked [196]. Godoy et al. demonstrated that, in a rat model of PD, LPS administration exacerbates neurodegeneration in the SN and induces increased production of IL-1β, an effect detrimental for adult neurogenesis [197].

Alzheimer's disease (AD) is a neurodegenerative disorder characterized by the presence of beta-amyloid plaques and neurofibrillary tangles [198]. Deposition of beta-amyloid peptide induces a chronic inflammatory response leading to activation of microglial cells and astrocytes, releasing different cytokines and chemokines into the brain parenchyma [199]. Despite some discrepancies, most studies carried out on AD mouse models show reduced SVZ and DG adult neurogenesis [reviewed in 200]. Conditioned media from microglial cells of mice carrying presenilin variants, proteins linked to familiar AD, have been shown to impair proliferation and neuronal differentiation of hippocampal NPCs. In vivo, EE is unable to increase adult neurogenesis in these presenilin transgenic mice, an effect attributed to microglial dysfunction [201]. Moreover, microglial cells exposed to aggregated beta-amyloid induce a decrease in the survival of SVZ NPCs [202]. Treatment with diverse cytokines that affect adult neurogenesis, by acting on microglial cells, like IL-10, IL-4 and IFN-γ, are effective in increasing neurogenesis and recover memory deficits in AD mouse models [117, 131, 202]. Furthermore, induction of a T cell response (Th2) using glatiramer acetate decreases the accumulation of amyloid plaques, enhances DG adult neurogenesis, and reduces spatial memory deficits in a mouse model of AD. Interestingly, the effect of glatiramer was suggested to be mediated by a change in microglia which become dendritic-like cells (expressing CD11c) and produce IGF-1 [202]. Thus, modulation of T cell response and its interaction with microglial cells could serve as a useful therapeutic target to improve cognitive capacities, promote adult neurogenesis and regulate the inflammatory, and phagocytic activity of microglia on AD.

14.10 Conclusions

The analysis of current data about microglia interaction with the adult neurogenic niche reveals the need for more intensive research on this field. Although some studies address the effect of microglia on adult neurogenesis during brain inflammation and neurodegeneration, there is still a lack of understanding about the role of microglia in healthy neurogenic niches. Involvement of microglia on dendritic and axonal development, synaptic maturation and cell death, as well as in injury-stimulated neurogenesis has not been deeply studied. Importantly, effects of microglia on neuroblast migration through the RMS and in the OB have not been properly evaluated. Despite this, we know that microglial cells express different molecules to communicate with NPCs and other components of the neurogenic niche, which endows these cells with the potential to regulate and maintain normal neurogenesis. New exciting data suggest that alteration of the relationship between microglia and neurogenic niche elements underlies the outcome of diverse CNS pathologies. In fact, regulation of microglial activity seems to be efficient against some pathological situations, but could trigger side effects on adult neurogenesis. We should consider whether such strategies also change adult neurogenesis in a positive or negative manner, and how this impact brain functions.

Microglial cells hold an immense, but poorly explored, potential as proneurogenic agents. Importantly, microglia possess the ability to produce/release and unleash signaling cascades that will inevitably play a role in neurogenesis. Modulating the release of these extrinsic factors, that regulate defined features of the neurogenic process, and the interaction between microglia and other cellular partners could be an important strategy in the amelioration or treatment of CNS disorders.

Acknowledgments Our work has been supported by Fundação para a Ciência e a Tecnologia (FCT). MFE is part of the MIT-Portugal PhD Program, focused in Bioengineering.

References

1. Alliot F, Godin I, Pessac B (1999) Microglia derive from progenitors, originating from the yolk sac, and which proliferate in the brain. Brain Res Dev Brain Res 117(2):145–152
2. Guillemin GJ, Brew BJ (2004) Microglia, macrophages, perivascular macrophages, and pericytes: a review of function and identification. J Leukoc Biol 75(3):388–397
3. Walter L, Neumann H (2009) Role of microglia in neuronal degeneration and regeneration. Semin Immunopathol 31(4):513–525
4. Castellano F, Chavrier P, Caron E (2001) Actin dynamics during phagocytosis. Semin Immunol 13(6):347–355
5. Myers KR, Casanova JE (2008) Regulation of actin cytoskeleton dynamics by Arf-family GTPases. Trends Cell Biol 18(4):184–192
6. Tricker E, Cheng G (2008) With a little help from my friends: modulation of phagocytosis through TLR activation. Cell Res 18(7):711–712
7. Koizumi S, Shigemoto-Mogami Y, Nasu-Tada K, Shinozaki Y, Ohsawa K, Tsuda M, Joshi BV, Jacobson KA, Kohsaka S, Inoue K (2007) UDP acting at P2Y6 receptors is a mediator of microglial phagocytosis. Nature 446(7139):1091–1095
8. Davalos D, Grutzendler J, Yang G, Kim JV, Zuo Y, Jung S, Littman DR, Dustin ML, Gan WB (2005) ATP mediates rapid microglial response to local brain injury in vivo. Nat Neurosci 8(6):752–758
9. Nimmerjahn A, Kirchhoff F, Helmchen F (2005) Resting microglial cells are highly dynamic surveillants of brain parenchyma in vivo. Science 308(5726):1314–1318
10. Biber K, Neumann H, Inoue K, Boddeke HW (2007) Neuronal 'On' and 'Off' signals control microglia. Trends Neurosci 30(11):596–602
11. Garden GA, Moller T (2006) Microglia biology in health and disease. J Neuroimmune Pharmacol 1(2):127–137

12. Graeber MB, Streit WJ (2010) Microglia: biology and pathology. Acta Neuropathol 119(1): 89–105
13. Danton GH, Dietrich WD (2003) Inflammatory mechanisms after ischemia and stroke. J Neuropathol Exp Neurol 62(2):127–136
14. Choi J, Koh S (2008) Role of brain inflammation in epileptogenesis. Yonsei Med J 49(1): 1–18
15. Block ML, Zecca L, Hong JS (2007) Microglia-mediated neurotoxicity: uncovering the molecular mechanisms. Nat Rev Neurosci 8(1):57–69
16. Forstreuter F, Lucius R, Mentlein R (2002) Vascular endothelial growth factor induces chemotaxis and proliferation of microglial cells. J Neuroimmunol 132(1–2):93–98
17. Kawai T, Akira S (2010) The role of pattern-recognition receptors in innate immunity: update on Toll-like receptors. Nat Immunol 11(5):373–384
18. Czeh M, Gressens P, Kaindl AM (2011) The yin and yang of microglia. Dev Neurosci 33(3–4):199–209
19. Chavarria A, Alcocer-Varela J (2004) Is damage in central nervous system due to inflammation? Autoimmun Rev 3(4):251–260
20. Turrin NP, Rivest S (2006) Molecular and cellular immune mediators of neuroprotection. Mol Neurobiol 34(3):221–242
21. Olah M, Biber K, Vinet J, Boddeke HW (2011) Microglia phenotype diversity. CNS Neurol Disord Drug Targets 10(1):108–118
22. Pont-Lezica L, Bechade C, Belarif-Cantaut Y, Pascual O, Bessis A (2011) Physiological roles of microglia during development. J Neurochem 119(5):901–908
23. Wakselman S, Bechade C, Roumier A, Bernard D, Triller A, Bessis A (2008) Developmental neuronal death in hippocampus requires the microglial CD11b integrin and DAP12 immunoreceptor. J Neurosci 28(32):8138–8143
24. Bessis A, Bechade C, Bernard D, Roumier A (2007) Microglial control of neuronal death and synaptic properties. Glia 55(3):233–238
25. Fantin A, Vieira JM, Gestri G, Denti L, Schwarz Q, Prykhozhij S, Peri F, Wilson SW, Ruhrberg C (2010) Tissue macrophages act as cellular chaperones for vascular anastomosis downstream of VEGF-mediated endothelial tip cell induction. Blood 116(5):829–840
26. Rymo SF, Gerhardt H, Wolfhagen SF, Lang R, Uv A, Betsholtz C (2011) A two-way communication between microglial cells and angiogenic sprouts regulates angiogenesis in aortic ring cultures. PLoS One. 6(1):e15846
27. Tremblay ME, Majewska AK (2011) A role for microglia in synaptic plasticity? Commun Integr Biol 4(2):220–222
28. Altman J (1969) Autoradiographic and histological studies of postnatal neurogenesis. IV. Cell proliferation and migration in the anterior forebrain, with special reference to persisting neurogenesis in the olfactory bulb. J Comp Neurol 137(4):433–457
29. Quinones-Hinojosa A, Sanai N, Soriano-Navarro M, Gonzalez-Perez O, Mirzadeh Z, Gil-Perotin S, Romero-Rodriguez R, Berger MS, Garcia-Verdugo JM, Alvarez-Buylla A (2006) Cellular composition and cytoarchitecture of the adult human subventricular zone: a niche of neural stem cells. J Comp Neurol 494(3):415–434
30. Doetsch F, García-Verdugo JM, Álvarez-Buylla A (1997) Cellular composition and three-dimensional organization of the subventricular germinal zone in the adult mammalian brain. J Neurosci 17(13):5046–5061
31. Curtis MA, Kam M, Nannmark U, Anderson MF, Axell MZ, Wikkelso C, Holtas S, van Roon-Mom WM, Bjork-Eriksson T, Nordborg C, Frisen J, Dragunow M, Faull RL, Eriksson PS (2007) Human neuroblasts migrate to the olfactory bulb via a lateral ventricular extension. Science 315(5816):1243–1249
32. Altman J, Das GD (1965) Autoradiographic and histological evidence of postnatal hippocampal neurogenesis in rats. J Comp Neurol 124(3):319–335
33. van Praag H, Schinder AF, Christie BR, Toni N, Palmer TD, Gage FH (2002) Functional neurogenesis in the adult hippocampus. Nature 415(6875):1030–1034

34. Eriksson PS, Perfilieva E, Björk E, Alborn AM, Nordborg C, Peterson DA, Gage FH (1998) Neurogenesis in the adult human hippocampus. Nat Med 4(11):1313–1317
35. Weiss S, Reynolds BA, Vescovi AL, Morshead C, Craig CG, van der Kooy D (1996) Is there a neural stem cell in the mammalian forebrain? Trends Neurosci 19(9):387–393
36. Bonaguidi MA, Wheeler MA, Shapiro JS, Stadel RP, Sun GJ, Ming GL, Song H (2011) In vivo clonal analysis reveals self-renewing and multipotent adult neural stem cell characteristics. Cell 145(7):1142–1155
37. Doetsch F, Caille I, Lim DA, García-Verdugo JM, Álvarez-Buylla A (1999) Subventricular zone astrocytes are neural stem cells in the adult mammalian brain. Cell 97(6):703–716
38. Morshead CM, van der Kooy D (2004) Disguising adult neural stem cells. Curr Opin Neurobiol 14(1):125–131
39. Chiasson BJ, Tropepe V, Morshead CM, van der Kooy D (1999) Adult mammalian forebrain ependymal and subependymal cells demonstrate proliferative potential, but only subependymal cells have neural stem cell characteristics. J Neurosci 19(11):4462–4471
40. Aimone JB, Wiles J, Gage FH (2009) Computational influence of adult neurogenesis on memory encoding. Neuron 61(2):187–202
41. Breton-Provencher V, Lemasson M, Peralta MR III, Saghatelyan A (2009) Interneurons produced in adulthood are required for the normal functioning of the olfactory bulb network and for the execution of selected olfactory behaviors. J Neurosci 29(48):15245–15257
42. Eisch AJ, Cameron HA, Encinas JM, Meltzer LA, Ming GL, Overstreet-Wadiche LS (2008) Adult neurogenesis, mental health, and mental illness: hope or hype? J Neurosci 28(46): 11785–11791
43. Lois C, García-Verdugo JM, Álvarez-Buylla A (1996) Chain migration of neuronal precursors. Science 271(5251):978–981
44. Bonfanti L, Peretto P, Merighi A, Fasolo A (1997) Newly-generated cells from the rostral migratory stream in the accessory olfactory bulb of the adult rat. Neuroscience 81(2): 489–502
45. Rochefort C, Gheusi G, Vincent JD, Lledo PM (2002) Enriched odor exposure increases the number of newborn neurons in the adult olfactory bulb and improves odor memory. J Neurosci 22(7):2679–2689
46. Oboti L, Schellino R, Giachino C, Chamero P, Pyrski M, Leinders-Zufall T, Zufall F, Fasolo A, Peretto P (2011) Newborn interneurons in the accessory olfactory bulb promote mate recognition in female mice. Front Neurosci 5:113
47. Lagace DC, Whitman MC, Noonan MA, Ables JL, DeCarolis NA, Arguello AA, Donovan MH, Fischer SJ, Farnbauch LA, Beech RD, DiLeone RJ, Greer CA, Mandyam CD, Eisch AJ (2007) Dynamic contribution of nestin-expressing stem cells to adult neurogenesis. J Neurosci 27(46):12623–12629
48. Platel JC, Gordon V, Heintz T, Bordey A (2009) GFAP-GFP neural progenitors are antigenically homogeneous and anchored in their enclosed mosaic niche. Glia 57(1):66–78
49. Mirzadeh Z, Merkle FT, Soriano-Navarro M, Garcia-Verdugo JM, Alvarez-Buylla A (2008) Neural stem cells confer unique pinwheel architecture to the ventricular surface in neurogenic regions of the adult brain. Cell Stem Cell 3(3):265–278
50. Shen Q, Wang Y, Kokovay E, Lin G, Chuang SM, Goderie SK, Roysam B, Temple S (2008) Adult SVZ stem cells lie in a vascular niche: a quantitative analysis of niche cell–cell interactions. Cell Stem Cell 3(3):289–300
51. Tavazoie M, Van der Veken L, Silva-Vargas V, Louissaint M, Colonna L, Zaidi B, Garcia-Verdugo JM, Doetsch F (2008) A specialized vascular niche for adult neural stem cells. Cell Stem Cell 3(3):279–288
52. Kerever A, Schnack J, Vellinga D, Ichikawa N, Moon C, Arikawa-Hirasawa E, Efird JT, Mercier F (2007) Novel extracellular matrix structures in the neural stem cell niche capture the neurogenic factor fibroblast growth factor 2 from the extracellular milieu. Stem Cells 25(9):2146–2157
53. Whitman MC, Fan W, Rela L, Rodriguez-Gil DJ, Greer CA (2009) Blood vessels form a migratory scaffold in the rostral migratory stream. J Comp Neurol 516(2):94–104

54. Lois C, Álvarez-Buylla A (1994) Long-distance neuronal migration in the adult mammalian brain. Science 264(5162):1145–1148
55. Jankovski A, Sotelo C (1996) Subventricular zone-olfactory bulb migratory pathway in the adult mouse: cellular composition and specificity as determined by heterochronic and heterotopic transplantation. J Comp Neurol 371(3):376–396
56. Valero J, Weruaga E, Murias AR, Recio JS, Curto GG, Gomez C, Alonso JR (2007) Changes in cell migration and survival in the olfactory bulb of the pcd/pcd mouse. Dev Neurobiol 67(7):839–859
57. Cameron HA, McKay RD (2001) Adult neurogenesis produces a large pool of new granule cells in the dentate gyrus. J Comp Neurol 435(4):406–417
58. Filippov V, Kronenberg G, Pivneva T, Reuter K, Steiner B, Wang LP, Yamaguchi M, Kettenmann H, Kempermann G (2003) Subpopulation of nestin-expressing progenitor cells in the adult murine hippocampus shows electrophysiological and morphological characteristics of astrocytes. Mol Cell Neurosci 23(3):373–382
59. Han YG, Spassky N, Romaguera-Ros M, Garcia-Verdugo JM, Aguilar A, Schneider-Maunoury S, Alvarez-Buylla A (2008) Hedgehog signaling and primary cilia are required for the formation of adult neural stem cells. Nat Neurosci 11(3):277–284
60. Amador-Arjona A, Elliott J, Miller A, Ginbey A, Pazour GJ, Enikolopov G, Roberts AJ, Terskikh AV (2011) Primary cilia regulate proliferation of amplifying progenitors in adult hippocampus: implications for learning and memory. J Neurosci 31(27):9933–9944
61. Kronenberg G, Reuter K, Steiner B, Brandt MD, Jessberger S, Yamaguchi M, Kempermann G (2003) Subpopulations of proliferating cells of the adult hippocampus respond differently to physiologic neurogenic stimuli. J Comp Neurol 467(4):455–463
62. Plumpe T, Ehninger D, Steiner B, Klempin F, Jessberger S, Brandt M, Romer B, Rodriguez GR, Kronenberg G, Kempermann G (2006) Variability of doublecortin-associated dendrite maturation in adult hippocampal neurogenesis is independent of the regulation of precursor cell proliferation. BMC Neurosci 7:77
63. Hastings NB, Gould E (1999) Rapid extension of axons into the CA3 region by adult-generated granule cells. J Comp Neurol 413(1):146–154
64. Forster E, Jossin Y, Zhao S, Chai X, Frotscher M, Goffinet AM (2006) Recent progress in understanding the role of Reelin in radial neuronal migration, with specific emphasis on the dentate gyrus. Eur J Neurosci 23(4):901–909
65. Zhao C, Teng EM, Summers RG Jr, Ming GL, Gage FH (2006) Distinct morphological stages of dentate granule neuron maturation in the adult mouse hippocampus I. J Neurosci 26(1):3–11
66. Ide Y, Fujiyama F, Okamoto-Furuta K, Tamamaki N, Kaneko T, Hisatsune T (2008) Rapid integration of young newborn dentate gyrus granule cells in the adult hippocampal circuitry. Eur J Neurosci 28(12):2381–2392
67. Biebl M, Cooper CM, Winkler J, Kuhn HG (2000) Analysis of neurogenesis and programmed cell death reveals a self-renewing capacity in the adult rat brain. Neurosci Lett 291(1):17–20
68. Kee N, Teixeira CM, Wang AH, Frankland PW (2007) Preferential incorporation of adult-generated granule cells into spatial memory networks in the dentate gyrus. Nat Neurosci 10(3):355–362
69. Mu Y, Lee SW, Gage FH (2010) Signaling in adult neurogenesis. Curr Opin Neurobiol 20(4):416–423
70. Ming GL, Song H (2011) Adult neurogenesis in the mammalian brain: significant answers and significant questions. Neuron 70(4):687–702
71. Sun J, Sun J, Ming GL, Song H (2011) Epigenetic regulation of neurogenesis in the adult mammalian brain. Eur J Neurosci 33(6):1087–1093
72. Ninkovic J, Gotz M (2007) Signaling in adult neurogenesis: from stem cell niche to neuronal networks. Curr Opin Neurobiol 17(3):338–344
73. Llorens-Martin M, Torres-Aleman I, Trejo JL (2009) Mechanisms mediating brain plasticity: IGF1 and adult hippocampal neurogenesis. Neuroscientist 15(2):134–148

74. Olson AK, Eadie BD, Ernst C, Christie BR (2006) Environmental enrichment and voluntary exercise massively increase neurogenesis in the adult hippocampus via dissociable pathways. Hippocampus 16(3):250–260
75. Park HR, Lee J (2011) Neurogenic contributions made by dietary regulation to hippocampal neurogenesis. Ann N Y Acad Sci 1229:23–28
76. Gheusi G, Ortega-Perez I, Murray K, Lledo PM (2009) A niche for adult neurogenesis in social behavior. Behav Brain Res 200(2):315–322
77. Schoenfeld TJ, Gould E (2011) Stress, stress hormones, and adult neurogenesis. Exp Neurol 233(1):12–21
78. Martoncikova M, Lievajova K, Orendacova J, Blasko J, Racekova E (2011) Odor enrichment influences neurogenesis in the rostral migratory stream of young rats. Acta Histochem 113(3):326–332
79. Bovetti S, Veyrac A, Peretto P, Fasolo A, De MS (2009) Olfactory enrichment influences adult neurogenesis modulating GAD67 and plasticity-related molecules expression in newborn cells of the olfactory bulb. PLoS One 4(7):e6359
80. Lee SW, Clemenson GD, Gage FH (2011) New neurons in an aged brain. Behav Brain Res 227(2):497–507
81. Yoneyama M, Shiba T, Hasebe S, Ogita K (2011) Adult neurogenesis is regulated by endogenous factors produced during neurodegeneration. J Pharmacol Sci 115(4):425–432
82. Russo I, Barlati S, Bosetti F (2011) Effects of neuroinflammation on the regenerative capacity of brain stem cells. J Neurochem 116(6):947–956
83. Whitney NP, Eidem TM, Peng H, Huang Y, Zheng JC (2009) Inflammation mediates varying effects in neurogenesis: relevance to the pathogenesis of brain injury and neurodegenerative disorders. J Neurochem 108(6):1343–1359
84. Sierra A, Encinas JM, Deudero JJ, Chancey JH, Enikolopov G, Overstreet-Wadiche LS, Tsirka SE, Maletic-Savatic M (2010) Microglia shape adult hippocampal neurogenesis through apoptosis-coupled phagocytosis. Cell Stem Cell 7(4):483–495
85. Blakemore WF (1969) The ultrastructure of the subependymal plate in the rat. J Anat 104(Pt 3): 423–433
86. Peretto P, Merighi A, Fasolo A, Bonfanti L (1999) The subependymal layer in rodents: a site of structural plasticity and cell migration in the adult mammalian brain. Brain Res Bull 49(4):221–243
87. Mercier F, Kitasako JT, Hatton GI (2002) Anatomy of the brain neurogenic zones revisited: fractones and the fibroblast/macrophage network. J Comp Neurol 451(2):170–188
88. Goings GE, Kozlowski DA, Szele FG (2006) Differential activation of microglia in neurogenic versus non-neurogenic regions of the forebrain. Glia 54(4):329–342
89. Marshall GP, Demir M, Steindler DA, Laywell ED (2008) Subventricular zone microglia possess a unique capacity for massive in vitro expansion. Glia 56(16):1799–1808
90. Thored P, Heldmann U, Gomes-Leal W, Gisler R, Darsalia V, Taneera J, Nygren JM, Jacobsen SE, Ekdahl CT, Kokaia Z, Lindvall O (2009) Long-term accumulation of microglia with proneurogenic phenotype concomitant with persistent neurogenesis in adult subventricular zone after stroke. Glia 57(8):835–849
91. Bulloch K, Miller MM, Gal-Toth J, Milner TA, Gottfried-Blackmore A, Waters EM, Kaunzner UW, Liu K, Lindquist R, Nussenzweig MC, Steinman RM, McEwen BS (2008) CD11c/EYFP transgene illuminates a discrete network of dendritic cells within the embryonic, neonatal, adult, and injured mouse brain. J Comp Neurol 508(5):687–710
92. Rasmussen S, Imitola J, Ayuso-Sacido A, Wang Y, Starossom SC, Kivisakk P, Zhu B, Meyer M, Bronson RT, Garcia-Verdugo JM, Khoury SJ (2011) Reversible neural stem cell niche dysfunction in a model of multiple sclerosis. Ann Neurol 69(5):878–891
93. Goncalves MB, Williams EJ, Yip P, Yanez-Munoz RJ, Williams G, Doherty P (2010) The COX-2 inhibitors, meloxicam and nimesulide, suppress neurogenesis in the adult mouse brain. Br J Pharmacol 159(5):1118–1125
94. Walton NM, Sutter BM, Laywell ED, Levkoff LH, Kearns SM, Marshall GP, Scheffler B, Steindler DA (2006) Microglia instruct subventricular zone neurogenesis. Glia 54(8):815–825

95. Lledo PM, Saghatelyan A, Lemasson M (2004) Inhibitory interneurons in the olfactory bulb: from development to function. Neuroscientist 10(4):292–303
96. Wu CH, Chien HF, Chang CY, Ling EA (1997) Heterogeneity of antigen expression and lectin labeling on microglial cells in the olfactory bulb of adult rats. Neurosci Res 28(1): 67–75
97. Ziv Y, Ron N, Butovsky O, Landa G, Sudai E, Greenberg N, Cohen H, Kipnis J, Schwartz M (2006) Immune cells contribute to the maintenance of neurogenesis and spatial learning abilities in adulthood. Nat Neurosci 9(2):268–275
98. Muzio L, Cavasinni F, Marinaro C, Bergamaschi A, Bergami A, Porcheri C, Cerri F, Dina G, Quattrini A, Comi G, Furlan R, Martino G (2010) Cxcl10 enhances blood cells migration in the sub-ventricular zone of mice affected by experimental autoimmune encephalomyelitis. Mol Cell Neurosci 43(3):268–280
99. Ziv Y, Schwartz M (2008) Orchestrating brain-cell renewal: the role of immune cells in adult neurogenesis in health and disease. Trends Mol Med 14(11):471–478
100. Butovsky O, Ziv Y, Schwartz A, Landa G, Talpalar AE, Pluchino S, Martino G, Schwartz M (2006) Microglia activated by IL-4 or IFN-gamma differentially induce neurogenesis and oligodendrogenesis from adult stem/progenitor cells. Mol Cell Neurosci 31(1):149–160
101. Li L, Walker TL, Zhang Y, Mackay EW, Bartlett PF (2010) Endogenous interferon gamma directly regulates neural precursors in the non-inflammatory brain. J Neurosci 30(27): 9038–9050
102. Iosif RE, Ahlenius H, Ekdahl CT, Darsalia V, Thored P, Jovinge S, Kokaia Z, Lindvall O (2008) Suppression of stroke-induced progenitor proliferation in adult subventricular zone by tumor necrosis factor receptor 1. J Cereb Blood Flow Metab 28(9):1574–1587
103. Ben-Hur T, Cialic R, Weidenfeld J (2003) Virus and host factors that mediate the clinical and behavioral signs of experimental herpetic encephalitis. A short auto-review. Acta Microbiol Immunol Hung 50(4):443–451
104. Bernardino L, Agasse F, Silva B, Ferreira R, Grade S, Malva JO (2008) Tumor necrosis factor-alpha modulates survival, proliferation, and neuronal differentiation in neonatal subventricular zone cell cultures. Stem Cells 26(9):2361–2371
105. Wong G, Goldshmit Y, Turnley AM (2004) Interferon-gamma but not TNF alpha promotes neuronal differentiation and neurite outgrowth of murine adult neural stem cells. Exp Neurol 187(1):171–177
106. Widera D, Mikenberg I, Elvers M, Kaltschmidt C, Kaltschmidt B (2006) Tumor necrosis factor alpha triggers proliferation of adult neural stem cells via IKK/NF-kappaB signaling. BMC Neurosci 7:64
107. Wu JP, Kuo JS, Liu YL, Tzeng SF (2000) Tumor necrosis factor-alpha modulates the proliferation of neural progenitors in the subventricular/ventricular zone of adult rat brain. Neurosci Lett 292(3):203–206
108. Iosif RE, Ekdahl CT, Ahlenius H, Pronk CJ, Bonde S, Kokaia Z, Jacobsen SE, Lindvall O (2006) Tumor necrosis factor receptor 1 is a negative regulator of progenitor proliferation in adult hippocampal neurogenesis. J Neurosci 26(38):9703–9712
109. Bowen KK, Dempsey RJ, Vemuganti R (2011) Adult interleukin-6 knockout mice show compromised neurogenesis. NeuroReport 22(3):126–130
110. Widera D, Holtkamp W, Entschladen F, Niggemann B, Zanker K, Kaltschmidt B, Kaltschmidt C (2004) MCP-1 induces migration of adult neural stem cells. Eur J Cell Biol 83(8):381–387
111. Liu XS, Zhang ZG, Zhang RL, Gregg SR, Wang L, Yier T, Chopp M (2007) Chemokine ligand 2 (CCL2) induces migration and differentiation of subventricular zone cells after stroke. J Neurosci Res 85(10):2120–2125
112. Olah M, Ping G, de Haas AH, Brouwer N, Meerlo P, Van Der Zee EA, Biber K, Boddeke HW (2009) Enhanced hippocampal neurogenesis in the absence of microglia T cell interaction and microglia activation in the murine running wheel model. Glia 57(10):1046–1061
113. Sasaki T, Kitagawa K, Sugiura S, Omura-Matsuoka E, Tanaka S, Yagita Y, Okano H, Matsumoto M, Hori M (2003) Implication of cyclooxygenase-2 on enhanced proliferation

of neural progenitor cells in the adult mouse hippocampus after ischemia. J Neurosci Res 72(4):461–471
114. McPherson CA, Kraft AD, Harry GJ (2011) Injury-induced neurogenesis: consideration of resident microglia as supportive of neural progenitor cells. Neurotox Res 19(2):341–352
115. Bachstetter AD, Morganti JM, Jernberg J, Schlunk A, Mitchell SH, Brewster KW, Hudson CE, Cole MJ, Harrison JK, Bickford PC, Gemma C (2009) Fractalkine and CX(3)CR1 regulate hippocampal neurogenesis in adult and aged rats. Neurobiol Aging 32(11):2030–2044
116. Wolf SA, Steiner B, Akpinarli A, Kammertoens T, Nassenstein C, Braun A, Blankenstein T, Kempermann G (2009) CD4-positive T lymphocytes provide a neuroimmunological link in the control of adult hippocampal neurogenesis. J Immunol 182(7):3979–3984
117. Kiyota T, Ingraham KL, Swan RJ, Jacobsen MT, Andrews SJ, Ikezu T (2011) AAV serotype 2/1-mediated gene delivery of anti-inflammatory interleukin-10 enhances neurogenesis and cognitive function in APP + PS1 mice. Gene Ther. doi: 10.1038/gt.2011.126 (In Press)
118. Beck RD Jr, Wasserfall C, Ha GK, Cushman JD, Huang Z, Atkinson MA, Petitto JM (2005) Changes in hippocampal IL-15, related cytokines, and neurogenesis in IL-2 deficient mice. Brain Res 1041(2):223–230
119. Basu A, Krady JK, Levison SW (2004) Interleukin-1: a master regulator of neuroinflammation. J Neurosci Res 78(2):151–156
120. Gemma C, Bachstetter AD, Cole MJ, Fister M, Hudson C, Bickford PC (2007) Blockade of caspase-1 increases neurogenesis in the aged hippocampus. Eur J Neurosci 26(10): 2795–2803
121. Kuzumaki N, Ikegami D, Imai S, Narita M, Tamura R, Yajima M, Suzuki A, Miyashita K, Niikura K, Takeshima H, Ando T, Ushijima T, Suzuki T, Narita M (2010) Enhanced IL-1beta production in response to the activation of hippocampal glial cells impairs neurogenesis in aged mice. Synapse 64(9):721–728
122. Koo JW, Duman RS (2008) IL-1beta is an essential mediator of the antineurogenic and anhedonic effects of stress. Proc Natl Acad Sci U S A 105(2):751–756
123. Mathieu P, Battista D, Depino A, Roca V, Graciarena M, Pitossi F (2010) The more you have, the less you get: the functional role of inflammation on neuronal differentiation of endogenous and transplanted neural stem cells in the adult brain. J Neurochem 112(6): 1368–1385
124. Maggi L, Scianni M, Branchi I, D'Andrea I, Lauro C, Limatola C (2011) CX(3)CR1 deficiency alters hippocampal-dependent plasticity phenomena blunting the effects of enriched environment. Front Cell Neurosci 5:22
125. Pitossi F, del RA, Kabiersch A, Besedovsky H (1997) Induction of cytokine transcripts in the central nervous system and pituitary following peripheral administration of endotoxin to mice. J Neurosci Res 48(4):287–298
126. Arguello AA, Fischer SJ, Schonborn JR, Markus RW, Brekken RA, Eisch AJ (2009) Effect of chronic morphine on the dentate gyrus neurogenic microenvironment. Neuroscience 159(3):1003–1010
127. Spulber S, Oprica M, Bartfai T, Winblad B, Schultzberg M (2008) Blunted neurogenesis and gliosis due to transgenic overexpression of human soluble IL-1ra in the mouse. Eur J Neurosci 27(3):549–558
128. Vallieres L, Rivest S (1997) Regulation of the genes encoding interleukin-6, its receptor, and gp130 in the rat brain in response to the immune activator lipopolysaccharide and the proinflammatory cytokine interleukin-1beta. J Neurochem 69(4):1668–1683
129. Vallieres L, Campbell IL, Gage FH, Sawchenko PE (2002) Reduced hippocampal neurogenesis in adult transgenic mice with chronic astrocytic production of interleukin-6. J Neurosci 22(2):486–492
130. Seguin JA, Brennan J, Mangano E, Hayley S (2009) Proinflammatory cytokines differentially influence adult hippocampal cell proliferation depending upon the route and chronicity of administration. Neuropsychiatr Dis Treat 5:5–14

131. Baron R, Nemirovsky A, Harpaz I, Cohen H, Owens T, Monsonego A (2008) IFN-gamma enhances neurogenesis in wild-type mice and in a mouse model of Alzheimer's disease. FASEB J 22(8):2843–2852
132. Harry GJ (2008) Lefebvre, d.C., McPherson, C.A., Funk, J.A., Aoyama, M., Wine, R.N.: Tumor necrosis factor p55 and p75 receptors are involved in chemical-induced apoptosis of dentate granule neurons. J Neurochem 106(1):281–298
133. Cacci E, Claasen JH, Kokaia Z (2005) Microglia-derived tumor necrosis factor-alpha exaggerates death of newborn hippocampal progenitor cells in vitro. J Neurosci Res 80(6): 789–797
134. Bravo JA, Parra CS, Arancibia S, Andres S, Morales P, Herrera-Marschitz M, Herrera L, Lara HE, Fiedler JL (2006) Adrenalectomy promotes a permanent decrease of plasma corticoid levels and a transient increase of apoptosis and the expression of Transforming Growth Factor beta1 (TGF-beta1) in hippocampus: effect of a TGF-beta1 oligo-antisense. BMC Neurosci 7:40
135. Bye N, Zieba M, Wreford NG, Nichols NR (2001) Resistance of the dentate gyrus to induced apoptosis during ageing is associated with increases in transforming growth factor-beta1 messenger RNA. Neuroscience 105(4):853–862
136. Buckwalter MS, Yamane M, Coleman BS, Ormerod BK, Chin JT, Palmer T, Wyss-Coray T (2006) Chronically increased transforming growth factor-beta1 strongly inhibits hippocampal neurogenesis in aged mice. Am J Pathol 169(1):154–164
137. Wachs FP, Winner B, Couillard-Despres S, Schiller T, Aigner R, Winkler J, Bogdahn U, Aigner L (2006) Transforming growth factor-beta1 is a negative modulator of adult neurogenesis. J Neuropathol Exp Neurol 65(4):358–370
138. Graciarena M, Depino AM, Pitossi FJ (2010) Prenatal inflammation impairs adult neurogenesis and memory related behavior through persistent hippocampal TGFbeta1 downregulation. Brain Behav Immun 24(8):1301–1309
139. Battista D, Ferrari CC, Gage FH, Pitossi FJ (2006) Neurogenic niche modulation by activated microglia: transforming growth factor beta increases neurogenesis in the adult dentate gyrus. Eur J Neurosci 23(1):83–93
140. Valero J, España J, Parra-Damas A, Martín E, Rodríguez-Álvarez J, Saura CA (2011) Short-term Environmental Enrichment Rescues Adult Neurogenesis and Memory Deficits in $APP_{Sw,Ind}$ Transgenic Mice. PLoS One 6(2):e16832
141. Kempermann G, Kuhn HG, Gage FH (1997) More hippocampal neurons in adult mice living in an enriched environment. Nature 386(6624):493–495
142. Kronenberg G, Bick-Sander A, Bunk E, Wolf C, Ehninger D, Kempermann G (2006) Physical exercise prevents age-related decline in precursor cell activity in the mouse dentate gyrus. Neurobiol Aging 27(10):1505–1513
143. Llorens-Martin M, Torres-Aleman I, Trejo JL (2010) Exercise modulates insulin-like growth factor 1-dependent and -independent effects on adult hippocampal neurogenesis and behaviour. Mol Cell Neurosci 44(2):109–117
144. Brown J, Cooper-Kuhn CM, Kempermann G, van PH, Winkler J, Gage FH, Kuhn HG (2003) Enriched environment and physical activity stimulate hippocampal but not olfactory bulb neurogenesis. Eur J Neurosci 17(10):2042–2046
145. Lazarov O, Robinson J, Tang YP, Hairston IS, Korade-Mirnics Z, Lee VM, Hersh LB, Sapolsky RM, Mirnics K, Sisodia SS (2005) Environmental enrichment reduces Abeta levels and amyloid deposition in transgenic mice. Cell 120(5):701–713
146. Lahiani-Cohen I, Lourbopoulos A, Haber E, Rozenstein-Tsalkovich L, Abramsky O, Grigoriadis N, Rosenmann H (2011) Moderate environmental enrichment mitigates tauopathy in a neurofibrillary tangle mouse model. J Neuropathol Exp Neurol 70(7):610–621
147. Lyons A, Lynch AM, Downer EJ, Hanley R, O'Sullivan JB, Smith A, Lynch MA (2009) Fractalkine-induced activation of the phosphatidylinositol-3 kinase pathway attentuates microglial activation in vivo and in vitro. J Neurochem 110(5):1547–1556

148. Bachstetter AD, Pabon MM, Cole MJ, Hudson CE, Sanberg PR, Willing AE, Bickford PC, Gemma C (2008) Peripheral injection of human umbilical cord blood stimulates neurogenesis in the aged rat brain. BMC Neurosci 9:22
149. Trejo JL, Llorens-Martin MV, Torres-Aleman I (2008) The effects of exercise on spatial learning and anxiety-like behavior are mediated by an IGF-I-dependent mechanism related to hippocampal neurogenesis. Mol Cell Neurosci 37(2):402–411
150. Kohman RA, Deyoung EK, Bhattacharya TK, Peterson LN, Rhodes JS (2011) Wheel running attenuates microglia proliferation and increases expression of a proneurogenic phenotype in the hippocampus of aged mice. Brain Behav Immun (In Press)
151. Choi YS, Cho HY, Hoyt KR, Naegele JR, Obrietan K (2008) IGF-1 receptor-mediated ERK/MAPK signaling couples status epilepticus to progenitor cell proliferation in the subgranular layer of the dentate gyrus. Glia 56(7):791–800
152. Singh AK, Jiang Y (2004) How does peripheral lipopolysaccharide induce gene expression in the brain of rats? Toxicology 201(1–3):197–207
153. Mori K, Kaneko YS, Nakashima A, Nagatsu T, Nagatsu I, Ota A (2010) Peripherally injected lipopolysaccharide induces apoptosis in the subventricular zone of young adult mice. Neurosci Lett 481(2):126–130
154. Monje ML, Toda H, Palmer TD (2003) Inflammatory blockade restores adult hippocampal neurogenesis. Science 302(5651):1760–1765
155. Cacci E, Ajmone-Cat MA, Anelli T, Biagioni S, Minghetti L (2008) In vitro neuronal and glial differentiation from embryonic or adult neural precursor cells are differently affected by chronic or acute activation of microglia. Glia 56(4):412–425
156. Keene CD, Chang R, Stephen C, Nivison M, Nutt SE, Look A, Breyer RM, Horner PJ, Hevner R, Montine TJ (2009) Protection of hippocampal neurogenesis from toll-like receptor 4-dependent innate immune activation by ablation of prostaglandin E2 receptor subtype EP1 or EP2. Am J Pathol 174(6):2300–2309
157. Packer MA, Stasiv Y, Benraiss A, Chmielnicki E, Grinberg A, Westphal H, Goldman SA, Enikolopov G (2003) Nitric oxide negatively regulates mammalian adult neurogenesis. Proc Natl Acad Sci U S A 100(16):9566–9571
158. Carreira BP, Morte MI, Inacio A, Costa G, Rosmaninho-Salgado J, Agasse F, Carmo A, Couceiro P, Brundin P, Ambrosio AF, Carvalho CM, Araujo IM (2010) Nitric oxide stimulates the proliferation of neural stem cells bypassing the epidermal growth factor receptor. Stem Cells 28(7):1219–1230
159. Sun Y, Jin K, Childs JT, Xie L, Mao XO, Greenberg DA (2005) Neuronal nitric oxide synthase and ischemia-induced neurogenesis. J Cereb Blood Flow Metab 25(4):485–492
160. Hoffmann O, Mahrhofer C, Rueter N, Freyer D, Bert B, Fink H, Weber JR (2007) Pneumococcal cell wall-induced meningitis impairs adult hippocampal neurogenesis. Infect Immun 75(9):4289–4297
161. Ekdahl CT, Kokaia Z, Lindvall O (2009) Brain inflammation and adult neurogenesis: the dual role of microglia. Neuroscience 158(3):1021–1029
162. Qiu L, Zhu C, Wang X, Xu F, Eriksson PS, Nilsson M, Cooper-Kuhn CM, Kuhn HG, Blomgren K (2007) Less neurogenesis and inflammation in the immature than in the juvenile brain after cerebral hypoxia-ischemia. J Cereb Blood Flow Metab 27(4):785–794
163. Liu Z, Fan Y, Won SJ, Neumann M, Hu D, Zhou L, Weinstein PR, Liu J (2007) Chronic treatment with minocycline preserves adult new neurons and reduces functional impairment after focal cerebral ischemia. Stroke 38(1):146–152
164. Hoehn BD, Palmer TD, Steinberg GK (2005) Neurogenesis in rats after focal cerebral ischemia is enhanced by indomethacin. Stroke 36(12):2718–2724
165. Kim BJ, Kim MJ, Park JM, Lee SH, Kim YJ, Ryu S, Kim YH, Yoon BW (2009) Reduced neurogenesis after suppressed inflammation by minocycline in transient cerebral ischemia in rat. J Neurol Sci 279(1–2):70–75
166. Deierborg T, Roybon L, Inacio AR, Pesic J, Brundin P (2010) Brain injury activates microglia that induce neural stem cell proliferation ex vivo and promote differentiation of

neurosphere-derived cells into neurons and oligodendrocytes. Neuroscience 171(4): 1386–1396

167. Yan YP, Sailor KA, Vemuganti R, Dempsey RJ (2006) Insulin-like growth factor-1 is an endogenous mediator of focal ischemia-induced neural progenitor proliferation. Eur J Neurosci 24(1):45–54
168. Yan YP, Lang BT, Vemuganti R, Dempsey RJ (2009) Galectin-3 mediates post-ischemic tissue remodeling. Brain Res 1288:116–124
169. Yan YP, Sailor KA, Lang BT, Park SW, Vemuganti R, Dempsey RJ (2007) Monocyte chemoattractant protein-1 plays a critical role in neuroblast migration after focal cerebral ischemia. J Cereb Blood Flow Metab 27(6):1213–1224
170. Yan YP, Lang BT, Vemuganti R, Dempsey RJ (2009) Osteopontin is a mediator of the lateral migration of neuroblasts from the subventricular zone after focal cerebral ischemia. Neurochem Int 55(8):826–832
171. Parent JM, Yu TW, Leibowitz RT, Geschwind DH, Sloviter RS, Lowenstein DH (1997) Dentate granule cell neurogenesis is increased by seizures and contributes to aberrant network reorganization in the adult rat hippocampus. J Neurosci 17(10):3727–3738
172. Bonde S, Ekdahl CT, Lindvall O (2006) Long-term neuronal replacement in adult rat hippocampus after status epilepticus despite chronic inflammation. Eur J Neurosci 23(4): 965–974
173. Ekdahl CT, Claasen JH, Bonde S, Kokaia Z, Lindvall O (2003) Inflammation is detrimental for neurogenesis in adult brain. Proc Natl Acad Sci U S A 100(23):13632–13637
174. Yang F, Liu ZR, Chen J, Zhang SJ, Quan QY, Huang YG, Jiang W (2010) Roles of astrocytes and microglia in seizure-induced aberrant neurogenesis in the hippocampus of adult rats. J Neurosci Res 88(3):519–529
175. Jung KH, Chu K, Lee ST, Kim J, Sinn DI, Kim JM, Park DK, Lee JJ, Kim SU, Kim M, Lee SK, Roh JK (2006) Cyclooxygenase-2 inhibitor, celecoxib, inhibits the altered hippocampal neurogenesis with attenuation of spontaneous recurrent seizures following pilocarpine-induced status epilepticus. Neurobiol Dis 23(2):237–246
176. Zhang HJ, Sun RP, Lei GF, Yang L, Liu CX (2008) Cyclooxygenase-2 inhibitor inhibits hippocampal synaptic reorganization in pilocarpine-induced status epilepticus rats. J Zhejiang Univ Sci B 9(11):903–915
177. Rola R, Raber J, Rizk A, Otsuka S, VandenBerg SR, Morhardt DR, Fike JR (2004) Radiation-induced impairment of hippocampal neurogenesis is associated with cognitive deficits in young mice. Exp Neurol 188(2):316–330
178. Mizumatsu S, Monje ML, Morhardt DR, Rola R, Palmer TD, Fike JR (2003) Extreme sensitivity of adult neurogenesis to low doses of X-irradiation. Cancer Res 63(14):4021–4027
179. Ramanan S, Kooshki M, Zhao W, Hsu FC, Riddle DR, Robbins ME (2009) The PPARalpha agonist fenofibrate preserves hippocampal neurogenesis and inhibits microglial activation after whole-brain irradiation. Int J Radiat Oncol Biol Phys 75(3):870–877
180. Hellstrom NA, Lindberg OR, Stahlberg A, Swanpalmer J, Pekny M, Blomgren K, Kuhn HG (2011) Unique gene expression patterns indicate microglial contribution to neural stem cell recovery following irradiation. Mol Cell Neurosci 46(4):710–719
181. Enwere E, Shingo T, Gregg C, Fujikawa H, Ohta S, Weiss S (2004) Aging results in reduced epidermal growth factor receptor signaling, diminished olfactory neurogenesis, and deficits in fine olfactory discrimination. J Neurosci 24(38):8354–8365
182. Drapeau E, Mayo W, Aurousseau C, Le MM, Piazza PV, Abrous DN (2003) Spatial memory performances of aged rats in the water maze predict levels of hippocampal neurogenesis. Proc Natl Acad Sci U S A 100(24):14385–14390
183. Hattiangady B, Shetty AK (2008) Aging does not alter the number or phenotype of putative stem/progenitor cells in the neurogenic region of the hippocampus. Neurobiol Aging 29(1): 129–147
184. Jin K, Sun Y, Xie L, Batteur S, Mao XO, Smelick C, Logvinova A, Greenberg DA (2003) Neurogenesis and aging: FGF-2 and HB-EGF restore neurogenesis in hippocampus and subventricular zone of aged mice. Aging Cell 2(3):175–183

185. Lichtenwalner RJ, Forbes ME, Bennett SA, Lynch CD, Sonntag WE, Riddle DR (2001) Intracerebroventricular infusion of insulin-like growth factor-I ameliorates the age-related decline in hippocampal neurogenesis. Neuroscience 107(4):603–613
186. Kempermann G, Kuhn HG, Gage FH (1998) Experience-induced neurogenesis in the senescent dentate gyrus. J Neurosci 18(9):3206–3212
187. Encinas JM, Michurina TV, Peunova N, Park JH, Tordo J, Peterson DA, Fishell G, Koulakov A, Enikolopov G (2011) Division-coupled astrocytic differentiation and age-related depletion of neural stem cells in the adult hippocampus. Cell Stem Cell 8(5): 566–579
188. Gemma C, Bachstetter AD, Bickford PC (2010) Neuron-Microglia Dialogue and Hippocampal Neurogenesis in the Aged Brain. Aging Dis 1(3):232–244
189. Luo XG, Ding JQ, Chen SD (2010) Microglia in the aging brain: relevance to neurodegeneration. Mol Neurodegener 5:12
190. Klempin F, Kempermann G (2007) Adult hippocampal neurogenesis and aging. Eur Arch Psychiatry Clin Neurosci 257(5):271–280
191. Sierra A, Gottfried-Blackmore AC, McEwen BS, Bulloch K (2007) Microglia derived from aging mice exhibit an altered inflammatory profile. Glia 55(4):412–424
192. Petry KG, Brochet B, Dousset V, Vignes JR, Boiziau C (2010) Inflammation induced neurological handicap processes in multiple sclerosis: new insights from preclinical studies. J Neural Transm 117(8):907–917
193. Curtis MA, Penney EB, Pearson AG, van Roon-Mom WM, Butterworth NJ, Dragunow M, Connor B, Faull RL (2003) Increased cell proliferation and neurogenesis in the adult human Huntington's disease brain. Proc Natl Acad Sci U S A 100(15):9023–9027
194. Gil-Mohapel J, Simpson JM, Ghilan M, Christie BR (2011) Neurogenesis in Huntington's disease: can studying adult neurogenesis lead to the development of new therapeutic strategies? Brain Res 1406:84–105
195. Hsiao HY, Chern Y (2010) Targeting glial cells to elucidate the pathogenesis of Huntington's disease. Mol Neurobiol 41(2–3):248–255
196. Depino AM, Earl C, Kaczmarczyk E, Ferrari C, Besedovsky H, del RA, Pitossi FJ, Oertel WH (2003) Microglial activation with atypical proinflammatory cytokine expression in a rat model of Parkinson's disease. Eur J Neurosci 18(10):2731–2742
197. Pott Godoy MC, Tarelli R, Ferrari CC, Sarchi MI, Pitossi FJ (2008) Central and systemic IL-1 exacerbates neurodegeneration and motor symptoms in a model of Parkinson's disease. Brain 131(Pt 7):1880–1894
198. Selkoe DJ (2002) Alzheimer's disease is a synaptic failure. Science 298(5594):789–791
199. McGeer EG, McGeer PL (2010) Neuroinflammation in Alzheimer's disease and mild cognitive impairment: a field in its infancy. J Alzheimers Dis 19:355–361
200. Rodriguez JJ, Verkhratsky A (2011) Neurogenesis in Alzheimer's disease. J Anat 219(1): 78–89
201. Choi SH, Veeraraghavalu K, Lazarov O, Marler S, Ransohoff RM, Ramirez JM, Sisodia SS (2008) Non-cell-autonomous effects of presenilin 1 variants on enrichment-mediated hippocampal progenitor cell proliferation and differentiation. Neuron 59(4):568–580
202. Butovsky O, Koronyo-Hamaoui M, Kunis G, Ophir E, Landa G, Cohen H, Schwartz M (2006) Glatiramer acetate fights against Alzheimer's disease by inducing dendritic-like microglia expressing insulin-like growth factor 1. Proc Natl Acad Sci U S A 103(31): 11784–11789

Chapter 15
Immunosuppressive Properties of Mesenchymal Stromal Cells

Francesco Lanza, Diana Campioni, Endri Mauro, Annalisa Pasini and Roberta Rizzo

Abstract The main interest in mesenchymal stromal cells (MSCs) is correlated with their ability to suppress the proliferation of T lymphocytes induced by mitogenic agents and alloantigens which regulate the transplantation rejection. Moreover, MSCs are resistant to the CD8+ T lymphocyte cytotoxicity, they are able to inhibit the differentiation of dendritic cells responsible for the antigen presentation, the proliferation, and antibody production of B lymphocytes and they stimulate the formation of regulatory T cells. The mechanisms at the basis of MSCs activity need cell–cell interaction and the secretion of soluble molecules induced by the micro-environment. The inhibitory functions of MSCs involve several soluble molecules as hepatocyte growth factor, transforming growth factor-beta, interleukin-10 and -2, tumor necrosis factor-alpha, prostaglandin E2, indoleamine 2,3-dioxygenase (IDO), and soluble HLA-G antigens. A large consensus has been obtained on the immuno-modulatory role of IDO and HLA-G molecules.

15.1 Introduction

Human multipotent mesenchymal stromal cells (hMSCs), first described by Friedenstein in 1970 [1] as nonhematopoietic cell precursors with osteogenic potential, represent stem cells for nonhematopoietic tissues [2]. The functional and

F. Lanza (✉) · E. Mauro · A. Pasini
Hematology and Stem Cell Unit, Cremona, Italy
e-mail: f.lanza@ospedale.cremona.it

D. Campioni
Hematology Section, University of Ferrara, Ferrara, Italy

R. Rizzo
Microbiology Section, University of Ferrara, Ferrara, Italy

H. Baharvand and N. Aghdami (eds.), *Advances in Stem Cell Research*,
Stem Cell Biology and Regenerative Medicine, DOI: 10.1007/978-1-61779-940-2_15,

immunophenotic characteristics of stromal cells have been recently reviewed and redefined by International Society for Cellular Therapy (ISCT) [3]: they must be defined as hMSCs, which are a plastic adherent cell population that retain in vitro clonogenic potential (defined by the presence of the fibroblast-colony forming unit, (CFU-F) [4, 5] (Fig. 15.1a, b), capable of supporting hematopoiesis [6, 7] and having a differentiation capacity toward a number of different cell types (osteoblasts, chondrocytes, adipocytes, myocytes) [8–11]. Bone marrow (BM)-derived hMSCs have been described to form in vitro stromal layers also with other adherent nonhemopoietic cells such as endothelial cells or with the transiently adherent macrophages (Fig. 15.1b, c). In vitro and in vivo studies also showed that hMSC could differentiate into cells of nonmesodermal origin such as neurons, skin and gut epithelial cells, hepatocytes, and pneumocytes [12].

By using multicolor flow cytometry, the existence of morphologically distinct hMSCs subsets has been proposed, further allowing the identification of a functional subpopulation of rapidly in vitro self renewing (RS1) MSC [13]. Recent data using carboxyfluorescein succimidyl ester (CFDA-SE) [14], that is a lipophilic dye that irreversibly couples to cellular proteins and determine the division history of cell populations undergoing proliferation, partially confirmed Colter findings, and suggested the presence of different MSC subpopulations having a heterogeneous proliferative pattern. Moreover, these functionally defined hMSC subsets displayed a distinct gene profile [15].

MSC are also capable of expressing genes of embryonic origin, cell–cell contact molecules, different extracellular matrix proteins, collagen, and fibronectin. hMSC may also secrete chemokine ligands (including CCL2,CCL4, CCL5, CCL20, CXCL8.) and interleukins (SDF-1, IL-6, IL-7, IL-8, IL-10, IL-11, IL-12, SCF, etc.) that regulates microenvironmental homeostasis. hMSC can produce leukemia inhibitory factor, granulocyte-colony stimulating factor, granulocyte–macrophage colony stimulating factor [16, 17].

hMSC are initially isolated and characterized from the BM, but could be also obtained from other different sources such as amniotic membrane, skin, adipose tissue, cord blood, fetal liver, placenta, and synovium [18]. Whereas MSC have historically been thought to be the same general population of cells regardless of the tissue source, recent data suggest that hMSC gene expression profile reflects their tissue origin, indicating that hMSC tissue heterogeneity is biologically relevant [19].

15.2 Immunophenotypic Characteristics of MSC

The isolation of hMSC from primary tissues is hampered by the limited selectivity of available markers while cultured hMSC may be defined only by a combination of nonspecific immunophenotypic markers. The in vivo phenotype of hMSC has been difficult to establish because of the exceptionally low frequency estimated on about 0.01–0.1%. Moreover, there are no universally accepted antigenic

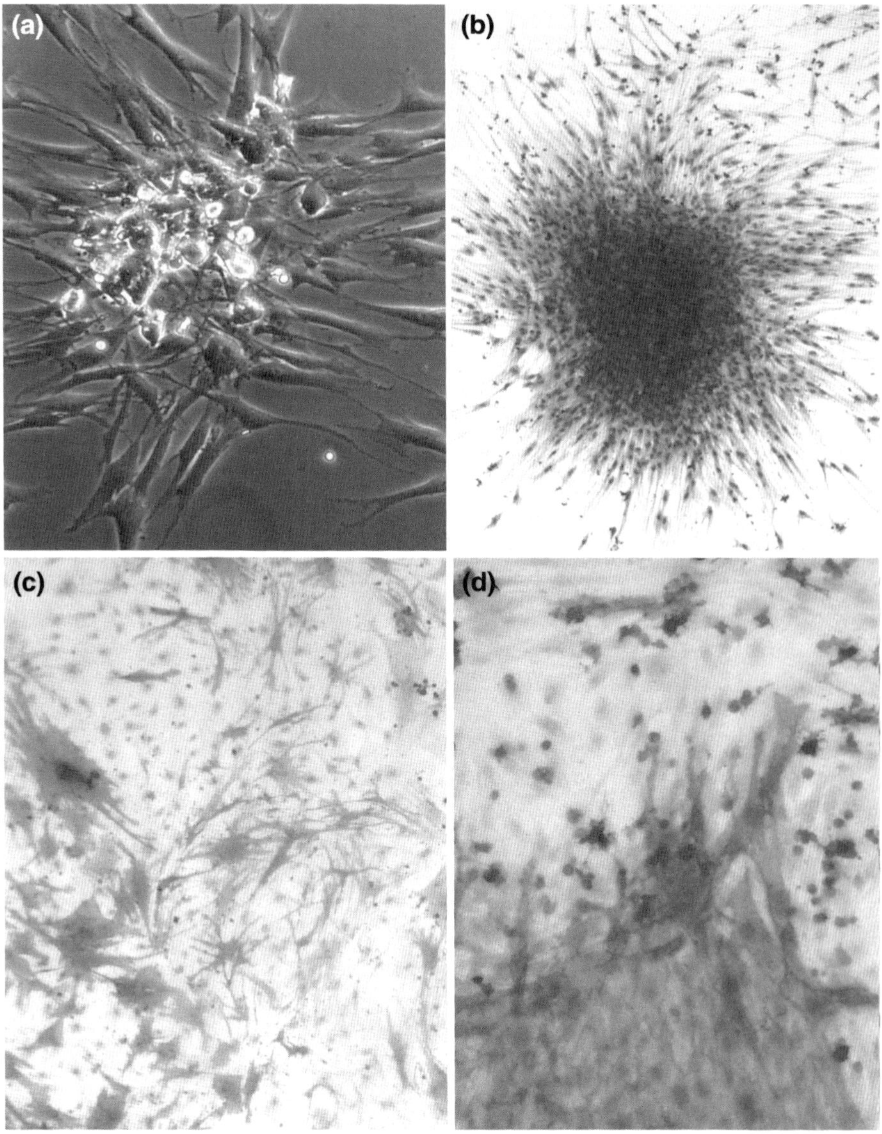

Fig. 15.1 A classical BM-derived CFU-F growing in culture (**a**) and stained with May-Grünwald Giemsa is reported (**b**). Alkaline-phosphatase (ALP) positive BM-MSC (*blue* cells) growing in mixed culture after ex vivo expansion with BM-derived endothelial cells that resulted ALP negative (**c**). BM-derived endothelial cells (*red* cells positive to CD31 antigen) have been observed in ex vivo expanded MSC monolayers (*blue* cells) (**d**)

determinants for the phenotypic characterization and isolation of hMSC. Only few markers have been developed and proved to be suitable for hMSC isolation. Functionally isolated MSC do not provide any information about the antigenic composition of the starting population. A variety of surface markers are exclusively found on cultured hMSC but not on their primary counterparts (such as CD271 or CD56 that are highly expressed on primary hMSC progenitor cells but rapidly downregulated in cultured hMSC). In 1990, Simmons and colleagues identified a stromal cell precursors in human BM by a "novel" monoclonal antibody, STRO-1, in a murine model [20]. Most of the monoclonal antibodies now available recognize also normal fibroblasts and mesenchymal derived stromal cells [21, 22]. No standardized protocols exist, nevertheless several protocols have been used to facilitate the prospective isolation of hMSC [23]. To identify "naïve" hMSC, successfully positive selection with antibodies against surface molecules such as CD49 [24] frizzled-9 (FZD-9, Wnt receptor) CD349, CD56, CDCP1, CD73, [25, 26] and in particular CD105, D7Fib [21], were used. Isolation and characterization of CD146 positive hMSC have been also analyzed revealing their specific miRNA expression profile [26]. Some studies showed that hMSC may be enriched by negative selection based on the depletion of CD45, CD14, CD34, and/or glycophorin A (CD235) markers. The CD271 seems to be one of the most specific markers [27] especially in combination with the CD56, and MSCA-1 [24] to purifty nonexpanded MSC progenitor cells (Fig. 15.2). Using high sensitive flow cytometry, other authors have demonstrated that SSEA-1 and SSEA-4 could identify BM-hMSCs [28], suggesting how these markers may be prospectively used to identify the most primitive progenitors with embryonic stem cell-like-features. More recently, it has been shown that mesenchymal and hemopoietic cells form a unique BM niche in which nestin positive cells are hMSC [29] and that CD146/CD271 expression on primary nonhematopoietic BM stem cells is correlated with in situ localization [30]. BM-MSC precursors express also a neural ganglioside (GD2) [31] that is not expressed by hemopoietic cells.

The immunophenotype of ex vivo expanded hMSC is currently defined by ISCT that between 2005 and 2006 has established the nomenclature and the minimal criteria for defining mesenchymal multipotent stromal cells [3] as showed in Table 15.1.

So far, immunophenotypic analysis of cultured hMSC from most of the laboratories demonstrated that hMSC fulfilled general immunophenotypical criteria as stated by ISCT as above reported [3] hMSC are reported to be uniformly positive for CD90, CD105, CD73, and negative for stem cell antigen CD34 and CD45 panleukocyte marker, and also for monocytic and lymphocytic cell antigens such as CD14, CD19, CD79a, HLA-DR by using flow cytometry. Other studies demonstrated the hMSC positivity for other antigens such as adhesion molecules (CD29, CD106, CD105, CD166, CD36), extracellular matrix protein (CD90, CD44), hemopoietic markers (CD10, CD59), histocompatibility antigens (HLA-I and II class) [21, 22], chemokine receptors (CD210, IL-10 receptor, CD184-SDF-1 receptor) [32], neural and endothelial receptors (CD271, CD146). Only few studies analysed hMSC phenotype by a multiparametric cytofluorimetric approach

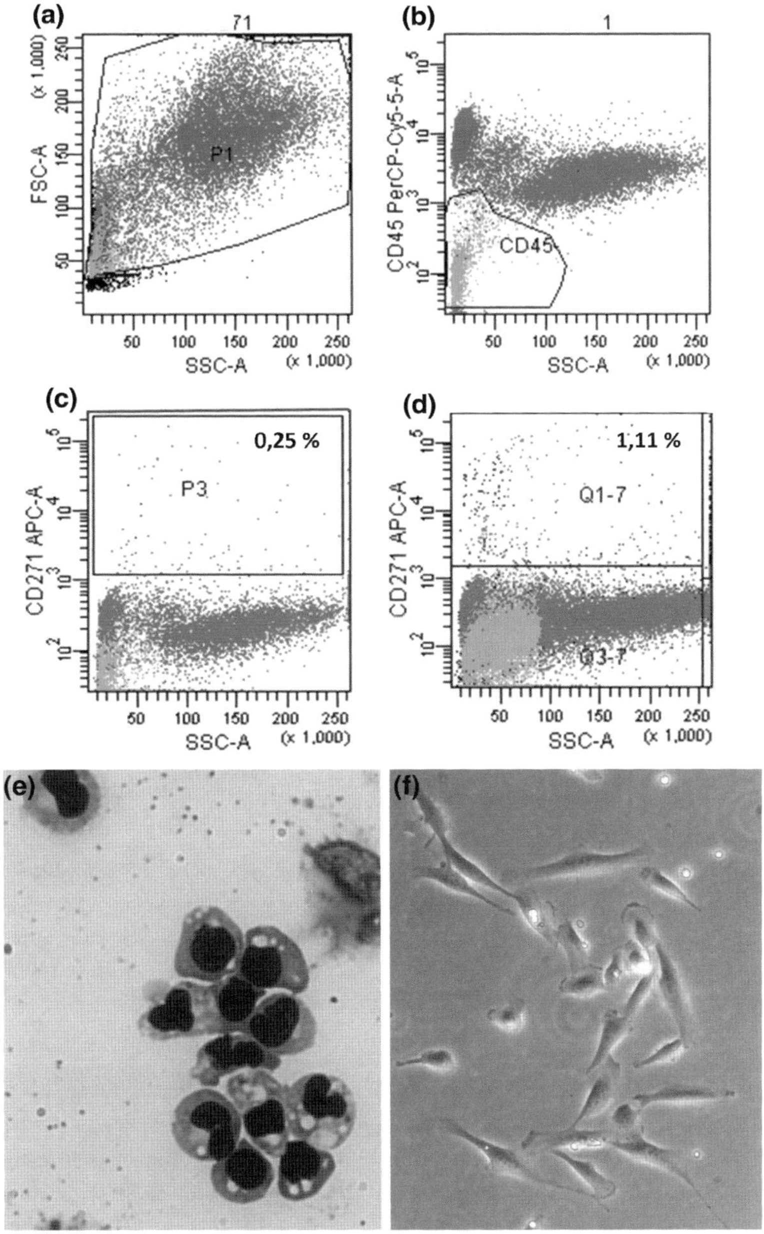

Fig. 15.2 A flow cytometric protocol for the study of nonexpanded MSC is shown. The putative MSC progenitor cells in BM fresh samples were gated on viable cells (P1), CD45medlow (P2), and CD271 (P3) positive fraction (**a**, **b**). **c** and **d** show two distinct patients affected by HM having a different percentage of CD271 positive MSC progenitor cells. Selected CD271+ positive cells from BM fresh samples are visible after staining with May-Grünwald Giemsa (**e**) and during the in vitro culture (**f**)

Table 15.1 Minimal criteria to define MSC as established by ISCT

Phenotype positive (≥95%)	Phenotype negative (≤2%)
CD105	CD45
CD73	CD34
CD90	CD14 o CD11b
	CD79α o CD19
	HLA-DR

Adherence to plastic
Specific surface antigen (Atg) expression
Multipotent differentiation potential: osteoblasts, adipocytes, chondrocytes (demonstrated by staining of in vitro cell culture)

[33, 34]. Standardized protocol in functional and phenotypic hMSC analysis are needed. A 4 colors cytofluorimetric protocol to study the immunophenotype of hMSC has been recently proposed [35, 36]. From these studies it was stressed that CD45+ positive cells as well as endothelial cells should be gated out from hMSC analysis (Fig. 15.1). Furthermore, in hematological malignancies (HM) blast cells resulted sometimes strictly adherent to in vitro expanded BM-derived MSC, thus rendering difficult the assessment of hMSC phenotype. (Fig. 15.3). Moreover, the use of 7-AAD (7-aminoactinomycin) or Cytox 13 to identify nonviable cells is recommended [37].

An emerging paradigm is that hMSCs could have key functional roles in the tissue in which they reside but phenotypic and functional characteristics under their ex vivo expansion conditions including different isolation methods, culture protocols: use of serum free, platelets lysate, additional cytokines, density, growth supplements, passages, effects of cryopreservation are still not elucidated [38–41]. For example the hMSC tissue source (normal vs. pathologic, and different tissues, donors variability, age in culture) may be also critically important in determining their biological activity (heterogeneity of gene expression) [42]. All these variables might have implications on the selection of cell type generating possible functionally and phenotypically distinct hMSC subsets and altering the hMSC clonogenic and plastic potential [43]. These observations suggest that the cell processing protocols can be modified to enhance or repress expression of specific gene in order to optimize the cytokine profile for a given clinical indication. Standardized protocol has to be developed assuring that the manufactured cells behave solely in the clinical intended purpose [44, 45] and do not exert adverse effects such as for example, uncontrolled differentiation or transformation.

The challenge for scientist is translating research into clinical scale manufacturing of mesenchymal stromal cells [46] even though this is still completely not achieved.

Present definitions of hMSC emphasize generic functional properties of these cells and fail to distinguish for example these cells from generic fibroblasts and to detect subsets of stromal cells with specialized niche functions [47, 48].

Emerging data focused on hMSC immunophenotype with special regards to the expression of Thy-1 (CD90) molecule, and in a lesser extent to the LNGFR

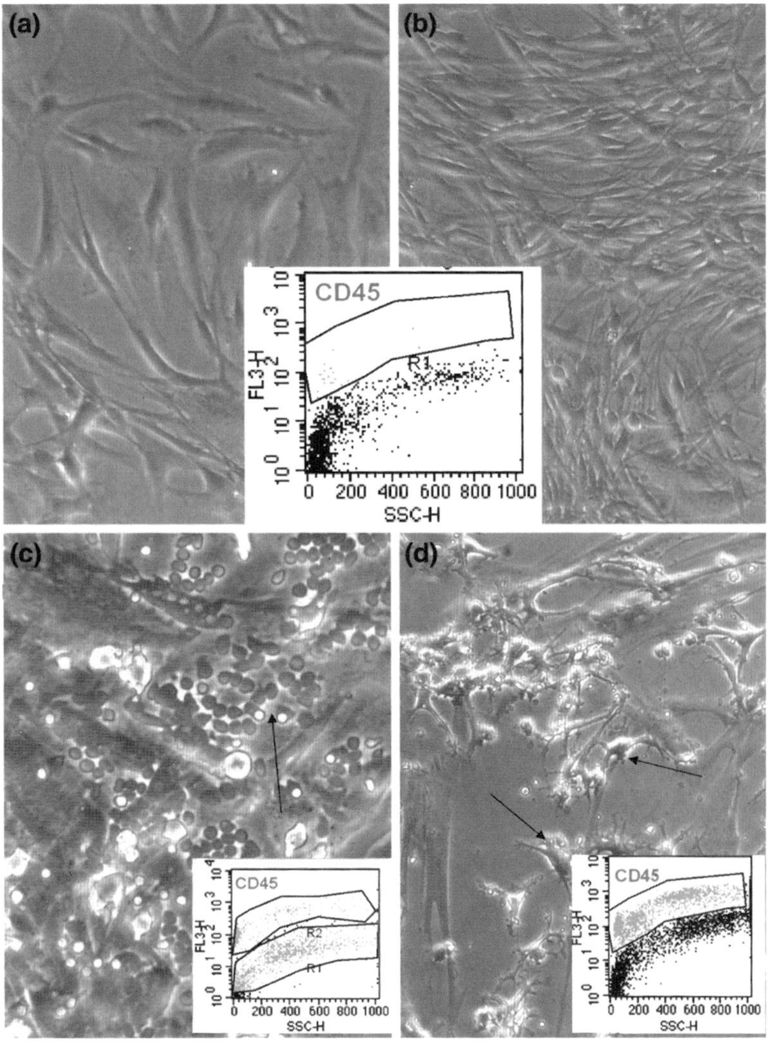

Fig. 15.3 In **a** and **b** the same BM-derived MSC from a normal sample are cultured in two different media: in **a** with serum and in **b** without serum but with different type of cytokines. After 20 days of in vitro culture the percentage of CD45+ positive hematopoietic cells is reduced as showed in the little cytometric *plot*. In **c** and **d** two different examples of in vitro expanded BM-derived MSC from patients with respectively a chronic lymphoprolipherative disorder and acute myeloid leukemia where blasts are strictly adherent to the autologous MSC layer after 30 days of culture and different changes of the medium

(CD271), CD105, CD44, CD10, CD146 molecules expression since they could give informations about different hMSC specialization of function. For example differences related to the expression of CD105, CD106, CD44 by cultured human BM-derived hMSCs from hematological patients (HM-MSCs) as compared to normal samples (NS) and Skin fibroblasts (SF) seem to be correlated to a different differentiation, and adhesion capacity by hMSC [49, 50].

These data may have therapeutic implication, since the use of autologous hMSCs has been proposed in a wide range of clinical applications including those in the area of the regenerative medicine and cell therapies.

15.3 Immunological Functions of MSC

In the last years several studies have demonstrated the peculiar immunological characteristics of hMSCs, as low antigenicity and high immunomodulatory ability, suggesting their clinical use to counteract graft versus host disease (GvHD), one of the most frequent and severe complication in allogeneic BM transplantations [51–53]. hMSCs share surface markers with thymic epithelial cells and express adhesion molecules for T lymphocyte interaction, as vascular adhesion molecule 1 and intracellular adhesion molecule 2. The low antigenicity of hMSCs is mainly caused by the low expression of classical human leukocyte antigen (HLA) class I molecules and the complete absence of HLA class II antigens and co-stimulatory molecules as CD80 (B7-1), CD86 (B7-2), and CD40 [52]. The main interest in hMSCs is correlated with their ability to suppress the proliferation of T lymphocytes induced by mitogenic agents and alloantigens which regulate the transplantation rejection [54]. Moreover, hMSCs are resistant to the CD8+ T lymphocyte cytotoxicity, they are able to inhibit the differentiation of dendritic cells responsible for the antigen presentation, the proliferation, and antibody production of B lymphocytes and they stimulate the formation of regulatory T cells. These hematopoietic precursors are able to mediate a dose-dependent tolerogenic activity toward both innate and adaptative immunity. The mechanisms at the basis of hMSCs activity need cell–cell interaction and the secretion of soluble molecules induced by the micro-environment. The inhibitory functions of hMSCs involve several soluble molecules as hepatocyte growth factor, transforming growth factor-beta (TGF-beta), interleukin-10 and -2 (IL-10, IL-2), tumor necrosis factor-alpha (TNF-alpha), prostaglandin E2 (PGE2), indoleamine 2,3-dioxygenase (IDO) [55, 56], and soluble HLA-G antigens [53]. A large consensus has been obtained on the immuno-modulatory role of IDO and HLA-G molecules (Fig. 15.4) described as follows:

15.3.1 Indoleamine 2,3-dioxygenase

IDO is an immuno-modulatory enzyme that, with the epathic enzyme tryptophan 2,3 dioxygenase, catalyzes the degradation of the essential amino acid L-tryptophan to N-formylkynurenine [55]. The physiological role of IDO is not yet well understood, but its ubiquitary distribution, its up-modulation by several cytokines,

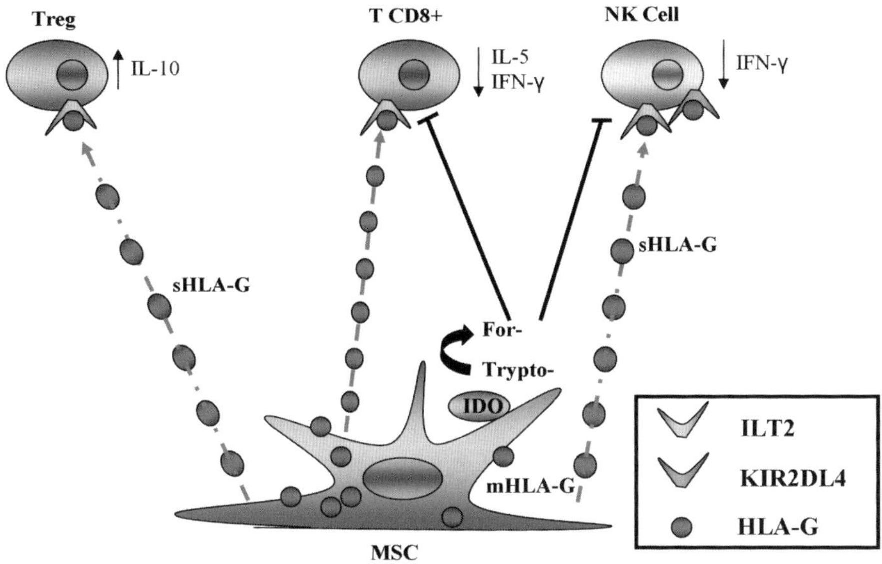

Fig. 15.4 MSCs effect on immune system cells mediated by IDO and HLA-G molecules. Membrane (m) and soluble (s) HLA-G molecules from MSCs is able to inhibit the lytic activity of natural killer (NK) cells and their secretion of interferon-gamma (IFN-γ), the cytotoxic activity of CD8+ T cells and their secretion of IFN-γ and interleukin-5 (IL-5), to induce regulatory T cells (T reg) and their secretion of interleukin-10 (IL-10). IDO from MSCs creates an immunosuppressive micro-environment inhibiting NK cell and CD8+ T lymphocyte cytotoxicity and cytokine secretion

as IFN-γ and the ability to control the tryptophan levels, an essential amino acid for cellular survival, increases the interest for this enzyme. In antigen presenting cells the expression of IFN-γ-induced IDO converts tryptophan into formylkynurenine with an immuno-suppressive effect toward T lymphocytes inducing the inhibition of their proliferation and activation [56]. MSCs present a costitutive expression of IDO. The presence of hMSCs and IDO in mixed lymphocyte reactions inhibits lymphocyte proliferation, which is partially reactivated adding tryptophan to cell cultures [57]. The partial response of lymphocyte after tryptophan treatment is due to the production of metabolites by IDO activity which contrasts T cell proliferation [58]. The immuno-modulatory role of IDO is confirmed by the abrogation of rat heart transplant survival after administration of 1-methyl tryptophan which inhibits IDO activity [59]. These data suggest IDO as a soluble factor implicated in the complex immuno-modulatory system of hMSCs. Spaggiari et al. [60] have demonstrated that hMSCs are able to inhibit the proliferation of natural killer (NK) cells, the expression of activatory NK cell receptors (NKp44, NKp30, NKG2D) and the induction of cytotoxic activity and cytokine production. The inhibition of IFN-γ receptor 1 (IFN-γ R1) on hMSCs and the consequent down-modulation of IDO production is not able to affect completely the hMSCs ability to control lympho-proliferation, suggesting the presence of alternative ways for the hMSC immuno-modulatory functions [61].

15.3.2 HLA-G Molecule

HLA-G antigen is a nonclassical HLA class I molecule with immuno-modulatory functions. It is characterized by seven mRNA splicing isoforms, four membrane-bound (HLA-G1,G2, G3, G4) and three soluble (HLA-G5, G6, G7). Both soluble and membrane-bound isoforms are able to inhibit several immune functions as lytic and cytotoxic activity of NK cells and T CD8+ lymphocytes, the maturation of dendritic cells, the alloproliferation of CD4+ T lymphocytes, and to induce the formation of regulatory T cells. HLA-G molecules differ from classical HLA antigens for their lower allelic polymorphism (47 alleles) and the limited tissue distribution which is restricted to cytotrophoblasts and thymus in physiological conditions. HLA-G expression is induced in pathological conditions, as tumors, organ transplantation, viral infections [62]. Taking into consideration the tolerogenic functions of HLA-G molecules, they could be a good candidate as a soluble factor for hMSCs immuno-modulatory activity. Götherström et al. [63] have demonstrated the presence of HLA-G mRNA in both fetal and adult hMSCs. The modulation of HLA-G has been documented on the surface of hMSCs [64] and in their culture supernatants [53]. Several studies have confirmed that the cell–cell contact between hMSCs and T lymphocyte induces the secretion of HLA-G molecules and IL-10, a cytokine that is able to up-modulate HLA-G production. The use of neutralizing antibodies against HLA-G and IL-10 has demonstrated the importance of soluble HLA-G, induced by IL-10, for the suppressive effect of hMSCs toward cell proliferation [53], cytotoxic activity, and IFN-γ secretion by NK cells, and the ability to induce regulatory T $CD4^{+}CD25^{high}FoxP3^{+}$ cells formation [65]. These data suggest a fundamental role for HLA-G antigens in the tolerogenic function of hMSCs. Interestingly Ning et al. [66] have reported a correlation between co-transplantation of hMSCs and the increase of hematologic disease relapse. It is known that the modulation of HLA-G molecules is considered as a favorable event in pregnancy and organ transplantation, where the regulation of the cell-mediated immune response is essential for a positive outcome. On the contrary, the expression of HLA-G molecules is associated with a negative follow-up in tumors and viral infections, where the tolerogenic function of HLA-G molecules allows the mutated/infected cells to escape the immune system. The association between hMSCs co-transplant and tumor relapse could be due to the ability of HLA-G to increase the risk of cancer [67]. These considerations support the necessity to evaluate "a priori" the immuno-modulatory activity of hMSCs and the production of HLA-G could be a good biomarker for this purpose [68].

Different factors could be responsible for the induction of HLA-G expression by hMSCs. Progesterone, a steroid hormone involved in the female menstrual cycle, pregnancy, and embryogenesis, is able to up-regulate the HLA-G expression by hMSCs [69]. These data could explain the ability of progesterone in regulating uterine immune function in ways that allows for inhibition of immune responses at the utero-placental interface without systemic immunosuppression.

hMSCs are newly detected targets for immunomodulatory activity and HLA-G molecules could be an important factor of this mechanism (Fig. 15.4). The available data suggest that there is a redundancy in the mechanisms of immunosuppression by hMSCs and only a complete comprehension of these suppressor mechanisms will offer an insight into the clinical use of these cells in human therapy.

15.4 Immunophenotype and Immunoregulation

Since differences in molecules' expression could give informations about different specialization of function and hMSC heterogeneity, some authors have focused on the hMSC immunophenotype in relation to their immunomodulant properties. The low antigenicity of hMSCs is mainly caused by the low expression of classical HLA class I molecules and the complete absence of HLA class II antigens and co-stimulatory molecules as CD80 (B7-1), CD86 (B7-2), and CD40 [52]. Nevertheless the expression of class II HLA-DR molecule on BM-derived hMSCs from patients with lymphoprolipherative disorders has been described [70]. Moreover, other studies reported up-regulation of MHC class II molecules only after pretreatment with IFNy and TNF-a or FGF [71]. On the other hand, the down-regulation of some antigens expression seems to be associated with multilineage differentiation in cord blood-derived hMSC [49]. Very interesting is the finding that (BM) derived hMSCs from HM isolated and cultured under in vitro angiogenic conditions exhibit a depressed expression of CD90 antigens that resulted associated with high in vitro proliferative rate but with a diminished immunosuppressive cell activity on T cell proliferation [72] suggesting the Thy-1 molecule may be considered a marker implicated in the hMSC inhibitory ability and might cooperate with HLA-G molecule in regulating suppressive versus stimulatory properties of hMSC.

The immunomodulant role of some hMSC immunophenotypic subsets has been recently demonstrated. For example, the CD105+, Stro-1, and CD271+ positive fraction displayed immunomodulant properties [73, 74].

15.5 MSC Clinical Application: an Overview

Based on their multipotential capabilities and immunosuppressive effects on T lymphocytes proliferation [53, 75], hMSCs could be used and expanded for clinical applications in cell therapy as well as in regenerative medicine [76–80]. However, a number of issues still need to be solved before its widespread use in the clinics, namely a very low percentage of engraftment after systemic infusion, and the absence of definitive evidences about a therapeutic effect. In fact, it is still unclear whether hMSCs contribute to tissue repair by differentiation into tissue-specific cell types, or mediate a paracrine effect [81]. Secretion of soluble mediators seems to be the predominant mechanism of action after systemic infusion of hMSC. Recent evidences [81] sustain the hypothesis about a paracrine role of hMSC that are able to: (1) differentiate and populate the resident tissue, giving them a therapeutic potential for regenerative medicine; (2) secrete cytokines or other soluble mediators, (3) serve as a vehicle for delivery of proteins and gene therapy.

To date, administration of hMSC proved to be safe and also efficacious in a variety of disorders. hMSC are increasingly used in many preclinical as well as clinical settings. Many animal models have been used to assess the in vivo hMSC effect on tissue transplantation and autoimmunity.

Ectodermal regeneration has been described for hMSC that can differentiate into **epidermal** cells.

BM-derived hMSC have shown **neural** differentiation by different experimental protocols (chemical agents, growth factors, co-culture with other neuronal cells) [82]. When used in vivo hMSC can survive and differentiate into neural–glial cells. Neurological recovery has been shown in animal models of Parkinson's disease [83], hypoxic–ischemic neural damage. Recent data provide remarkable cues regarding the potential of hMSC in promoting endogenous reparative mechanisms that may prove applicable and promising therapeutic applications in Parkinson's disease [84, 85]. hMSC regenerative activity in humans has been described in the treatment of autoimmune diseases characterized by autoimmune axonal demyelination associated with inflammation [multiple sclerosis and experimental encephalomyelitis (EAE)]. Nevertheless, in these cases the beneficial effect seems to be related to the inhibition of autoreactive cells in lymphoid organs while the direct contribution to neuroregeneration in chronic EAE is still controversial [86].

The role of hMSC in *mesodermal regeneration*, especially bones, cartilage, tendons, and muscles has been described. hMSC in vivo differentiation toward **osteoblastic and chondrogenic** pathway was demonstrated by use of combination of hMSC plus biomaterials and heterotopic implantation and directly in human after infusion, suggesting an important role for these cells in bone and cartilage repair and in the treatment of osteoarthritis [76, 78].

BM-derived hMSC can differentiate in vitro in **cardiomyocytes** if treated with 5-azacytidine and once injected in an infarcted heart, they reduce infarct size. Spontaneous differentiation of adipose tissue-derived hMSC into beating cardiomyocytes has been also observed [87]. Moreover, human cardiac and BM-derived stromal cells exhibit distinctive properties in cardiac repair related to their origin [88] and their pre-activation demonstrated greater cytoprotection than unmodified hMSCs [89]. They could be also implanted in vascular prosthesis or scaffold ameliorating the restoration of vascular wall [90]. The beneficial effects of hMSC therapy in cardiovascular disease may involve multiple mechanisms: hMSC secrete a large amount of factors that could help resident cells to survive and could promote angiogenesis. hMSC can also differentiate and/or stimulate endothelial cell proliferation suggesting that paracrine effects are more likely responsible despite infrequent cellular fusion or differentiation or limited engraftment. Recently, induced pluripotent stem cell-derived cardiovascular progenitor cells represent a suitable autologous cell source for myocardial regeneration as they have the capability to form myocardial cells and to contribute to revascularization [91].

hMSC are also involved in *endodermal regeneration* since they can be induced into **hepatic differentiation** and may serve as alternative for hepatocyte transplantation, cell-based therapy for liver injury, and preclinical drug testing [92]. hMSC may help in inducing regeneration and/or proliferation of resident insulin-producing cells in **diabetes** disease models [93].

The use of hMSC in many **kidney disorders** (involving both ischemic/inflammatory and immunological injury) has been considered as a clinically relevant solution as alternative to pharmacologic agents that target only a single event or pathway in the pathophysiology of a given disease. In contrast to conventional therapy hMSC use can promote cellular repair and tissue remodeling. Infusion of hMSC enhanced recovery and renal function [94, 95].

15.5.1 hMSC and Tissue Engineering

hMSCs-mediated tissue regeneration is a promising approach for tissue repair. The use of hMSC in three-dimensional (3D) scaffolds is promising even though it is limited by the need for an ideal scaffolds. The ideal scaffold is expected to be biocompatible and biodegradable mimicking the structure and function of the native extracellular matrix. The hMSC loaded on 3D scaffold are allowed to interact with the damaged area, wherein they generate and replace the tissue especially in the presence of specific molecules that could promote cell signaling, proliferation, and differentiation. For example, BMP-2 or bFGF growth factor are often associated with ceramics supports, or injected in a carrier material such as poly-lactic-co-glycolic acid, or gelatin, and collagen thus improving bone and tendon formation [96].

15.5.2 MSC and Clinical Application in Gene Therapy

The ability to transfer genes into hMSC and the ability of MSC to migrate to a lesion provided evidences for a possible use of hMSC in clinical application. MSC was successfully transfected with BMP-2 and injected in articular fractures [97]. Transplantation of IL-7 gene-engineered hMSC into lethally irradiated mice led to significant increase in tymopoiesis and protected from GvHD [98]. For example, cell-mediated delivery of TRAIL in BM-MSC to metastatic rhabdomyosarcoma tumor sites can repress the growth of tumor. Targeted delivery of TRAIL to tumors may allow systemic exposure of patients to drugs that may overcome resistance for TRAIL-induced apoptosis in RMS cells [99].

15.5.3 *MSC for Improvement of Engraftment and Treatment/ Prevention GVHD*

Several disorders are characterized by both inflammation and tissue defects and hMSC has been shown to provide a valid alternative to pharmacological therapy to treat this clinical conditions. Nevertheless, it is sometimes unclear whether efficacy of hMSC is due to their production of trophic factors which stimulate endogenous repair mechanism through an immunomodulatory effect, or a putative direct differentiation into various cell types.

Recent studies have demonstrated that cotransplantation of BM-derived HSC and adipose tissue-derived hMSC induced a rapid and efficient hematopoietic reconstitution [100, 101]. Relevant preclinical studies with hMSC have shown that hMSC exert potent immunosuppressive effect in vitro and in vivo reducing the risk of graft failure and incidence of acute GvHD [102]. Allogeneic hMSC have been used to treat severe aplastic anemia refractory to conventional treatment and not eligible for allogeneic HSC transplantation, for inflammatory and autoimmune diseases such as Chron's disease [103, 104]. However, treating GVHD with hMSC might be a double-edged sword [67, 105] since has been demonstrated that the risk of the relapse was increased and the survival decreased in the hMSC treated group [66]. It was also shown that when hMSC were infused at the onset of GVHD there was no therapeutic effect, suggesting in this model that hMSC have a greater role in prophylaxis than treatment [106]. These data support a previous observation that immunosuppressive effects are exerted more on T cell proliferation than on effector function. Moreover, not all studies give concordant or positive results and several factors are likely to contribute to these discrepancies, including the tissue sources and method of isolation and expansion of hMSC, the immunodepletion process, the dose and scheduling of hMSC administration. Another possible explanation for the lack of an anti-GVHD effect is that, despite the putatively immunoprivileged features of hMSC the cells could be rejected. Furthermore, in the context of a steroid-resistant acute GVHD where a patient could have a fatal outcome weeks or months later, the administration of hMSC to manage this life-threatening condition resulted feasible, despite the limited and not standardized preclinical data available [107].

15.6 Potential Risks of Use of MSC

Human MSC maintained under standard culture conditions were shown to be nontumorigenic, however, several reports presented their capability to modulate tumor microenvironment thus having an impact on the tumor behavior [108].

The complex relationship between hMSC and tumors should be better clarified and the potential risks related to the use of hMSC should be better investigated. Furthermore, hMSC may provide a support for leukemia blasts growth and relapse, and more studies are required to better understand the factors affecting malignant cell quiescence versus proliferation during hMSC-tumor cells interactions [109, 110]. Recent concerns have been expressed about the potential transformation of hMSC during the culture process before infusion. hMSC were found rarely to help tumor development due to their immunosuppressive capacity [111]. However, the nature and the role of hMSC in their microenvironmental niche should be further explored. Moreover, although it has been demonstrated that ex vivo expansion processing of hMSC is feasible and should be obtained according to GMP and regulation constraints [46], definitive standards to produce clinical grade hMSC are still lacking.

In conclusion, the emerging functional and immunophenotypic heterogeneity of hMSCs in relation either to anatomical sites, or donor type (normal vs. pathological) or the different culture expansion conditions, should be carefully considered as an important step before using hMSC for clinical purposes before reinfusion after ex vivo expansion.

References

1. Friedenstein AJ, Chailakhyan RK, Lalykina KS (1970) The development of fibroblast colonies in monolayer cultures of guinea pig bone marrow and spleen colonies. Cell Tissue Kinet 3:393–403
2. Prockop DJ (1997) Marrow stromal cells as stem cells for non-hematopoietic tissues. Science 276:71
3. Dominici M, Le Blank K, Prockop DI, Howartz E (2006) Position paper: minimal criteria for defining multipotent mesenchymal stromal cells. The International Society for Cellular Therapy position statement. Cytotherapy 8:315–317
4. Dexter TM (1979) Cell interactions in vitro. Cin Haematol 8:453–468
5. Castro-Malaspina H, Gay RE, Resnik G, Kapoor N, Meyers P, Chiarieri D, McKenzie S, Broxmeyer HE, Moore MA (1980) Characterization of human bone marrow fibroblast colony-forming cells (CFU-F) and their progeny. Blood 56:289–301
6. Tavassoli M, Friedenstein A (1983) Hemopoietic stromal microenvironment. Am J Haematol 15:196–203
7. Eaves AC, Eaves CJ (1988) Maintenance and proliferation control of primitive hemopoietic progenitors in long-term cultures of human marrow cells. Blood Cells 14:355–368
8. Owen M, Friedenstein AJ (1988) Stromal stem cells: marrow-derived osteogenic precursors. Ciba Found Symp 136:42
9. Wulf GG, Jackson KA, Goodell MA (2001) Somatic stem cell plasticity: current evidence and emerging concepts. Exp Hematol 29:1361–1370
10. Pittenger MF, Mackay AM, Beck SC, Jaiswal RK, Douglas R, Mosca JD, Moorman MA, Simonetti DW, Craig S, Marshak DR (1999) Multilineage potential of adult human mesenchymal stem cells. Science 284:143
11. Di Girolamo CM, Stokes D, Colter D, Phinney DG, Class R, Prockop DJ (1999) Propagation and senescence of human marrow stromal cells in culture: a simple colony-forming assay identifies samples with the greatest potential to propagate and differentiate. Br J Haematol 107:275–281

12. Reyes M, Lund T, Lenvik T, Aquiar D, Koodie L, Verfaillie CM (2001) Purification and ex vivo expansion of postnatal human marrow mesodermal progenitor cells. Blood 98: 2615–2625
13. Colter DC, Sekiya I, Prockop DJ (2001) Identification of a subpopulation of rapidly self-renewing and multipotential adult stem cells in colonies of human marrow stromal cells. PNAS 98:7841–7845
14. Urbani S, Caporale R, Lombardini L, Bosi A, Saccaridi R (2006) Use of CFDA-SE for evaluating the in vitro proliferation pattern of human MSC. Cytotherapy 8:243–253
15. Taormin A, Brune JC, Olsson E, Valcich J, Neuman U, Olofsson T, Jacobsen SE, Scheding S (2009) Characterization of bone marrow derived MSC based on gene expression profiling of functionally defined MSC subsets. Cytotherapy 11:114–128
16. Majumdar MK, Keane-Moore M, Buyaner D et al (2003) Characterization and functionality of cell surface molecules on human mesenchymal stem cells. J Biomed Sci 10:228–241
17. Conget PA, Minguell JJ (1999) Phenotypical and functional properties of human bone marrow mesenchymal progenitor cells. J Cell Physiol 181:67–73
18. Beltrami AP, Cesselli D, Bergamin N, Marcon P, Rigo S, Puppato E, D'Aurizio F, Verardon R, Piazza S, Pignatelli A, Poz A, Baccaroni U, Damiani D, Fanin R, Mariuzzi L, Finato N, Masolini P, Burelli S, Belluzzi O, Schneider C, Beltrami CA (2007) Multipotent cells can be generated in vitro from several adult human organs (heart, liver, bone marrow). Blood 110:3438
19. Götherström C, West A, Liden J, Uzunel M, Lahesmaa R, Le Blanc K (2005) Differences in gene expression between human fetal liver and adult bone marrow mesenchymal stem cells. Haematologica 90:1017–1026
20. Simmons PJ, Torok-Storb B (1991) Identification of stromal precursors in human bone marrow by a novel monoclonal antibody, STRO-1. Blood 78:55–62
21. Campioni D, Lanza F, Moretti S, Dominaci M, Punturieri M, Pauli S, Hofmann T, Horwitz E, Castaldi GL (2003) Functional and immunophenotypic characteristics of isolated CD105+ and fibroblasts + mesenchymal cells from acute myeloid leukaemia: implication for their plasticity along endothelial lineage. Cytotherapy 5:66–79
22. Buhring HJ, Battula VL, Treml S, Schewe B, Kanz L, Vogel W (2007) Novel markers for the prospective isolation of human MSC. Ann N Y Acad Sci 1106:262–271
23. Deschaseux F, Charbord P (2000) Human marrow stromal precursors are $\alpha 1$ integrin subunit-positive. J Cell Physiol 184:319–325
24. Battula VL, Treml S, Bareiss PM, Gieseke F, Roelofs H, de Zwart P, Muller I, Schewe B, Skutella T, Fibbe WE, Kanz L, Buhring HJ (2009) Isolation of functionally distinct MSC cell subsets using antibodies aginst CD56, CD271 and mesenchymal stem cell antigen-1 (MSCA-1). Haematologica 94:173–184
25. Bhuring HJ, Kuci S, Conze T, Rathke G, Scherl-Mostageer M, Brummendorf TH, Schweifer N, Lammers R (2004) CDCP1 identifies a broad spectrum of normal and malignant stem/progenitor cell subsets of hematopoietic and non-hematopoietic origin. Stem Cells 22:334–343
26. SorrentinoA, Ferracin M, Castelli G, Biffoni M, Tomaselli G, Baiocchi M, Fatica A, Negrini M, Peschle C, Valtieri M (2008) Isolation and characterization of CD146+ multipotent mesenchymal stromal cells. Exp Haematol 36:1035–1046
27. Quirici N, Soligo D, Bossolasco P, Servida F, Lumini C, Delivers GL (2002) Isolation of bone marrow mesenchymal stem cells by anti-nerve growth factor receptor antibodies. Exp Haematol 30:783–791
28. Gang EJ, Bosnakovsky D, Figueiredo CA et al (2007) SSEA-4 identifies MSC from bone marrow. Blood 109:1743–1751
29. Mendez-Ferrer S, Michurina TV, Ferraro F, Mazloom AR, Macarthur BD, Lira SA, Scadden DT, Ma'ayan A, Enikopolov GN, Frenette PS (2010) Mesenchymal and haematopoietic stem cells form a unique bone marrow niche. Nature 466(7308):829–834

30. Tormin A, Li O, Brune JC, Walsh S, Schtz B, Ehinger M, Ditzel N, Kassem M, Scheding S (2011) CD146 expression on primary non-hematopoietic bone marrow stem cells is correlated with in situ localization. Blood 117:5067–5077
31. Martinez C, Hofman TJ, Marino R, Dominaci M, Horwitz EM (2007) Human bone marrow mesenchymal cells express the neural ganglioside GD2: a novel surface markers for the identification of MSC. Blood 109:4245–4248
32. Sordi V, Malosio ML, Marchesi F, Mercalli A, Melzi R, Giordano T, Belmonte N, Ferrari G, Leone BE, Bertuzzi F, Zerbini G, Allavena P, Bonifacio E, Piemonti L (2005) Bone marrow mesenchymal stem cells express a restricted set of functionally active chemokine receptors capable of promoting migration to pancreatic islet. Blood 106:419–425
33. Sanchez-Guijo FM, Blanco JF, Cruz G, Muntion S, Gomez M, Carrancio S, Lopez-Villar O, Barbado MV, Sanchez-Abarca LI, Blanco B, Brinon JG, Del Canizo MC (2009) Multiparametric comparison of mesenchymal stromal cells obtained from trabecular bone by using a novel isolation method with those obtained by iliac crest aspiration from the subjects. Cell Tissue Res 336(3):501–507
34. Landuyt KBV, Jones EA, McGonagle D, Luyten FP, Lories RJ (2010) Flow cytometric characterization of freshly isolated and cultured expanded human synovial cell population in patients with chronic arthritis. Arthr Res Ther 12:R15
35. Jones E, English A, Kinsey S et al (2006) Optimization of a flow cytometry-based protocol for detection and phenotypic characterization of multipotent mesenchymal stromal cells from human bone marrow. Cytometry Part B 70B:391–399
36. Bradford JA, Clarke ST (2011) Panel development for multicolor flow cytometry testing of proliferation and immunophenotype in h MSCs. Methods Mol Biol 698:367–385
37. Tabera S, Perez-Simon JA, Diez-Campelo M, Sanchez-Abarca LI, Blanco B, Lopez A, Benito A, Ocio E, Sanchez-Guijo FM, Canizo C, San Miguel JF (2008) The effect of mesenchymal stem cells on the viability, proliferation and differentiation of B-Lymphocytes. Haematologica 93:1301–1309
38. Meuleman N, Tondreau T, Delforge A, Dejeneffe M, Massy M, Libertalis M, Bron D, Lagneaux L (2006) Human marrow mesenchymal stem cell culture : serum-free medium allows better expansion than classical a-MEM medium. Eur J Haematol 76:309–316
39. Mueller I, Kordowich S, Holzwarth C, Spano C, Isensee G, Staiber A, Viebahn S, Gleseke F, Langer H, Gawaz MP, Horwitz EM, Conte P (2006) Animal serum-free culture conditions for isolation and expansion of multipotent mesenchymal stromal cells from human BM. Cytotherapy 8:437–444
40. Esposito MT, Di Noto R, Mirabelli P, Garrese M, Parisi S, Montanaro D, Del Vecchio L, Pastore L (2009) Culture conditions allow selection of different MSC progenitors from adult mouse bone marrow. Tissue Eng Part A 15:124
41. Sotiropoulou PA, Perez SA, Salagianni M, Baxevanis CN, Papamichail M (2006) Characterization of the optimal culture conditions for clinical scale production of human mesenchymal stem cells. Stem Cells 24:74–85
42. Kalz N, Ringe J, Holzwarth C, Chrabord P, Niemeyer M, Jacobs VR, Peschel C (2010) Novel markers of mesenchymal stem cells defined by genome-wide gene expression analysis of stromal cells from different sources. Exp Cell Res 316:2609–2617
43. Lee RH, Hsu SC, Munoz J, Jung JS, Lee NR, Pochampally R, Prockop DJ (2006) A subset of human rapidly-self renewing marrow stromal cells (MSCs) preferentially engraft in mice. Blood 107:2153–2161
44. Samuelsson H, Ringden O, Lönnies L, Le Blanc K (2009) Optimizing in vitro conditions for immunomodulation and expansion of mesenchymal stem cells. Cytotherapy 11:129–136
45. Grisendi G, Anneren C, Cafarelli L, Sternieri R, Veronesi E, Cervo GL, Luminari S, Maur M, Frassoldati A, Palazzi G, Otsuru S, Bambi F, Paolucci P, Conte PF, Horwitz E, Dominici M (2010) GMP-manufactured density gradient media for optimized mesenchymal stromal stem cell isolation and expansion. Cytotherapy 12:466–477
46. Bieback K, Kinzebach S, Karagianni M (2010) Translating research into clinical scale manufacturing of mesenchymal stromal cells. Stem Cells Int 2010:193519

47. Haniffa MA, Collin MP, Buckley CD, Dazzi F (2009) Mesenchymal stem cells: the fibroblasts' new clothes? Haematologica 94:258–263
48. Chang HY, Chi JT, Dudoit S, Bondre C, van de Rijn M, Botstein D, Brown PO (2002) Diversity, topographic differentiation, and positional memory in human fibroblasts. Proc Natl Acad Sci U S A 99:12877–12882
49. Jin HJ, Park SK, Oh W, Yang YS, Kim SW, Choi SJ (2009) Down-regulation of CD105 is associated with multilineage differentiation in human umbilical cord derived-MSC. Biochem Biophys Res Comm 381:676–681
50. Kemp K, Morse R, Wexler S, Cox C, Mallam E, Hows J, Donaldson C (2010) Chemotherapy induce mesenchymal stem cell damage in patients with haematological malignancy. Ann Hematol 89:701–713
51. Di Nicola M, Carlo-Stella C, Magni M, Milanesi M, Longoni PD, Matteucci P, Grisanti S, Gianni AM (2000) Human bone marrow stromal cells suppress T-lymphocyte proliferation induced by cellular or nonspecific mitogenic stimuli. Blood 99:3838–3843
52. Le Blanc K, Tammik C, Rosendahl K, Zetterbergie E, Ringdén O (2003) HLA expression and immunologic properties of differentiated and undifferentiated mesenchymal cells. Exp Hematol 31:890–896
53. Rizzo R, Campioni D, Stignani M, Melchiorri L, Bagnara GP, Bonsi L, Alviano F, Lanzoni G, Moretti S, Cuneo A, Lanza F, Baricordi OR (2008) A functional role for soluble HLA-G antigens in immune modulation mediated by mesenchymal stromal cells. Cytotherapy 10:364–375
54. Tse WT, Pendleton JD, Beyer WM, Egalka MC, Guinan EC (2003) Suppression of allogeneic T-cell proliferation by human marrow stromal cells: implications in transplantation. Transplant 75:389–397
55. Meisel R, Zibert A, Laryea M, Gobel U, Daubener W, Dilloo D (2004) Human bone marrow stromal cells inhibit allogeneic T-cell responses by indoleamine 2,3-dioxygenase-mediated tryptophan degradation. Blood 103:4619–4621
56. Taylor MW, Feng G (1991) Relationship between interferon-gamma, indoleamine 2,3-dioxygenase, and tryptophan catabolism. FASEB J 5(11):2516–2522
57. Munn DH, Sharma MD, Lee JR, Jhaver KG, Johnson TS, Keskin DB, Marshall B, Chandler P, Antonia S, Burgess R, Slingwff CL Jr, Mellor AL (2002) Potential regulatory function of human dendritic cells expressing indoleamine 2,3-dioxygenase. Science 297:1867–1870
58. Terness P, Bauer TM, Rose L, Dufter C, Watzlik A, Simon H, Opelz G (2002) Inhibition of allogeneic T cell proliferation by indoleamine 2,3-dioxygenase-expressing dendritic cells: mediation of suppression by tryptophan metabolites. J Exp Med 196:447–457
59. Popp FC, Eggenhofer E, Renner P, Slowik P, Lang SA, Kaspar H, Geissler EK, Piso P, Schlitt HJ, Dahlke MH (2008) Mesenchymal stem cells can induce long-term acceptance of solid organ allografts in synergy with low-dose mycophenolate. Transpl Immunol 20(1–2): 55–60
60. Spaggiari GM, Capobianco A, Abdelrazik H, Becchetti F, Mingari MC, Moretta L (2008) Mesenchymal stem cells inhibit natural killer-cell proliferation, cytotoxicity, and cytokine production: role of indoleamine 2,3-dioxygenase and prostaglandin E2. Blood 111(3): 1327–1333
61. Spoerri PE, Caballero S, Wilson SH, Shaw LC, Grant MB (2003) Expression of IGFBP-3 by human retinal endothelial cell cultures: IGFBP-3 involvement in growth inhibition and apoptosis. Invest Ophthalmol Vis Sci 44(1):365–369
62. Baricordi OR, Stignani M, Melchiorri L, Rizzo R (2008) HLA-G and inflammatory diseases. Inflamm Allergy Drug Targets 7:67–74
63. Götherström C, West A, Liden J, Uzunel M, Lahesmaa R, Le Blanc K (2005) Difference in gene expression between human fetal liver and adult bone marrow mesenchymal stem cells. Haematol 90:1017–1026
64. Nasef A, Mathieu N, Chapel A, Frick J, François S, Mazurier C, Boutarfa A, Bouchet S, Gorin NC, Thierry D, Fouillard L (2007) Immunosuppressive effects of mesenchymal stem cells: involvement of HLA-G. Transplant 84(2):231–237

65. Selmani Z, Naji A, Zidi I, Favier B, Gaiffe E, Obert L, Borg C, Saas P, Tiberghien P, Rouas-Freiss N, Carosella ED, Deschaseaux F (2008) Human leukocyte antigen-G5 secretion by human mesenchymal stem cells is required to suppress T lymphocyte and natural killer function and to induce CD4+CD25highFOXP3+ regulatory T cells. Stem Cells 2(1):212–222
66. Ning H, Yang F, Jiang M, Hu L, Feng K, Zhang J, Yu Z, Li B, Xu C, Li Y, Wang J, Hu J, Lou X, Chen H (2008) The correlation between cotransplantation of mesenchymal stem cells and higher recurrence rate in hematologic malignancy patients: outcome of a pilot clinical study. Leukemia 22:593–599
67. Rizzo R, Campioni D, Lanza F, Baricordi OR (2008) Cotransplantation of mesenchymal cells and a higher relapse rate: a role for HLA-G molecules? Leukemia 22:2273
68. Rizzo R, Lanzoni G, Stignani M, Campioni D, Alviano F, Ricci F, Tazzari PL, Melchiorri L, Scalinci SZ, Cuneo A, Bonsi L, Lanza F, Bagnara GP, Baricordi OR (2011) A simple method for identifying bone marrow mesenchymal stromal cells with a high immunosuppressive potential. Cytotherapy 13:523–527
69. Ivanova-Todorova E, Mourdjeva M, Kyurkchiev D, Bochev I, Stoyanova E, Dimitrov R, Timeva T, Yunakova M, Bukarev D, Shterev A, Tivchev P, Kyurkchiev S (2009) HLA-G expression is up-regulated by progesterone in mesenchymal stem cells. Am J Reprod Immunol 62(1):25–33
70. Lanza F, Campioni D, Moretti S, Ferrari L, Rizzo R, Baricordi R, Cuneo A (2007) Aberrant expression of HLA-DR antigen by bone marrow-derived mesenchymal stromal cells from patients affected by acute lymphoproliferative disorders. Leukemia 21:378–381
71. Bocelli-Tyndall C, Zajac P, Di Maggio N, Trella E, Benvenuto F, Iezzi G, Scherberich A, Barbero A, Schaeren S, Pistoia V, Spagnoli G, Vukcevik M, Martin I, Tyndall A (2010) Fibroblast growth factor 2 and Platelet-derived growth factor, but not platelet lysete, induce proliferation-dependent, functional class II major histocompatibility complex antigen in human mesenchymal stem cells. Arthritis Rheum 62:3815–3825
72. Campioni D, Rizzo R, Stignani M, Melchiorri L, Ferrari L, Rizzo R, Baricordi R, Cuneo A, Lanza F, Moretti S, Ferrari L, Alviano F, Russo A (2009) A decreased positivity for CD90 on human mesenchymal stromal cells (MSCs) is associated with a loss of immunosuppressive activity by MSCs. Cytometry 76(3):225–230
73. Kuci S, Kuci Z, Kreyemberg H, Deak E, Putsch K, Huenecke S, Amara C, Koller S, Rettinger E, Grez M, Koehl U, Latifi-Pupovci H, Henschler R, Tonn T, von Laer D, Kliengebiel T, Bader P (2010) CD271 antigen define a subset of multipotent stromal cells with immunosuppressive and lymphohematopoietic engraftment-promoting properties. Haematologica 95(4):651–659
74. Nasef A, Zhang YZ, Mazurier C, Bouchet S, Bensidhoum M, Francois S, Gorin NC, Lopez M, Thierry D, Fouillard L, Chapel A (2009) Selected STRO-1 enriched bone marrow stromal cells display a major suppressive effect on lymphocyte proliferation. Int J Lab Hematol 31:9–19
75. Krampera M, Franchini M, Pizzolo G, Aprili G (2007) Mesenchymal stem cells: from biology to clinical use. Blood Transfus 5:120–129
76. Schäfer R, Dominici M, Müller I, Horwitz E, Asahara T, Bulte JW, Bieback K, le Blanc K, Buhring HJ, Capogrossi MC, Dazzi F, Gorodetsky R, Henschler R, Handgretinger R, Kajstura J, K Luger P, Lange C, Luettichau I, Mertsching H, Schrezenmejer H, Sievert KD, Strunk D, Verfaillie C, Northoff H (2008) Basic research and clinical applications of non-hematopoietic stem cells. Cytotherapy 11:245–255 (4–5 April 2008, Tübingen, Germany)
77. Le Blanc K, Frassoni F, Ball L, Locatelli F, Roelofs H, Lewis I, Lanino E, Sundberg B, Bernardo ME, Remberger M, Dini G, Egeler RM, Bacigalupo A, Fibbe W, Ringden O (2008) Mesenchymal stem cells for treatment of steroid-resistant, severe, acute graft-versus host disease: a phase II study. Lancet 371:1579–1586
78. Horwitz EM, Gordon PL, Koo WKK et al (2002) Isolated allogeneic bone marrow-derived mesenchymal cells engraft and stimulate growth in children with osteogenesis imperfecta: implications for cell therapy of bone. Proc Natl Acad Sci U S A 99:8932–8937

79. Almeida-Porada G, Flake AW, Glimp HA, Flake AW, Glimp HA, Zanjani ED (1999) Cotransplantation of stroma results in enhancement of engraftment and early expression of donor hematopoietic stem cells in utero. Exp Hematol 27:1569–1575
80. Devine SM, Cobbs C, Jennings M (2003) Mesenchymal stem cells distribute to a wide range of tissue following systemic infusion into nonhuman primates. Blood 101:2999–3001
81. Horwitz EM, Dominici M (2008) How do mesenchymal stromal cells exert their therapeutic benefit? Cytotherapy 10:771–774
82. Woodbury D, Schwarz EJ, Prockop DJ, Black IB (2000) Adult rat and human bone marrow stem cells differentiate into neurons. J Neurosci Res 61:223–229
83. Bouchez G, Sensebè L, Vourc'h P, Garreau L, Bodard S, Rico A, Guilloteau D, Charbord P, Besnard JC, Chalon S (2008) Partial recovery of dopaminergic pathway after graft of adult MSC in a rat model of Parkinson's disease. Neurochem Int 52:1332–1342
84. Kalbermatten DF, Pierer G, Terenghi G, Kingham PJ (2010) Regeneration potential and survival of transplanted undifferentiated adipose tissue-derived stem cells in peripheral nerve conduits. J Plast Reconstr Aesthet Surg 12:e811–e817
85. Kingham PJ, Mantovani C, Terenghi G (2011) Stem cell and neuron co-cultures for the study of nerve regeneration. Methods Mol Biol 695:115–127
86. Lu P, Blesch A, Tuszynski MH (2004) Induction of bone marrow stem cells into neurons: differentiation, transdifferentiation or artefact? J Neurosci Res 77:174–191
87. Planat-Bénard V, Menard C, Andrè M, Puceat M, Perez A, Garcia-Verdugo JM, Penicaud L, Casteilla L (2004) Spontaneous cardiomyocyte differentiation from adipose tissue stroma cells. Circ Res 94:223–229
88. Rossini A, Frati C, Lagrasta C, Graiani G, Scopece A,Cavalli S, Musso E, Baccarini M, Di Segni M, Fagnoni F, Germani A, Quaini E, Mayr M, Qingbo Xu, Barbuti A, Di Francesco, Pompilio G, Quaini F, Gaetano C, Capogrossi MC (2011) Human cardiac and bone marrow stromal cells exhibit distinctive properties related to their origin. Cardiovascular Res 89:650–660
89. Gnecchi M, He H, Melo GM, Noiseux N, Morello F, De Boer R, Zhang L, Pratt RE, Dzau VJ, Ingwall JS (2009) Early beneficial effects of bone marrow derived mesenchymal cells overexpressing Akt on cardiac metabolism after myocardial infarction. Stem Cells 27:971–979
90. Mirza A, Hyvelin JM, Rochefort GY, Lermusiaux P, Antier D, Awede B, Bonnet P, Domenech J, Eder V (2008) Undifferentiated MSC seeded on a vascular prosthesis contribute to the restoration of vascular wall. J Vasc Surg 47:1313–1321
91. Mauritz C, Martens A, Rojas SV, Schnick T, Rathert C, Schecker N, Menke S, Glage S, Zweigerdt R, Haverich A, Martin U, Kutschaka I (2011) Induced pluripotent stem cell (iPSC)-derived Flk-1 progenitor cells engraft, differentiate, and improve heart function in a mouse model of acute myocardial infarction. Eur Heart J. doi:10.1093/eurheart/ehr166
92. Lee KD, Kuo TK, Whang-Peng J, Chung YF, Lin CT, Chou SH, Chen JR, Chen YP, Lee OK (2004) In vitro hepatic differentiation of human MSC. Hepatology 40:1275–1284
93. Wang HS, Shyu JF, Shen WS, Hsu HC, Chi TC, Chen CP, Huang SW, Shyr YM, Tang KT, Chen TH (2011) Transplantation of insulin producing cells derived from umbilical cord stromal mesenchymal cells to treat NOD mice. Cell Transplant 20(3):455–466
94. Lange C, Togel F, Ittrich H, Clayton F, Nolte- Ernesting C, Zander AR, Westen Felder C (2005) Administered MSC enhance recovery from ischemia/reperfusion-induced acute renal failure in rats. Kidney Int 68:1613–1617
95. Bussolati B (2011) Stem cells for organ repair: support or replace? Biomatter 1:95
96. Kode JA, Mukherjee S, Joglekar MV, Hardikar AA (2009) Mesenchymal stem cells: immunobiology and in immunomodulation and tissue regeneration. Cytotherapy 11:377–391
97. Zachos T, Diggs A, Weisbrode S, Bartlett J, Bertone A (2007) Mesenchymal stem cells mediated gene delivery of bone morphogenetic protein-2 in an articular fracture model. Mol Ther 15:1543–1550
98. Caplan AI (2007) Adult MSC for tissue engineering versus regenerative medicine. J Cell Physiol 213:341–347

99. Barti-Juhasz H, Mihalik R, Nagy K, Grisendi G, Dominici M, Petak I (2011) Bone marrow derived MSC transduced with full length human TRAIL repress the growth of rhabdomyosarcoma cells in vitro. Haematologica 96:e21
100. Liechty KW, MacKenzie TC, Shaaban AF, Radu A, Moseley AM, Deans R, Marshak DR, Flake AW (2000) Human mesenchymal stem cells engraft and demonstrate site-specific differentiation after in utero transplantation in sheep. Nat Med 6:1282–1286
101. Ball LM, Bernardo ME, Roelofs H, Lankester A, Cometa A, Egeler RM, Locatelli F, Fibbe WE (2007) Cotransplantation of ex vivo expanded MSC accelerates lymphocyte recovery and may reduce risk of graft failure in haploidentical hematopoietic stem cell transplantation. Blood 110:2764–2767
102. Lazarus HM, Koc ON, Devine SM, Curtin P, Marziaz RT, Holland HK, Shpall EJ, McCarthy P, Atkinson K, Cooper BW, Gerson SL, Laughlin MJ, Loberizia FR, Moseley AB, Bacigalupo A (2005) Cotransplantation of HLA-identical sibling culture expanded MSC and HSC in hematologic malignancy patients. Biol Blood Marrow Transpl 11:389–398
103. Le Blanc K, Ringden O (2005) Immunobiology of human mesenchymal stem cells and future use in hematopoietic stem cell transplantation. Biol Blood Marrow Transpl 11:321–334
104. Sensebè L, Krampera M, Schrezenmeier H, Bourin P, Giordano R (2010) Mesenchymal stem cell for clinical application. Vox Sang 98:93–107
105. Vianello F, Dazzi F (2008) Mesenchymal stem cells for graft-versus host disease: a doible-egde sword? Leukemia 22:463–465
106. Tolar J, Villeneuve P, Keating A (2011) Mesenchymal stromal cells for graft-versus-host disease. Hum Gene Ther 22:257–262
107. Le Blanc K, Frassoni F, Ball L, Locatelli F, Roelofs H, Lewis I, lanino E, Sundberg B, Bernardo ME, Remberger M, Dini G, Egeler RM, Bacigalupo A, Fibbe W, Ringden O (2008) Mesenchymal stem cells for treatment of steroid resistant, severe, acut GVHD: a phase II study. Lancet 363:1439–1441
108. Kucerova L, Matuskova M, Hlubinova K, Altanerova V, Altaner C (2010) Tumor cell behaviour modulation by mesenchymal stromal cells. Mol Cancer 9:129
109. Prockop D, Brenner M, Fibbe W (2010) Defining the risk of therapies with MSC. Cytotherapy 12:576–578
110. Marini FC (2009) Commentary: the complex love-hate relationship between mesenchymal stromal cells and tumors. Cytotherapy 11:375–376
111. Djouad F, Plence P, Bony C (2003) Immunosuppressive effect of mesenchymal stem cells favours tumor growth in allogeneic animals. Blood 102:3837–3844

Chapter 16
Cellular Therapy for Hematology Malignancies: Allogeneic Hematopoietic Stem Transplantation, Graft-Versus-Host Disease, and Graft Versus Leukemia Effects

James L. M. Ferrara and Pavan Reddy

Abstract The ability of allogeneic hematopoietic cell transplantation (HCT) to cure hematologic malignancies is widely recognized. An important therapeutic aspect of HCT in eradicating malignant cells is the graft-versus-leukemia (GVL) effect. But the GVL effect is closely associated with graft-versus-host disease (GVHD) the major complication of HCT. GVHD remains the major barrier to the wider application of allogeneic HCT for a variety of diseases. GVHD occurs in two forms, acute and chronic, and both are associated with GVL effects. Recent advances in the understanding of genetic polymorphisms, the chemo-cytokine networks, several novel cellular subsets including regulatory T cells, and of the direct mediators of cellular cytotoxicity have led to improved understanding of these complex processes. Animal studies show that modulating several mediators of the complex GVHD cascade may be able to reduce the undesirable inflammatory aspects of GVHD while preserving the benefits of GVL. However, most of the laboratory observations remain to be studied in well-controlled clinical trials. Multiple cellular effectors may be involved in GVL, although donor T cell recognition of host antigens is an important element of this process. Cellular immunotherapy such as donor leukocyte infusion offers a strategy for separating GVHD and the GVL effect. Both experimental and clinical data suggest that post-transplantation cellular immunotherapy can be performed relatively safely and effectively, and optimization of patient selection, cell dose, and timing of administration may all serve to limit toxicity and enhance the potential GVL effects.

J. L. M. Ferrara (✉)
Medicine and Pediatrics, Combined Bone Marrow Transplant Program,
University of Michigan Cancer Center, Ann Arbor, MI, USA
e-mail: Ferrara@umich.edu

P. Reddy
Department of Internal Medicine, Bone Marrow Transplant Program,
University of Michigan Cancer Center, Ann Arbor, MI, USA

H. Baharvand and N. Aghdami (eds.), *Advances in Stem Cell Research*,
Stem Cell Biology and Regenerative Medicine, DOI: 10.1007/978-1-61779-940-2_16,

16.1 Introduction

The ability of allogeneic hematopoietic cell transplantation (HCT) to cure certain hematologic malignancies is widely recognized. An important therapeutic aspect of HCT in eradicating malignant cells is the graft-versus-leukemia (GVL) effect. The importance of the GVL effect in allogeneic HCT has been recognized since the earliest experiments in stem cell transplantation. Forty years ago Barnes and colleagues noted that leukemic mice treated with a subtherapeutic dose of radiation and a syngeneic (identical twin) graft transplant were more likely to relapse than mice given an allogeneic stem cell transplant [1, 2]. They hypothesized that the allogeneic graft contained cells with immune reactivity necessary for eradicating residual leukemia cells. They also noted that recipients of allogeneic grafts, though less likely to relapse, died of a "wasting syndrome" now recognized as graft-versus-host disease (GVHD). Thus, in addition to describing GVL, these experiments highlighted for the first time the intricate relationship between GVL and GVHD. Since these early experiments, both GVHD and the GVL effect have been studied extensively [3]. This chapter reviews the pathophysiology, clinical features, treatment of GVHD, and summarizes current understanding of the relationships between GVHD and the GVL effect.

16.2 Graft-Versus-Host Disease: Clinical and Pathologic Aspects

Ten years after the work of Barnes and Loutit, Billingham formulated the requirements for the development of GVHD: the graft must contain immunologically competent cells, the recipient must express tissue antigens that are not present in the transplant donor, and the recipient must be incapable of mounting an effective response to destroy the transplanted cells [4]. According to these criteria, GVHD can develop in various clinical settings when tissues containing immunocompetent cells (blood products, bone marrow, and some solid organs) are transferred between persons. The most common setting for the development of GVHD is following allogeneic HCT; without prophylactic immunosuppression, most allogeneic HCTs will be complicated by GVHD. GVHD occurs secondary to mismatches between histocompatibility antigens between the donor and recipient. Matching of the major histocompatibility complex (MHC) antigens speeds engraftment and reduces the severity of GVHD [5]. The MHC contains the genes that encode tissue antigens and were first identified functionally in murine models as transplantation antigens responsible for rejection of tissue grafts. In humans, the MHC region lies on the short arm of chromosome 6 and is called the human leukocyte antigen (HLA) region [6]. The HLA region includes many genes, not all of which are involved in immune activation. It is divided into two classes, Class I and Class II, each containing numerous gene loci that encode a large number of polymorphic alleles. MHC class I molecules are involved in the presentation of peptides to $CD8^+$ T cells, and class II molecules present peptides to $CD4^+$ T cells [6].

Each MHC antigen is composed of two polypeptide chains. Class I antigens are made up of a heavy chain that contains the polymorphic regions and the nonpolymorphic light chain, beta-2 microglobulin. The class I HLA antigens include HLA A, B, and C antigens. These are expressed on almost all cells of the body at varying densities [6]. Both chains of class II antigens contain polymorphic regions and are encoded in the MHC. The class II antigens are further divided into DR, DQ, and DP antigens. Class II antigens are expressed on B cells, dendritic cells, and monocytes and their expression can be induced on many other cell types following inflammation or injury [6, 7]. The determination of HLA types has become much more accurate with molecular techniques that replace earlier serologic or cellular methods. In patients whose ancestry involves extensive interracial mixing, the chances of identifying an HLA-identical donor are diminished [8].

Despite HLA identity between a patient and donor, substantial numbers of patients still develop GVHD due to differences in minor histocompatibility antigens that lie outside the HLA loci. Most minor antigens are expressed on the cell surface as degraded peptides bound to specific HLA molecules, but the precise elucidation of many human minor antigens is yet to be accomplished [9]. In the United States, the average patient has a 25% chance of having an HLA match within his immediate family [8]. Patients who lack an HLA-identical family member donor must seek unrelated donor volunteers or cord blood donations.

16.2.1 Acute Graft-Versus-Host Disease

Acute GVHD can occur within days (in recipients who are not HLA-matched with the donor or in patients not given any prophylaxis) or as late as 2 months after transplantation. The incidence ranges from less than 10 to more than 80%, depending on the degree of histoincompatibility between donor and recipient, the number of T cells in the graft, the patient's age, and the GVHD prophylactic regimen [9]. The principal target organs include the immune system, skin, liver, and intestine. GVHD occurs first and most commonly in the skin as a pruritic maculopapular rash, often involoving the palms, soles, and ears; it can progress to total-body erythroderma, with bullae formation, rupture along the epidermal-dermal border, and desquamation in severe cases [9]. Gastrointestinal (GI) and liver manifestations often appear later and rarely represent the first and only findings. Intestinal symptoms include anorexia, nausea, diarrhea (sometimes bloody), abdominal pain, and paralytic ileus [9]. Liver dysfunction includes hyperbilirubinemia and increased serum alkaline phosphatase and aminotransferase values. Coagulation studies may become abnormal, and hepatic failure with ascites and encephalopathy may develop in severe cases [9–11]. Hepatic GVHD can be distinguished from hepatic veno-occlusive disease (VOD) by weight gain or pain in the right upper quadrant in the latter [11]. Acute GVHD also results in the delayed recovery of immunocompetence [9]. The clinical result is profound immunodeficiency and susceptibility to infections, often further accentuated by the immunosuppressive agents used to treat GVHD [9].

Pathologically, the sine qua non of acute GVHD is selective epithelial damage of target organs [12, 13]. The epidermis and hair follicles are damaged and sometimes destroyed. Small bile ducts are profoundly affected, with segmental disruption. The destruction of intestinal crypts results in mucosal ulcerations that may be either patchy or diffuse. Other epithelial surfaces, such as the conjunctivae, vagina, and esophagus, are less commonly involved. A peculiarity of GVHD histology is the frequent paucity of mononuclear cell infiltrates; however, as the disease progresses the inflammatory component may be substantial. Recent studies have identified inflammatory cytokines as soluble mediators of GVHD and have suggested that direct contact between target cells and lymphocytes may not be required for target cell destruction (see following sections). GVHD lesions are not evenly distributed in the target tissues. In the skin, damage is prominent at the tip of rete ridges; in the intestine, at the base of the crypts; and in the liver, in the periductular epithelium. These areas contain a high proportion of stem cells, giving rise to the idea that GVHD targets may be undifferentiated epithelial cells with primitive surface antigens [14].

The histologic severity of a given lesion is at best semiquantitative and, consequently, the severity of pathologic findings are not used in the grading of GVHD. As it is often difficult to obtain an adequate tissue biopsy, the physician is left to use clinical judgment. It can be very difficult to distinguish GVHD from other post-BMT complications such as drug eruptions, infectious complication, even on histology.

A recent multicenter phase III trial used an independent committee to assess the presence and severity of GVHD. The incidence of GVHD as determined by investigators was substantially higher than the review committee could confirm [15]. Nevertheless, for an experienced clinician a combination of physical and laboratory findings in the appropriate context provides a working diagnosis of GVHD that is satisfactory to produce a meaningful prognostic scale based on clinical grading system [12, 17]. Standard grading systems generally include clinical changes in the skin, GI tract, liver, and performance status (Table 16.2) [16]. Although the severity of GVHD is sometimes difficult to quantify, the overall grade correlates with disease outcome. While mild GVHD (grade I or II) is associated with little morbidity and almost no mortality, higher grades are associated with significantly decreased survival [16, 17]. With grade IV GVHD, the mortality is almost 100% [17].

16.2.1.1 Clinical Features of Acute GVHD

The clinical features, staging and grading of acute GVHD are summarized in Tables 16.1 and 16.2.

In a comprehensive review of patients receiving therapy for acute GVHD, it was found that 81% had skin involvement, 54% had GI involvement, and 50%

Table 16.1 Clinical manifestations and staging of acute GVHD

Organ	Clinical manifestations	Staging
Skin	Erythematous, maculopapular rash involving palms and soles may become confluent.	Stage 1: <25% rash Stage 2: 25–50% rash Stage 3: generalized erytroderma
	Severe disease: bullae	Stage 4: bullae
Liver	Painless jaundice with conjugated hyperbilirubinemia and increased alkaline phosphatase	Stage 1: bili 2–3 mg/dL Stage 2: bili 3.1–6 mg/dL Stage 3: bili 6.1–15 mg/dL Stage 4: bili > 15 mg/dL
Gastro-intestinal (GI) tract	Upper: nausea, vomiting, anorexia	Stage 1: diarrhea > 500 mL/day Stage 2: diarrhea > 1000 mL/day
	Lower: diarrhea, abdominal cramps, distention, ileus, and bleeding	Stage 3: diarrhea > 1500 mL/day Stage 4: ileus, bleeding

Table 16.2 Glucksberg criteria for staging of acute GVHD[a]

Overall grade	Skin	Liver		Gut
I	1–2	0		0
II	1–3	1	And/or	1
III	2–3	2–4	And/or	2–3
IV	2–4	2–4	And/or	2–4

[a] See Table 16.1 for individual organ staging. Traditionally, individual organs are staged without regard to attribution. The overall grade of GVHD, however, reflects the actual extent of GVHD. To achieve each overall grade, skin disease plus liver and/or gut involvement are required

had liver involvement at the initiation of therapy [18]. After high-intensity (conventional) conditioning, acute GVHD generally occurs within 14–35 days of stem cell infusion. The time of onset may depend upon the degree of histocompatibility, the number of donor T cells infused, and the prophylactic regimen for GVHD. A "hyperacute" form of GVHD may occur in patients with severe HLA mismatches and in patients who receive T cell replete transplants without or with inadequate in vivo GVHD prophylaxis [19]. It is, however, important to note that this "hyperacute" form is pathophysiolocally distinct from the hyperacute rejection after solid organ allo-grafting. This form of GVHD is manifested by fever, generalized erythroderma and desquamation, and often edema. It typically occurs about 1 week after stem cell infusion and may be rapidly fatal. In patients receiving more conventional (in vivo) GVHD prophylaxis, such as a combination of cyclosporine (CSP) and methotrexate, the median onset of GVHD is typically 21–25 days after transplantation; however, after in vitro T cell depletion of the graft the onset may be much later [19]. Thus, the findings of rash and diarrhea by 1 week after transplantation would very likely be hyperacute manifestations of GVHD if minimal or ineffective prophylaxis were administered; the same kinetics would be very unlikely with the use of calcineurin inhibitors or in vitro T cell depletion of the stem cell inoculum. A less ominous syndrome of fever, rash, and fluid retention occurring in the first 1–2 weeks after stem cell infusion is the "engraftment syndrome". These manifestations may be seen with either allogeneic or autologous transplantation. While the pathophysiology is poorly understood, it is thought to be due to a wave of cytokine production as the graft starts to recover. This is related to, but distinct from, the "cytokine storm" that is thought to contribute to acute GVHD in which there is no concomitant T cell-mediated attack [19, 20]. This syndrome responds immediately to steroids in most patients and it typically presents earlier than acute GVHD [16]. In autologous transplantation the differential diagnosis is of little relevance, but in allogeneic transplant recipients it must be distinguished from the hyperacute manifestations of GVHD. A prompt response to steroids would argue in favor of an engraftment syndrome, although some patients with GVHD will also respond.

Skin is the most commonly affected organ. In patients receiving transplants after myeloablative conditioning, the skin is usually the first organ involved, and GVHD often coincides with engraftment. However, the presentation of GVHD is more varied following nonmyeloablative transplants or donor lymphocyte infusions [21]. The characteristic maculopapular rash can spread throughout the rest of the body but usually spares the scalp, and is often described as feeling like a sunburn, tight, or pruritic. In severe cases the skin may blister and ulcerate [22]. Histologic confirmation is critical to rule out drug reactions, viral infections, etc. Apoptosis at the base of dermal crypts is characteristic. Other features include dyskeratosis, exocytosis of lymphocytes, satellite lymphocytes adjacent to dyskeratotic epidermal keratinocytes, and dermal perivascular lymphocytic infiltration [21, 23].

Gastrointestinal tract involvement of GVHD may present as nausea, vomiting, anorexia, diarrhea, and/or abdominal pain [23]. It is a pan-intestinal process, often with differences in severity between the upper and lower GI tracts. Gastric involvement gives rise to postprandial vomiting that is not always preceded by nausea. Although gastroparesis is seen after bone marrow transplant, it is usually not associated with GVHD. The diarrhea of GVHD is secretory and significant GI blood loss may occur as a result of mucosal ulceration and is associated with a poor prognosis [24]. In advanced disease, diffuse, severe abdominal pain and distention is accompanied by voluminous diarrhea (>2 L/day) [17, 25].

Radiologic findings of the GI tract include lumenal dilatation with thickening of the wall of the small bowel and air/fluid levels suggestive of an ileus on abdominal flat plates or small bowel series. Abdominal computed tomography may show the "ribbon" sign of diffuse thickening of the small bowel wall [22]. Little correlation exists between the extent of disease and the appearance of mucosa on endoscopy, but mucosal sloughing is pathognomonic for severe disease [26]. Nevertheless, some studies have shown that antral biopsies correlate well with the severity of GVHD in the duodenum and in the colon even when the presenting symptom is diarrhea [26]. Histologic analysis of tissue is imperative to establish the diagnosis. The histologic features of GI GVHD are the presence of apoptotic bodies in the base of crypts, crypt abscesses, crypt loss, and flattening of the surface epithelium [25, 27].

Liver function test abnormalities are common after bone marrow transplant and occur secondary to VOD, drug toxicity, viral infection, sepsis, iron overload, and other causes of extrahepatic biliary obstruction [11]. The exact incidence of hepatic GVHD is probably underreported because many patients do not undergo liver biopsies. The development of jaundice or an increase in the alkaline phosphatase and bilirubin are the initial features of acute GVHD of the liver. The histologic features of hepatic GVHD are endothelialitis, lymphocytic infiltration of the portal areas, pericholangitis, and bile duct destruction and loss [19, 30].

16.2.1.2 Other Organs

Whether GVHD affects organs other than the classic triad of skin, liver, and gut has remained a matter of debate. However, numerous reports suggest additional organ manifestations. The most likely candidate is the lung. Lung toxicity, including interstitial pneumonitis and diffuse alveolar hemorrhage, may occur in 20–60% of allogeneic transplant recipients but in fewer autologous transplant recipients. Causes of pulmonary damage other than GVHD include engraftment syndrome (see below), infection, radiation pneumonitis, and chemotherapy-related toxicity (e.g., methotrexate, busulfan) [19, 28]. At least one retrospective analysis failed to link severe pulmonary complications to clinical acute GVHD per se [29]. The mortality due to pneumonia increases with the severity of GVHD, but this association does not necessarily imply that GVHD, as opposed to immunosuppression given for therapy, is causative [19]. A particular histopathologic syndrome of lymphocytic bronchitis has been attributed directly to GVHD, although this association has not been confirmed by others [28].

Despite the fact that kidneys and heart can be targets for allogeneic damage as evidenced by their rejection after renal and cardiac transplants respectively, there is no convincing evidence for direct renal or cardiac damage from acute GVHD that is not secondary to drugs or infection. Similarly, neurologic complications are also common after transplantation but most can be attributed to drug toxicity, infection, or vascular insults.

16.2.1.3 Differential Diagnosis

Acute GVHD ought to be distinguished from any process that causes a constellation of fever, erythematous skin rash with/or without low-pressure, and pulmonary edema that may occur during neutrophil recovery. This picture may reflect the dysregulated production of inflammatory cytokines and cellular responses to these molecules, and has been termed engraftment or capillary leak syndrome [30, 31]. The picture is most clearly recognized after autologous transplantation where, theoretically, GVHD should not occur. In allogeneic transplant recipients distinction from acute GVHD is difficult. This engraftment syndrome is thought to reflect cellular and cytokine activities during early recovery of (donor-derived) blood cell counts and/or homeostatic proliferation of lymphocytes, but a precise delineation of the offending cells and mechanisms has not been accomplished. Engraftment syndromes may be associated with increased mortality, primarily from pulmonary failure and also (other) multi-organ dysfunction. Corticosteroid therapy may be effective particularly for the treatment of pulmonary manifestations [32]. The differential diagnosis of skin rashes, diarrhea, and liver function abnormalities can be difficult to resolve. Skin rashes may reflect delayed reactions to the conditioning regimen, antibiotics, or infections and furthermore histopathologic skin changes consistent with acute GVHD can be mimicked by chemoradiotherapy and drug reactions [19, 33]. Diarrhea can be a consequence of TBI, viral infections, especially with CMV and other herpes viruses, parasites, *Clostridium difficile*, nonspecific gastritis, narcotic withdrawal, and drug reactions, all of which mimic GVHD of the gut. Liver dysfunction can be due to parenteral nutrition, VOD, viral or drug-induced hepatitis.

16.3 Pathophysiology of Acute Graft-Versus-Host Disease

It is helpful to remember two important principles when considering the pathophysiology of acute GVHD. First, acute GVHD represents exaggerated but normal inflammatory responses against foreign antigens (allo-antigens) that are ubiquitously expressed in a setting where they are undesirable. The donor lymphocytes that have been infused into the recipient function appropriately, given the foreign environment they encounter. Second, donor lymphocytes encounter tissues in the recipient that have been often profoundly damaged. The effects of the underlying disease, prior infections, and the intensity of conditioning regimen all result in substantial changes not only in the immune cells but also in the endothelial and epithelial cells. Thus the allogeneic donor cells rapidly encounter not only a foreign environment, but one that has been altered to promote the activation and proliferation of inflammatory cells. Thus, the pathophysiology of acute GVHD may be considered a distortion of the normal inflammatory cellular responses that, in addition to the absolute requirement of donor T cells, involves multiple other innate and adaptive cells and mediators [34]. The development and evolution of acute GVHD can be conceptualized in three sequential phases (Fig. 16.1) to provide a unified perspective on the complex cellular interactions and inflammatory cascades that lead to acute GVHD: (1) activation of the antigen-presenting cells (APCs); (2) donor T cell activation, differentiation, and migration; and (3) effector phase [34]. It is important to note that the three-phase description as discussed below allows for a unified perspective in understanding the biology. It is, however, not meant to suggest that all three phases are of equal importance or that GVHD occurs in a step-wise and sequential manner. The spatio-temporal relationships between the biological processes described below, depending on the context, are more likely to be chaotic and of varying intensity and relevance in the induction, severity, and maintenance of GVHD.

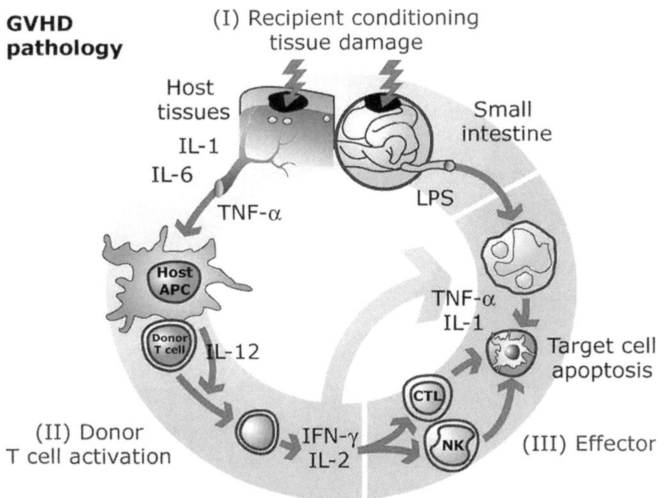

Fig. 16.1 Pathophysiology of GVHD: during step 1, irradiation and chemotherapy both damage and activate host tissues, including intestinal mucosa, liver, and the skin. Activated cell hosts then secrete inflammatory cytokines (e.g., TNF-α and IL-1), which can be measured in the systemic circulation. The cytokine release has important effects on APCs of the host, including increased expression of adhesion molecules (e.g., ICAM-1, VCAM-1) and of MHC class II antigens. These changes in the APCs enhance the recognition of host MHC and/or minor H antigens by mature donor T cells. During step 2, donor T cell activation is characterized by proliferation of GVHD T cells and secretion of multiple cytokines, including IL-2 and IFN-α. Induction of CTL and NK cell responses, and the priming of mononuclear phagocytes. In step 3, mononuclear cells primed by IFN-α and possibly other cytokines are triggered by a second signal such as endotoxin (LPS) to secrete cytopathic amounts of IL-I and TNF-α. LPS can leak through the intestinal mucosa damaged by the conditioning regimen to stimulate gut-associated lymphoid tissue or Kupffer cells in the liver; LPS that penetrate the epidermis may stimulate keratinocytes, dermal fibroblasts, and macrophages to produce similar cytokines in the skin. This mechanism results in the amplification of local tissue injury and further production of inflammatory effectors such as nitric oxide, which, together with CTL and NK effectors, leads to the observed target tissue destruction in the stem cell transplant host. CTL effectors use Fas/FasL, perforin/granzyme B, and membrane-bound cytokines to lyse target cells

A. Phase 1: Activation of antigen-presenting cells

The earliest phase of acute GVHD is initiated by the profound damage caused by the underlying disease and infections and further exacerbated by the BMT conditioning regimens (which include total-body irradiation (TBI) and/or chemotherapy) that are administered even before the infusion of donor cells [35–39]. This first step results in activation of the APCs [7]. Specifically, damaged host tissues respond with multiple changes, including the secretion of proinflammatory cytokines, such as TNF-α and IL-1, described as the "cytokine storm" [37, 38, 40]. Such changes increase expression of adhesion molecules, costimulatory molecules, MHC antigens, and chemokines gradients that alert the residual host and the infused donor immune cells [38]. These "danger signals" activate host

APCs [41, 42]. Damage to the GI tract from the conditioning is particularly important in this process because it allows for systemic translocation of immunostimulatory microbial products such as lipopolysaccaride (LPS) that further enhance the activation of host APCs and the secondary lymphoid tissue in the GI tract is likely the initial site of interaction between activated APCs and donor T cells [38, 43, 44]. This scenario accords with the observation that an increased risk of GVHD is associated with intensive conditioning regimens that cause extensive injury to epithelial and endothelial surfaces with a subsequent release of inflammatory cytokines and increases in expression of cell surface adhesion molecules [38, 39]. The relationship among conditioning intensity, inflammatory cytokine, and GVHD severity has been supported by elegant murine studies [40]. Furthermore, the observations from these experimental studies have led to two recent clinical innovations to reduce clinical acute GVHD: (a) reduced-intensity conditioning to decrease the damage to host tissues and thus limit activation of host APC; and (b) KIR mismatches between donor and recipients to eliminate the host APCs by the alloreactive NK cells [45, 46].

Host-type APCs that are present and have been primed by conditioning are critical for the induction of this phase; recent evidence suggests that donor-type APCs exacerbate GVHD, but, in certain experimental models, donor-type APC chimeras also induce GVHD [42, 47–49]. In clinical situations, if donor-type APCs are present in sufficient quantity and have been appropriately primed, they too might play a role in the initiation and exacerbation of GVHD [50–52]. Among the cells with antigen-presenting capability, DCs are the most potent and play an important role in the induction of GVHD [53]. Experimental data suggest that GVHD can be regulated by qualitatively or quantitatively modulating distinct DC subsets [54–59]. Langerhans cells were also shown to be sufficient for the induction of GVHD when all other APCs were unable to prime donor T cells, although the role for Langerhans cells when all APCs are intact is unknown [60]. Studies are yet to define roles for other DC subsets. In one clinical study, persistence of host DC after day 100 correlated with the severity of acute GVHD while elimination of host DCs was associated with reduced severity of acute GVHD [51]. The allo-stimulatory capacity of mature monocyte-derived DCs (mDCs) after reduced-intensity transplants was lower for up to 6 months compared to the mDCs from myeloablative transplant recipients, thus suggesting a role for host DCs and the reduction in "danger signals" secondary to less intense conditioning in acute GVHD [61]. Nonetheless, this concept of enhanced host APC activation explains a number of clinical observations such as increased risks of acute GVHD associated with advanced stage malignancy, conditioning intensity, and histories of viral infections.

Other professional APCs such as monocytes/macrophages or semi-professional APCs might also play a role in this phase [7]. For example, recent data suggest that host-type B cells might play a regulatory role under certain contexts [62]. Also host-/or donor-type nonhematopoietic stem cells, such as mesenchymal stem cells or stromal cells, when acting as APCs, have been shown to reduce T cell allogeneic responses, although the mechanism for such inhibition remains unclear.

The relative contributions of various APCs, professional or otherwise, remain to be elucidated.

B. Phase 2: Donor-T-Cell Activation, differentiation, and migration

The infused donor T cells interact with the primed APCs leading to the initiation of the second phase of acute GVHD. This phase includes antigen presentation by primed APCs, the subsequent activation, proliferation, differentiation, and migration of alloreactive donor T cells. After allogeneic HSC transplants, both host- and donor-derived APCs are present in secondary lymphoid organs [63, 64]. The T cell receptor (TCR) of the donor T cells can recognize alloantigens either on host APCs (direct presentation) or on donor APCs (indirect presentation) [65, 66]. In direct presentation, donor T cells recognize either the peptide bound to allogeneic MHC molecules or allogeneic MHC molecules without peptide [66, 67]. During indirect presentation, T cells respond to the peptide generated by degradation of the allogeneic MHC molecules presented on self-MHC [67]. Experimental studies demonstrated that APCs derived from the host, rather than from the donor, are critical in inducing GVHD across MiHA mismatch [7, 65]. Recent data suggest that presentation of distinct target antigens by the host and donor-type APCs might play a differential role in mediating target organ damage [7, 68, 69]. In humans, most cases of acute GVHD developed when both host DCs and donor dendritic cells (DCs) are present in peripheral blood after BMT [51].

1. **Co-stimulation**: The interaction of donor lymphocyte TCR with the host allopeptide presented on the MHC of APCs alone is insufficient to induce T cell activation [7, 70]. Both TCR ligation and co-stimulation via a "second" signal through interaction between the T cell co-stimulatory molecules and their ligands on APCs are required to achieve T proliferation, differentiation, and survival [71]. The danger signals generated in phase 1 augment these interactions and significant progress has been made on the nature and impact of these "second" signals [72, 73]. Co-stimulatory pathways are now known to deliver both positive and negative signals and molecules from two major families, the B7 family and the TNF receptor (TNFR) family play pivotal roles in GVHD [74]. Interruption of the second signal by blockade of various positive co-stimulatory molecules (CD28, ICOS, CD40, CD30, 4-1BB, and OX40) reduces acute GVHD in several murine models while antagonism of the inhibitory signals (PD-1 and CTLA-4) exacerbates the severity of acute GVHD [75–81]. The various T cell and APC co-stimulatory molecules and the impact on acute GVHD are summarized in Table 16.3. The specific context and the hierarchy in which each of these signals play a dominant role in the modulation of GVHD remain to be determined.

2. **T cell subsets**: T cells consist of several subsets whose responses differ based on antigenic stimuli, activation thresholds, and effector functions. The alloantigen composition of the host determines which donor T cell subsets proliferate and differentiate.

$CD4^+$ and $CD8^+$ cells: CD4 and CD8 proteins are co-receptors for constant portions of MHC class II and class I molecules, respectively [82]. Therefore, MHC class I (HLA-A, -B, -C) differences stimulate $CD8^+$ T cells and MHC class II (HLA-DR -DP, -DQ) differences stimulate $CD4^+$ T cells [82–85]. But clinical

Table 16.3 T cell-APC interactions

T cell	APC
Adhesion	
ICAMs	LFA-1
LFA-1	ICAMs
CD2 (LFA-2)	LFA-3
Recognition	
TCR/CD4	MHC II
TCR/CD8	MHC I
Costimulation	
CD28	CD80/86
CD152 (CTLA-4)	CD80/86
ICOS	B7H/B7RP-1
PD-1	PD-L1, PD-L2
Unknown	B7-H3
CD154 (CD40L)	CD40
CD134 (OX 40)	CD134L (OX40L)
CD137 (4-1BB)	CD137L (4-1BBL)
HVEM	LIGHT

HVEM, HSV glycoprotein D for herpesvirus entry mediator; LIGHT, homologous to lymphotoxins, shows inducible expression, and competes with herpes simplex virus glycoprotein D for herpes virus entry mediator (HVEM), a receptor expressed by T lymphocytes

trials of $CD4^+$ or $CD8^+$ depletion have been inconclusive [86]. This perhaps is not surprising, because GVHD is induced by MiHAs in the majority of HLA-identical BMT, which are peptides derived from polymorphic cellular proteins that are presented by MHC molecules [87]. Because the manner of protein processing depends on genes of the MHC, two siblings will have many different peptides in the MHC groove [87]. Thus in the majority of HLA-identical BMT, acute GVHD may be induced by either or both $CD4^+$ and $CD8^+$ subsets in response to minor histocompatibility antigens [86]. The peptide repertoire for class I or class II MHC remains unknown and might even be different in different individuals [88]. But it is plausible that only a few of the many of these peptides might behave as immunodominant "major minor" antigens that can potentially induce GVHD. In any event, such antigens remain to be identified and validated in large patient population.

Naïve and Memory subsets: Several independent groups have intriguingly found that the na ($CD62L^+$) T cells were alloreactive and caused acute GVHD but not the memory ($CD62L^-$) T cells across different donor/recipient strain combinations [89–92]. Furthermore, expression of na T cell marker CD62L was also found to be critical for regulation of GVHD by donor natural regulatory T cells [96]. By contrast, another recent study demonstrated that alloreactive memory T cells and their precursor cells (memory stem cells) caused robust GVHD [93, 94]. It remains as yet unknown whether the reduced GVHD potential of memory-type T cells from a na murine donor, in contrast to their ability to cause greater solid organ allo-rejection, is due to a consequence of the intense conditioning regimen and/or altered trafficking or from a restricted repertoire and/or from T cell intrinsic defect [99].

Regulatory T cells: Recent advances indicate that distinct subsets of regulatory $CD4^+CD25^+$, $CD4^+CD25^-IL10^+$ Tr cells, $\gamma\delta$T cells, DN^- T cells, NK T cells, and regulatory DCs control immune responses by induction of anergy or active suppression of alloreactive T cells [55, 56, 95–103]. Several studies have demonstrated a critical role for the natural donor $CD4^+CD25^+$ $Foxp3^+$ regulatory T (Treg) cells, obtained from na animals or generated ex vivo, in the outcome of acute GVHD. Donor $CD4^+CD25^+$ T cells suppressed the early expansion of alloreactive donor T cells and their capacity to induce acute GVHD without abrogating GVL effector function against these tumors [104, 105]. $CD4^+CD25^+$ T cells induced/generated by immature or regulatory host-type DCs and by regulatory donor-type myeloid APCs were also able to suppress acute GVHD [55]. One of the clinical studies that evaluated the relationship between donor $CD4^+CD25^+$ cells and acute GVHD in humans after matching sibling donor grafts and found that in contrast to the murine studies, donor grafts containing larger numbers of $CD4^+$ $CD25^+$ T cells developed more severe acute GVHD [106]. These data suggest that co-expression of $CD4^+$ and $CD25^+$ is insufficient because an increase in $CD25^+$ T cells in donor grafts is associated with greater risks of acute GVHD after clinical HCT. Another recent study found that Foxp3 mRNA expression (considered a specific marker for naturally occurring $CD4^+CD25^+$ Tregs) was significantly decreased in peripheral blood mononuclear cells from patients with acute GVHD [107, 108]. But Foxp3 expression in humans, unlike mice, may not be specific for T cells with a regulatory phenotype [109]. It is likely that the precise role of regulatory T cells in clinical acute GVHD will, therefore, not only depend upon identification of specific molecular markers in addition to Foxp3, but also on the ability for ex vivo expansion of these cells in sufficient numbers. Several clinical trials are underway in the US and Europe with attempts to substantially expand these cells ex vivo and use for prevention of GVHD.

Host $NK1.1^+$ T cells are another T cell subsets with suppressive functions have also been shown to suppress acute GVHD in an IL-4 dependent manner [102, 103, 110]. By contrast, donor NKT cells were found to reduce GVHD and enhance perforin mediated GVL in an IFN-γ dependent manner [111, 112]. Recent clinical data suggests that enhancing recipient NKT cells by repeated TLI conditioning promoted Th2 polarization and dramatically reduced GVHD [103]. Experimental data also show that activated donor NK cells can reduce GVHD through the elimination of host APCs or by secretion of transforming growth factor-β (TGF-β) secretion [112]. A murine BMT study using mice lacking SH2-containing inositol phosphatase (SHIP), in which the NK compartment is dominated by cells that express two inhibitory receptors capable of binding either self or allogeneic MHC ligands, suggests that host NK cells may play a role in the initiation of GVHD [113].

3. **T cell apoptosis**: Deletional mechanisms of tolerance fall into two categories: (1) central (thymic) deletion; and (2) peripheral deletion [114]. Central deletion is an effective way to eliminate continued thymic production of

alloreactive T cells. To this end, lymphoablative treatments have been used as a condition to create a mixed hematopoietic chimeric state in murine BMT models [120]. In this strategy, donor cells seed the thymus and maturing donor-reactive T cell clones are deleted through intrathymic apoptosis [115, 116]. The proportion of the peripheral T cell repertoire that can respond to allogeneic MHC antigens can play a critical role in the development of tolerance [117]. In the case of MHC-mismatched transplantation, the frequency of alloreactive T cells is at least five orders of magnitude greater than the frequency of peptide-specific T cells responding to a nominal antigen [117, 118]. The pathways of T cell apoptosis by which peripheral deletion occurs can be broadly categorized into activation-induced cell death (AICD) and passive cell death (PCD) [117]. An important mediator of AICD in T cells is the Fas receptor [119]. Activated T cells expressing the Fas molecule undergo apoptotic cell death when brought into contact with cells expressing Fas ligand. A critical role for Fas-mediated AICD has been clearly demonstrated in attenuation of acute GVHD by several Th1 cytokines [120–124].

PCD or "death by neglect" illustrates the exquisite dependence of activated T cells upon growth factors (e.g., IL-2, IL-4, IL-7, and/or IL-15) for survival; apoptotic cell death in this instance is largely due to rapid downregulation of Bcl-2 [125–127]. Transplantation of Bcl-xL T cells into nonirradiated recipients significantly exacerbates GVHD; however, no difference in GVHD mortality is observed in animals that have been lethally irradiated [128]. Selective elimination of donor T cells in vivo after BMT using transgenic T cells in which a thymidine kinase suicide gene is targeted to T cells has also been shown to attenuate the severity of acute GVHD [128–131]. Another recent approach to prevent GVHD is the selective depletion of alloantigen-specific donor T cells by photodynamic cell purging process wherein donor T cells are treated with photoactive 4, 5-dibromorhodamine 123 and subsequently exposed to visible light [132]. Thus, several deletional mechanisms have been shown to reduce acute GVHD but the conditions under which one or another of these deletional mechanisms predominate role remain to be determined.

4. **Cytokines and T cell differentiation**: APC and T cell activation result in rapid intracellular biochemical cascades that induce transcription of many genes including cytokines and their receptors. The Th1 cytokines (IFN-γ, IL-2 and TNF-α) have been implicated in the pathophysiology of acute GVHD [133–135]. IL-2 production by donor T cells remains the main target of many current clinical therapeutic and prophylactic approaches, such as CSP, tacrolimus, and monoclonal antibodies (mAbs) against the IL-2 and its receptor to control acute GVHD [136, 137]. But emerging data indicate an important role for IL-2 in the generation and maintenance of $CD4^{+}CD25^{+}$ $Foxp3^{+}$ T regs, suggesting that prolonged interference with IL-2 may have an unintended consequence in the prevention of the development of long-term tolerance after allogeneic HCT [138–141]. Similarly the role of other Th1 cytokines IFN-γ or their inducers as regulators or inducers of GVHD severity depends on the degree of allo-mismatch, the intensity of conditioning, and the T cell subsets that are involved after BMT [142–144]. Thus although the "cytokine storm" initiated in phase 1 and amplified by the Th1

cytokines correlates with the development of acute GVHD, early Th1 polarization of donor T cells to HCT recipients can attenuate acute GVHD suggesting that physiological and adequate amounts of Th1 cytokines are critical for GVHD induction, while inadequate production (extremely low or high) could modulate acute GVHD through a breakdown of negative feedback mechanisms for activated donor T cells [124, 135, 144–146]. Several different cytokines that polarize donor T cells to Th2 such as IL-4, G-CSF, IL-18, IL-11, rapamycin, and the secretion of IL-4 by NK1.1$^+$ T cells can reduce acute GVHD [147–154]. But Th1 and Th2 subsets cause injury of distinct acute GVHD target tissues and some studies failed to show a beneficial effect of Th2 polarization on acute GVHD [155]. Thus, the Th1/Th2 paradigm of donor T cells in the immunopathogenesis of acute GVHD has evolved over the last few years and its causal role in acute GVHD is complex and incompletely understood. IL-10 plays a key role in suppression of immune responses and its role in regulating experimental acute GVHD is unclear [156]. Recent clinical data demonstrate an unequivocal association of IL-10 polymorphisms with the severity of acute GVHD [157]. TGF-β, another suppressive cytokine was shown to suppress acute GVHD but to exacerbate chronic GVHD [158]. The roles of some other cytokines, such as IL-7 (that promotes immune reconstitution) and IL-13, remain unclear [159–162]. The role for Th17 cells, a recently described novel T cell differentiation in many immunological processes, is not yet known [163]. In any case, all of the experimental data so far collectively suggest that the timing of administration, the production of any given cytokine, the intensity of the conditioning regimen, and the donor-recipient combination may all be critical to the eventual outcome of acute GVHD.

5. **Leukocyte migration**: Donor T cells migrate to lymphoid tissues, recognize alloantigens on either host or donor APCs, and become activated. They then exit the lymphoid tissues and traffic to the target organs and cause tissue damage [164]. The molecular interactions necessary for T cell migration and the role of lymphoid organs during acute GVHD have recently become the focus of a growing body of research. Chemokines play a critical role in the migration of immune cells to secondary lymphoid organs and target tissues [165]. T-lymphocyte production of macrophage inflammatory protein-1alpha is critical to the recruitment of CD8$^+$ but not CD4$^+$ T cells to the liver, lung, and spleen during acute GVHD [166]. Several chemokines such as CCL2-5, CXCL2, CXCL9-11, CCL17, and CCL27 are overexpressed and might play a critical role in the migration of leukocyte subsets to target organs liver, spleen, skin, and lungs during acute GVHD [164, 167]. CXCR3$^+$ T and CCR5$^+$ T cells cause acute GVHD in the liver and intestine [164, 168–170]. CCR5-expression has also been found to be critical for Treg migration in GVHD [171]. In addition to chemokines and their receptors, expression of selectins and integrins and their ligands also regulate the migration of inflammatory cells to target organs [165]. For example, interaction between $\alpha4\beta7$ integrin and its ligand MadCAM-1 are important for homing of donor T cells to Peyer's patches and in the initiation of intestinal GVHD [43, 172]. $\alpha L\beta2$/ICAM1, 2, 3 and $\alpha4\beta1$/VCAM-2 interactions are important for homing to the lung and liver after experimental HCT [164]. The expression of CD62L on donor Tregs is critical for

their regulation of acute GVHD suggesting that their migration in secondary tissues is critical for their regulatory effects [64]. The migratory requirement of donor T cells to specific lymph nodes (e.g., Peyer's patches) for the induction of GVHD might depend on other factors such as the conditioning regimen, inflammatory milieu, etc. [43, 173]. Furthermore, FTY720, a pharmacologic sphingosine-1-phosphate receptor agonist, inhibited GVHD in murine but not in canine models of HCT [174, 175]. Thus, there might also be significant species differences in the ability of these molecules to regulate GVHD.

C. Phase 3: Effector Phase

The effector phase that leads to the GVHD target organ damage is a complex cascade of multiple cellular and inflammatory effectors that further modulate each reciprocal response, either simultaneously or successively. Effector mechanisms of acute GVHD can be grouped into cellular effectors [e.g., Cytotoxic T cells (CTLs)] and inflammatory effectors such as cytokines. Inflammatory chemokines expressed in inflamed tissues upon stimulation by proinflammatory effectors such as cytokines are specialized for the recruitment of effector cells, such as CTLs [176]. Furthermore, the spatio-temporal expression of the cytochemokine gradients might determine not only the severity but also the unusual cluster of GVHD target organs (skin, gut, and liver) [164, 177].

1: **Cellular effectors**: CTLs are the major cellular effectors of GVHD [178, 179]. The Fas–Fas ligand (FasL), the perforin–grazyme (or granule exocytosis), and TNFR-like death receptors (DR), such as TNF-related apoptosis-inducing ligand (TRAIL: DR4, 5 ligand) and TNF-like weak inducers of apoptosis (TWEAK: DR3 ligand) are the principle CTL effector pathways that have been evaluated after allogeneic BMT [179–184]. The involvement of each of these molecules in GVHD has been testing by utilizing donor cells that are unable to mediate each pathway. Perforin is stored in cytotoxic granules of CTLs and NK cells, together with granzymes and other proteins. Although the exact mechanisms remain unclear, following the recognition of a target cell through the TCR-MHC interaction, perforin is secreted and inserted into the cell-membrane, forming "perforin pores" that allow granzymes to enter the target cells and induce apoptosis through various downstream effector pathways such as caspases [185]. Ligation of Fas results in the formation of the death-inducing signaling complex and also activates caspases [186, 187].

Transplantation of perforin-deficient T cells results in a marked delay in the onset of GVHD in transplants across MiHA disparities only, both MHC and MiHA disparities, and across isolated MHC I or II disparities [179, 188–192]. However, mortality and clinical and histological signs of GVHD were still induced in these studies, even in the absence of perforin-dependent killing, demonstrating that the perforin–granzyme pathways play little role in target-organ damage. A role for the perforin–granzyme pathway for GVHD induction is also evident in studies employing donor-T cell subsets. Perforin- or granzyme B-deficient $CD8^+$ T cells caused less mortality than wild-type T cells in experimental transplants across a single MHC Class I mismatch. This pathway, however, seems to be less important compared to Fas/FasL pathway in CD4-mediated GVHD [191–193]. Thus, it seems that $CD4^+$ CTLs preferentially use the Fas-FasL pathway, whereas $CD8^+$ CTLs primarily use the perforin–granzyme pathway.

Fas, a TNF-receptor family member, is expressed by many tissues, including GVHD target organs [194]. Its expression can be upregulated by inflammatory cytokines such as IFN-γ and TNF-α during GVHD, and the expression of FasL is also increased on donor T cells, indicating that FasL-mediated cytotoxicity may be a particularly important effector pathway in GVHD [179, 195]. FasL-defective T cells cause less GVHD in the liver, skin and lymphoid organs [190, 193, 195]. The Fas–FasL pathway is particularly important in hepatic GVHD, consistent with the keen sensitivity of hepatocytes to Fas-mediated cytotoxicity in experimental models of murine hepatitis [179]. Fas-deficient recipients are protected from hepatic GVHD, but not from other organ GVHD, and administration of anti-FasL (but not anti-TNF) MAbs significantly blocked hepatic GVHD damage occurring in murine models [179, 196, 197]. Although the use of FasL-deficient donor T cells or the administration of neutralizing FasL MAbs had no effect on the development of intestinal GVHD in several studies, the Fas–FasL pathway may play a role in this target organ, because intestinal epithelial lymphocytes exhibit increased FasL-mediated killing potential [198]. Elevated serum levels of soluble FasL and Fas have also been observed in at least some patients with acute GVHD [199, 200].

The utilization of a perforin–granzyme and FasL cytotoxic double-deficient (cdd) mouse provides an opportunity to address whether other effector pathways are capable of inducing GVHD target organ pathology. An initial study demonstrated that cdd T cells were unable to induce lethal GVHD across MHC class I and class II disparities after sublethal irradiation [189]. However, subsequent studies demonstrated that cytotoxic effector mechanisms of donor T cells are critical in preventing host resistance to GVHD [183, 201]. Thus, when recipients were conditioned with lethal doses of irradiation, cdd $CD4^+$ T cells produced similar mortality to wild-type $CD4^+$ T cells [183]. These results were confirmed by a recent study demonstrating that GVHD target damage can occur in mice that lack alloantigen expression on the epithelium, preventing direct interaction between CTLs and target cells [184].

The participation of another death ligand receptor signaling pathway, TNF/TNFRs, has also been evaluated. Experimental data suggest that this pathway is crucial for GI GVHD (discussed below). Recently, several additional TNF family apoptosis-inducing receptors/ligands have been identified, including TWEAK, TRAIL, and LTβ/LIGHT; all three have all been proposed to play a role in GVHD and GVL responses [75, 202–208]. However, whether these distinct pathways play a more specific role for GVHD mediated by distinct T cell subsets in certain situations remains unknown. Intriguingly, recent data suggest that none of these pathways may be critical for mediating the rejection of donor grafts [202, 209]. Thus it is likely that their role in GVHD may be modulated by the intensity of conditioning and by the recipient T cell subsets. Existing experimental data suggest that perforin and TRAIL cytotoxic pathways are associated with $CD8^+$ T cell-mediated GVL [179]. The available experimental data are strongly skewed toward $CD8^+$ T cell-mediated GVL, based on the dominant role of this effector population in most murine GVT models; however, $CD4^+$ T cells can mediate GVL and might be crucial in clinical BMT depending on the type of malignancy and the expression of immunodominant antigens.

Taken together, although experimental data suggest there might be some distinction between the use of different lytic pathways for the specific GVHD target organs and GVL, the clinical applicability of these observations is as yet largely unknown.

2. **Inflammatory effectors**: Inflammatory cytokines synergize with CTLs resulting in the amplification of local tissue injury and further promotion of an inflammation, which ultimately leads to the observed target tissue destruction in the transplant recipient [20]. Macrophages which had been primed with IFN-γ during step 2 produce inflammatory cytokines TNF-α and IL-1, when stimulated by a secondary triggering signal [210]. This stimulus may be provided through Toll-like receptors (TLRs) by microbial products such as LPS and other microbial particles, which can leak through the intestinal mucosa damaged by the conditioning regimen and gut GVHD [211, 212]. It is now apparent that immune recognition through both TLR and non-TLRs (such as NOD) by the innate immune system also controls activation of adaptive immune responses [211, 213]. Recent clinical studies of GVHD suggested the possible association with TLR/NOD polymorphisms and severity of GVHD [214–216]. LPS and other innate stimuli may stimulate gut-associated lymphocytes, keratinocytes, dermal fibroblasts, and macrophages to produce pro-inflammatory effectors that play a direct role in causing target organ damage. Indeed, experimental data with MHC-mismatched BMT suggest that, under certain circumstances, these inflammatory mediators are sufficient in causing GVHD damage even in the absence of direct CTL-induced damage [47]. The severity of GVHD appears to be directly related to the level of innate and adaptive immune cell priming and release of pro-inflammatory cytokines such as TNF-α, IL-1, and nitric oxide (NO) [47, 212, 217–219].

The cytokines TNF-α and IL-1 are produced by an abundance of cell types during processes of both innate and adaptive immunity; they often have synergistic, pleiotrophic, and redundant effects on both activation and effector phases of GVHD [135]. A critical role for TNF-α in the pathophysiology of acute GVHD was first suggested over 20 years ago, because mice transplanted with mixtures of allogeneic BM and T cells developed severe skin, gut, and lung lesions that were associated with high levels of TNF-α mRNA in these tissues [220]. Target organ damage could be inhibited by infusion of anti-TNF-α MAbs, and mortality could be reduced from 100 to 50% by the administration of the soluble form of the TNF-α receptor (sTNFR), an antagonist of TNF-α [37, 40, 218]. Accumulating experimental data further suggests that TNF-α is involved in a multistep process of GVHD pathophysiology. TNF-α can: (1) cause cachexia, a characteristic feature of GVHD; (2) induce maturation of DCs, thus enhancing alloantigen presentation; (3) recruit effector T cells, neutrophilis, and monocytes into target organs through the induction of inflammatory chemokines; and (4) cause direct tissue damage by inducing apoptosis and necrosis. TNF-α is also involved in donor-T cell activation directly through its signaling via TNFR1 and TNFR2 on T cells. TNF-TNF1 interactions on donor T cells promote alloreactive T cell responses and TNF–TNFR2 interactions are critical for intestinal GVHD [205, 221]. TNF-α also seems to be important effector molecules in GVHD in skin and lymphoid tissue [220, 222]. Additionally, TNF-α might also be involved in hepatic GVHD, probably by enhancing effector cell migration to the liver via the induction of inflammatory chemokines [223]. An important role for TNF-α in clinical acute GVHD has been suggested by studies demonstrating elevated serum levels or TNF-α or elevated TNF-α mRNA expression in peripheral blood mononuclear cells in patients with acute GVHD and other endothelial complications, such as hepatic VOD [223–226]. Phase I-II trials using TNF-α antagonists reduced the severity of GVHD suggesting that it is a relevant effector in causing target organ damage [227, 228].

The second major pro-inflammatory cytokine that appears to play an important role in the effector phase of acute GVHD is IL-1 [229]. Secretion of IL-1 appears to occur predominantly during the effector phase of GVHD of the spleen and skin, two major GVHD target organs [230]. A similar increase in mononuclear cell IL-1 mRNA has been shown during clinical acute GVHD. Indirect evidence of a role for IL-1 in GVHD was obtained with administration of this cytokine to recipients in an allogeneic murine BMT model. Mice receiving IL-1 displayed a wasting syndrome and increased mortality that appeared to be an accelerated form of disease. By contrast, intra-peritoneal administration of IL-1ra starting upon d-10 post-transplant was able to reverse the development of GVHD in the majority of animals, providing a significant survival advantage to treated animals [231]. However, the attempt to use IL-1ra to prevent acute GBHD in a randomized trial was not successful [232].

As a result of activation during GVHD, macrophages also produce NO, which contributes to the deleterious effects on GVHD target tissues, particularly immunosuppression [219, 233]. NO also inhibits the repair mechanisms of target tissue destruction by inhibiting proliferation of epithelial stem cells in the gut and skin [234]. In humans and rats, the development of GVHD is preceded by an increase in serum levels of NO oxidation products [235–238].

Existing data demonstrate important role for various inflammatory effectors in GVHD. The relevance of currently studied or as yet unknown specific effectors could still be determined by other factors, including the intensity of preparatory regimens, the type of allograft, the T cell subsets, and the duration of BMT. In any event, both experimental and clinical data suggest an important role for both the cellular and inflammatory mediators in GVHD-induced target organ damage.

16.4 Prevention of Acute GVHD

Elimination of T cells with mAbs, immunotoxins, lectins, CD34 columns, or physical techniques are effective at reducing GVHD. A typical unmanipulated marrow transplant entails the infusion of $\sim 10^7$ T cells per kg of recipient weight. A T cell dose $\leq 10^5$/kg has been associated with complete control of GVHD [239].

More recently the combination of very high stem cell numbers and $<3 \times 10^4$ CD3 cells/kg allowed haploidentical transplantation without GVHD [240]. Presumably, host immune cells that survive the initial conditioning are responsible for graft rejection. When the stem cell source contains a large number of T cells, the GVHD reaction further reduces the residual population capable of alloreactivity, thus decreasing graft rejection. To some degree, the higher graft failure rates may be controlled by increasing the intensity of the immunosuppression of the conditioning regimen, or adding T cells back in [241–243]. Overall, there has been no improvement in survival that can be definitively attributed to T cell depletion.

Treatment of established GVHD with specific T cell antibodies has produced mixed results. While anti-thymocyte globulin has definite activity in established GVHD, the nonspecific clearance of T cells may result in increased opportunistic infections and no improvement in survival [18, 244, 245]. More specific therapy with humanized anti-IL-2 receptor antibody, the humanized anti-CD3 antibody, is promising since it offers the potential to selectively remove the activated T cells [246–250]. However, an increased risk of infection may still be observed [251].

The first generally prescribed GVHD preventive regimen was the administration of intermittent low dose methotrexate as developed in a dog model by Thomas and Storb [252]. The principle of this approach was to administer a cell-cycle-specific chemotherapeutic agent immediately after the transplant, when the T cells have started to divide after exposure to allogeneic antigens. Subsequently, the addition of anti-thymocyte globulin, prednisone, or both resulted in incremental improvement in the GVHD rate but no improvement in survival [253, 254]. Ultimately the course of methotrexate was abbreviated and combined with a T cell activation inhibitor, such as CSP or tacrolimus. The introduction of CSP in the late 1970s was a significant advance in GVHD prevention. A similar agent, tacrolimus, has been shown to provide similar control of GVHD [255]. As a single agent, CSP was about as effective as methotrexate [256]. However, in combination with methotrexate, there was a significant reduction in the incidence of GVHD and an improvement in survival [257]. Subsequent trials of tacrolimus and methotrexate, as compared with CSP and methotrexate, showed no advantage for either combination [255]. The addition of prednisone to the conventional two-drug regimen resulted in similar rates of GVHD and no improvement in survival [258].

Sirolimus (rapamycin) is a macrocyclic lactone immunosuppressant that is similar in structure to tacrolimus and CSP. All three drugs bind to immunophilins; however, sirolimus complexed with FKBP12 inhibits T cell proliferation by interfering with signal transduction and cell-cycle progression and can prevent GVHD [259]. Because Sirolimus acts through a separate mechanism from the tacrolimus-FKBP complex (and cyclosporine–cyclophilin complex), it is synergistic with both tacrolimus and CSP. More recently, mycophenolate mofetil (MMF) has been studied. It is the prodrug of mycophenolic acid (MPA) a selective inhibitor of inosine monophosphate dehydrogenase, an enzyme critical to the de novo synthesis of guanosine nucleotide. Since T lymphocytes are more dependent on such synthesis than myeloid or mucosal cells, MPA preferentially inhibits proliferative responses of T cells [260].

One hypothesis that flows from the three-step model of GVHD is that reduction of intestinal colonization with bacteria could prevent GVHD. Animal studies in germ-free environments support this notion, where GVHD was not observed until mice were colonized with gram-negative organisms [261]. Later, gut decontamination and use of a laminar air flow environment was associated with less GVHD and better survival in patients with severe aplastic anemia [262]. Similarly, studies of intestinal decontamination in patients with malignancies have shown less GVHD in some, but not all, studies [263–265]. Finally, another recent approach to GVHD prevention has been the use of nonmyeloablative conditioning transplants. This less intensive preparative regimen decreases the tissue toxicity and subsequent release of cytokines in animal models [40, 266]. Patients generally experience mild toxicity in the initial peritransplant period and develop little or no GVHD, although many develop GVHD later, especially after donor lymphocyte infusions. In fact, the rates of GVHD are often higher than with conventional transplants, and GVHD is associated with a significant portion of the GVL effect [267, 268].

An important role for TNF-α in clinical acute GVHD has been suggested by studies demonstrating elevated levels of TNF-α in the serum of patients with acute GVHD and other endothelial complications such as VOD [269–272]. Therapy of GVHD with humanized anti-TNF-α (infliximab) or a dimeric fusion protein consisting of the extracellular ligand-binding portion of the human TNF-α receptor (TNFR) linked to the Fc portion of human IgG1 (etanercept) have shown some promise [273–275]. The second major proinflammatory cytokine that appears to play an important role in the effector phase of acute GVHD is IL-1. Secretion of IL-1 appears to occur predominantly during the effector phase of GVHD in the spleen and skin, two major GVHD target organs [276]. IL-1 receptor antagonist (IL-1RA) is a naturally occurring pure competitive inhibitor of IL-1 that is produced by monocytes/macrophages and keratinocytes. Interestingly, the IL-1RA gene is polymorphic and the presence in the donor of the allele that is linked to higher secretion of IL-1RA was associated with less acute GVHD [277]. Two phase I/II trials showed promising data that specific inhibition of IL-1 with either the soluble receptor or IL-1RA could result in remissions in 50–60% of patients with steroid-resistant GVHD [278, 279]. But a subsequent randomized trial of the addition of IL-1RA or placebo to CSP and methotrexate beginning at the time of conditioning and continuing through day 14 after stem cell infusion did not show any protective effect of the drug, despite the attainment of very high plasma levels [232, 280]. Thus, at least as administered in this study, IL-1 inhibition was insufficient to prevent GVHD in humans. IL-11 was also able to protect the GI tract in animal models and prevent GVHD, but it did not prevent clinical GVHD [280]. Thus not all preclinical strategies successfully translate into new therapies.

Therapy for acute GVHD

Glucocorticoid steroids are the initial therapy for acute GVHD. The mechanisms by which steroids work are multifactorial, they act as lympholytic agents and inhibit the release of inflammatory cytokines such as IL-1, IL-2, IL-6, gamma interferon, and TNF-α. Because of its intravenous availability, methylprednisolone is the steroid most commonly given for acute GVHD. Various dosing regimens have been used, none of which is clearly superior. High-bolus doses (10–20 mg/kg or 500 mg/m^2) have higher initial response rates, but flares on tapering and opportunistic infections are common. Both the Seattle and Minnesota transplant groups have found that treatment with steroids was as effective as, or more effective than, other therapies or combination of therapies, with 20–40% of patients having durable long-term responses [281, 282]. Long-term salvage rates for patients who did not respond to steroids were 20% or less; most patients eventually died from infection, acute GVHD, and/or chronic GVHD. More recently, a randomized trial demonstrated that topical therapy with oral budosenide can have prednisone-sparing effects and is efficacious in treatment of GI GVHD [283]. Clinically, two types of failure of corticosteroid treatment of acute GVHD can be distinguished: true steroid resistance, i.e. progression of GVHD symptoms and manifestations while patients are receiving full dose corticosteroid treatment; and steroid dependence, i.e. re-occurrence (or flare) of GVHD during/or after tapering of steroid treatment [284]. In general, the prognosis with true steroid-resistant GVHD is worse than the prognosis of steroid-dependent patients [284]. A comparison of trials dealing with steroid resistant GVHD is hampered by variable inclusion of both patient groups in many of these trials. A variety of agents has been tested, including chemical immunosuppressants such as MMF, ATG, anti-CD3 anti-T cell antibodies, as well as more specific agents directed against activation or adhesion molecules anti-CD25, anti-CD147 or cytokines, or extracorporeal photopheresis [284, 285]. To date, there are no randomized trails testing one versus the other in this clinical situation.

Other Supportive Approaches

Infections are the main cause of death in patients with steroid-refractory acute GVHD, so careful surveillance and control of infections is mandatory in patients with acute GVHD. Fungal infections, especially aspergillosis, are the leading complication. Prophylaxis and early aggressive treatment should be facilitated by the introduction of new azoles (voriconazole and posiconozole) or echinocandins (caspofungin, micafungin), which broaden therapeutic efficacy with acceptable toxicity. A variety of further supplementary approaches have been suggested, such as large volumes of diarrhea, might be controlled by the use of octreotide and oral beclomethasone (or budesonide) [283, 286].

16.4.1 Chronic Graft-Versus-Host Disease

Chronic GVHD was initially defined as a GVHD syndrome presenting more than 100 days after transplant; its onset occurred either as an extension of acute GVHD (progressive), after a disease-free interval (quiescent), or with no precedent (de novo) [287, 288]. Chronic GVHD may be limited or extensive (Table 16.3). Any grade of acute GVHD increases the probability of chronic GVHD, although no singular pathologic feature of the former predicts the development of the latter. Its incidence ranges from 30 to 60% after transplantation with the bone marrow, although it may be higher after peripheral blood progenitor transplants [289].

As with acute GVHD, the immune system appears to be affected in all patients, who are highly susceptible to bacterial, viral, fungal, and opportunistic infections. Specific abnormalities of cellular immunity include decreases in the production of antibodies against specific antigens, defects in the number and function of $CD4^+$ T cells, and increases in the number of nonspecific suppressor cells, which further diminish lymphocyte responses. About 80% of patients have skin changes resembling widespread lichen planus with papulosquamous dermatitis, plaques, desquamation, dyspigmentation, and vitiligo [268, 290]. Destruction of dermal appendages leads to alopecia and onychodysplasia. Severe chronic GVHD of the skin can resemble scleroderma, with induration, joint contractures, atrophy, and chronic skin ulcers. Chronic cholestatic liver disease occurs in 80% of patients and often resembles acute GVHD; it rarely progresses to cirrhosis. Severe mucositis of the mouth and esophagus can result in weight loss and malnutrition. Intestinal involvement, however, is infrequent [268, 290]. Chronic GVHD also produces a sicca syndrome, with atrophy and dryness of mucosal surfaces caused by lymphocytic destruction of exocrine glands, usually affecting the eyes, mouth, airways, skin, and esophagus [22, 290, 291]. The hematopoietic system may also be affected, and thrombocytopenia is an unfavorable prognostic factor in patients with chronic GVHD [268]. Important predictors of unfavorable outcome are progressive onset, lichenoid skin changes, elevated serum bilirubin level, continued thrombocytopenia, and failure to respond in 9 months of therapy [268, 292–294]. Among patients with none of these risk factors, 70% are expected to survive, compared with less than 20% survival in patients with two or more of these risk factors [294].

Histologic examination of the immune system reveals involution of thymic epithelium, disappearance of Hassall's corpuscles, depletion of lymphocytes, and absence of secondary germinal centers in lymph nodes [295]. Pathologic skin findings include epidermal atrophy with changes characteristic of lichen planus and striking inflammation around eccrine units. Sclerosis of the dermis and fibrosis of the hypodermis subsequently develop. GI lesions include localized inflammation of the mucosa and stricture formation in the esophagus and small intestine [290]. Histologic findings in the liver are often similar to those that occur in acute GVHD but are more intense, with chronic changes such as fibrosis with hyalinization of portal triads, obliteration of bile ducts, and hepatocellular cholestasis [268]. The endocrine glands of the eyes, mouth, esophagus, and bronchi show destruction focused on centrally draining ducts, with secondary involvement of alveolar components [295]. Findings of bronchiolitis obliterans, similar to those that occur in rejection of lung transplants, are now generally considered a pulmonary manifestation of chronic GVHD, although the pathogenesis of this process remains unclear [295].

Table 16.4 Commonly administered drugs for GVHD prophylaxis and treatment

Drug	Mechanism	Adverse effects
Corticosteroids	Direct lymphocyte toxicity; suppress pro-inflammatory cytokines such as TNF-α	Hyperglycemia; acute psychosis; severe myopathy; neuropathy; osteoporosis; cataract development
Methotrexate (MTX)	Antimetabolite: inhibit T cell proliferation	Significant renal, hepatic, and gastrointestinal toxicities
Cyclosporine A (CSA)	IL-2 suppressor; blocks $Ca2^{+}$ dependent signal transduction distal to TCR engagement	Renal and hepatic insufficiency; hypertension; hyperglycemia; headache; nausea and vomiting; hirsutism; gum hypertrophy; seizure with severe toxicity
Tacrolimus (FK506)	IL-2 receptor; blocks $Ca2^{+}$ dependent signal transduction distal to TCR engagement	Similar to CSA
Mycophenolate mofetil (MMF)	Inhibits de novo purine synthesis	Body aches; abdominal pain; nausea and vomiting; diarrhea; neutropenia
Sirolimus	mTOR inhibitor	Thrombocytopenia; hyperlipidemia; TTP
Antithymocyte globulin (ATG)	Polyclonal immunoglobulin	Anaphylaxis; serum sickness

16.4.1.1 Clinical Manifestations of Chronic GVHD

Chronic GVHD can present with a plethora of clinical manifestations. Because of its unpredictable pattern and the late onset when patients are no longer receiving care at their transplant center, the diagnosis is often delayed or not recognized. The staging of chronic GVHD is summarized in Table 16.4. However, the recently developed NIH Consensus Criteria, upon validation, might soon become the standard for diagnosing and evaluating responses for chronic GVHD [296, 297].

Dermatological: Skin involvement in chronic GVHD presents with varied features. Lichenoid chronic GVHD presents as an erythematous, papular rash that resembles lichen planus with no typical distribution pattern [22]. Scleradermatous GVHD may involve the dermis and/or the muscular fascia and clinically resembles systemic sclerosis. The skin is thickened, tight, and fragile, with very poor wound-healing. Either hypopigmentation or hyperpigmentation may occur. In severe cases, the skin may become blistered and ulcerate. Hair changes can include increased brittleness, premature graying, and alopecia. Fingernails and toenails may also be affected by chronic GVHD. Destruction of sweat glands can cause hyperthermia [298].

Ocular: Ocular GVHD usually presents with xerophthalmia or dry eyes. Irreversible destruction of the lacrimal glands results in dryness, photophobia, and burning. Local therapy with preservative-free tears and ointment or the placement of punctal plugs by an ophthalmologist might be required. Conjunctival GVHD, a rare manifestation of severe chronic GVHD, has a poor prognosis [22, 298].

Oral: Oral GVHD causes xerostomia and/or food sensitivity [298]. More advanced disease may cause odynophagia due to esophageal damage and strictures, although esophageal involvement occurs rarely without oral disease. Physical examination may reveal only erythema with a few white plaques, prompting a misdiagnosis of thrush or herpetic infections. Lichenoid changes in advanced disease can cause extensive plaque formation [22].

Gastrointestinal: Patients with chronic GVHD have GI complaints that mimic other disease states, including acute GVHD, infection, dysmotility, lactose intolerance, pancreatic insufficiency, and drug-related side effects. In one retrospective review of the intestinal biopsies of patients with chronic GVHD and persistent GI symptoms, a majority of patients had evidence of both acute and chronic GVHD, and only 7% of the patients had isolated chronic GVHD [22, 290]. Thus, although chronic GVHD may involve the GI tract alone, it may be difficult to diagnose in those circumstances without concurrent acute GVHD.

Hepatic: Hepatic disease typically presents as cholestasis with elevated serum levels of alkaline phosphatase and bilirubin. Isolated hepatic chronic GVHD has become more common with the increasing use of donor lymphocyte infusions [10]. Liver biopsy is required to confirm the diagnosis of chronic hepatic GVHD in patients with no other target organ involvement.

Pulmonary: Bronchiolitis obliterans is a late and serious manifestation of chronic GVHD. Patients typically present with cough or dyspnea [298]. Severe sclerotic disease of the chest wall may also give rise to similar symptoms with no intrinsic pulmonary disease. Pulmonary function tests demonstrate obstructive physiology and a reduction in DLCO. Chest computed tomography results may be normal or may show hyperinflation with a ground glass appearance. Overall, patients with bronchiolitis obliterans have minimal response to therapy and a very poor prognosis. Patients with chronic GVHD are also at risk for chronic sinopulmonary infections but symptoms may be minimal [22].

Hematopoietic: Cytopenias in chronic GVHD are common. This may be a result of stromal damage, but autoimmune neutropenia, anemia, and/or thrombocytopenia are also seen. Thrombocytopenia at the time of chronic GVHD diagnosis is associated with poor prognosis. However, thrombocytopenia post-transplant is a poor prognostic factor regardless of GVHD and eosinophilia is occasionally seen with chronic GVHD [298].

Immunologic: Chronic GVHD is inherently immunosuppressive. Functional asplenia with an increased susceptibility to encapsulated bacteria is common, and circulating Howell-Jolly bodies can be seen on peripheral blood smear. Patients are also at risk for invasive fungal infections and *Pneumocystis carinii* pneumonia (PCP) [298]. Hypoglobulinemia is common and patients with levels below 500 mg/dL should be supplemented with intravenous immunoglobulin [298].

Musculoskeletal: Fascial involvement in sclerodermatous GVHD is usually associated with skin changes. Fasciitis in joint areas can cause severe restriction of range of motion. Muscle cramps are a common complaint in patients with chronic GVHD, but myositis with elevated muscle enzymes is rare. Many patients with chronic GVHD are on steroid therapy and have low levels of sex hormone post-transplant, avascular necrosis, osteopenia, and osteoporosis are thus frequent complications [298].

Although several cases have been described, whether kidneys, which are are primary targets in some animal models of chronic GVHD, are also involved is yet to be determined in large studies [299]. Among the myriad clinical features of chronic GVHD, three definitive signs appear to be risk factors for increased mortality: (1) extensive skin GVHD involving >50% of the body surface area; (2) platelet count of <100,000/μL; and (3) progressive onset and acute GVHD that continues uninterrupted beyond day 100 [300]. However, chronic GVHD remains and, except in those with obvious features, it is a hard diagnosis and even harder to assess in terms of response to therapy. Recent criteria established by NIH consensus conference might prove to be beneficial in establishing uniform criteria for diagnosis, treatment, and response [297]. The NIH consensus criteria are currently in the process being evaluated.

16.4.1.2 Differential Diagnosis

The distinction between chronic and acute GVHD has traditionally been based on the time of onset. However, with the advent of low intensity HCT that distinction has become somewhat moot. The NIH Working Group has, in addition to the two main categories of GVHD, added two subcategories. The broad category of acute GVHD includes: (1) classic acute GVHD (maculopapular rash, nausea, vomiting, anorexia, profuse diarrhea, ileus, or cholestatic hepatitis) occurring within 100 days after transplantation or donor leukocyte infusion (DLI) (without diagnostic or distinctive signs of chronic GVHD); and (2) persistent, recurrent, or late acute GVHD: features of classic acute GVHD without diagnostic or distinctive manifestations of chronic GVHD occurring beyond 100 days of transplantation or DLI (often seen after withdrawal of immune suppression). The broad category of chronic GVHD includes: (1) classic chronic GVHD without features characteristic of acute GVHD; and (2) an overlap syndrome in which features of chronic and acute GVHD appear together. In the absence of histologic or clinical signs or symptoms of chronic GVHD, the persistence, recurrence, or new onset of characteristic skin, GI tract, or liver abnormalities should be classified as acute GVHD, regardless of the time after transplantation. With appropriate stratification, patients with persistent, recurrent, or late/acute GVHD or overlap syndrome can be included in clinical trials with patients who have chronic GVHD.

16.4.2 Chronic GVHD: Pathophysiology

The pathophysiology of chronic GVHD is generally much less well understood than that of acute GVHD and has undergone less intensive experimental modeling [22]. It is important to recognize that chronic GVHD was originally defined as a temporal rather than a clinical or pathophysiological entity. The initial clinical reports of chronic GVHD described abnormalities that occurred at least 150 days after stem cell infusion [301, 302]. By convention, day 100 after stem cell infusion is used as an arbitrary divider between acute and chronic GVHD. But some manifestations of acute GVHD occur after day 100 and some manifestations of chronic GVHD may occur before day 100 [22, 298]. Thus, it is preferable to consider the clinical symptoms and signs *per se* rather than their timing of onset.

Relatively little is known about the pathophysiology of chronic GVHD. This is due in part to the absence of appropriate animal models that can capture the kinetics and the protean manifestation of chronic GVHD [303]. Based on certain clinical features, chronic GVHD has been considered to be an "autoimmune" disease, with some experimental data suggesting that chronic GVHD results from defective central negative selection leading to the generation of autoreactive clones that escape tolerogenic mechanisms operating in the periphery [304, 305]. This would indicate that the nonpolymorphic antigens expressed in both the donor and recipient, rather than MiHA antigens, are the likely targets. In contrast to T cells from animals with acute GVHD that are specific for host alloantigens, T cells from animals with chronic GVHD are specific for a public (common) determinant of MHC Class II molecules [306, 307]. These T cells are considered autoreactive because they recognize public MHC Class II determinants that are common to both donor and recipient, rather than polymorphic histocompatibility antigens that are specific for the host. The autoreactive cells of chronic GVHD are associated with a damaged thymus, which can be injured by several mechanisms including acute GVHD, the conditioning regimen, or age-related involution and atrophy. In chronic GVHD the ability of the thymus to delete autoreactive T cells (negative selection) and to induce tolerance is impaired [22, 308, 309]. However, no clear data exist on the isolation of "autoreactive" donor-derived T cell clones that equally recognize nonpolymorphic antigens from the donor and recipient cells. This, though, does not directly preclude the existence or a causative role for "autoreactive" T cells. Chronic GVHD could also be a product of T cells that have undergone relatively chronic antigen stimulation due the presence of inexhaustible and ubiquitous MiHA antigens. Allo-T cells under circumstances of chronic MiHA antigen stimulation can induce syndromes resembling those induced by the chronic antigen stimulation in autoimmune diseases. This concept is also consistent with the proposal of acute GVHD as a risk factor for chronic GVHD. The antigens targeted in chronic GVHD could be the same dominant ones targeted in acute GVHD, but the reactive T cells could be different; for example, they may secrete TGF-β. But antigens other than those that were initially immunodominant, even those not initially targeted but introduced through epitope spread—either the nonpolymorphic "auto-antigens" or now distinct MiHAs—could be important. It is also conceivable that regulatory mechanisms may frequently fail in allo-HCT, resulting in activation and expansion of T cells that recognize both nonpolymorphic and MiHA epitopes. One murine study suggested that development of chronic GVHD-like syndrome is target antigen-dependent [68]. Furthermore, chronic GVHD pathogenesis could, in part, be a consequence of T cell priming by donor-derived APCs [69]. In some patient subsets, responses to rituximab, presence of MiHA-specific antibodies, and the presence of chronic GVHD after T cell-depleted (TCD) allo-BMT would indicate that, in addition to donor T cells, donor B cells might either have direct effector role or prime T cells as APCs [310, 311]. It is important to understand that, given the myriad clinical presentation of chronic GVHD that tend to occur at variable time after HCT, it is possible that separate pathogenic mechanisms might be involved in causing distinct manifestations and that no single putative mechanism might be sufficient to cause chronic GVHD.

16.5 Therapy for Chronic GVHD

Chronic GVHD has a major impact on both QOL and survival, frequently involves multiple organs, and necessitates prolonged immunosuppressive therapy [312]. One report noted that 15% of cancer-free patients were still receiving immunosuppressive therapy after 7 years [313]. The more severe forms of chronic GVHD are clearly associated with less disease-free survival. Thus, the potential benefit of a graft-versus-leukemia effect is shadowed by significant treatment-related mortality [313].

Current therapies for chronic GVHD are of limited efficacy and there is no long-term satisfactory regimen for patients who fail front-line steroid-based therapy. Indeed, there is no Food and Drug Administration-approved medication for use in chronic GVHD. The lack of standardized response criteria to measure therapeutic efficacy poses a major obstacle to pursuing therapeutic trials in chronic GVHD. Overall survival and/or discontinuation of systemic immunosuppression are accepted long-term endpoints in chronic GVHD trials. The recent National Institutes of Health (NIH)-sponsored consensus project provided for the first time a set of standardized measures and definitions to use as response criteria in chronic GVHD [296, 297]. Nonetheless, these recommendations are yet be tested and validated in prospective studies. The NIH Consensus Conference has defined response measures classified in two main groups: clinician-assessed and patient-reported (see below) [303].

Measure	Clinician-assessed	Patient-reported
I. *Chronic GVHD-specific core measures*		
Signs	Organ-specific measures	Not applicable
Symptoms	Clinician-assessed symptoms	Patient-reported symptoms
Global rating	Mild, moderate, or severe	Mild, moderate, or severe
	0–10 Severity scale	0–10 Severity scale
	Seven-point change scale	Seven-point change scale
II. *Chronic GVHD nonspecific ancillary measures*		
Function	Grip strength	Patient-reported function
	Two-minute walk time	
Quality of life	–	Patient-reported health-related quality of life

The prevention of acute GVHD has not consistently resulted in a lower incidence of chronic GVHD. A clear example is the use of reduced-intensity transplants, consistently associated with a lower incidence of acute GVHD but with no major impact on chronic GVHD [314, 315]. The extended use of GVHD prophylaxis with CSP, or variations in the CSP dosage used, showed no beneficial effects on the incidence of chronic GVHD [312, 316]. The addition of thalidomide to CSP and methotrexate prophylaxis, the administration of IVIg, or early treatment based on biopsy findings of subclinical GVHD in an attempt to preemptively treat chronic GVHD were unsuccessful [312]. The most commonly used therapies to treat chronic GVHD are cyclosporine A (CSA) and prednisone. Sullivan et al. reported that prednisone alone is superior to prednisone plus azathioprine for primary treatment of patients with chronic GVHD [317]. However, in patients classified as high risk on the basis of platelet counts below 100,000/μL, treatment with prednisone alone resulted in only a 26% 5-year survival rate. When a similar group of patients was treated with alternating-day CSA and prednisone, 5-year survival exceeded 50% [318]. A recent randomized study of 287 patients with extensive GVHD found no statistically significant difference in non-relapse death at 5 years or in cumulative incidence of secondary therapy at 5 years when prednisone alone was compared to prednisone plus CSA [22]. For chronic GVHD that recurs or fails to respond to initial therapy, there is no standard treatment. Experimental therapies currently under evaluation include psoralen plus ultraviolet light A, MMF, thalidomide, total lymphoid irradiation, plaquenil, extracorporeal photopheresis, pentostatin, and acetretin (Table 16.4 lists the commonly used GVHD drugs and their common side-effects) [292, 319].

16.5.1 Syngeneic Graft-Versus-Host Disease

A GVHD-like syndrome that is usually self-limited and predominantly affects the skin can occur in recipients of syngeneic or autologous transplants [320]. It is also possible that some of the cases of syngeneic GVHD reflected a mistaken assumption that the donor was syngeneic without extensive molecular confirmation. It is primarily manifested as a rash that usually responds promptly to corticosteroid therapy. Although the level of severity may be grade II or III, the disease generally resolves promptly with the administration of glucocorticoids and is not life-threatening. Virtually all patients in whom a GVHD-like syndrome develops after syngeneic transplantation have been prepared with intensive conditioning regimens, usually involving irradiation. Experimental studies suggest that such conditioning is essential for the induction of thymic dysfunction, which is necessary for the development of the disease. Generation of autoreactive cells (a defect in thymic negative selection) as well as elimination of regulatory cells appear to be the requirements for the development of this disease [309]. An additional hypothesis is that in some individuals, maternal cells transmitted to them during their fetal development remain present throughout adult life [321]. These observations suggest the possibility that in some instances small numbers of HLA- incompatible cells (derived from the donor's mother) may be transmitted with HLA-identical transplants. Transplacentally transferred maternal cells may also play a role in the development of neonatal GVHD [321].

16.5.1.1 Transfusion-Associated GVHD

Most blood products administered to immunocompromised patients are now irradiated or at least leukocyte-depleted to avoid the transfusion of viable alloreactive T cells. With most homologous blood products, the MHC incompatibility between donor and recipient results in rapid clearance of transfused T cells by the recipient's immune system. However, occasionally transfusions from donors who are homozygous for one of the recipient's MHC haplotypes are not recognized as foreign by the recipient [322–324]. These cells can survive, "engraft", and may mount an immunologic attack against the unshared haplotype in the patient, resulting in transfusion-induced GVHD [324]. This syndrome of transfusion-associated GVHD differs from GVHD occurring after transplantation in regard to its kinetics but also in regard to its manifestations insofar as the recipient marrow is a major target [323]. Since the number of stem cells in the offending blood product is inadequate, there is no hemopoietic recovery from donor cells. This syndrome is generally fatal due to refractory pancytopenia in addition to other organ involvement.

16.6 Graft-Versus-Leukemia Responses

The GVL response after allogeneic HCT comes from the immunological attack of the host tissue and, by extension, the leukemia/tumor. This response represents a potent form of immunotherapy that circumvents some of the "immunoediting" mechanisms used by tumor cells to develop in the hosts. The power of the allo-immune response to eliminate malignancy was first reported more than 50 years ago in experimental models by Barnes and Loutit [1]. But the entity "GVL" and its close association with GVHD was not established until another 15 years later [325].

16.7 Clinical Features

Clinical evidence that the donor graft mediates important antileukemic effect comes from higher relapse rates for recipients of syngeneic stem cells than recipients of HLA-matched sibling grafts [326]. These findings have also been confirmed in a multicenter analysis of HCT recipients with acute myelogenous leukemia (AML) in first remission and subsequent retrospective analyses by the International Bone Marrow Transplant Registry (IBMTR) [327–329]. The second IBMTR analysis also showed that the magnitude of this GVL effect is greater for patients with CML and AML and not statistically significant for patients with ALL in first remission [330, 331].

Several case reports of patients with relapse of leukemia after allogeneic HCT noted remissions of the malignancy either after abrupt withdrawal of immunosuppression or during a flare of acute GVHD [332–335]. Patients who develop GVHD after allogeneic HCT experience relapse less frequently than similar patients who do not develop clinical disease. GVHD is protective against relapse both for HCT recipients with advanced leukemia and for patients who receive transplants in earlier stages of malignancy [336–339]. Additional analyses also suggest that the the magnitude of the GVL effect appears to be disease- and stage-specific [339–341]. Initial reports suggested that chronic GVHD was most protective against relapse, but other analyses demonstrate that acute GVHD is also protective [338, 339]. Based on these reports, newer trials of immunotherapy are designed to include cessation of immunosuppressive therapy (without taper) to induce a GVL reaction for patients whose malignancy has relapsed after HCT. Furthermore, Childs et al. [342] demonstrated that the graft-versus-tumor effect also plays an important role in inducing remissions from a non-hematologic malignancy, renal cell carcinoma.

Another line of clinical evidence regarding the GVL effect of allogeneic HCT and its tight linkage to GVHD comes from the studies employing T cell depletion of the donor graft. Donor T cells included in the stem cell graft are critical for acute GVHD, and T cell depletion by various strategies is one of the most successful means of reducing the incidence and severity of GVHD after allogeneic HCT [343–349]. Unfortunately, although T cell depletion results in less treatment-related morbidity and mortality, improved overall survival rates have not been reliably demonstrated. This failure is due in large part to a reciprocal increase in the subsequent relapse rate after T cell depletion, as well as to graft failure and other complications [341, 350, 351]. T cell depletion increases relapse rates, particularly in CML [339, 341, 352]. This observation provides further strong, albeit indirect, evidence that allogeneic donor T cells are important mediators not only of GVHD, but also of the GVL properties of the allogeneic stem cell graft. Finally the most compelling evidence of donor T cells in mediating GVL comes from the observations made from donor lymphocyte infusions (discussed below).

The induction of GVL is a complex process.

Genetic basis: The immunotherapeutic effect that occurs in allo-BMT setting is primarily mediated by allogeneic donor T or NK cells directed against the alloantigens shared by the recipient tumor and target tissues and/or tumor-specific antigens that have the advantage of not being subjected to tolerance mechanisms by the host tumor [353–355]. Understanding of the exquisite specificity of T cell responses led to attempts to identify specific antigens that are responsible for the GVL effect. Much of the focus has been on the identification of: (1) certain oncogenic viral proteins because these are absent in normal cells but expressed by transformed tumor cells (certain EBV peptides such as EBNA-1, LMP-1, LMP, LCL); (2) antigens that are expressed in a tissue-specific fashion (melanoma specific proteins); (3) proteins that are overexpressed in tumors (WT1, proteinase 3, survivin, telomerase reverse transcriptase, CYPB1 and Her-2/neu) [87, 354]. Although these antigens are specific, most T cell responses to these antigens are limited because of the poor immunogenicity of these proteins, expression on normal cells, defects in the processing or presentation of tumor antigens, or by production of factors that disable T cell responses. Thus, clinical attempts to obtain high specificity of T cell responses have been offset with difficulties in obtaining enough sensitivity and vice versa. Furthermore, given the current concepts of "stem cell" origins of leukemia and cancers, identification of the immunogenic proteins that are specifically expressed in the malignant stem cells and harnessing T cell responses to those antigens will be needed for the optimal GVL effect to cure malignancy [356, 357].

In contrast to the tumor-specific or associated antigens discussed above, alloantigens are not subjected to tolerance mechanisms. Vaccination strategies with autologous T cells utilizing tumor-associated or specific antigens have yielded disappointing clinical anti-tumor responses [358]. By contrast, allogeneic HCT has met with remarkable GVL responses, perhaps due to recognition of minor alloantigens, in addition to the tumor-associated antigens [359]. This concept has been demonstrated by recent murine studies, which showed that alloantigen on the tumor cells is required for GVL responses and that the principal targets of GVL are the immuno-dominant allogeneic MiHAs rather than the tumor-associated antigens [49, 360]. Thus T cells specific for MiHA antigens could provide for a potent GVL effect. Significant progress has been made in the identification of MiHAs that are specifically expressed in the host hematopoietic tissues and, therefore, might allow for a GVL response without causing GVHD [87]. Together, these results suggest that, in addition to tumor-specific proteins, expression of alloantigens and cognate interactions between donor T cells and the tumor tissues are required for the effective induction of the majority of GVL responses. However, T cells specific for some MiHAs are also responsible for GVHD and a means of consistently separating the beneficial GVL effect from GVHD has not yet been clinically achieved.

KIR polymorphisms: The two competing models, as described above, "mismatched ligand" and the "missing ligand" models for HLA-KIR allorecognition have been supported by clinical observations of GVL responses in different patient and transplant populations [46, 361]. The former model has been shown to separate GVL and GVHD responses in the context of TCD haplo-identical HCT for AML [362, 363]. Even though this model is supported by elegant laboratory studies, it was found to be invalid for ALL and also for AML after unrelated donor HCT with immunosuppression [362]. Recent retrospective clinical data suggest that GVHD and GVL can be separated by the "missing ligand" model in CML/AML and MDS patients after TCD HLA-identical sibling HCT [361, 363, 364]. Further validation of either models by clinical prospective studies and the a better understanding of the balance between the inhibitory and activating receptor-ligand interaction of the NK cells are needed to adequately exploit the interface between HLA and KIR genetics to separate GVHD from GVL [365, 366].

16.8 Immunobiology of GVL Responses

Given the tight association of clinical GVHD and GVL and the common biological principles governing these responses after allogeneic HCT, we will discuss below the similarities and distinctions between them in the context of the three cellular phases of GVHD [367].

Phase 1: Activation of APCs

The concept that tumor eradication after allogeneic HCT might not require toxic chemo-radiotherapy and could be achieved primarily by the immunotherapeutic effect from the GVL responses has led to the clinical development of non-myeloablative HCT for hematological and non-hematological malignancies [368]. Phase 1 is characterized by the development of "cytokine storm" generated "danger signals" from the conditioning regimen and the subsequent activation of APCs [367]. Experimental data suggested that the reduction in conditioning would attenuate the "cytokine storm", lead to the development of mixed donor-host chimerism, and confine the GVH response primarily to secondary lymphoid organs, thus cause less severe GVHD without impairing GVL responses [369–371]. However, non-myeloablative HCT has delayed the kinetics but did not reduce the overall incidence of GVHD and a significant number of patients either failed to respond or relapsed [19]. Furthermore, recent murine and human studies have suggested that homeostatic expansion of T cells in a lymphopenic environment induced by conditioning (as opposed to mere immunosuppression) improves the antitumor efficacy of adoptively transferred syngeneic or autologous T cells by increasing the availability of space, enhancing the memory responses, and reducing the competition for homeostatic cytokines (such as IL-7 and IL-15) for transferred T cells while eliminating regulatory T cells [372–374]. Thus low intensity HCT clearly demonstrates the principle of GVL effect, but the role of "cytokine storm", and homeostatic expansion of allogeneic T cells in shaping the intensity of GVL responses is not known.

Host and donor APCs are critical for the induction and severity of GVHD [7]. Activation of APCs is the key step in Phase 1 of GVHD [367]. Significant progress has been made in understanding the role of APCs in GVL. Recent experimental evidence demonstrates a crucial role for professional host APCs in the induction of GVL responses mediated by donor T cells even when the tumor cells showed some features of APCs [49, 375]. Tumors that merely express costimulatory molecules may still be unable to stimulate an effective immune response because of their various "immunoediting" processes that cause ineffective antigen presentation [376]. However, when the tumor cell itself functions as a professional APC, as with CML, it can generate an effective GVL response [377, 378]. By contrast, cancers such as acute leukemias that seldom differentiate into APCs generate poor GVL responses. Data also demonstrated that given sufficient time and a low tumor burden, cross-presentation of tumor-associated antigens and/or alloantigens by professional donor APCs can occur, and may promote or sustain GVL responses by maintaining or expanding alloreactive T cells after initial priming on host APCs [49, 378]. This concept is consistent with clinical GVL responses in CML in which the final stage of a GVL response to CML may be the result of donor T cells responding indirectly to the small number of CML stem cells or progenitors (which would be undifferentiated and therefore poor APCs) but to tumor antigens cross-presented on professional donor APCs. But it is unclear if such cross-presentation is sufficient to elicit effective GVL responses against acute or advanced leukemia. These data, however, suggest that GVL responses generated after low intensity conditioning may not be as robust as those after full intensity HCT and highlight the need for a clearer understanding of the effects of "cytokine storm" and lymphopenia generated danger signals on the activation of APCs in mediating GVL.

Phase 2: Donor T cell activation

The core of GVL responses, as with GVHD, is also dependent on the activation of appropriate numbers of T cells. The "second signals" from professional APCs (or certain tumor cells that function as effective APCs) are critical for generating an effective GVL response [75]. Several of the co-stimulatory pathways that modulate GVHD have also been have been evaluated in mediating GVL responses [75]. Blockade of CD28 co-stimulation preserved GVL responses but reduced GVHD in murine studies [379]. However, when the tumor cells also expressed B7 molecules, such blockade reduced he GVL responses [377]. Ex vivo blockade of CD40-CD40L interaction has been shown to reduce GVHD by generating Tregs but preserving GVL [75, 95, 380]. By contrast, blockade of the 4-1BB pathway reduced both GVHD and GVL [79]. The other co-stimulatory molecules (OX40 and ICOS) and the inhibitory molecules (CTLA-4 and PD-1) also modulate anti-tumor responses [80, 81]. A better understanding of the context (i.e. low intensity or DLI) and the hierarchy of timing, duration, and extent of co-stimulatory requirements of donor T cell subsets might allow for balancing the intensity of GVL and GVHD responses. Clinical and experimental evidence suggest that donor T cells numbers correlate with the severity of GVHD and GVL responses. TCD had reduced GVHD but increased disease relapse suggesting a role for T cell numbers in GVL responses as well [381]. Clinical attempts to separate GVHD and GVL by regulating allogeneic T cell dose have met with limited success. For example, administration of 1×10^5 T cells/kg after HLA-matched sibling transplantation did not mediate GVL effects and yet was associated with a measurable incidence of GVHD. Therefore, infusion of right numbers of donor T cell effectors is crucial for GVL responses [381]. This has been demonstrated by durable responses that are observed in CML and other malignancies after DLI (see below) despite the experimental evidence that host APCs stimulate a stronger GVL response than donor APCs [49, 378]. This could be because, clinically, DLI is almost always given without immunosuppression to an individual who has not developed GVHD either from the either chemical immunosuppression or physical removal of donor T cells from the allograft. This lack of immunosuppression after DLI increases the likelihood of a GVH response and DLI are almost always associated with clinical GVHD. The delivery of additional allogeneic effector cells in DLI also increases the effector:target ratio compared to the time of initial HCT. The latter is also clinically demonstrated by a more effective GVL response to DLI against minimal residual disease (bcr-abl positivity by PCR) compared to the response against high leukemic burden (e.g., blast crisis) in CML patients [378]. Thus DLI provides the proof, in principle, for the concepts that sufficient T cell numbers and appropriate antigen presentation are required for both GVHD and an effective GVL response.

T cell subsets: Most experimental studies have implicated donor $CD8^{+}$ T cells as the primary mediators of GVL, but there are no clinical data for $CD8^{+}$-mediated GVL responses in the absence of CD4 T cells [179, 354, 378]. Moreover, some clinical data suggest a role for greater CD4 mediated GVL responses without an increase in GVHD after allogeneic HCT and DLI [375, 382–385]. But it is unclear whether $CD4^{+}$ T cell-initiated GVL responses occur in the absence of generation of MiHA-specific $CD8^{+}$ T cells. Given the critical requirement of alloantigens for most GVL responses, the specific requirement of either CD4 and/or CD8 T cells for GVL and GVHD is likely to be determined by the expression of the relevant immuno-dominant MiHAs and/or tumor-associated proteins. Therefore, it is unlikely that GVHD and GVL responses can be separated under all circumstances merely by depletion of either subset of alloreactive T cells. However, experimental data suggest it might be possible to separate GVHD and GVL when certain donor T cell subsets are either depleted or infused (DLI) at appropriate intervals after transplant [375]. But the optimal time interval, if any, after clinical HCT is yet to be determined.

Because of recent identification and understanding of the role of various T cell subsets in mediating immune responses, depletion of specific T cell subsets to separate GVHD and GVL remains an area of active investigation. For example, recent experimental data suggest that CD62L expressing na T cells home to secondary lymph nodes are critical for initiating GVHD [90]. By contrast, CD62L negative effector memory T cells with enhanced reactivity to recall antigen-mediated GVL responses with minimal GVHD [90]. An important caveat to these data is the lack of a priori knowledge of the repertoire of human memory T cells, which would make it difficult to predict if their cells might cross-react only with tumor-associated antigens and/or with the recipient's alloantigens. Using CD62L status alone as a determinant of GVHD potential can also have other unintended consequences as recent studies have demonstrated, since its expression is critical for the regulation of GVHD by Tregs (see below). Moreover, it is not known whether the behavior of human memory T cells parallels that of murine memory T cells in their migratory, functional and cytolytic capabilities. Although Tregs reduce anti-tumor immunity in murine models and in human subjects, experimental data suggest that administration of donor-type Tregs, either at the time of HCT or when delayed, reduced GVHD but preserved $CD8^{+}$ mediated perforin-dependent GVL responses [104, 386]. Similar preservation of experimental GVL was also observed by harnessing donor NKT function with GCSF analogs [111, 112]. However, whether these observations are valid after clinical HCT, when the GVL responses might not be entirely dependent on CD8 T cells, remains unclear.

T cell migration: It is conceivable that manipulation of these interactions to focus the alloimmune response to lympho-hematopoietic tissues would enhance GVL responses but not GVHD. For example, blockade of CCR9 ligand TECK or CCR5, and CCL17 may prevent the migration of donor T cells to GI tract and skin, respectively, but preserve GVL [164]. Pharmacological manipulation with the immunosuppressive agent FTY720 has recently provided the proof in principle for this approach [174]. Given this redundancy, strategies to modulate the chemokine biology for separation of GVHD and GVL will require greater understanding of these networks in modulating the migration of not only specific T cell subsets but also of the other immune cells in the context of different conditioning regimens.

Phase 3: Effector phase of GVL

The effector arm of GVL is also characterized primarily by the antigen-specific cellular and less by the inflammatory components of allo-response. Experimental data demonstrate that neutralization of IL-1-α reduced GVHD preserved GVL [276]. By contrast, donor TNF-α secretion contributes to both GVHD and GVL effects and in some cases, antagonism of TNF-α, reduced GVHD and GVL responses [218, 387, 388]. Nonetheless antagonism of non-specific inflammatory effectors such as either IL-1 or TNF-α appears to regulate GVHD to a greater extent than GVL responses after experimental allogeneic HCT [388].

Several lines of experimental and clinical data demonstrate that antigen-specific donor T cell subsets and NK cells are the key effectors of GVL [354]. The cytotoxic pathways that are operative in the NK and T cell-mediated anti-tumor responses have been well characterized [354]. Fas ligand-mediated CTL of tumor targets is utilized by both NK and T (mostly Th1) cells. But most murine experiments with FasL-deficient donor T cells suggested that FasL is a key effector molecule for causing GVHD but not for GVL [179]. However, one study found that FasL is required for $CD4^+$ mediated GVL against myeloid leukemia [389, 390]. By contrast, even though perforin-mediated CTL pathways are also utilized by T (mostly Th2) and NK cells, experimental data with perforin-deficient donor T cells demonstrated a loss of GVL with a diminution in the severity of GVHD [179]. In some other experimental models, perforin was required only for GVL but not GVHD [179]. Recent data showed that TRAIL-mediated CTL had no effect on GVHD severity but was required for optimal GVL [203]. Therefore strategies that increase donor T cell TRAIL expression or enhance the susceptibility of tumors to TRAIL-mediated CTL (such as HDAC inhibitors) may promote a robust GVL effect without exacerbating GVHD [391–393]. Thus a significant progress has been made recently in understanding of the CTL pathways employed by donor T cells for GVL responses, but the role and context of utilization of these pathways by donor NK and NKT cells after allogeneic HCT are unknown.

16.9 Donor Leukocyte Infusions

Until recently the evidence for an important GVL effect in clinical transplantation was strong but largely circumstantial. The use of DLI to treat relapses after allogeneic HCT has now provided direct evidence of the GVL effect [394–396]. Kolb et al. [397] first reported three patients with relapsed CML who achieved complete cytogenetic remission after treatment with IFN-γ and DLI from the original donors. Subsequently these findings have been confirmed in several reports [398–400]. Two large retrospective studies of DLI have been reported from Europe and North America [396, 401]. Although the treatment protocols varied slightly among institutions, the results have been remarkably consistent. When results from these trials are combined, the complete remission rate for patients treated for relapsed CML is consistently 60–80% [394]. In many patients with CML, the response to DLI is not immediate. The average time to obtain a molecular remission is between 4 and 6 months, but disease-free survival after DLI depends on the stage of CML [402, 403]. Complete cytogenetic and molecular responses are achieved in almost 80% of patients treated with either early relapse (cytogenetic or molecular) or hematologic relapse of chronic phase [331]. Patients with more advanced CML (accelerated phase or blast crisis) are less likely to respond. Two recent analyses suggest that the outcome after unrelated DLI is similar to matched sibling DLI in CML patients [396, 401].

DLI has also produced remissions in acute leukemias. Several retrospective studies have reported respone rates in AML ranging from 20 to 65% [331, 401, 404–406]. A prospective study of 65 patients with advanced myeloid leukemia who received cytarabine and GCSF-primed DLI showed that 47% of the patients achieved complete remission with an overall survival of 19% at 2 years [407]. A recent EBMT analysis of DLI in patients with relapsed AML showed a response rate of 41% that did not change if chemotherapy was used prior to DLI [408]. Generally, DLI produces lower response rates and higher relapse rates in patients with AML than with CML [331, 378, 401, 404–406]. Extramedullary relapses at multiple regions that are usually not considered sanctuary sites for leukemia have been observed after DLI [406].

Although GVL responses to ALL after allogeneic HCT have been noted, DLI for ALL has generally been ineffective [409]. Reports both from Europe and from America found little benefit for DLI in ALL patients [378, 401, 404, 405]. Furthermore, Ruggeri et al. [362, 410] found that following TCD haploidentical HCT where the donors had anti-recipient NK cells, the probability of relapse was 85% for ALL in contrast to 0% for AML.

The experience with DLI for other hematologic malignancies such as NHL and multiple myeloma (MM) is much more limited. Case reports and small series demonstrate responses to DLI in patients with post-transplant relapses of low grade lymphoma and CLL and also after low intensity HCT [411, 412]. In MM, complete responses to DLI were observed in 25% of the patients in two small series [378, 413, 414]. DLI has also been shown to induce complete remission in a majority of patients with post-transplant lymphoproliferative disorders after allogeneic HCT [398, 415]. Viral infections may be treatable with DLI, and adoptive immune therapy with T cells specific for EBV and CMV have been shown to both treat and prevent these complications [416–420].

Complications of DLI: Adoptive immunotherapy with DLI causes significant morbidity. The major complications are myelosuppression and GVHD. Myelosuppression with anemia/thrombocytopenia/leucopenia or pancytopenia occurs in 34% of the patients [403]. Marrow aplasia presumably results from the destruction of host leukemia cells before recovery of normal donor hematopoiesis. This idea is supported by the observation that patients treated with donor MNC infusions for cytogenetic or molecular relapse rarely experience pancytopenia [396]. Occasionally, marrow aplasia has been persistent [398, 399, 421], although this toxicity has been successfully reversed with the infusion of additional donor stem cells in some patients [398, 399]. If pancytopenia is associated with chronic GVHD then immunosuppression might be the most appropriate therapy [422]. Acute and chronic GVHD have been the major direct complications from DLI. In retrospective and prospective studies, acute or chronic GVHD has been reported in 40–60% of evaluable patients. In most studies, GVHD correlates with GVL activity and response [396, 401]. In the North American analysis, over 90% of complete responders developed acute GVHD and 88% of responders developed chronic GVHD. Of 23 patients who did not experience GVHD, only three achieved a complete remission. In 92 patients who had no response, only 35% had acute GHVD and only 13% had chronic GVHD. In the EBMT analysis, 41% of DLI recipients developed grade II–IV acute GVHD [396]. It should be noted, however, that many patients who fail to respond to GVL induction will die shortly of progressive disease and may not survive long enough for GVHD to develop. This is particularly important for patients with acute leukemia. It is also important to emphasize that a number of complete responses were seen in patients without any sign of GVHD. The GVL effect in the absence of clinical GVHD provides important evidence for GVL activity separate from GVHD.

In general, GVHD that occurs after DLI has been mild-to-moderate and has been responsive to immunosuppressive therapy. This observation is important since the dose of T cells often administered with DLI may be 10- to 100-fold higher than the T cell dose administered at the time of transplant, and DLI is given without additional immunosuppression. At the time of transplant, if immunosuppression is withheld or a similar dose of donor T cells is given as is used for DLI (to augment GVL), toxicity is unacceptable [423, 424]. This may be because the effects of GVHD are more tolerable when separated from the acute transplant toxicity. GVHD can be stimulated and exacerbated by the "cytokine storm" that can accompany transplantation. Tissue damage from the intensive conditioning regimen, infections, and other physiological insults result in a cascade of events that ultimately augment the GVHD reaction [229]. When GVHD is induced independently of other transplant-related toxicity, it may be more manageable with appropriate immunosuppression. Nevertheless, in some cases DLI-induced GVHD may be quite severe; as shown in Table 16.2, 20-35% of DLI recipients can be anticipated to develop grade III–IV acute GVHD. Furthermore, acute GVHD has contributed to death in almost 10% of patients [425]. Chronic GVHD occurs in 30−60% of recipients of unmanipulated DLI.

It is notable that the clinical presentation of GVHD after DLI is somewhat different than after myeloablative allogeneic HCT. For instance, the onset of GVHD may be later/after DLI. The median time to onset of acute GVHD is approximately 32–42 days [401, 426] compared to a median time to onset of 16–20 days following myeloablative T-replete transplantation [255, 427]. Therefore, not only can the severity of GVHD be influenced by the use of intensive conditioning, the tissue damage and inflammatory cytokine release may influence the pace of GVHD development as well. This possibility is supported by the finding that time to onset of acute GVHD is also delayed following reduced-intensity transplantation [428]. The target organs of acute GVHD following DLI are the same as those seen following HCT, but the clinical manifestations may differ. A hepatitic variant of liver GVHD characterized by marked elevations of serum aminotransferase levels more than 10 times the upper limit of normal was observed in 11/73 (15%) patients who received DLI at Johns Hopkins University [268]. Characteristic skin, liver, and intestinal acute GVHD manifestations can be seen following DLI, but their frequency and severity has not been well described. In a study of 81 patients who received DLI for relapse, mixed chimerism, or prophylaxis following reduced-intensity HCT, skin, liver, and intestinal GVHD developed in 26, 8, and 14%, respectively [429]. Likewise, characteristic findings of chronic GVHD can be seen following DLI [429] and on occasion both acute and chronic GVHD can develop simultaneously following DLI [430].

In recent years, DLI has been utilized in the context of reduced-intensity allogeneic HCT. When conditioning is minimal and GVHD has not already occurred, DLI is often required to either reverse mixed chimerism or treat persistent disease [429, 431, 432]. GVHD after DLI, following non-myeloablative setting, develops 19–45% of the time, while chronic GVHD develops 28–34% of the time. In two studies of a combined 134 subjects who received DLI after reduced-intensity transplantation, 37 (28%) patients developed acute GVHD, 15 (11%) developed grade III–IV GVHD, and 8 (6%) died from GVHD [429, 431]. There was no statistically significant relationship between DLI dose and GVHD in either study. The clinical features of GVHD after DLI, following reduced-intensity HCT, are similar to the GVHD that occurs after DLI is used to treat relapse following myeloablative HCT. The similarity between GVHD after DLI in either setting suggests that the reduction of tissue damage associated with either no chemotherapy (as in DLI for relapse following prior myeloablative HCT) or reduced dose chemotherapy (as in DLI after reduced-intensity HCT) may be an important common theme.

It is likely that effector cells responsible for the GVL and GVHD effects of HCT will similarly be responsible for the GVL effect associated with DLI, although this assumption has not been formally proven. The administration of select subsets of donor mononuclear cell fractions is the ideal setting in which to dissect the cellular mechanisms responsible for GVL induction and strategies that delay the infusion of these various cellular subsets will help define the mechanisms and the enhance the efficacy of DLI.

References

1. Barnes D, Corp M, Loutit J, Neal F (1956) Treatment of murine leukaemia with X-rays and homologous bone marrow: preliminary communication. Br Med J 2:626–630
2. Barnes D, Loutit J (1957) Treatment of murine leukaemia with X-rays and homologous bone marrow. Br J Haematol 3:241–252
3. Truitt R, Johnson B, McCabe C, Weiler M (1997) Graft versus leukemia. In: Ferrara J, Deeg H, Burakoff S (eds) Graft-vs-host disease (edn 2). Marcel Dekker, Inc., New York, pp 385–424
4. Billingham RE (1966) The biology of graft-versus-host reactions. Harvey Lec 62:21–78
5. Petersdorf EW (2006) Immunogenomics of unrelated hematopoietic cell transplantation. Curr Opin Immunol 18(5):559–564
6. Erlich HA, Opelz G, Hansen J (2001) HLA DNA typing and transplantation. Immunity 14(4):347–356
7. Shlomchik WD (2007) Graft-versus-host disease. Nat Rev Immunol 7(5):340–352
8. Mori M, Beatty PG, Graves M, Boucher KM, Milford EL (1997) HLA gene and haplotype frequencies in the North American population: the National Marrow Donor Program Donor Registry. Transplantation 64(7):1017–1027
9. Deeg HJ, Cottler-Fox M (1990) Clinical spectrum and pathophysiology of acute graft-vs.-host disease. In: Burakoff SJ, Deeg HJ, Ferrara J, Atkinson K (eds) Graft-vs.-host disease: immunology, pathophysiology, and treatment. Marcel Dekker, Inc., New York, pp 539–555
10. Arai S, Lee LA, Vogelsang GB (2002) A systematic approach to hepatic complications in hematopoietic stem cell transplantation. J Hematother Stem Cell Res 11(2):215–229
11. Arai S, Margolis J, Zahurak M, Anders V, Vogelsang GB (2002) Poor outcome in steroid-refractory graft-versus-host disease with antithymocyte globulin treatment. Biol Blood Marrow Transplant 8(3):155–160
12. Sale GE, Shulman HM (1984) The pathology of bone marrow transplantation. Masson, New York
13. Sale GE (1996) Does graft-versus-host disease attack epithelial stem cells? Mol Med Today 2:114–119
14. Sale GE, Shulman HM, Gallucci BB, Thomas ED (1985) Young rete ridge keratinocytes are preferred targets in cutaneous graft-vs-host disease. Am J Pathol 118:278–287
15. Weisdorf DJ, Hurd D, Carter S, Howe C, Jensen LA, Wagner J, Stablein D, Thompson J, Kernan NA (2003) Prospective grading of graft-versus-host disease after unrelated donor marrow transplantation: a grading algorithm versus blinded expert panel review. Biol Blood Marrow Transplant 9(8):512–518
16. Glucksberg H, Storb R, Fefer A, Buckner C, Neiman P, Clift R, Lerner K, Thomas E (1974) Clinical manifestations of graft-versus-host disease in human recipients of marrow from HLA matched sibling donors. Transplantation 18:295–304
17. Leisenring WM, Martin PJ, Petersdorf EW, Regan AE, Aboulhosn N, Stern JM, Aker SN, Salazar RC, McDonald GB (2006) An acute graft-versus-host disease activity index to predict survival after hematopoietic cell transplantation with myeloablative conditioning regimens. Blood 108(2):749–755
18. Martin P, Schoch G, Fisher L, Byers V, Anasetti C, Applebaum F, Beatty P, Doney K, McDonald G, Sanders J, Sullivan K, Storb R, Thomas E, Witherspoon R, Lomen P, Hannigan J, Hansen J (1990) A retrospective analysis of therapy for acute graft-versus-host diesase: initial Treatment. Blood 76:1464–1472

19. Deeg HJ, Antin JH (2006) The clinical spectrum of acute graft-versus-host disease. Semin Hematol 43(1):24–31
20. Antin JH, Ferrara JL (1992) Cytokine dysregulation and acute graft-versus-host disease. Blood 80(12):2964–2968
21. Goker H, Haznedaroglu IC, Chao NJ (2001) Acute graft-vs-host disease: pathobiology and management. Exp Hematol 29(3):259–277
22. Vogelsang GB, Lee L, Bensen-Kennedy DM (2003) Pathogenesis and treatment of graft-versus-host disease after bone marrow transplant. Annu Rev Med 54:29–52
23. Ferrara JLM, Deeg HJ (1991) Graft versus host disease. N Engl J Med 324:667–674
24. Nevo S, Enger C, Swan V, Wojno KJ, Fuller AK, Altomonte V, Braine HG, Noga SJ, Vogelsang GB (1999) Acute bleeding after allogeneic bone marrow transplantation: association with graft versus host disease and effect on survival. Transplantation 67(5):681–689
25. Iqbal N, Salzman D, Lazenby AJ, Wilcox CM (2000) Diagnosis of gastrointestinal graft-versus-host disease. Am J Gastroenterol 95(11):3034–3038
26. Ponec RJ, Hackman RC, McDonald GB (1999) Endoscopic and histologic diagnosis of intestinal graft-versus-host disease after marrow transplantation. Gastrointest Endosc 49(5):612–621
27. Snover DC, Weisdorf SA, Vercellotti GM, Rank B, Hutton S, McGlave P (1985) A histopathological study of gastric and small intestinal graft-versus-host disease following allogeneic bone marrow transplantation. Hum Pathol 16:387–392
28. Cooke KR, Yanik G (2004) Acute lung injury after allogeneic stem cell transplantation: is the lung a target of acute graft-versus-host disease? Bone Marrow Transplant 34(9):753–765
29. Fukuda T, Hackman RC, Guthrie KA, Sandmaier BM, Boeckh M, Maris MB, Maloney DG, Deeg HJ, Martin PJ, Storb RF, Madtes DK (2003) Risks and outcomes of idiopathic pneumonia syndrome after nonmyeloablative and conventional conditioning regimens for allogeneic hematopoietic stem cell transplantation. Blood 102(8):2777–2785
30. Lee CK, Gingrich RD, Hohl RJ, Ajram KA (1995) Engraftment syndrome in autologous bone marrow and peripheral stem cell transplantation. Bone Marrow Transplant 16(1):175–182
31. Gorak E, Geller N, Srinivasan R, Espinoza-Delgado I, Donohue T, Barrett AJ, Suffredini A, Childs R (2005) Engraftment syndrome after nonmyeloablative allogeneic hematopoietic stem cell transplantation: incidence and effects on survival. Biol Blood Marrow Transplant 11(7):542–550
32. Spitzer TR (2001) Engraftment syndrome following hematopoietic stem cell transplantation. Bone Marrow Transplant 27(9):893–898
33. Sviland L, Pearson AD, Eastham EJ, Hamilton PJ, Proctor SJ, Malcolm AJ (1988) Histological features of skin and rectal biopsy specimens after autologous and allogeneic bone marrow transplantation. J Clin Pathol 41(2):148–154
34. Reddy P, Ferrara JL (2003) Immunobiology of acute graft-versus-host disease. Blood Rev 17(4):187–194
35. Gale RP, Bortin MM, van Bekkum DW, Biggs JC, Dicke KA, Gluckman E, Good RA, Hoffman RG, Kay HEM, Kersey JH, Marmont A, Masaoka T, Rimm AA, van Rood JJ, Zwaan FE (1987) Risk factors for acute graft-versus-host disease. Br J Haematol 67:397–406
36. Clift RA, Buckner CD, Appelbaum FR, Bearman SI, Petersen FB, Fisher LB, Anasetti C, Beatty P, Bensigner WI, Doney K, Hill RS, McDonald GB, Martin P, Sanders J, Singer J, Stewart P, Sullivan KM, Witherspoon R, Storb R, Hansen JA, Thomas ED (1990) Allogeneic marrow transplantation in patients with acute myeloid leukemia in first remission: a randomized trial of two irradiation regimens. Blood 76:1867–1871
37. Xun CQ, Thompson JS, Jennings CD, Brown SA, Widmer MB (1994) Effect of total body irradiation, busulfan-cyclophosphamide, or cyclophosphamide conditioning on inflammatory cytokine release and development of acute and chronic graft-versus-host disease in H-2-incompatible transplanted SCID mice. Blood 83:2360–2367
38. Hill G, Ferrara J (2000) The primacy of the gastrointestinal tract as a target organ of acute graft-versus-host disease: rationale for the use of cytokine shields in allogeneic bone marrow transplantation. Blood 95(9):2754–2759

39. Paris F, Fuks Z, Kang A, Capodieci P, Juan G, Ehleiter D, Haimovitz-Friedman A, Cordon-Cardo C, Kolesnick R (2001) Endothelial apoptosis as the primary lesion initiating intestinal radiation damage in mice. Science 293(5528):293–297
40. Hill GR, Crawford JM, Cooke KR, Brinson YS, Pan L, Ferrara JL (1997) Total body irradiation and acute graft-versus-host disease: the role of gastrointestinal damage and inflammatory cytokines. Blood 90(8):3204–3213
41. Matzinger P (2002) The danger model: a renewed sense of self. Science 296(5566):301–305
42. Shlomchik WD, Couzens MS, Tang CB, McNiff J, Robert ME, Liu J, Shlomchik MJ, Emerson SG (1999) Prevention of graft versus host disease by inactivation of host antigen-presenting cells. Science 285(5426):412–415
43. Murai M, Yoneyama H, Ezaki T, Suematsu M, Terashima Y, Harada A, Hamada H, Asakura H, Ishikawa H, Matsushima K (2003) Peyer's patch is the essential site in initiating murine acute and lethal graft-versus-host reaction. Nat Immunol 4(2):154–160
44. Cooke K, Hill G, Crawford J, Bungard D, Brinson Y, Delmonte J Jr, Ferrara J (1998) Tumor necrosis factor-a production to lipopolysaccharide stimulation by donor cells predicts the severity of experimental acute graft versus host disease. J Clin Invest 102:1882–1891
45. Slavin S (2000) New strategies for bone marrow transplantation. Curr Opin Immunol 12(5):542–551
46. Velardi A, Ruggeri L, Alessandro Moretta, Moretta L (2002) NK cells: a lesson from mismatched hematopoietic transplantation. Trends Immunol 23(9):438–444
47. Teshima T, Ordemann R, Reddy P, Gagin S, Liu C, Cooke KR, Ferrara JL (2002) Acute graft-versus-host disease does not require alloantigen expression on host epithelium. Nat Med 8(6):575–581
48. Jones SC, Murphy GF, Friedman TM, Korngold R (2003) Importance of minor histocompatibility antigen expression by nonhematopoietic tissues in a CD4+ T cell-mediated graft-versus-host disease model. J Clin Invest 112(12):1880–1886
49. Reddy P, Maeda Y, Liu C, Krijanovski OI, Korngold R, Ferrara JL (2005) A crucial role for antigen-presenting cells and alloantigen expression in graft-versus-leukemia responses. Nat Med 11(11):1244–1249
50. Arpinati M, Green CL, Heimfeld S, Heuser JE, Anasetti C (2000) Granulocyte-colony stimulating factor mobilizes T helper 2-inducing dendritic cells. Blood 95(8):2484–2490
51. Auffermann-Gretzinger S, Lossos IS, Vayntrub TA, Leong W, Grumet FC, Blume KG, Stockerl-Goldstein KE, Levy R, Shizuru JA (2002) Rapid establishment of dendritic cell chimerism in allogeneic hematopoietic cell transplant recipients. Blood 99(4):1442–1448
52. MacDonald KP, Rowe V, Clouston AD, Welply JK, Kuns RD, Ferrara JL, Thomas R, Hill GR (2005) Cytokine expanded myeloid precursors function as regulatory antigen-presenting cells and promote tolerance through IL-10-producing regulatory T cells. J Immunol 174(4):1841–1850
53. Banchereau J, Steinman RM (1998) Dendritic cells and the control of immunity. Nature 392:245–252
54. Duffner UA, Maeda Y, Cooke KR, Reddy P, Ordemann R, Liu C, Ferrara JL, Teshima T (2004) Host dendritic cells alone are sufficient to initiate acute graft-versus-host disease. J Immunol 172(12):7393–7398
55. Sato K, Yamashita N, Baba M, Matsuyama T (2003) Regulatory dendritic cells protect mice from murine acute graft-versus-host disease and leukemia relapse. Immunity 18(3):367–379
56. Maeda Y, Reddy P, Lowler KP, Liu C, Bishop DK, Ferrara JL (2005) Critical role of host gammadelta T cells in experimental acute graft-versus-host disease. Blood 106(2):749–755
57. Chorny A, Gonzalez-Rey E, Fernandez-Martin A, Ganea D, Delgado M (2006) Vasoactive intestinal peptide induces regulatory dendritic cells that prevent acute graft-versus-host disease while maintaining the graft-versus-tumor response. Blood 107(9):3787–3794
58. Macdonald KP, Kuns RD, Rowe V, Morris ES, Banovic T, Bofinger H, O'Sullivan B, Markey KA, Don AL, Thomas R, Hill GR (2007) Effector and regulatory T cell function is differentially regulated by RelB within antigen presenting cells during GVHD. Blood 109:5049–5057

59. Paraiso KH, Ghansah T, Costello A, Engelman RW, Kerr WG (2007) Induced SHIP deficiency expands myeloid regulatory cells and abrogates graft-versus-host disease. J Immunol 178(5):2893–2900
60. Merad M, Hoffmann P, Ranheim E, Slaymaker S, Manz MG, Lira SA, Charo I, Cook DN, Weissman IL, Strober S, Engleman EG (2004) Depletion of host Langerhans cells before transplantation of donor alloreactive T cells prevents skin graft-versus-host disease. Nat Med 10(5):510–517
61. Nachbaur D, Kircher B, Eisendle K, Latzer K, Haun M, Gastl G (2003) Phenotype, function and chimaerism of monocyte-derived blood dendritic cells after allogeneic haematopoietic stem cell transplantation. Br J Haematol 123(1):119–126
62. Rowe V, Banovic T, MacDonald KP, Kuns R, Don AL, Morris ES, Burman AC, Bofinger HM, Clouston AD, Hill GR (2006) Host B cells produce IL-10 following TBI and attenuate acute GVHD after allogeneic bone marrow transplantation. Blood 108(7):2485–2492
63. Korngold R, Sprent J (1980) Negative selection of T cells causing lethal graft-versus-host disease across minor histocompatibility barriers. Role of the H-2 complex. J Exp Med 151(5):1114–1124
64. Beilhack A, Schulz S, Baker J, Beilhack GF, Wieland CB, Herman EI, Baker EM, Cao YA, Contag CH, Negrin RS (2005) In vivo analyses of early events in acute graft-versus-host disease reveal sequential infiltration of T-cell subsets. Blood 106(3):1113–1122
65. Shlomchik WD (2003) Antigen presentation in graft-vs-host disease. Exp Hematol 31(12):1187–1197
66. Lechler R, Ng WF, Steinman RM (2001) Dendritic cells in transplantation—friend or foe? Immunity 14(4):357–368
67. Sayegh MH, Carpenter CB (1996) Role of indirect allorecognition in allograft rejection. Int Rev Immunol 13(3):221–229
68. Kaplan DH, Anderson BE, McNiff JM, Jain D, Shlomchik MJ, Shlomchik WD (2004) Target antigens determine graft-versus-host disease phenotype. J Immunol 173(9):5467–5475
69. Anderson BE, McNiff JM, Jain D, Blazar BR, Shlomchik WD, Shlomchik MJ (2005) Distinct roles for donor- and host-derived antigen-presenting cells and costimulatory molecules in murine chronic graft-versus-host disease: requirements depend on target organ. Blood 105(5):2227–2234
70. Appleman LJ, Boussiotis VA (2003) T cell anergy and costimulation. Immunol Rev 192:161–180
71. Sharpe AH, Freeman GJ (2002) The B7-CD28 superfamily. Nat Rev Immunol 2(2):116–126
72. Bromley SK, Iaboni A, Davis SJ, Whitty A, Green JM, Shaw AS, Weiss A, Dustin ML (2001) The immunological synapse and CD28-CD80 interactions. Nat Immunol 2(12):1159–1166
73. Dustin ML (2001) Role of adhesion molecules in activation signaling in T lymphocytes. J Clin Immunol 21(4):258–263
74. Greenwald RJ, Freeman GJ, Sharpe AH (2005) The B7 family revisited. Annu Rev Immunol 23:515–548
75. Welniak LA, Blazar BR, Murphy WJ (2006) Immunobiology of allogeneic hematopoietic stem cell transplantation. Annu Rev Immunol 25:139–170
76. Blazar BR, Taylor PA, Linsley PS, Vallera DA (1994) In vivo blockade of CD28/CTLA4: B7/BB1 interaction with CTLA4-Ig reduces lethal murine graft-versus-host disease across the major histocompatibility complex barrier in mice. Blood 83:3815–3825
77. Blazar BR, Taylor PA, Panoskaltsis-Mortari A, Buhlman J, Xu J, Flavell RA, Korngold R, Noelle R, Vallera DA (1997) Blockade of CD40 ligand-CD40 interaction impairs CD4+ T cell-mediated alloreactivity by inhibiting mature donor T cell expansion and function after bone marrow transplantation. J Immunol 158(1):29–39
78. Blazar BR, Taylor PA, Panoskaltis-Mortari A, Gray GS, DA V (1995) Co-blockade of the LFA1:ICAM and CD28/CTLA4:B7 pathways is a highly effective means of preventing acute lethal graft-versus host disease induced by fully major histocompatibility complex-disparate donor grafts. Blood 85:2607–2618

79. Blazar BR, Kwon BS, Panoskaltsis-Mortari A, Kwak KB, Peschon JJ, Taylor PA (2001) Ligation of 4-1BB (CDw137) regulates graft-versus-host disease, graft-versus-leukemia, and graft rejection in allogeneic bone marrow transplant recipients. J Immunol 166(5):3174–3183
80. Blazar BR, Sharpe AH, Chen AI, Panoskaltsis-Mortari A, Lees C, Akiba H, Yagita H, Killeen N, Taylor PA (2003) Ligation of OX40 (CD134) regulates graft-versus-host disease (GVHD) and graft rejection in allogeneic bone marrow transplant recipients. Blood 101(9):3741–3748
81. Blazar BR, Carreno BM, Panoskaltsis-Mortari A, Carter L, Iwai Y, Yagita H, Nishimura H, Taylor PA (2003) Blockade of programmed death-1 engagement accelerates graft-versus-host disease lethality by an IFN-gamma-dependent mechanism. J Immunol 171(3):1272–1277
82. Csencsits KL, Bishop DK (2003) Contrasting alloreactive CD4+ and CD8+ T cells: there's more to it than MHC restriction. Am J Transplant 3(2):107–115
83. Korngold R, Sprent J (1982) Features of T cells causing H-2-restricted lethal graft-vs.-host disease across minor histocompatibility barriers. J Exp Med 155(3):872–883
84. Korngold R, Sprent J (1985) Surface markers of T cells causing lethal graft-vs-host disease to class I vs class II H-2 differences. Journal of Immunology 135:3004–3010
85. Korngold R, Sprent J (1987) T cell subsets and graft-versus-host disease. Transplantation 44(3):335–339
86. Wu CJ, Ritz J (2006) Induction of tumor immunity following allogeneic stem cell transplantation. Adv Immunol 90:133–173
87. Goulmy E (2006) Minor histocompatibility antigens: from transplantation problems to therapy of cancer. Hum Immunol 67(6):433–438
88. Spierings E, Drabbels J, Hendriks M, Pool J, Spruyt-Gerritse M, Claas F, Goulmy E (2006) A uniform genomic minor histocompatibility antigen typing methodology and database designed to facilitate clinical applications. PLoS ONE 1:e42
89. Anderson BE, McNiff J, Yan J, Doyle H, Mamula M, Shlomchik MJ, Shlomchik WD (2003) Memory CD4+ T cells do not induce graft-versus-host disease. J Clin Invest 112(1):101–108
90. Chen BJ, Cui X, Sempowski GD, Liu C, Chao NJ (2004) Transfer of allogeneic CD62L- memory T cells without graft-versus-host disease. Blood 103(4):1534–1541
91. Zhang Y, Joe G, Zhu J, Carroll R, Levine B, Hexner E, June C, Emerson SG (2004) Dendritic cell-activated CD44hiCD8+ T cells are defective in mediating acute graft-versus-host disease but retain graft-versus-leukemia activity. Blood 103(10):3970–3978
92. Maeda Y, Tawara I, Teshima T, Liu C, Hashimoto D, Matsuoka K, Tanimoto M, Reddy P (2007) Lymphopenia-induced proliferation of donor T cells reduces their capacity for causing acute graft-versus-host disease. Exp Hematol 35(2):274–286
93. Zhang Y, Joe G, Hexner E, Zhu J, Emerson SG (2005) Alloreactive memory T cells are responsible for the persistence of graft-versus-host disease. J Immunol 174(5):3051–3058
94. Zhang Y, Joe G, Hexner E, Zhu J, Emerson SG (2005) Host-reactive CD8+ memory stem cells in graft-versus-host disease. Nat Med 11(12):1299–1305
95. Blazar BR, Taylor PA (2005) Regulatory T cells. Biol Blood Marrow Transplant 11(Suppl 2):46–49
96. Cohen JL, Boyer O (2006) The role of CD4(+)CD25(hi) regulatory T cells in the physiopathogeny of graft-versus-host disease. Curr Opin Immunol 18:580–585
97. Johnson BD, Konkol MC, Truitt RL (2002) CD25+ immunoregulatory T-cells of donor origin suppress alloreactivity after BMT. Biol Blood Marrow Transplant 8(10):525–535
98. Hoffmann P, Ermann J, Edinger M, Fathman CG, Strober S (2002) Donor-type CD4(+)CD25(+) regulatory T cells suppress lethal acute graft-versus-host disease after allogeneic bone marrow transplantation. J Exp Med 196(3):389–399
99. Cohen JL, Trenado A, Vasey D, Klatzmann D, Salomon BL (2002) CD4(+)CD25(+) immunoregulatory T Cells: new therapeutics for graft-versus-host disease. J Exp Med 196(3):401–406

100. Roncarolo MG (1997) The role of interleukin-10 in transplantation and GVHD. In: Ferrara JLM, Deeg HJ, Burakoff SJ (eds) Graft-vs.-host disease. Marcel Dekker, Inc., New York, pp 693–715
101. Young KJ, DuTemple B, Phillips MJ, Zhang L (2003) Inhibition of graft-versus-host disease by double-negative regulatory T cells. J Immunol 171(1):134–141
102. Zeng D, Lewis D, Dejbakhsh-Jones S, Lan F, Garcia-Ojeda M, Sibley R, Strober S (1999) Bone marrow NK1.1(-) and NK1.1(+) T cells reciprocally regulate acute graft versus host disease. J Exp Med 189(7):1073–1081
103. Lowsky R, Takahashi T, Liu YP, Dejbakhsh-Jones S, Grumet FC, Shizuru JA, Laport GG, Stockerl-Goldstein KE, Johnston LJ, Hoppe RT, Bloch DA, Blume KG, Negrin RS, Strober S (2005) Protective conditioning for acute graft-versus-host disease. N Engl J Med 353(13):1321–1331
104. Edinger M, Hoffmann P, Ermann J, Drago K, Fathman CG, Strober S, Negrin RS (2003) CD4(+)CD25(+) regulatory T cells preserve graft-versus-tumor activity while inhibiting graft-versus-host disease after bone marrow transplantation. Nat Med 9(9):1144–1150
105. Nguyen VH, Zeiser R, Dasilva DL, Chang DS, Beilhack A, Contag CH, Negrin RS (2007) In vivo dynamics of regulatory T-cell trafficking and survival predict effective strategies to control graft-versus-host disease following allogeneic transplantation. Blood 109(6):2649–2656
106. Stanzani M, Martins SL, Saliba RM, St John LS, Bryan S, Couriel D, McMannis J, Champlin RE, Molldrem JJ, Komanduri KV (2003) CD25 expression on donor CD4+ or CD8+ T cells is associated with an increased risk of graft-versus-host disease following HLA-identical stem cell transplantation in humans. Blood 103:1140–1146
107. Miura Y, Thoburn CJ, Bright EC, Phelps ML, Shin T, Matsui EC, Matsui WH, Arai S, Fuchs EJ, Vogelsang GB, Jones RJ, Hess AD (2004) Association of Foxp3 regulatory gene expression with graft-versus-host disease. Blood 104(7):2187–2193
108. Zorn E, Kim HT, Lee SJ, Floyd BH, Litsa D, Arumugarajah S, Bellucci R, Alyea EP, Antin JH, Soiffer RJ, Ritz J (2005) Reduced frequency of FOXP3+ CD4+ CD25+ regulatory T cells in patients with chronic graft-versus-host disease. Blood 106(8):2903–2911
109. Ziegler SF (2006) FOXP3: of mice and men. Annu Rev Immunol 24:209–226
110. Hashimoto D, Asakura S, Miyake S, Yamamura T, Van Kaer L, Tanimoto M, Teshima T (2003) Host NKT cells promote Th2 polarization of donor T cells and regulate acute GVHD after experimental BMT via a STAT6-dependent mechanism. Blood 102(Suppl 11):191a
111. Morris ES, MacDonald KP, Rowe V, Banovic T, Kuns RD, Don AL, Bofinger HM, Burman AC, Olver SD, Kienzle N, Porcelli SA, Pellicci DG, Godfrey DI, Smyth MJ, Hill GR (2005) NKT cell-dependent leukemia eradication following stem cell mobilization with potent G-CSF analogs. J Clin Invest 115(11):3093–3103
112. Morris ES, MacDonald KP, Hill GR (2006) Stem cell mobilization with G-CSF analogs: a rational approach to separate GVHD and GVL? Blood 107(9):3430–3435
113. Wang JW, Howson JM, Ghansah T, Desponts C, Ninos JM, May SL, Nguyen KH, Toyama-Sorimachi N, Kerr WG (2002) Influence of SHIP on the NK repertoire and allogeneic bone marrow transplantation. Science 295(5562):2094–2097
114. Kamradt T, Mitchison NA (2001) Tolerance and autoimmunity. N Engl J Med 344(9): 655–664
115. Wekerle T, Kurtz J, Sayegh M, Ito H, Wells A, Bensinger S, Shaffer J, Turka L, Sykes M (2001) Peripheral deletion after bone marrow transplantation with costimulatory blockade has features of both activation-induced cell death and passive cell death. J Immunol 166(4):2311–2316
116. Sykes M, Sachs DH (2001) Mixed chimerism. Philos Trans R Soc Lond B Biol Sci 356(1409):707–726
117. Li XC, Strom TB, Turka LA, Wells AD (2001) T cell death and transplantation tolerance. Immunity 14(4):407–416
118. Suchin EJ, Langmuir PB, Palmer E, Sayegh MH, Wells AD, Turka LA (2001) Quantifying the frequency of alloreactive T cells in vivo: new answers to an old question. J Immunol 166(2):973–981

119. Van Parijs L, Ibraghimov A, Abbas AK (1996) The roles of costimulation and Fas in T-cell apoptosis and peripheral tolerance. Immunity 4:321–328
120. Sykes M, Abraham VS (1992) The mechanism of IL-2-mediated protection against GVHD in mice. II. Protection occurs independently of NK/LAK cells. Transplantation 53:1063–1070
121. Sykes M, Abraham VS, Harty MW, Pearson DA (1993) IL-2 reduces graft-versus-host disease and preserves a graft-versus-leukemia effect by selectively inhibiting CD4+ T cell activity. J Immunol 150:197–205
122. Yang Y, Dey B, Sergio J, Sykes M (1997) Interleukin-12 prevents severe acute graft-versus-host disease (GVHD) and GVHD-associated immune dysfunction in a fully major histocompatibility complex haplotype-mismatched murine bone marrow transplantation model. Transplantation 64:1343–1352
123. Yang YG, Dey BR, Sergio JJ, Pearson DA, Sykes M (1998) Donor-derived interferon gamma is required for inhibition of acute graft-versus-host disease by interleukin 12. J Clin Invest 102(12):2126–2135
124. Reddy P, Teshima T, Kukuruga M, Ordemann R, Liu C, Lowler K, Ferrara JL (2001) Interleukin-18 regulates acute graft-versus-host disease by enhancing Fas-mediated donor T cell apoptosis. J Exp Med 194(10):1433–1440
125. Marrack P, Mitchell T, Bender J, Hildeman D, Kedl R, Teague K, Kappler J (1998) T-cell survival. Immunol Rev 165:279–285
126. Hildeman DA, Zhu Y, Mitchell TC, Bouillet P, Strasser A, Kappler J, Marrack P (2002) Activated T cell death in vivo mediated by proapoptotic bcl-2 family member bim. Immunity 16(6):759–767
127. Hildeman DA, Zhu Y, Mitchell TC, Kappler J, Marrack P (2002) Molecular mechanisms of activated T cell death in vivo. Curr Opin Immunol 14(3):354–359
128. Drobyski WR, Komorowski R, Logan B, Gendelman M (2002) Role of the passive apoptotic pathway in graft-versus-host disease. J Immunol 169(3):1626–1633
129. Drobyski WR, Gendelman M (2002) Regulation of alloresponses after bone marrow transplantation using donor T cells expressing a thymidine kinase suicide gene. Leuk Lymphoma 43(10):2011–2016
130. Drobyski WR, Klein J, Flomenberg N, Pietryga D, Vesole DH, Margolis DA, Keever-Taylor CA (2002) Superior survival associated with transplantation of matched unrelated versus one-antigen-mismatched unrelated or highly human leukocyte antigen-disparate haploidentical family donor marrow grafts for the treatment of hematologic malignancies: establishing a treatment algorithm for recipients of alternative donor grafts. Blood 99(3):806–814
131. Bonini C, Ferrari G, Verzeletti S, Servida P, Zappone E, Ruggieri L, Ponzoni M, Rossini S, Mavilio F, Traversari C, Bordignon C (1997) HSV-TK gene transfer into donor lymphocytes for control of allogeneic graft-versus-leukemia. Science 276(5319):1719–1724
132. Chen BJ, Cui X, Liu C, Chao NJ (2002) Prevention of graft-versus-host disease while preserving graft-versus-leukemia effect after selective depletion of host-reactive T cells by photodynamic cell purging process. Blood 99(9):3083–3088
133. Ferrara J, Krenger W (1998) Graft Versus Host Disease: the Influence of Type 1 and Type 2 T Cell Cytokines. Transfus Med Rev 12:1–17
134. Ferrara JLM (1994) The cytokine storm of acute graft-versus host disease. Haematol Rev 8:27
135. Reddy P (2003) Pathophysiology of acute graft-versus-host disease. Hematol Oncol 21(4):149–161
136. Ratanatharathorn V, Nash RA, Przepiorka D et al (1998) Phase III study comparing methotrexate and tacrolimus (prpgraf, FK506) with methotrexate and cyclosporine for graft-versus-host disease prophylaxis after HLA-identical sibling bone marrow transplantation. Blood 92:2303–2314
137. Liu EH, Siegel RM, Harlan DM, O'Shea JJ (2007) T cell-directed therapies: lessons learned and future prospects. Nat Immunol 8(1):25–30

138. Zeiser R, Nguyen VH, Beilhack A, Buess M, Schulz S, Baker J, Contag CH, Negrin RS (2006) Inhibition of CD4+CD25+ regulatory T-cell function by calcineurin-dependent interleukin-2 production. Blood 108(1):390–399
139. Zhang H, Chua KS, Guimond M, Kapoor V, Brown MV, Fleisher TA, Long LM, Bernstein D, Hill BJ, Douek DC, Berzofsky JA, Carter CS, Read EJ, Helman LJ, Mackall CL (2005) Lymphopenia and interleukin-2 therapy alter homeostasis of CD4+CD25+ regulatory T cells. Nat Med 11(11):1238–1243
140. Liston A, Rudensky AY (2007) Thymic development and peripheral homeostasis of regulatory T cells. Curr Opin Immunol 19(2):176–185
141. Gavin MA, Rasmussen JP, Fontenot JD, Vasta V, Manganiello VC, Beavo JA, Rudensky AY (2007) Foxp3-dependent programme of regulatory T-cell differentiation. Nature 445(7129):771–775
142. Yang Y-G, Sergio JJ, Pearson DA, Szot GL, Shimizu A, Sykes M (1997) Interleukin-12 preserves the graft-versus-leukemia effect of allogeneic CD8 T cells while inhibiting CD4-dependent graft-versus-host disease in mice. Blood 90:4651–4660
143. Sykes M, Pearson DA, Taylor PA, Szot GL, Goldman SJ, Blazar BR (1999) Dose and timing of interleukin (IL)-12 and timing and type of total-body irradiation: effects on graft-vs.-host disease inhibition and toxicity of exogenous IL-12 in murine bone marrow transplant recipients. Biol Blood Marrow Transplant 5(5):277–284. doi:S1083-8791(99)70002-9
144. Reddy P (2004) Interleukin-18: recent advances. Curr Opin Hematol 11(6):405–410
145. Sykes M, Harty MW, Szot GL, Pearson DA (1994) Interleukin-2 inhibits graft-versus-host disease-promoting activity of CD4+ cells while preserving CD4- and CD8-mediated graft-versus-leukemia effects. Blood 83:2560–2569
146. Sykes M, Szot GL, Nguyen PL, Pearson DA (1995) Interleukin-12 inhibits murine graft-versus-host disease. Blood 86:2429–2438
147. Fowler DH, Kurasawa K, Smith R, Eckhaus MA, Gress RE (1994) Donor CD4-enriched cells of Th2 cytokine phenotype regulate graft-versus-host disease without impairing allogeneic engraftment in sublethally irradiated mice. Blood 84:3540
148. Krenger W, Snyder KM, Byon CH, Falzarano G, Ferrara JLM (1995) Polarized type 2 alloreactive CD4+ and CD8+ donor T cells fail to induce experimental acute graft-versus-host disease. J Immunol 155:585–593
149. Pan L, Delmonte J, Jalonen C, Ferrara J (1995) Pretreatment of donor mice with granulocyte colony-stimulating factor polarizes donor T-lymphocytes toward type-2 cytokine production and reduces severity of experimental graft-versus-host disease. Blood 86:4422–4429
150. Hill GR, Cooke KR, Teshima T, Crawford JM, Keith JC Jr, Brinson YS, Bungard D, Ferrara JL (1998) Interleukin-11 promotes T cell polarization and prevents acute graft-versus-host disease after allogeneic bone marrow transplantation. J Clin Invest 102(1):115–123
151. Reddy P, Teshima T, Hildebrandt G, Williams DL, Liu C, Cooke KR, Ferrara JL (2003) Pretreatment of donors with interleukin-18 attenuates acute graft-versus-host disease via STAT6 and preserves graft-versus-leukemia effects. Blood 101(7):2877–2885
152. Foley JE, Jung U, Miera A, Borenstein T, Mariotti J, Eckhaus M, Bierer BE, Fowler DH (2005) Ex vivo rapamycin generates donor Th2 cells that potently inhibit graft-versus-host disease and graft-versus-tumor effects via an IL-4-dependent mechanism. J Immunol 175(9):5732–5743
153. Jung U, Foley JE, Erdmann AA, Toda Y, Borenstein T, Mariotti J, Fowler DH (2006) Ex vivo rapamycin generates Th1/Tc1 or Th2/Tc2 effector T cells with enhanced in vivo function and differential sensitivity to post-transplant rapamycin herapy. Biol Blood Marrow Transplant 12(9):905–918
154. Fowler DH, Gress RE (2000) Th2 and Tc2 cells in the regulation of GVHD, GVL, and graft rejection: considerations for the allogeneic transplantation therapy of leukemia and lymphoma. Leuk Lymphoma 38(3–4):221–234
155. Nikolic B, Lee S, Bronson R, Grusby M, Sykes M (2000) Th1 and Th2 mediate acute graft-versus-host disease, each with distinct end-organ targets. J Clin Invest 105:1289–1298

156. Blazar BR, Taylor PA, Smith S, Vallera DA (1995) Interleukin-10 administration decreases survival in murine recipients of major histocompatibility complex disparate donor bone marrow grafts. Blood 85:842–851
157. Lin MT, Storer B, Martin PJ, Tseng LH, Gooley T, Chen PJ, Hansen JA (2003) Relation of an interleukin-10 promoter polymorphism to graft-versus-host disease and survival after hematopoietic-cell transplantation. N Engl J Med 349(23):2201–2210
158. Banovic T, MacDonald KP, Morris ES, Rowe V, Kuns R, Don A, Kelly J, Ledbetter S, Clouston AD, Hill GR (2005) TGF-beta in allogeneic stem cell transplantation: friend or foe? Blood 106(6):2206–2214
159. Alpdogan O, Schmaltz C, Muriglan SJ, Kappel BJ, Perales MA, Rotolo JA, Halm JA, Rich BE, van den Brink MR (2001) Administration of interleukin-7 after allogeneic bone marrow transplantation improves immune reconstitution without aggravating graft-versus-host disease. Blood 98(7):2256–2265
160. Alpdogan O, Muriglan SJ, Eng JM, Willis LM, Greenberg AS, Kappel BJ, van den Brink MR (2003) IL-7 enhances peripheral T cell reconstitution after allogeneic hematopoietic stem cell transplantation. J Clin Invest 112(7):1095–1107
161. Sinha ML, Fry TJ, Fowler DH, Miller G, Mackall CL (2002) Interleukin 7 worsens graft-versus-host disease. Blood 100(7):2642–2649
162. Gendelman M, Hecht T, Logan B, Vodanovic-Jankovic S, Komorowski R, Drobyski WR (2004) Host conditioning is a primary determinant in modulating the effect of IL-7 on murine graft-versus-host disease. J Immunol 172(5):3328–3336
163. Weaver CT, Murphy KM (2007) T-cell subsets: the more the merrier. Curr Biol 17(2):R61–63
164. Wysocki CA, Panoskaltsis-Mortari A, Blazar BR, Serody JS (2005) Leukocyte migration and graft-versus-host disease. Blood 105(11):4191–4199
165. Cyster JG (2005) Chemokines, sphingosine-1-phosphate, and cell migration in secondary lymphoid organs. Annu Rev Immunol 23:127–159
166. Serody JS, Burkett SE, Panoskaltsis-Mortari A, Ng-Cashin J, McMahon E, Matsushima GK, Lira SA, Cook DN, Blazar BR (2000) T-lymphocyte production of macrophage inflammatory protein-1alpha is critical to the recruitment of CD8(+) T cells to the liver, lung, and spleen during graft-versus-host disease. Blood 96(9):2973–2980
167. Mapara MY, Leng C, Kim YM, Bronson R, Lokshin A, Luster A, Sykes M (2006) Expression of chemokines in GVHD target organs is influenced by conditioning and genetic factors and amplified by GVHR. Biol Blood Marrow Transplant 12(6):623–634
168. Duffner U, Lu B, Hildebrandt GC, Teshima T, Williams DL, Reddy P, Ordemann R, Clouthier SG, Lowler K, Liu C, Gerard C, Cooke KR, Ferrara JL (2003) Role of CXCR3-induced donor T-cell migration in acute GVHD. Exp Hematol 31(10):897–902
169. Wysocki CA, Burkett SB, Panoskaltsis-Mortari A, Kirby SL, Luster AD, McKinnon K, Blazar BR, Serody JS (2004) Differential roles for CCR5 expression on donor T cells during graft-versus-host disease based on pretransplant conditioning. J Immunol 173(2):845–854
170. Murai M, Yoneyama H, Harada A, Yi Z, Vestergaard C, Guo B, Suzuki K, Asakura H, Matsushima K (1999) Active participation of CCR5(+)CD8(+) T lymphocytes in the pathogenesis of liver injury in graft-versus-host disease. J Clin Invest 104(1):49–57
171. Wysocki C, Burkett S, Chwastiak K, Taylor P, Blazar BR, Serody JS (2003) CCR5 expression on CD4+CD25+ regulatory T cells plays a critical role in their ability to prevent GVHD. Blood 102(11):517
172. Waldman E, Lu SX, Hubbard VM, Kochman AA, Eng JM, Terwey TH, Muriglan SJ, Kim TD, Heller G, Murphy GF, Liu C, Alpdogan O, van den Brink MR (2006) Absence of beta7 integrin results in less graft-versus-host disease because of decreased homing of alloreactive T cells to intestine. Blood 107(4):1703–1711
173. Welniak LA, Kuprash DV, Tumanov AV, Panoskaltsis-Mortari A, Blazar BR, Sun K, Nedospasov SA, Murphy WJ (2005) Peyer's patches are not required for acute graft-versus-host disease after myeloablative conditioning and murine allogeneic bone marrow transplantation. Blood 107:410–412

174. Kim YM, Sachs T, Asavaroengchai W, Bronson R, Sykes M (2003) Graft-versus-host disease can be separated from graft-versus-lymphoma effects by control of lymphocyte trafficking with FTY720. J Clin Invest 111(5):659–669
175. Lee RS, Kuhr CS, Sale GE, Zellmer E, Hogan WJ, Storb R, Little MT (2003) FTY720 does not abrogate acute graft-versus-host disease in the dog leukocyte antigen-nonidentical unrelated canine model. Transplantation 76(8):1155–1158
176. Sallusto F, Mackay CR, Lanzavecchia A (2000) The role of chemokine receptors in primary, effector, and memory immune responses. Annu Rev Immunol 18:593–620
177. Sackstein R (2006) A revision of Billingham's tenets: the central role of lymphocyte migration in acute graft-versus-host disease. Biol Blood Marrow Transplant 12(Suppl 1):2–8
178. Kagi D, Vignaux F, Ledermann B, Burki K, Depraetere V, Nagata S, Hengartner H, Golstein P (1994) Fas and perforin pathways as major mechanisms of T cell-mediated cytotoxicity. Science 265:528–530
179. van den Brink MR, Burakoff SJ (2002) Cytolytic pathways in haematopoietic stem-cell transplantation. Nat Rev Immunol 2(4):273–281
180. Pan G, O'Rourke K, Chinnaiyan AM, Gentz R, Ebner R (1997) The receptor for the cytotoxic ligand TRAIL. Science 276(5309):111–113
181. Chicheportiche Y, Bourdon P, Xu H, Hsu Y, Scott H, Hession C, Garcia I, Browning J (1997) TWEAK, a new secreted ligand in the tumor necrosis factor family that weakly induces apoptosis. J Biol Chem 272:32401–34210
182. Jiang S, Herrera O, Lechler RI (2004) New spectrum of allorecognition pathways: implications for graft rejection and transplantation tolerance. Curr Opin Immunol 16(5):550–557
183. Jiang Z, Podack E, Levy R (1998) Donor T cells which cannot mediate perforin-dependent and FasL-dependent cytotoxicity can effect graft vs host reactivity following allogeneic bone marrow transplantation. Periodicum Biologorum 100:477–481
184. Maeda Y, Levy RB, Reddy P, Liu C, Clouthier SG, Teshima T, Ferrara JL (2005) Both perforin and Fas ligand are required for the regulation of alloreactive CD8+ T cells during acute graft-versus-host disease. Blood 105(5):2023–2027
185. Voskoboinik I, Smyth MJ, Trapani JA (2006) Perforin-mediated target-cell death and immune homeostasis. Nat Rev Immunol 6(12):940–952
186. Chinnaiyan A, O'Rourke K, Yu G, Lyons R, Garg M, Duan D, Xing L, Gentz R, Ni J, Dixit V (1996) Signal transduction by DR3, a death domain-containing receptor related to TNFR-1 and CD95. Science 274:990–992
187. Krammer PH (2000) CD95's deadly mission in the immune system. Nature 407(6805):789–795
188. Baker MB, Altman NH, Podack ER, Levy RB (1996) The role of cell-mediated cytotoxicity in acute GVHD after MHC-matched allogeneic bone marrow transplantation in mice. J Exp Med 183:2645–2656
189. Braun YM, Lowin B, French L, Acha-Orbea H, Tschopp J (1996) Cytotoxic T cells deficient in both functional Fas ligand and perforin show residual cytolytic activity yet lose their capacity to induce lethal acute graft-versus-host disease. J Exp Med 183:657–661
190. Baker MB, Riley RL, Podack ER, Levy RB (1997) GVHD-associated lymphoid hypoplasia and B cell dysfunction is dependent upon donor T cell-mediated Fas-ligand function, but not perforin function. Proc Natl Acad Sci U S A 94:1366–1371
191. Graubert TA, Russell JH, Ley T (1996) The role of granzyme B in murine models of acute graft-versus-host disease and graft rejection. Blood 87:1232–1237
192. Graubert TA, DiPersio JF, Russell JH, Ley TJ (1997) Perforin/granzyme-dependent and independent mechanisms are both important for the development of graft-versus-host disease after murine bone marrow transplantation. J Clin Invest 100(4):904–911
193. Ueno Y, Ishii M, Yahagi K, Mano Y, Kisara N, Nakamura N, Shimosegawa T, Toyota T, Nagata S (2000) Fas-mediated cholangiopathy in the murine model of graft versus host disease. Hepatology 31(4):966–974

194. Aggarwal BB (2003) Signalling pathways of the TNF superfamily: a double-edged sword. Nat Rev Immunol 3(9):745–756
195. Via C, Nguyen P, Shustov A, Drappa J, Elkon K (1996) A major role for the Fas pathway in acute graft-versus-host disease. J Immunol 157:5387–5393
196. Hattori K, Hirano T, Miyajima H, Yamakawa N, Tateno M, Oshimi K, Kayagaki N, Yagita H, Okumura K (1998) Differential effects of anti-Fas ligand and anti-tumor necrosis factor alpha antibodies on acute graft-versus-host disease pathologies. Blood 91(11):4051–4055
197. van den Brink M, Moore E, Horndasch E, Crawford J, Hoffman J, Murphy G, Burakoff S (2000) Fas-deficient *lpr* mice are more susceptible to graft-versus-host disease. J Immunol 164:469–480
198. Lin T, Brunner T, Tietz B, Madsen J, Bonfoco E, Reaves M, Huflejt M, Green DR (1998) Fas ligand-mediated killing by intestinal intraepithelial lymphocytes. Participation in intestinal graft-versus-host disease. J Clin Invest 101(3):570–577
199. Liem LM, van Lopik T, van Nieuwenhuijze AEM, van Houwelingen HC, Aarden L, Goulmy E (1998) Soluble fas levels in sera of bone marrow transplantation recipients are increased during acute graft-versus-host disease but not during infections. Blood 91:1464–1468
200. Das H, Imoto S, Murayama T, Kajimoto K, Sugimoto T, Isobe T, Nakagawa T, Nishimura R, Koizumi T (1999) Levels of soluble FasL and FasL gene expression during the development of graft-versus-host disease in DLT-treated patients. Br J Haematol 104(4):795–800
201. Martin PJ, Akatsuka Y, Hahne M, Sale G (1998) Involvement of donor T-cell cytotoxic effector mechanisms in preventing allogeneic marrow graft rejection. Blood 92(6):2177–2181
202. Zimmerman Z, Shatry A, Deyev V, Podack E, Mammolenti M, Blazar BR, Yagita H, Levy RB (2005) Effector cells derived from host CD8 memory T cells mediate rapid resistance against minor histocompatibility antigen-mismatched allogeneic marrow grafts without participation of perforin, Fas ligand, and the simultaneous inhibition of 3 tumor necrosis factor family effector pathways. Biol Blood Marrow Transplant 11(8):576–586
203. Schmaltz C, Alpdogan O, Kappel BJ, Muriglan SJ, Rotolo JA, Ongchin J, Willis LM, Greenberg AS, Eng JM, Crawford JM, Murphy GF, Yagita H, Walczak H, Peschon JJ, Van Den Brink MR (2002) T cells require TRAIL for optimal graft-versus-tumor activity. Nat Med 8(12):1433–1437
204. Sato K, Nakaoka T, Yamashita N, Yagita H, Kawasaki H, Morimoto C, Baba M, Matsuyama T (2005) TRAIL-transduced dendritic cells protect mice from acute graft-versus-host disease and leukemia relapse. J Immunol 174(7):4025–4033
205. Brown GR, Lee E, Thiele DL (2002) TNF-TNFR2 interactions are critical for the development of intestinal graft-versus-host disease in MHC class II-disparate (C57BL/6J–>C57BL/6J x bm12)F1 mice. J Immunol 168(6):3065–3071
206. Brown GR, Thiele DL (2000) Enhancement of MHC class I-stimulated alloresponses by TNF/TNF receptor (TNFR)1 interactions and of MHC class II-stimulated alloresponses by TNF/TNFR2 interactions. Eur J Immunol 30(10):2900–2907
207. Brown GR, Lee EL, El-Hayek J, Kintner K, Luck C (2005) IL-12-independent LIGHT signaling enhances MHC class II disparate CD4+ T cell alloproliferation, IFN-gamma responses, and intestinal graft-versus-host disease. J Immunol 174(8):4688–4695
208. Xu Y, Flies AS, Flies DB, Zhu G, Anand S, Flies SJ, Xu H, Anders RA, Hancock WW, Chen L, Tamada K (2006) Selective targeting of the LIGHT-HVEM co-stimulatory system for the treatment of graft-versus-host disease. Blood 109:4097–4104
209. Zimmerman Z, Jones M, Shatry A, Komatsu M, Mammolenti M, Levy R (2005) Cytolytic pathways used by effector cells derived from recipient naive and memory T cells and natural killer cells in resistance to allogeneic hematopoietic cell transplantation. Biol Blood Marrow Transplant 11(12):957–971
210. Ferrara JL, Levy R, Chao NJ (1999) Pathophysiologic mechanisms of acute graft-vs.-host disease. Biol Blood Marrow Transplant 5(6):347–356

211. Iwasaki A, Medzhitov R (2004) Toll-like receptor control of the adaptive immune responses. Nat Immunol 5(10):987–995
212. Hill GR, Ferrara JL (2000) The primacy of the gastrointestinal tract as a target organ of acute graft-versus-host disease: rationale for the use of cytokine shields in allogeneic bone marrow transplantation. Blood 95(9):2754–2759
213. Fritz JH, Ferrero RL, Philpott DJ, Girardin SE (2006) Nod-like proteins in immunity, inflammation and disease. Nat Immunol 7(12):1250–1257
214. Dickinson AM, Middleton PG, Rocha V, Gluckman E, Holler E (2004) Genetic polymorphisms predicting the outcome of bone marrow transplants. Br J Haematol 127(5):479–490
215. Lorenz E, Schwartz DA, Martin PJ, Gooley T, Lin MT, Chien JW, Hansen JA, Clark JG (2001) Association of TLR4 mutations and the risk for acute GVHD after HLA-matched-sibling hematopoietic stem cell transplantation. Biol Blood Marrow Transplant 7(7):384–387
216. Holler E, Rogler G, Brenmoehl J, Hahn J, Herfarth H, Greinix H, Dickinson AM, Socie G, Wolff D, Fischer G, Jackson G, Rocha V, Steiner B, Eissner G, Marienhagen J, Schoelmerich J, Andreesen R (2006) Prognostic significance of NOD2/CARD15 variants in HLA-identical sibling hematopoietic stem cell transplantation: effect on long-term outcome is confirmed in 2 independent cohorts and may be modulated by the type of gastrointestinal decontamination. Blood 107(10):4189–4193
217. Nestel FP, Price KS, Seemayer TA, Lapp WS (1992) Macrophage priming and lipopolysaccharide-triggered release of tumor necrosis factor alpha during graft-versus-host disease. J Exp Med 175(2):405–413
218. Hill GR, Teshima T, Gerbitz A, Pan L, Cooke KR, Brinson YS, Crawford JM, Ferrara JL (1999) Differential roles of IL-1 and TNF-alpha on graft-versus-host disease and graft versus leukemia. J Clin Invest 104(4):459–467
219. Krenger W, Falzarano G, Delmonte J, Snyder KM, Byon JCH, Ferrara JLM (1996) Interferon-g suppresses T-cell proliferation to mitogen via the nitric oxide pathway during experimental acute graft-versus-host disease. Blood 88:1113–1121
220. Piguet PF, Grau GE, Allet B, Vassalli PJ (1987) Tumor necrosis factor/cachectin is an effector of skin and gut lesions of the acute phase of graft-versus-host disease. J Exp Med 166:1280–1289
221. Hill GR, Teshima T, Rebel VI, Krijanovski OI, Cooke KR, Brinson YS, Ferrara JL (2000) The p55 TNF-alpha receptor plays a critical role in T cell alloreactivity. J Immunol 164(2):656–663
222. Piguet PF (1990) Tumor necrosis factor and graft-versus-host disease. In: Burakoff SJ, Deeg HJ, Ferrara JLM, Atkinson K (eds) Graft-vs.-host disease. Marcel Dekker Inc., New York, pp 255–276
223. Tanaka J, Imamura M, Kasai M, Masauzi N, Watanabe M, Matsuura A, Morii K, Kiyama Y, Naohara T, Higa T, Honke K, Gasa S, Sakurada K, Miyazaki T (1993) Rapid analysis of tumor necrosis factor-alpha mRNA expression during venooclusive disease of the liver after allogeneic bone marrow transplantation. Transplantation 55:430–432
224. Tanaka J, Imamura M, Kasai M, Masauzi N, Matsuura A, Ohizumi H, Morii K, Kiyama Y, Naohara T, Saitho M et al (1993) Cytokine gene expression in peripheral blood mononuclear cells during graft-versus-host disease after allogeneic bone marrow transplantation. Br J Haematol 85:558–565
225. Holler E, Kolb HJ, Moller A, Kempeni J, Lisenfeld S, Pechumer H, Lehmacher W, Ruckdeschel G, Gleixner B, Riedner C, Ledderose G, Brehm G, Mittermuller J, Wilmanns W (1990) Increased serum levels of tumor necrosis factor alpha precede major complications of bone marrow transplantation. Blood 75:1011–1016
226. Holler E, Kolb HJ, Hintermeier-Knabe R, Mittermuller J, Thierfelde S, Kaul M, Wilmanns W (1993) The role of tumor necrosis factor alpha in acute graft-versus-host disease and complications following allogeneic bone marrow transplantation. Transplant Proc 25:1234–1236

227. Herve P, Flesch M, Tiberghien P, Wijdenes J, Racadot E, Bordigoni P, Plouvier E, Stephan JL, Bourdeau H, Holler E, Lioure B, Roche C, Vilmer E, Demeocq F, Kuentz M, Cahn YJ (1992) Phase I-II trial of a monoclonal anti-tumor necrosis factor alpha antibody for the treatment of refractory severe acute graft-versus-host disease. Blood 81:1993–1999
228. Uberti JP, Ayash L, Ratanatharathorn V, Silver S, Reynolds C, Becker M, Reddy P, Cooke KR, Yanik G, Whitfield J, Jones D, Hutchinson R, Braun T, Ferrara JL, Levine JE (2005) Pilot trial on the use of etanercept and methylprednisolone as primary treatment for acute graft-vs.-host disease. Biol Blood Marrow Transplant 11(9):680–687
229. Antin JH, Ferrara JLM (1992) Cytokine dysregulation and acute graft-versus-host disease. Blood 80:2964–2968
230. Abhyankar S, Gilliland DG, Ferrara JLM (1993) Interleukin 1 is a critical effector molecule during cytokine dysregulation in graft-versus-host disease to minor histocompatibility antigens. Transplantation 56:1518–1523
231. McCarthy PL, Abhyankar S, Neben S, Newman G, Sieff C, Thompson RC, Burakoff SJ, Ferrara JLM (1991) Inhibition of interleukin-1 by an interleukin-1 receptor antagonist prevents graft-versus-host disease. Blood 78:1915–1918
232. Antin JH, Weisdorf D, Neuberg D, Nicklow R, Clouthier S, Lee SJ, Alyea E, McGarigle C, Blazar BR, Sonis S, Soiffer RJ, Ferrara JL (2002) Interleukin-1 blockade does not prevent acute graft-versus-host disease: results of a randomized, double-blind, placebo-controlled trial of interleukin-1 receptor antagonist in allogeneic bone marrow transplantation. Blood 100(10):3479–3482
233. Falzarano G, Krenger W, Snyder KM, Delmonte J, Karandikar M, Ferrara JLM (1996) Suppression of B cell proliferation to lipopolysaccharide is mediated through induction of the nitric oxide pathway by tumor necrosis factor-a in mice with acute graft-versus-host disease. Blood 87:2853–2860
234. Nestel FP, Greene RN, Kichian K, Ponka P, Lapp WS (2000) Activation of macrophage cytostatic effector mechanisms during acute graft-versus-host disease: release of intracellular iron and nitric oxide-mediated cytostasis. Blood 96(5):1836–1843
235. Weiss G, Schwaighofer H, Herold M, Nachbaur D, Wachter H, Niederwieser D, Werner ER (1995) Nitric oxide formation as predictive parameter for acute graft-versus-host disease after human allogeneic bone marrow transplantation. Transplantation 60(11):1239–1244
236. Langrehr JM, Murase N, Markus PM, Cai X, Neuhaus P, Schraut W, Simmons RL, Hoffmann RA (1992) Nitric oxide production in host-versus-graft and graft-versus-host reactions in the rat. J Clin Invest 90:679–683
237. Langrehr JM, Muller AR, Bergonia HA, Jacob TD, Lee TK, Schraut WH, Lancaster JR, Hoffman RA, Simmons RL (1992) Detection of nitric oxide by electron paramagnetic resonance spectroscopy during rejection and graft-versus-host disease after small-bowel transplantation in the rat. Surgery 112:395–401
238. Bogdan C (2001) Nitric oxide and the immune response. Nat Immunol 2(10):907–916
239. Kernan NA, Collins NH, Juliano L, Cartagena T, Dupont B, O'Reilly RJ (1986) Clonable T lymphocytes in T cell-depleted bone marrow transplants correlate with development of graft-v-host disease. Blood 68(3):770–773
240. Aversa F, Tabilio A, Velardi A, Cunningham I, Terenzi A, Falzetti F, Ruggeri L, Barbabietola G, Aristei C, Latini P, Reisner Y, Martelli MF (1998) Treatment of high-risk acute leukemia with T-cell-depleted stem cells from related donors with one fully mismatched HLA haplotype. N Engl J Med 339(17):1186–1193
241. Champlin R, Ho W, Gajewski J, Feig S, Burnison M, Holley G, Greenberg P, Lee K, Schmid I, Giorgi J et al (1990) Selective depletion of CD8+ T lymphocytes for prevention of graft-versus-host disease after allogeneic bone marrow transplantation. Blood 76(2):418–423
242. Papadopoulos EB, Carabasi MH, Castro-Malaspina H, Childs BH, Mackinnon S, Boulad F, Gillio AP, Kernan NA, Small TN, Szabolcs P, Taylor J, Yahalom J, Collins NH, Bleau SA, Black PM, Heller G, O'Reilly RJ, Young JW (1998) T-cell-depleted allogeneic bone

marrow transplantation as postremission therapy for acute myelogenous leukemia: freedom from relapse in the absence of graft-versus-host disease. Blood 91(3):1083–1090

243. Wagner J, Santos G, Noga S (1990) Bone marrow graft engineering by counterflow centrifugal elutration: results of a phase I-II clinical trial. Blood 75(6):1370–1377
244. Martin PJ, Schoch G, Fisher L et al (1991) A retrospective analysis of therapy for acute graft-versus-host disease: secondary treatment. Blood 77:1821–1828
245. Cragg L, Blazar BR, Defor T, Kolatker N, Miller W, Kersey J, Ramsay M, McGlave P, Filipovich A, Weisdorf D (2000) A randomized trial comparing prednisone with antithymocyte globulin/prednisone as an initial systemic therapy for moderately severe acute graft-versus-host disease. Biol Blood Marrow Transplant 6(4A):441–447
246. Anasetti C, Hansen JA, Waldmann TA, Appelbaum FR, Davis J, Deeg HJ, Doney K, Martin PJ, Nash R, Storb R (1994) Treatment of acute graft-versus-host disease with humanized anti-Tac: an antibody that binds to the interleukin-2 receptor. Blood 84(4):1320–1327
247. Anasetti C, Hansen JA, Waldmann TA, Appelbaum FR, Davis J, Deeg HJ, Doney K, Martin PJ, Nash R, Storb R et al (1994) Treatment of acute graft-versus-host disease with humanized anti-Tac: an antibody that binds to the interleukin-2 receptor. Blood 84(4):1320–1327
248. Przepiorka D, Kernan NA, Ippoliti C, Papadopoulos EB, Giralt S, Khouri I, Lu JG, Gajewski J, Durett A, Cleary K, Champlin R, Andersson BS, Light S (2000) Daclizumab, a humanized anti-interleukin-2 receptor alpha chain antibody, for treatment of acute graft-versus-host disease. Blood 95(1):83–89
249. Carpenter PA, Appelbaum FR, Corey L et al (2002) A humanized non FcR-biding anti-CD3 antibody, visilizumb, for treatment of steroid-refractory acute graft-versus-host disease. Blood 99:2712–2719
250. Carpenter PA, Lowder J, Johnston L, Frangoul H, Khoury H, Parker P, Jerome KR, McCune JS, Storer B, Martin P, Appelbaum F, Abonour R, Westervelt P, Anasetti C (2005) A phase II multicenter study of visilizumab, humanized anti-CD3 antibody, to treat steroid-refractory acute graft-versus-host disease. Biol Blood Marrow Transplant 11(6):465–471
251. Willenbacher W, Basara N, Blau IW, Fauser AA, Kiehl MG (2001) Treatment of steroid refractory acute and chronic graft-versus-host disease with daclizumab. Br J Haematol 112(3):820–823
252. Storb R, Epstein RB, Graham TC, Thomas ED (1970) Methotrexate regimens for control of graft-versus-host disease in dogs with allogeneic marrow grafts. Transplantation 9(3):240–246
253. Ramsay NK, Kersey JH, Robison LL, McGlave PB, Woods WG, Krivit W, Kim TH, Goldman AI, Nesbit ME Jr (1982) A randomized study of the prevention of acute graft-versus-host disease. N Engl J Med 306(7):392–397
254. Blume KG, Beutler E, Bross KJ, Chillar RK, Ellington OB, Fahey JL, Farbstein MJ, Forman SJ, Schmidt GM, Scott EP, Spruce WE, Turner MA, Wolf JL (1980) Bone-marrow ablation and allogeneic marrow transplantation in acute leukemia. N Engl J Med 302(19):1041–1046
255. Ratanatharathorn V, Nash RA, Przepiorka D, Devine SM, Klein JL, Weisdorf D, Fay JW, Nademanee A, Antin JH, Christiansen NP, van der Jagt R, Herzig RH, Litzow MR, Wolff SN, Longo WL, Petersen FB, Karanes C, Avalos B, Storb R, Buell DN, Maher RM, Fitzsimmons WE, Wingard JR (1998) Phase III study comparing methotrexate and tacrolimus (prograf, FK506) with methotrexate and cyclosporine for graft-versus-host disease prophylaxis after HLA-identical sibling bone marrow transplantation. Blood 92(7):2303–2314
256. Storb R, Deeg HJ, Whitehead J, Appelbaum F, Beatty P, Bensinger W, Buckner CD, Clift R, Doney K, Farewell V et al (1986) Methotrexate and cyclosporine compared with cyclosporine alone for prophylaxis of acute graft versus host disease after marrow transplantation for leukemia. N Engl J Med 314(12):729–735
257. Storb R, Deeg HJ, Pepe M, Appelbaum F, Anasetti C, Beatty P, Bensinger W, Berenson R, Buckner CD, Clift R et al (1989) Methotrexate and cyclosporine versus cyclosporine alone

for prophylaxis of graft-versus-host disease in patients given HLA-identical marrow grafts for leukemia: long-term follow-up of a controlled trial. Blood 73(6):1729–1734
258. Chao NJ, Snyder DS, Jain M et al (2000) Equivalence of 2 effective graft-versus-host disease prophylaxis regimens: results of a prospective double-blind randomized trial. Biol Blood Marrow Transplant 6:254–261
259. Blazar BR, Taylor PA, Panoskaltsis-Mortari A, Vallera DA (1998) Rapamycin inhibits the generation of graft-versus-host disease- and graft-versus-leukemia-causing T cells by interfering with te production of Th1 or Th1 cytotoxic cytokines. J Immunol 160:5355–5365
260. Allison AC, Eugui EM (2000) Mycophenolate mofetil and its mechanisms of action. Immunopharmacology 47(2–3):85–118
261. van Bekkum DW, Roodenburg J, Heidt PJ, van der Waaij D (1974) Mitigation of secondary disease of allogeneic mouse radiation chimeras by modification of the intestinal microflora. J Nat Cancer Inst 52(2):401–404
262. Storb R, Prentice RL, Buckner CD, Clift RA, Appelbaum F, Deeg J, Doney K, Hansen JA, Mason M, Sanders JE, Singer J, Sullivan KM, Witherspoon RP, Thomas ED (1983) Graft-versus-host disease and survival in patients with aplastic anemia treated by marrow grafts from HLA-identical siblings Beneficial effect of a protective environment. N Engl J Med 308:302–307
263. Beelen DW, Haralambie E, Brandt H, Linzenmeier G, Muller K-D, Quabeck K, Sayer HG, Graeven U, Mahmoud HK, Schaefer UW (1992) Evidence that sustained growth suppression of intestinal anaerobic bacteria reduces the risk of acute graft-versus-host disease after sibling marrow transplantation. Blood 80:2668–2676
264. Beelen D, Elmaagacli A, Muller K, Hirche H, Schaefer U (1999) Influence of intestinal bacterial decontamination using metronidazole and ciprofloxacin or ciprofloxacin alone on the development of acute graft-versus-host disease after marrow transplantation in patients with hematologic malignancies: final results and long term follow-up of an open-label prospective randomized trial. Blood 93:3267–3275
265. Passweg J, Rowlings P, Atkinson K et al (1998) Influence of protective isolation on outcome of allogeneic bone marrow transplantation for leukemia. Bone Marrow Tansplant 21:1231–1238
266. Xun CQ, Thompson JS, Jennings CD, Brown SA, Widmer MB (1994) Effect of total body irradiation, busulfan-cyclophosphamide, or cyclophosphamide conditioning on inflammatory cytokine release and development of acute and chronic graft-versus-host disease in H-2-incompatible transplanted SCID mice. Blood 83(8):2360–2367
267. Akpek G, Lenz G, Lee SM, Sanchorawala V, Wright DG, Colarusso T, Waraska K, Lerner A, Vosburgh E, Skinner M, Comenzo RL (2001) Immunologic recovery after autologous blood stem cell transplantation in patients with AL-amyloidosis. Bone Marrow Transplant 28(12):1105–1109
268. Akpek G, Boitnott JK, Lee LA, Hallick JP, Torbenson M, Jacobsohn DA, Arai S, Anders V, Vogelsang GB (2002) Hepatitic variant of graft-versus-host disease after donor lymphocyte infusion. Blood 100(12):3903–3907
269. Holler E, Kolb HJ, Moller A, Kempeni J, Liesenfeld S, Pechumer H, Lehmacher W, Ruckdeschel G, Gleixner B, Riedner C (1990) Increased serum levels of tumor necrosis factor alpha precede major complications of bone marrow transplantation. Blood 75(4):1011–1016
270. Holler E, Kolb HJ, Hintermeier-Knabe R, Mittermuller J, Thierfelder S, Kaul M, Wilmanns W (1993) Role of tumor necrosis factor alpha in acute graft-versus-host disease and complications following allogeneic bone marrow transplantation. Transplant Proc 25(1 Pt 2):1234–1236
271. Holler E, Kolb HJ, Wilmanns W (1993) Treatment of GVHD–TNF-antibodies and related antagonists. Bone Marrow Transplant 12(Suppl 3):S29–31
272. Huang XJ, Wan J, Lu DP (2001) Serum TNFalpha levels in patients with acute graft-versus-host disease after bone marrow transplantation. Leukemia 15(7):1089–1091

273. Kobbe G, Schneider P, Rohr U, Fenk R, Neumann F, Aivado M, Dietze L, Kronenwett R, Hunerliturkoglu A, Haas R (2001) Treatment of severe steroid refractory acute graft-versus-host disease with infliximab, a chimeric human/mouse antiTNFalpha antibody. Bone Marrow Transplant 28(1):47–49
274. Couriel DR, Hicks K, Giralt S, Champlin RE (2000) Role of tumor necrosis factor-alpha inhibition with inflixiMAB in cancer therapy and hematopoietic stem cell transplantation. Curr Opin Oncol 12(6):582–587
275. Chiang KY, Abhyankar S, Bridges K, Godder K, Henslee-Downey JP (2002) Recombinant human tumor necrosis factor receptor fusion protein as complementary treatment for chronic graft-versus-host disease. Transplantation 73(4):665–667
276. Abhyankar S, Gilliland DG, Ferrara JL (1993) Interleukin-1 is a critical effector molecule during cytokine dysregulation in graft versus host disease to minor histocompatibility antigens. Transplantation 56(6):1518–1523
277. Cullup H, Dickinson A, Jackson G, Taylor P, Cavet J, Middleton P (2000) Donor interleukin 1 receptor antagonist genotype associated with acute graft-versus-host disease in human leucocyte antigen-matched sibling allogeneic transplants. Br J Haematol 113:807–813
278. McCarthy PL, Williams L, Harris-Bacile M, Yen J, Przepiorka D, Ippoliti C, Champlin R, Fay J, Blosch C, Jacobs C, Anasetti C (1996) A clinical phase I/II study of recombinant human interleukin-1 receptor in glucocorticoid-resistant graft-versus-host disease. Transplantation 62:626–631
279. Antin JH, Weinstein HJ, Guinan EC, McCarthy P, Bierer BE, Gilliland DG, Parsons SK, Ballen KK, Rimm IJ, Falzarano G, Bloedow DC, Abate L, Lebsack M, Burakoff SJ, Ferrara JLM (1994) Recombinant human interleukin-1 receptor antagonist in the treatment of steroid-resistant graft-versus-host disease. Blood 84:1342–1348
280. Antin JH, Lee SJ, Neuberg D, Alyea E, Soiffer RJ, Sonis S, Ferrara JL (2002) A phase I/II double-blind, placebo-controlled study of recombinant human interleukin-11 for mucositis and acute GVHD prevention in allogeneic stem cell transplantation. Bone Marrow Transplant 29(5):373–377
281. Martin PJ, Schoch G, Fisher L et al (1991) A retrospective analysis of therapy for acute graft-versus-host disease: secondary treatment. Blood 77:1821–1828
282. Weisdorf D, Haake R, Blazar B, Miller W, McGlave P, Ramsay N, Kersey J, Filipovich A (1990) Treatment of moderate/severe acute graft-versus-host disease after allogeneic bone marrow transplantation: an analysis of clinical risk features and outcome. Blood 75(4):1024–1030
283. Hockenbery DM, Cruickshank S, Rodell TC, Gooley T, Schuening F, Rowley S, David D, Brunvand M, Berryman B, Abhyankar S, Bouvier M, McDonald GB (2007) A randomized, placebo-controlled trial of oral beclomethasone dipropionate as a prednisone-sparing therapy for gastrointestinal graft-versus-host disease. Blood 109(10):4557–4563
284. Deeg HJ (2007) How I treat refractory acute GVHD. Blood 109:4119–4126
285. Greinix HT, Socie G, Bacigalupo A, Holler E, Edinger MG, Apperley JF, Schwarz T, Ullrich SE, Albert ML, Knobler RM, Peritt D, Ferrara JL (2006) Assessing the potential role of photopheresis in hematopoietic stem cell transplant. Bone Marrow Transplant 38(4):265–273
286. Ely P, Dunitz J, Rogosheske J, Weisdorf D (1991) Use of a somatostatin analogue, octreotide acetate, in the management of acute gastrointestinal graft-versus-host disease. Am J Med 90(6):707–710
287. Sullivan KM, Agura E, Anasetti C, Appelbaum F, Badger C, Bearman S, Erickson K, Flowers M, Hansen J, Loughran T et al (1991) Chronic graft-versus-host disease and other late complications of bone marrow transplantation. Semin Hematol 28(3):250–259
288. Vogelsang GB, Arai S (2001) Mycophenolate mofetil for the prevention and treatment of graft-versus-host disease following stem cell transplantation: preliminary findings. Bone Marrow Transplant 27(12):1255–1262
289. Flowers ME, Parker PM, Johnston LJ, Matos AV, Storer B, Bensinger WI, Storb R, Appelbaum FR, Forman SJ, Blume KG, Martin PJ (2002) Comparison of chronic graft-

versus-host disease after transplantation of peripheral blood stem cells versus bone marrow in allogeneic recipients: long-term follow-up of a randomized trial. Blood 100(2):415–419
290. Akpek G, Chinratanalab W, Lee LA, Torbenson M, Hallick JP, Anders V, Vogelsang GB (2003) Gastrointestinal involvement in chronic graft-versus-host disease: a clinicopathologic study. Biol Blood Marrow Transplant 9(1):46–51
291. Ghayur T, Seemayer T, Lapp WS (1990) Histologic correlates of immune functional deficits in graft-vs-host disease. In: Burakoff SJ, Deeg HJ, Ferrara J, Atkinson K (eds) Graft-vs.-host disease: immunology, pathophysiology, and treatment. Marcel Dekker, New York, pp 109132
292. Arai S, Vogelsang GB (2000) Management of graft-versus-host disease. Blood Rev 14(4):190–204
293. Wingard JR, Piantadosi S, Vogelsgang GB, Farmer ER, Jabs DA, Levin LS, Beschorner WE, Cahill RA, Miller DF, Harrison D, Saral R, Santos GW (1989) Predictors of death from chronic graft-versus-host disease after bone marrow transplantation. Blood 74:1428–1435
294. Wingard JR, Vogelsang GB, Deeg HJ (2002) Stem cell transplantation: supportive care and long-term complications. Hematology (Am Soc Hematol Educ Program):422-444
295. Ghayur T, Seemayer T, Lapp WS et al (1990) Histologic correlates of immune functional deficits in graft-vs-host disease. In: Burakoff SJ, Deeg HJ, Ferrara J, Atkinson K (eds) Graft-vs.-host disease: immunology, pathophysiology, and treatment. Marcel Dekker, New York, pp 109–132
296. Filipovich AH, Weisdorf D, Pavletic S, Socie G, Wingard JR, Lee SJ, Martin P, Chien J, Przepiorka D, Couriel D, Cowen EW, Dinndorf P, Farrell A, Hartzman R, Henslee-Downey J, Jacobsohn D, McDonald G, Mittleman B, Rizzo JD, Robinson M, Schubert M, Schultz K, Shulman H, Turner M, Vogelsang G, Flowers ME (2005) National Institutes of Health consensus development project on criteria for clinical trials in chronic graft-versus-host disease: I. Diagnosis and staging working group report. Biol Blood Marrow Transplant 11(12):945–956
297. Martin PJ, Weisdorf D, Przepiorka D, Hirschfeld S, Farrell A, Rizzo JD, Foley R, Socie G, Carter S, Couriel D, Schultz KR, Flowers ME, Filipovich AH, Saliba R, Vogelsang GB, Pavletic SZ, Lee SJ (2006) National Institutes of Health Consensus development project on criteria for clinical trials in chronic graft-versus-host disease: VI. Design of clinical trials working group report. Biol Blood Marrow Transplant 12(5):491–505
298. Ratanatharathorn V, Ayash L, Lazarus HM, Fu J, Uberti JP (2001) Chronic graft-versus-host disease: clinical manifestation and therapy. Bone Marrow Transplant 28(2):121–129
299. Reddy P, Johnson K, Uberti JP, Reynolds C, Silver S, Ayash L, Braun TM, Ratanatharathorn V (2006) Nephrotic syndrome associated with chronic graft-versus-host disease after allogeneic hematopoietic stem cell transplantation. Bone Marrow Transplant 38(5):351–357
300. Akpek G, Zahurak ML, Piantadosi S, Margolis J, Doherty J, Davidson R, Vogelsang GB (2001) Development of a prognostic model for grading chronic graft-versus-host disease. Blood 97(5):1219–1226
301. Shulman HM, Sharma P, Amos D, Fenster LF, McDonald GB (1988) A coded histologic study of hepatic graft-versus-host disease after human bone marrow transplantation. Hepatology 8(3):463–470
302. Shulman HM, Sale GE, Lerner KG, Barker EA, Weiden PL, Sullivan K, Gallucci B, Thomas ED, Storb R (1978) Chronic cutaneous graft-versus-host disease in man. Am J Pathol 91(3):545–570
303. Shlomchik WD, Lee SJ, Couriel D, Pavletic SZ (2007) Transplantation's greatest challenges: advances in chronic graft-versus-host disease. Biol Blood Marrow Transplant 13(Suppl 1):2–10
304. Parkman R (1993) Is chronic graft versus host disease an autoimmune disease? Curr Opin Immunol 5(5):800–803

305. Sakoda Y, Hashimoto D, Asakura S, Takeuchi K, Harada M, Tanimoto M, Teshima T (2007) Donor-derived thymic-dependent T cells cause chronic graft-versus-host disease. Blood 109(4):1756–1764
306. Hess AD, Horwitz L, Beschorner WE, Santos GW (1985) Development of graft-vs.-host disease-like syndrome in cyclosporine-treated rats after syngeneic bone marrow transplantation. I. Development of cytotoxic T lymphocytes with apparent polyclonal anti-Ia specificity, including autoreactivity. J Exp Med 161(4):718–730
307. Hess A, Thoburn C, Bright E, Horwitz L (1997) Specificity of effector mechanisms in syngeneic graft-vs-host disease: recognition of the MHC class II invariant chain peptide (CLIP). Transplant Proc 29(1–2):725–727
308. Holländer GA, Widmer B, Burakoff SJ (1994) Loss of normal thymic repertoire selection and persistence of autoreactive T cells in graft vs. host disease. J Immunol 152:1609–1617
309. Teshima T, Reddy P, Liu C, Williams D, Cooke KR, Ferrara JL (2003) Impaired thymic negative selection causes autoimmune graft-versus-host disease. Blood 102(2):429–435
310. Ratanatharathorn V, Ayash L, Reynolds C, Silver S, Reddy P, Becker M, Ferrara JL, Uberti JP (2003) Treatment of chronic graft-versus-host disease with anti-CD20 chimeric monoclonal antibody. Biol Blood Marrow Transplant 9(8):505–511
311. Cutler C, Antin JH (2006) Chronic graft-versus-host disease. Curr Opin Oncol 18(2):126–131
312. Lee SJ (2005) New approaches for preventing and treating chronic graft-versus-host disease. Blood 105(11):4200–4206
313. Stewart BL, Storer B, Storek J, Deeg HJ, Storb R, Hansen JA, Appelbaum FR, Carpenter PA, Sanders JE, Kiem HP, Nash RA, Petersdorf EW, Moravec C, Morton AJ, Anasetti C, Flowers ME, Martin PJ (2004) Duration of immunosuppressive treatment for chronic graft-versus-host disease. Blood 104(12):3501–3506
314. Mielcarek M, Martin PJ, Leisenring W, Flowers ME, Maloney DG, Sandmaier BM, Maris MB, Storb R (2003) Graft-versus-host disease after nonmyeloablative versus conventional hematopoietic stem cell transplantation. Blood 102(2):756–762
315. Couriel D, Carpenter PA, Cutler C, Bolanos-Meade J, Treister NS, Gea-Banacloche J, Shaughnessy P, Hymes S, Kim S, Wayne AS, Chien JW, Neumann J, Mitchell S, Syrjala K, Moravec CK, Abramovitz L, Liebermann J, Berger A, Gerber L, Schubert M, Filipovich AH, Weisdorf D, Schubert MM, Shulman H, Schultz K, MittelmanB, Pavletic S, Vogelsang GB, Martin PJ, Lee SJ, Flowers ME (2006) Ancillary therapy and supportive care of chronic graft-versus-host disease: national institutes of health consensus development project on criteria for clinical trials in chronic Graft-versus-host disease: V. Ancillary therapy and supportive care working group report. Biol Blood Marrow Transplant 12(4):375–396
316. Lee SJ, Klein JP, Barrett AJ, Ringden O, Antin JH, Cahn JY, Carabasi MH, Gale RP, Giralt S, Hale GA, Ilhan O, McCarthy PL, Socie G, Verdonck LF, Weisdorf DJ, Horowitz MM (2002) Severity of chronic graft-versus-host disease: association with treatment-related mortality and relapse. Blood 100(2):406–414
317. Sullivan KM, Witherspoon RP, Storb R, Weiden P, Flournoy N, Dahlberg S, Deeg HJ, Sanders JE, Doney KC, Appelbaum FR et al (1988) Prednisone and azathioprine compared with prednisone and placebo for treatment of chronic graft-v-host disease: prognostic influence of prolonged thrombocytopenia after allogeneic marrow transplantation. Blood 72(2):546–554
318. Sullivan KM, Witherspoon RP, Storb R, Deeg HJ, Dahlberg S, Sanders JE, Appelbaum FR, Doney KC, Weiden P, Anasetti C et al (1988) Alternating-day cyclosporine and prednisone for treatment of high-risk chronic graft-v-host disease. Blood 72(2):555–561
319. Investigator's Brochure (1997) Recombinant human keratinocyte growth (r-HuKGF), 3.0 ed. Amgen, Inc.
320. Hess AD (1997) The immunobiology of syngeneic/autologous graft-versus-host disease. In: Ferrara JLM, Deeg HJ, Burakoff SJ (eds) Graft-vs.-host disease. Marcel Dekker, Inc., New York, pp 561–586

321. Ichinohe T, Teshima T, Matsuoka K, Maruya E, Saji H (2005) Fetal-maternal microchimerism: impact on hematopoietic stem cell transplantation. Curr Opin Immunol 17(5):546–552
322. Anderson KC, Weinstein HJ (1990) Transfusion-associated graft-versus-host disease. N Engl J Med 323:315–321
323. Orlin JB, Ellis MH (1997) Transfusion-associated graft-versus-host disease. Curr Opin Hematol 4(6):442–448
324. Schroeder ML (2002) Transfusion-associated graft-versus-host disease. Br J Haematol 117(2):275–287
325. Bortin MM, Rimm AA, Saltzstein E (1973) Graft-versus-leukemia: quantification of adoptive immunotherapy in murine leukemia. Science 173:811–813
326. Weiden PL, Flournoy N, Thomas ED, Prentice R, Fefer A, Buckner CD, Storb R (1979) Antileukemic effect of graft-versus-host disease in human recipients of allogeneic-marrow grafts. N Engl J Med 300(19):1068–1073
327. Gale RP, Champlin RE (1984) How does bone-marrow transplantation cure leukaemia? Lancet 2(8393):28–30
328. Gale RP, Butturini A (1994) Stem cells, clonality and leukemia. Gematol Transfuziol 39(6):3–6
329. Horowitz MM, Bortin MM (1990) Current status of allogeneic bone marrow transplantation. Clin Transplant 4(2):41–52
330. Gale R, Horowitz M, Ash R (1994) Identical-twin bone marrow transplants for leukemia. Ann Intern Med 120:646–652
331. Porter DL, Antin JH (1999) The graft-versus-leukemia effects of allogeneic cell therapy. Annu Rev Med 50:369–386
332. Collins R, Rogers Z, Bennett M et al (1992) Hematologic relapse of chronic myelogenous leukemia following allogeneic bone marrow transplantation: apparent graft-versus-leukemia effect following abrupt discontinuation of immunosuppression. Bone Marrow Transplant 10:391–395
333. Higano C, Brixey M, Bryant E et al (1990) Durable complete remission of acute nonlymphocytic leukemia associated with discontinuation of immunosuppression following relapse after allogeneic bone marrow transplantation: a case report of a probable graft-versus-leukemia effect. Transplantation 50:175–177
334. Sullivan K, Storb R, Buckner D, Fefer A (1989) Graft-versus-host disease as adoptive immunotherapy in patients with advanced hematologic neoplasms. N Engl J Med 320:828–834
335. Odom L, August C, Githens J (1978) Remission of relapsed leukaemia during a graft-versus-host reaction. A "graft-versus-leukaemia" reaction in man? Lancet 2:537–540
336. Weiden P, Flournoy N, Donnall Thomas E et al (1979) Antileukemic effect of graft-versus-host disease in human recipients of allogeneic-marrow grafts. N Engl J Med 300:1068–1073
337. Sullivan K, Weiden P, Storb R (1989) Influence of acute and chronic graft-versus-host disease on relapse and survival after bone marrow transplantation from HLA-identical siblings as treatment of acute and chronic leukemia. Blood 73:1720–1728
338. Weiden P, Sullivan K, Flournoy N et al (1981) Antileukemic effect of chronic graft-versus-host disease: Contribution to improved survival after allogeneic marrow transplantation. N Engl J Med 304:1529–1533
339. Horowitz MM, Gale RP, Sondel PM, Goldman JM, Kersey J, Kolb HJ, Rimm AA, Ringden O, Rozman C, Speck B et al (1990) Graft-versus-leukemia reactions after bone marrow transplantation. Blood 75(3):555–562
340. Weiden PL, Zuckerman N, Hansen JA, Sale GE, Remlinger K, Beck TM, Buckner CD (1981) Fatal graft-versus-host disease in a patient with lymphoblastic leukemia following normal granulocyte transfusion. Blood 57(2):328–332
341. Goldman JM, Gale RP, Horowitz MM, Biggs JC, Champlin RE, Gluckman E, Hoffmann RG, Jacobsen SJ, Marmont AM, McGlave PB et al (1988) Bone marrow transplantation for

chronic myelogenous leukemia in chronic phase: increased risk for relapse associated with T-cell depletion. Ann Intern Med 108(6):806–814
342. Childs R, Chernoff A, Contentin N, Bahceci E, Schrump D, Leitman S, Read E, Tisdale J, Dunbar C, Linehan W, Neal S, Barrett A (2000) Regression of metastatic renal-cell carcinoma after nonmyeloablative allogeneic periphereal-blood stem-cell transplantation. N Engl J Med 343:750–758
343. Antin JH, Bierer BE, Smith BR, Ferrara J, Guinan EC, Sieff C, Golan DE, Macklis RM, Tarbell NJ, Lynch E et al (1991) Selective depletion of bone marrow T lymphocytes with anti-CD5 monoclonal antibodies: effective prophylaxis for graft-versus-host disease in patients with hematologic malignancies. Blood 78(8):2139–2149
344. Martin PJ, Hansen JA, Storb R, Thomas ED (1985) T cell depletion of donor marrow for prevention of acute graft-versus-host disease. Haematol Blood Transfus 29:42–43
345. Mitsuyasu RT, Champlin RE, Gale RP, Ho WG, Lenarsky C, Winston D, Selch M, Elashoff R, Giorgi JV, Wells J et al (1986) Treatment of donor bone marrow with monoclonal anti-T-cell antibody and complement for the prevention of graft-versus-host disease. A prospective, randomized, double-blind trial. Ann Intern Med 105(1):20–26
346. Soiffer RJ, Murray C, Mauch P, Anderson KC, Freedman AS, Rabinowe SN, Takvorian T, Robertson MJ, Spector N, Gonin R et al (1992) Prevention of graft-versus-host disease by selective depletion of CD6-positive T lymphocytes from donor bone marrow. J Clin Oncol 10(7):1191–1200
347. Waldmann H, Polliak A, Hale G, Or R, Cividalli G, Weiss L, Weshler Z, Samuel S, Manor D, Brautbar C et al (1984) Elimination of graft-versus-host disease by in vitro depletion of alloreactive lymphocytes with a monoclonal rat anti-human lymphocyte antibody (CAMPATH-1). Lancet 2(8401):483–486
348. Wagner JE, Santos GW, Noga SJ, Rowley SD, Davis J, Vogelsang GB, Farmer ER, Zehnbauer BA, Saral R, Donnenberg AD (1990) Bone marrow graft engineering by counterflow centrifugal elutriation: results of a phase I–II clinical trial. Blood 75(6):1370–1377
349. Young JW, Papadopoulos EB, Cunningham I, Castro-Malaspina H, Flomenberg N, Carabasi MH, Gulati SC, Brochstein JA, Heller G, Black P et al (1992) T-cell-depleted allogeneic bone marrow transplantation in adults with acute nonlymphocytic leukemia in first remission. Blood 79(12):3380–3387
350. Apperley JF, Jones L, Hale G, Waldmann H, Hows J, Rombos Y, Tsatalas C, Marcus RE, Goolden AW, Gordon-Smith EC et al (1986) Bone marrow transplantation for patients with chronic myeloid leukaemia: T-cell depletion with Campath-1 reduces the incidence of graft-versus-host disease but may increase the risk of leukaemic relapse. Bone Marrow Transplant 1(1):53–66
351. Goldman JM, Apperley JF, Jones L, Marcus R, Goolden AW, Batchelor R, Hale G, Waldmann H, Reid CD, Hows J et al (1986) Bone marrow transplantation for patients with chronic myeloid leukemia. N Engl J Med 314(4):202–207
352. Ho VT, Soiffer RJ (2002) The history and future of T-cell depletion as graft-versus-host disease prophylaxis for allogeneic hematopoietic stem cell transplantation. Blood 98:3192–3204
353. Farag SS, Fehniger TA, Becknell B, Blaser BW, Caligiuri MA (2003) New directions in natural killer cell-based immunotherapy of human cancer. Expert Opin Biol Ther 3(2):237–250
354. Bleakley M, Riddell SR (2004) Molecules and mechanisms of the graft-versus-leukaemia effect. Nat Rev Cancer 4(5):371–380
355. Swann JB, Smyth MJ (2007) Immune surveillance of tumors. J Clin Invest 117(5):1137–1146
356. Reya T, Morrison SJ, Clarke MF, Weissman IL (2001) Stem cells, cancer, and cancer stem cells. Nature 414(6859):105–111

357. Yilmaz OH, Valdez R, Theisen BK, Guo W, Ferguson DO, Wu H, Morrison SJ (2006) Pten dependence distinguishes haematopoietic stem cells from leukaemia-initiating cells. Nature 441(7092):475–482
358. Gilboa E (2007) DC-based cancer vaccines. J Clin Invest 117(5):1195–1203
359. Appelbaum FR (2001) Haematopoietic cell transplantation as immunotherapy. Nature 411(6835):385–389
360. Fontaine P, Roy-Proulx G, Knafo L, Baron C, Roy DC, Perreault C (2001) Adoptive transfer of minor histocompatibility antigen-specific T lymphocytes eradicates leukemia cells without causing graft-versus-host disease. Nat Med 7(7):789–794
361. Hsu KC, Gooley T, Malkki M, Pinto-Agnello C, Dupont B, Bignon JD, Bornhauser M, Christiansen F, Gratwohl A, Morishima Y, Oudshoorn M, Ringden O, van Rood JJ, Petersdorf E (2006) KIR ligands and prediction of relapse after unrelated donor hematopoietic cell transplantation for hematologic malignancy. Biol Blood Marrow Transplant 12(8):828–836
362. Ruggeri L, Capanni M, Urbani E, Perruccio K, Shlomchik WD, Tosti A, Posati S, Rogaia D, Frassoni F, Aversa F, Martelli MF, Velardi A (2002) Effectiveness of donor natural killer cell alloreactivity in mismatched hematopoietic transplants. Science 295(5562):2097–2100
363. Petersdorf EW, Malkki M (2006) Genetics of risk factors for graft-versus-host disease. Semin Hematol 43(1):11–23
364. Hsu KC, Keever-Taylor CA, Wilton A, Pinto C, Heller G, Arkun K, O'Reilly RJ, Horowitz MM, Dupont B (2005) Improved outcome in HLA-identical sibling hematopoietic stem-cell transplantation for acute myelogenous leukemia predicted by KIR and HLA genotypes. Blood 105(12):4878–4884
365. Farag SS, Bacigalupo A, Eapen M, Hurley C, Dupont B, Caligiuri MA, Boudreau C, Nelson G, Oudshoorn M, van Rood J, Velardi A, Maiers M, Setterholm M, Confer D, Posch PE, Anasetti C, Kamani N, Miller JS, Weisdorf D, Davies SM (2006) The effect of KIR ligand incompatibility on the outcome of unrelated donor transplantation: a report from the center for international blood and marrow transplant research, the European blood and marrow transplant registry, and the Dutch registry. Biol Blood Marrow Transplant 12(8):876–884
366. Miller JS, Cooley S, Parham P, Farag SS, Verneris MR, McQueen KL, Guethlein LA, Trachtenberg EA, Haagenson M, Horowitz MM, Klein JP, Weisdorf DJ (2007) Missing KIR-ligands is associated with less relapse and increased graft versus host disease (GVHD) following unrelated donor allogeneic HCT. Blood 109:5058–5061
367. Ferrara JL, Reddy P (2006) Pathophysiology of graft-versus-host disease. Semin Hematol 43(1):3–10
368. Slavin S, Morecki S, Weiss L, Shapira MY, Resnick I, Or R (2004) Nonmyeloablative stem cell transplantation: reduced-intensity conditioning for cancer immunotherapy—from bench to patient bedside. Semin Oncol 31(1):4–21
369. Mapara MY, Kim YM, Wang SP, Bronson R, Sachs DH, Sykes M (2002) Donor lymphocyte infusions mediate superior graft-versus-leukemia effects in mixed compared to fully allogeneic chimeras: a critical role for host antigen-presenting cells. Blood 100(5):1903–1909
370. Mapara MY, Sykes M (2004) Tolerance and cancer: mechanisms of tumor evasion and strategies for breaking tolerance. J Clin Oncol 22(6):1136–1151
371. Chakraverty R, Cote D, Buchli J, Cotter P, Hsu R, Zhao G, Sachs T, Pitsillides CM, Bronson R, Means T, Lin C, Sykes M (2006) An inflammatory checkpoint regulates recruitment of graft-versus-host reactive T cells to peripheral tissues. J Exp Med 203(8):2021–2031
372. Gattinoni L, Klebanoff CA, Palmer DC, Wrzesinski C, Kerstann K, Yu Z, Finkelstein SE, Theoret MR, Rosenberg SA, Restifo NP (2005) Acquisition of full effector function in vitro paradoxically impairs the in vivo antitumor efficacy of adoptively transferred CD8+ T cells. J Clin Invest 115(6):1616–1626
373. Gattinoni L, Finkelstein SE, Klebanoff CA, Antony PA, Palmer DC, Spiess PJ, Hwang LN, Yu Z, Wrzesinski C, Heimann DM, Surh CD, Rosenberg SA, Restifo NP (2005) Removal of

homeostatic cytokine sinks by lymphodepletion enhances the efficacy of adoptively transferred tumor-specific CD8+ T cells. J Exp Med 202(7):907–912
374. Wrzesinski C, Paulos CM, Gattinoni L, Palmer DC, Kaiser A, Yu Z, Rosenberg SA, Restifo NP (2007) Hematopoietic stem cells promote the expansion and function of adoptively transferred antitumor CD8 T cells. J Clin Invest 117(2):492–501
375. Chakraverty R, Eom HS, Sachs J, Buchli J, Cotter P, Hsu R, Zhao G, Sykes M (2006) Host MHC Class II+ antigen-presenting cells and CD4 cells are required for CD8-mediated graft-versus-leukemia responses following delayed donor leukocyte infusions. Blood 108:2106–2113
376. Dunn GP, Old LJ, Schreiber RD (2004) The immunobiology of cancer immunosurveillance and immunoediting. Immunity 21(2):137–148
377. Boyer MW, Vallera DA, Taylor PA, Gray GS, Katsanis E (1997) The role of B7 costimulation by murine acute myeloid leukemia in the generation and function of a CD8+ T-Cell line with potent in vivo graft-versus-leukemia properties. Blood 89(9):3477–3485
378. Kolb HJ, Schmid C, Barrett AJ, Schendel DJ (2004) Graft-versus-leukemia reactions in allogeneic chimeras. Blood 103(3):767–776
379. Blazar BR, Taylor PA, Boyer MW, Panoskaltsis-Mortari A, Allison JP, Vallera DA (1997) CD28/B7 interactions are required for sustaining the graft-versus-leukemia effect of delayed post-bone marrow transplantation splenocyte infusion in murine recipients of myeloid or lymphoid leukemia cells. J Immunol 159(7):3460–3473
380. Blazar B, Taylor P, Panoskaltsis-Mortari A, Buhlman J, Xu J, Flavell R, Korngold R, Noelle R, Vallera D (1997) Blockade of CD40 ligand-CD40 interaction impairs CD4+ T cell mediated alloreactivity by inhibiting mature donor T cell expansion and function after bone marrow transplantation. J Immunology 158:29–39
381. Porter DL, Antin JH (2006) Donor leukocyte infusions in myeloid malignancies: new strategies. Best Pract Res Clin Haematol 19(4):737–755
382. Jiang YZ, Mavroudis D, Dermime S, Hensel N, Couriel D (1996) Alloreactive CD4+ T lymphocytes can exert cytotoxicity to chronic myeloid leukemia cells processing and presenting exogenous antigen. Br J Haematol 93:606–612
383. Petersdorf EW, Kollman C, Hurley CK, Dupont B, Nademanee A, Begovich AB, Weisdorf D, McGlave P (2001) Effect of HLA class II gene disparity on clinical outcome in unrelated donor hematopoietic cell transplantation for chronic myeloid leukemia: the US National Marrow Donor Program Experience. Blood 98(10):2922–2929
384. Dodi IA, Van Rhee F, Forde HC, Roura-Mir C, Jaraquemada D, Goldman JM, Madrigal JA (2002) CD4(+) bias in T cells cloned from a CML patient with active graft versus leukemia effect. Cytotherapy 4(4):353–363
385. Madrigal JA, Travers PJ, Dodi IA (2005) Immunotherapeutic aspects of stem cell transplantation. Hematology 10(Suppl 1):289–292
386. Jones SC, Murphy GF, Korngold R (2003) Post-hematopoietic cell transplantation control of graft-versus-host disease by donor CD425 T cells to allow an effective graft-versus-leukemia response. Biol Blood Marrow Transplant 9(4):243–256
387. Schmaltz C, Alpdogan O, Muriglan SJ, Kappel BJ, Rotolo JA, Ricchetti ET, Greenberg AS, Willis LM, Murphy GF, Crawford JM, van den Brink MR (2003) Donor T cell-derived TNF is required for graft-versus-host disease and graft-versus-tumor activity after bone marrow transplantation. Blood 101(6):2440–2445
388. Korngold R, Marini JC, De Baca ME, Murphy GF, Giles-Komar J (2003) Role of tumor necrosis factor-alpha in graft-versus-host disease and graft-versus-leukemia responses. Biol Blood Marrow Transplant 9(5):292–303
389. Hsieh MH, Korngold R (2000) Differential use of FasL- and perforin-mediated cytolytic mechanisms by T-cell subsets involved in graft-versus-myeloid leukemia responses. Blood 96(3):1047–1055
390. Hsieh MH, Patterson AE, Korngold R (2000) T-cell subsets mediate graft-versus-myeloid leukemia responses via different cytotoxic mechanisms. Biol Blood Marrow Transplant 6(3):231–240

391. Nebbioso A, Clarke N, Voltz E, Germain E, Ambrosino C, Bontempo P, Alvarez R, Schiavone EM, Ferrara F, Bresciani F, Weisz A, de Lera AR, Gronemeyer H, Altucci L (2005) Tumor-selective action of HDAC inhibitors involves TRAIL induction in acute myeloid leukemia cells. Nat Med 11(1):77–84
392. Insinga A, Monestiroli S, Ronzoni S, Gelmetti V, Marchesi F, Viale A, Altucci L, Nervi C, Minucci S, Pelicci PG (2005) Inhibitors of histone deacetylases induce tumor-selective apoptosis through activation of the death receptor pathway. Nat Med 11(1):71–76
393. Reddy P, Maeda Y, Hotary K, Liu C, Reznikov LL, Dinarello CA, Ferrara JL (2004) Histone deacetylase inhibitor suberoylanilide hydroxamic acid reduces acute graft-versus-host disease and preserves graft-versus-leukemia effect. Proc Natl Acad Sci U S A 101(11):3921–3926
394. Porter D, Connors J, Van Deerlin V, Duffy K, McGarigle C, Saidman S, Leonard D, Antin J (1999) Graft-versus-tumor induction with donor leukocyte infusions as primary therapy for patients with malignancies. J Clin Oncol 17(4):1234–1243
395. van Rhee F, Lin F, Cullis JO, Spencer A, Cross NCP, Chase A, Garicochea B, Bungey J, Barrett J, Goldman JM (1994) Relapse of chronic myeloid leukemia after allogeneic bone marrow transplant: the case for giving donor leukocyte transfusions before the onset of hematologic relapse. Blood 83:3377–3383
396. Kolb H-J, Schattenberg A, Goldman JM, Hertenstein B et al (1995) Graft-versus-leukemia effect of donor lymphocyte transfusions in marrow grafted patients. Blood 86:2041–2050
397. Kolb H, Mittermuller J, Clemm C et al (1990) Donor leukocyte transfusions for treatment of recurrent chronic myelogenousleukemia in marrow transplant patients. Blood 76:2462–2465
398. Porter D, Roth M, McGarigle C et al (1994) Induction of graft versus host disease as immunotherapy for relapsed chronic myeloid leukemia. N Engl J Med 330:100–106
399. Drobyski W, Keever C, Roth M, Koeth S, Hanson G, McFadden P, Gottschall J, Ash RC, van Tuinen P, Horowitz MM (1993) Salvage immunotherapy using donor leukocyte transfusions as treatment for relapse chronic myelogenous leukemia after allogeneic bone marrow transplantation: efficacy and toxicity of a defined T-cell dose. Blood 82:2310–2318
400. Hertenstein B, Wiesneth M, Novotny J et al (1993) Interferon-alpha and donor buffy coat transfusions for treatment of relapsed chronic myeloid leukemia after allogeneic bone marrow transplantation. Transplantation 56:1114
401. Collins RH, Shpilberg WR, Drobyski WR, Porter DL, Giralt S (1997) Donor leukocyte infusions in 140 patients with relapsed malignancy after allogeneic bone marrow transplantation. J Clin Oncol 15(2):433–444
402. Dazzi F, Szydlo RM, Cross NC, Craddock C, Kaeda J, Kanfer E, Cwynarski K, Olavarria E, Yong A, Apperley JF, Goldman JM (2000) Durability of responses following donor lymphocyte infusions for patients who relapse after allogeneic stem cell transplantation for chronic myeloid leukemia. Blood 96(8):2712–2716
403. Dazzi F, Szydlo RM, Apperley JF, Goldman JM (2002) Prognostic factors for acute graft-versus-host disease after donor lymphocyte infusions. Blood 100(7):2673–2674
404. Collins RH Jr, Goldstein S, Giralt S, Levine J, Porter D, Drobyski W, Barrett J, Johnson M, Kirk A, Horowitz M, Parker P (2000) Donor leukocyte infusions in acute lymphocytic leukemia. Bone Marrow Transplant 26(5):511–516
405. Kolb HJ, Schattenberg A, Goldman JM, Hertenstein B, Jacobsen N, Arcese W, Ljungman P, Ferrant A, Verdonck L, Niederwieser D, van Rhee F, Mittermueller J, de Witte T, Holler E, Ansari H (1995) Graft-versus-leukemia effect of donor lymphocyte transfusions in marrow grafted patients. Blood 86(5):2041–2050
406. Singhal S, Powles R, Kulkarni S, Treleaven J, Saso R, Mehta J (1999) Long-term follow-up of relapsed acute leukemia treated with immunotherapy after allogeneic transplantation: the inseparability of graft-versus-host disease and graft-versus-leukemia, and the problem of extramedullary relapse. Leuk Lymphoma 32(5–6):505–512
407. Levine JE, Braun T, Penza SL, Beatty P, Cornetta K, Martino R, Drobyski WR, Barrett AJ, Porter DL, Giralt S, Leis J, Holmes HE, Johnson M, Horowitz M, Collins RH Jr (2002)

Prospective trial of chemotherapy and donor leukocyte infusions for relapse of advanced myeloid malignancies after allogeneic stem-cell transplantation. J Clin Oncol 20(2):405–412
408. Kolb HJ, Schmid C, Weisser M, Pihusch M, Tischer J, Pihusch R, Stoetzer O, Ledderose G, Schleuning M (2002) Cytoreduction, DLI, or mobilized peripheral blood progenitors. Ann Hematol 81(Suppl 2):S30–33
409. Slavin S, Morecki S, Weiss L, Or R (2002) Donor lymphocyte infusion: the use of alloreactive and tumor-reactive lymphocytes for immunotherapy of malignant and nonmalignant diseases in conjunction with allogeneic stem cell transplantation. J Hematother Stem Cell Res 11(2):265–276
410. Ruggeri L, Capanni M, Martelli MF, Velardi A (2001) Cellular therapy: exploiting NK cell alloreactivity in transplantation. Curr Opin Hematol 8(6):355–359
411. Rondon G, Giralt S, Huh Y, Khouri I, Andersson B, Andreeff M, Champlin R (1996) Graft-versus-leukemia effect after allogeneic bone marrow transplantation for chronic lymphocytic leukemia. Bone Marrow Transplant 18(3):669–672
412. Mandigers CM, Meijerink JP, Raemaekers JM, Schattenberg AV, Mensink EJ (1998) Graft-versus-lymphoma effect of donor leucocyte infusion shown by real-time quantitative PCR analysis of t(14;18). Lancet 352(9139):1522–1523
413. Salama M, Nevill T, Marcellus D, Parker P, Johnson M, Kirk A, Porter D, Giralt S, Levine JE, Drobyski W, Barrett AJ, Horowitz M, Collins RH (2000) Donor leukocyte infusions for multiple myeloma. Bone Marrow Transplant 26(11):1179–1184
414. Lokhorst HM, Schattenberg A, Cornelissen JJ, van Oers MH, Fibbe W, Russell I, Donk NW, Verdonck LF (2000) Donor lymphocyte infusions for relapsed multiple myeloma after allogeneic stem-cell transplantation: predictive factors for response and long-term outcome. J Clin Oncol 18(16):3031–3037
415. Papadopoulos EB, Ladanyi M, Emanuel D, Mackinnon S, Boulad F, Carabasi MH, Castro-Malaspina H, Childs BH, Gillio AP, Small TN, Young JW, Kernan NA, O'Reilly RJ (1994) Infusions of donor leukocytes to treat Epstein-Barr virus-associated lymphoproliferative disorders after allogeneic bone marrow transplantation. N Engl J Med 330:1185–1191
416. Rooney CM, Smith CA, Ng CY, Loftin SK, Sixbey JW, Gan Y, Srivastava DK, Bowman LC, Krance RA, Brenner MK, Heslop HE (1998) Infusion of cytotoxic T cells for the prevention and treatment of Epstein-Barr virus-induced lymphoma in allogeneic transplant recipients. Blood 92(5):1549–1555
417. Rooney CM, Aguilar LK, Huls MH, Brenner MK, Heslop HE (2001) Adoptive immunotherapy of EBV-associated malignancies with EBV-specific cytotoxic T-cell lines. Curr Top Microbiol Immunol 258:221–229
418. Riddell SR, Watanabe KS, Goodrich JM, Li CR, Agha ME, Greenberg PD (1992) Restoration of viral immunity in immunodeficient humans by the adoptive transfer of T cell clones. Science 257:238–241
419. Einsele H, Roosnek E, Rufer N, Sinzger C, Riegler S, Loffler J, Grigoleit U, Moris A, Rammensee HG, Kanz L, Kleihauer A, Frank F, Jahn G, Hebart H (2002) Infusion of cytomegalovirus (CMV)-specific T cells for the treatment of CMV infection not responding to antiviral chemotherapy. Blood 99(11):3916–3922
420. Boeckh M, Leisenring W, Riddell SR, Bowden RA, Huang ML, Myerson D, Stevens-Ayers T, Flowers ME, Cunningham T, Corey L (2003) Late cytomegalovirus disease and mortality in recipients of allogeneic hematopoietic stem cell transplants: importance of viral load and T-cell immunity. Blood 101(2):407–414
421. Frassoni F, Fagiloli F, Sessarego M et al (1992) The affect of donor leukocyte infusion in patients with leukemia following allogeneic bone marrow transplantation. Exp Hematol 20:712
422. Flowers ME, Kansu E, Sullivan KM (1999) Pathophysiology and treatment of graft-versus-host disease. Hematol Oncol Clin N Am 13(5):1091–1112, viii–ix
423. Elfenbein G, Graham-Pole J, Weiner R, Goedert T, Gross S (1987) Consequences of no prophylaxis for acute graft-versus-host disease after HLA-identical bone marrow transplantation (abstract). Blood 70(Suppl 1):305a

424. Sullivan K, Schulman H (1989) Chronic graft-versus-host disease, obliterative bronchiolitis, and graft-versus-leukemia effect: case histories. Transplant Proc 21:51–62
425. Porter D, Antin J (1997) Graft-versus-leukemia effect of allogeneic bone marrow transplantation and donor mononuclear cell infusions. In: Winter J (ed) Blood stem cell transplantation. Kluwer Academic Publishers, Norwell, pp 57–86
426. Porter D, Roth M, McGarigle C et al (1994) Induction of graft-vs-leukemia (GVL) reaction as therapy for relapsed leukemia after allogeneic bone marrow transplantation (BMT). J Cell Biochem 18B:94a
427. Nash A, Pepe MS, Storb R, Longton G, Pettinger M, Anasetti C, Appelbaum FR, Bowden RA, Deeg HJ, Doney K, Martin PJ, Sullivan KM, Sanders J, Witherspoon R (1992) Acute graft-versus-host disease analysis of risk factors after allogeneic marrow transpantation and prophylaxis with cyclosporine and methotrexate. Blood 80:1838–1845
428. Levine JE, Uberti JP, Ayash L, Reynolds C, Ferrara JL, Silver SM, Braun T, Yanik G, Hutchinson R, Ratanatharathorn V (2003) Lowered-intensity preparative regimen for allogeneic stem cell transplantation delays acute graft-versus-host disease but does not improve outcome for advanced hematologic malignancy. Biol Blood Marrow Transplant 9(3):189–197
429. Marks DI, Lush R, Cavenagh J, Milligan DW, Schey S, Parker A, Clark FJ, Hunt L, Yin J, Fuller S, Vandenberghe E, Marsh J, Littlewood T, Smith GM, Culligan D, Hunter A, Chopra R, Davies A, Towlson K, Williams CD (2002) The toxicity and efficacy of donor lymphocyte infusions given after reduced-intensity conditioning allogeneic stem cell transplantation. Blood 100(9):3108–3114
430. Alyea EP, Soiffer RJ, Canning C, Neuberg D, Schlossman R, Pickett C, Collins H, Wang Y, Anderson KC, Ritz J (1998) Toxicity and efficacy of defined doses of CD4(+) donor lymphocytes for treatment of relapse after allogeneic bone marrow transplant. Blood 91(10):3671–3680
431. Bethge WA, Hegenbart U, Stuart MJ, Storer BE, Maris MB, Flowers ME, Maloney DG, Chauncey T, Bruno B, Agura E, Forman SJ, Blume KG, Niederwieser D, Storb R, Sandmaier BM (2004) Adoptive immunotherapy with donor lymphocyte infusions after allogeneic hematopoietic cell transplantation following nonmyeloablative conditioning. Blood 103(3):790–795
432. Dey BR, McAfee S, Sackstein R, Colby C, Saidman S, Weymouth D, Poliquin C, Vanderklish J, Sachs DH, Sykes M, Spitzer TR (2001) Successful allogeneic stem cell transplantation with nonmyeloablative conditioning in patients with relapsed hematologic malignancy following autologous stem cell transplantation. Biol Blood Marrow Transplant 7(11):604–612

Chapter 17
Cardiac Versus Non-Cardiac Stem Cells to Repair the Heart: The Role of Autocrine/Paracrine Signals

Manlio Vinciguerra, Vincenzo Lionetti, Carlo Ventura and Nadia Rosenthal

Abstract The heart can grow and change to accommodate alterations in its workload in response to stress. Stress-induced pathological cardiac hypertrophy, characterized by an increase in the size of cardiac myocytes, is a major risk factor for the development of cardiomyopathies and heart failure. Restoration of damaged heart muscle through repair or regeneration is a promising strategy to cure cardiovascular diseases. The robust proliferative and differentiation capacity of stem cells holds the potential of an unlimited supply of functioning cardiomyocytes. A crucial issue in designing rational stem cell-based therapy approaches for cardiac diseases is the understanding of the exact mechanisms whereby each stem cell can affect the performance of the heart. Stem cells are self-renewing and can become functionally specialized cardiac cells with the proper milieu of locally acting growth and

M. Vinciguerra · N. Rosenthal
European Molecular Biology Laboratory (EMBL), Mouse Biology Unit, Campus A. Buzzati-Traverso, Monterotondo, Italy

M. Vinciguerra (✉)
Istituto EuroMEditerraneo di Scienza e Tecnologia (IEMEST), Palermo, Italy
e-mail: manliovinciguerra@iemest.eu

V. Lionetti
Laboratory of Medical Science, Scuola Superiore Sant'Anna, Institute of Life Sciences, Pisa, Italy

V. Lionetti · C. Ventura
Laboratory of Molecular Biology and Stem Cell Engineering, National Institute of Biostructures and Biosystems, University of Bologna, Bologna, Italy

C. Ventura
Cardiovascular Department, S. Orsola–Malpighi Hospital, University of Bologna, Bologna, Italy

N. Rosenthal
Australian Regenerative Medicine Institute (ARMI), Monash University, Melbourne, VIC, Australia

H. Baharvand and N. Aghdami (eds.), *Advances in Stem Cell Research*, Stem Cell Biology and Regenerative Medicine, DOI: 10.1007/978-1-61779-940-2_17,

signaling factors. In this chapter, we review the current advancements in our understanding of how adult stem cells within the myocardium, such as cardiac stem cells, or derived from other tissues, such as mesenchymal stem cells, may contribute to the repair of the damaged heart. Moreover we describe how this process could be experimentally boosted for therapy, by manipulating local autocrine and paracrine molecules regulating endogenous stem cell regenerative potential.

17.1 Introduction

The mammalian heart is a dynamic organ that can grow and change to accommodate alterations in its workload. In response to stress or dysfunction such as pressure/volume overload, congenital defects, myocardial infarction, or changes caused by adiposity (obesity) or diabetic cardiomyopathy, the heart undergoes adaptation processes, which lead to pathological hypertrophy, characterized by an increase in the size of individual cardiac myocytes [1]. This is the result of several biomechanical sensitive and neurohumoral mechanisms occurring simultaneously. These mechanisms converge on a limited number of intracellular signal-transduction circuits to mediate cardiac growth. At the cellular level, cardiac dysfunction and cardiomyocyte loss is often due to an imbalance between the generation of reactive oxygen species and the antioxidant defenses in favor of the former. Growing evidence demonstrates that metabolic stress, oxidative stress, and hypertrophy are mechanistically linked to each other in the long-term development of cardiomyopathies and heart failure (HF) [2, 3]. This may be accompanied also by the formation of scar tissue, an overload of blood flow, and pressure capacity.

Ultimately, stress-induced pathological cardiac hypertrophy is a major risk factor for the development of HF and related deaths. HF is defined as a deficiency in the capability of the heart to adequately pump blood in response to systemic demands. Stress-induced pathological cardiac hypertrophy is particularly relevant in the aging population. The social and economic burden is tremendous, with the incidence of HF being 1–2% in the total population and >10% in the elderly population in Europe [4]. Similarly, in the USA, about 5 million Americans are affected by HF, which is characterized by a 5-year survival of approximately 50% [5–7]. Numerous pathways have been implicated in the establishment of pathological hypertrophy, drawing a very complex and dynamic picture of this phenomenon [1]. Examples include the activation of G-protein coupled receptors [by Angiotensin II (Ang II), endothelin-1 and catecholamines], MAP kinase signaling, the calcineurin-NFAT circuit, the PI3K-AKT pathway (activated by insulin, IGF-1 and other growth factors), enzymes that regulate chromatin remodeling, and a plethora of transcription factors (NF-kB, GATA4, MEF2 and others) [8]. Despite these advances, a major challenge in the field is to translate our increased knowledge of these intricate signaling mechanisms into novel treatments for pathological cardiac hypertrophy and HF.

Restoration of damaged heart muscle, through repair or regeneration, is an exciting new strategy to treat HF and other cardiovascular diseases. The heart is classically considered a post-mitotic organ, i.e. cardiomyocytes do not undergo renewal once the organ is fully differentiated and developed, and cannot repair efficiently after a consistent damage such an infarction. Recently, this view has been challenged. It was shown for instance that, in patients from 20 to 100 years of age who died from causes other than cardiovascular diseases, the cardiomyocyte compartment is replaced 15 times in women and 11 times in men by a resident pool of cardiac stem cells (CSC) [9]. The regenerative capacity of the heart is however extremely slow, and this organ has among the lowest cell turnover rate in the body [10]. For these reasons, the heart muscle field has directed a great effort toward exploring the therapeutic potential of adult stem cells not only from the heart but also from different origins and with different approaches (cellular reprogramming, tissue engineering), which may provide a new therapeutic option for myocardial regeneration, although many challenges remain to be solved [11].

The outstanding proliferative and differentiation capacity of stem cells holds the potential of an unlimited supply of functioning cardiomyocytes. A key issue in designing rational stem cell-based therapy approaches for HF and other cardiac diseases is the understanding of the exact mechanisms whereby each stem cell or progenitor-cell type can affect the performance of the heart. Stem cells differ from other cell types by two unique features: first, they are self-renewing, even after long inactive periods; they can become tissue- or organ-specific functionally specialized cells with the proper physiologic or experimental setting (milieu of autocrine/paracrine growth and signaling factors). Second, stem cells can be pluripotent or multipotent. Pluripotent stem cells can differentiate theoretically to any cell type in the body whereas multipotent stem cells give rise to a restricted profile of cell types. Stem cells can be derived from three sources: embryonic, non-embryonic (somatic or adult), or induced pluripotent stem cells. Embryonic stem cells (ESCs) are derived from embryos that are 5–8 days old in humans, or from a blastocyst stage in mammals; however, they can be isolated even from earlier developmental stages. ESCs extracted in any of these stages are pluripotent. Non-embryonic stem cells, somatic, and adult are found in small quantities in most adult tissues in the body. Examples include the above-mentioned CSC, hematopoietic stem cells, endothelial progenitor cells, mesenchymal stem cells (MSC), and bone marrow-derived stem cells, which are widely studied for cellular transplantation therapies and cardiac recovery [12, 13]. Adult stem cells can also be found in placentas (placental stem cells, PSC) and umbilical cords: they have though a more limited proliferative capacity.

Between 2006 and 2007 the team of scientists led by Shinya Yamanaka identified conditions that both in mouse and in humans would allow somatic adult cells to be reprogrammed genetically to assume a stem cell-like condition (iPSCs [14–16]): accordingly, iPSCs can be defined as stem cell artificially derived from a non-pluripotent cell by inducing forced expression of specific key pluripotency genes (originally this was achieved by retroviral transfection of master regulators of transcription: Oct-3/4, SOX2, c-Myc, and Klf4). In other terms, iPSCs are adult cells that have been genetically rewired to an embryonic stem cell-like state by expressing genes and factors that define the properties of embryonic/pluripotent stem cells.

Current developments in the field of stem cells research and therapy run rapidly and hold great promise for the clinics, and the interested reader is referred to recent reviews on the functions and methods of ESCs and iPSCs to potentially treat HF and other cardiovascular disorders [17–19]. In this chapter, we will restrict our focus specifically on the use of adult stem cells in cardiovascular research and therapy. In particular, we will summarize the advancements in comprehension of how adult stem cells, resident in the heart or derived from other tissues, may contribute to the repair of the damaged myocardium. Since cardiomyocytes have a low turnover rate and a weak capacity to regenerate, we will explore the intriguing concept that this capacity could be experimentally boosted for therapy, by manipulating local autocrine/paracrine molecules regulating endogenous stem cell regenerative potential.

17.2 Modulation of the Cardiovascular Unit to Repair the Heart

To date, the cardiomyocyte has been traditionally viewed as the primary cell responsible for myocardial structure and function. This has led to an underestimation of the potential roles of other myocardial cell types, and has limited our understanding of the mechanisms underlying the pathophysiology of myocardial remodeling and regeneration.

The healthy myocardium appears as a highly organized tissue where each cardiomyocyte occupies its own space in the extracellular matrix and is served by three arterial capillaries, so the ordered architecture of the tissue results in an effective function. Ausoni et al. [20] have recently developed the concept of the "cardiovascular unit" (CVU), which is the dynamic morphofunctional unit of the myocardium and consists of cardiomyocytes, fibroblasts, adjacent capillaries enriched with several pericytes.

CVU cells are not isolated from each other but communicate through the release of humoral mediators acting in an autocrine and/or paracrine manner, or through different transmembrane ionic currents. CVU cells also have the capacity to communicate with the entire organism through the cardio-endothelial barrier and promote the recruitment and homing in the extracellular matrix of circulating inflammatory and/or stem cells. The multifaceted modulation of communication signals between different cells of the CVU may impact on the magnitude of myocardial remodeling and regeneration after an insult.

Recent studies have shown the presence of rare CSCs residing in the intercellular niches of the healthy myocardial of humans, rodents, and pigs. Although the existence of CSCs was confirmed in the adult rat myocardium [21], the role of CSCs is still under investigation. CSCs may participate in regenerative processes of the myocardium as producers of anti-apoptotic, vasculogenic, and/or anti-fibrotic humoral mediators, although, other investigators suggested that they may differentiate into new myocardial cells (cardiomyocytes, endothelial cells, fibroblasts, pericytes). However, a significant bottleneck in cardiac regeneration is the identification of a viable source of stem/progenitor cells that could contribute new muscle after ischemic heart disease and acute myocardial infarction. Cumulative evidence indicates an epicardial origin of at least a subset of the progenitor population, and embryonic reprogramming results in the mobilization of this population and concomitant differentiation to give rise to de novo cardiomyocytes. Smart et al. [22] revealed in mice the occurrence of transdifferentiation of epicardial stem cells to a myocyte fate in the absence of cell fusion. In contrast, Avitabile et al. [23] implicated cell fusion as the key mechanism whereby cardiac functional features are acquired by stem/progenitor cells in the heart.

The current debate is not likely to yield a reliable and definitive resolution on the mechanism behind the de novo cardiomyogenesis for some time. Indeed, Fortini et al. [24] demonstrated that circulating stem/progenitor cells progressively increased from NYHA class II to NYHA class IV in patients affected by HF without improving cardiac function and prognosis. Moreover, Cesselli et al. [25] showed that defects in the balance between cardiomyocyte mass and the pool of nonsenescent CSCs may condition the evolution of the decompensated myocardial dysfunction. These studies emphasize the role of cell age and function in the therapeutic context, and highlight the relevance of the intercellular cross-talk more than the number or the stem cells types available to replace the CVU. This cross-talk can be ameliorated by adjusting the biophysical or biochemical properties of the cardiac cells, including stem cell survival and activation in the heart. Variation in cell features can influence gene expression either directly or epigenetically, by conditioning the geometry and function of the chromatin. Transient exposure of cells to chemical and/or physical agents can determine stable phenotypic changes affecting CVU function and the magnitude of myocardial remodeling or regeneration [26, 27].

Recent studies showed that exposure of stem cells to electromagnetic fields induces de novo cardiomyogenesis at the transcriptional level [28]. In particular, Ventura et al. [29] demonstrated that the prodynorphin gene expression, and the synthesis and secretion of dynorphin B, an endorphin playing a major role in cardiogenesis, is upregulated by low frequency magnetic fields and can drive the cardiac differentiation of MSCs. Other authors proposed that ultrasound as well as static magnetic fields can modulate the function of other CVU cells, such as endothelial cells [30–32] and fibroblasts [33].

The survival, growth, and function of the CVU cells are modulated further in vitro and in vivo by natural or synthetic chemical compounds. In this regard, Lionetti et al. [34] have recently provided the evidence that hyaluronan mixed esters with retinoic and butyric acid (HBR) act as novel modulator of the CVU cells by eliciting de novo cardiomyogenesis in vivo without side effects. The rationale for the synthesis of these novel glycoconjugates is discussed in detail elsewhere [35]. HBR enhanced the transcription of vasculogenic, proliferative, and prosurvival factors, such as vascular endothelial growth factor (VEGF), hepatocyte growth factor (HGF), KDR, AKT, and pim-1 (Fig. 17.1). The gene transcription in cardiomyocytes, endothelial cells, and stem/progenitor cells was due to long-term acetylation of histone type 4. In particular, chromatin immune precipitation and transcriptional analyses showed that HBR increased the transcription of the cardiogenic gene Nkx-2.5 through the binding of Smad4, a well-known cardiac signal transducer, to its own consensus Smad site in human MSCs [36]. Ventura et al. [37] previously showed in vitro that HBR can redirect the gene profile of human MSCs to cardiovascular feature in a dose-dependent manner. Human MSC, mostly those of placental origin, after exposure to HBR appear to be epigenetically more sensitive to the extracellular environment and more oriented toward a more effective path to cardiovascular differentiation. Epigenetic activation of MSCs with the HBR is so stable and powerful as to limit significantly the post-ischemic myocardial remodeling in swine heart, which is ontogenetically more similar to the human heart, and with a plasticity lower than rodent heart [38].

In accord with the chemical strategy to modulate the epigenetics of CVU, Limana et al. [39] showed that the injection in murine peri-infarcted left ventricular (LV) region of High Mobility Group Box-1 protein, a highly conserved chromatin-binding factor promoting access of transcriptional complexes to the DNA, enhanced LV function, and attenuated LV remodeling, due to increased collagenolytic activity and miR-206-mediated inhibition of tissue inhibitor of metalloproteinase 3. This finding is important since it demonstrates the feasibility of reversing the remodeling of the failing heart by modulating the gene profile of the cardiac resident cells. In this context, Forini et al. [40] explored the possibility of preventing post-ischemic cardiac remodeling by low-dosage replacement of 3,5,3'-Levo-triiodothyronine (L-T3). L-T3 is a natural compound, which re-establishes the intercellular crosstalk into the heart and increases the resistance of cardiac cells to ischemic insult (Fig. 17.1).

17.3 Autocrine/Paracrine Signals to Empower Cardiac Stem Cells

The studies discussed above emphasize the role of intercellular crosstalk between stem cells and myocardial cells as crucial to the replacement of the damaged CVU, and illustrate the therapeutic impact of natural/synthetic compounds and physical forces in boosting this process. In the last years, a growing body of evidence has emerged supporting the hypothesis that endogenous paracrine mechanisms mediated by factors released by the adult stem cells play an essential role in the healing process observed after stem cell recruitment or injection into damaged hearts.

Paracrine mediators can be produced in a temporally and spatially controlled manner, exerting distinct effects depending on the cardiac microenvironment. Adult stem cells secrete a broad variety of signaling peptides and growth factors that could be involved in cardiac repair, including VEGF, fibroblast growth factor (FGF), HGF, and insulin growth factor I (IGF-I) [41–46]. These paracrine factors positively influence cardiac cells through several mechanisms: (i) myocardial

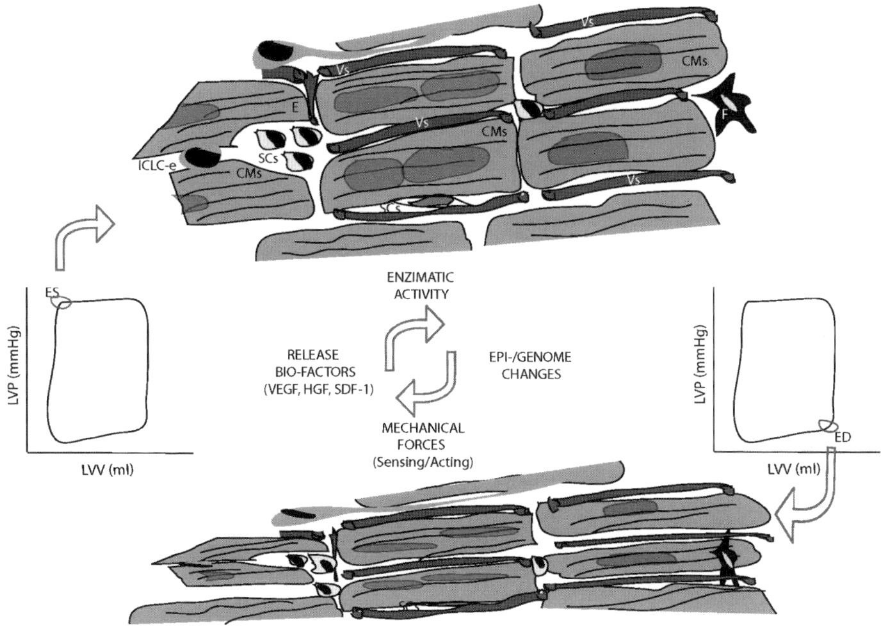

Fig. 17.1 Cardiovascular Unit (CVU) represented during the phases of a cardiac cycle in a normal heart: telesystole (ES) and telediastole (ED). This dynamic model lays the basis for the release of biofactors, changes in the enzymatic activities, changes in epigenetic patterns, and mechanical sensing systems. The model simplifies the complexity of the intramyocardial intercellular interactions during a cardiac cycle

protection and vascularization; (ii) improvement of cardiac metabolism, cardiac contractility, and cardiac regeneration; (iii) autocrine actions on the secreting stem cells themselves [47, 48] (Fig. 17.1). As proof-of-concept, we discuss below the role of paracrine factors in the rescue of damaged cardiac function by MSCs (non-cardiac adult stem cells) and CSCs (cardiac adult stem cells).

The role of paracrine factors in MSC function is illustrated by the direct cytoprotective action of conditioned medium from autologous MSCs, which reduces cell death of isolated rat ischemic cardiomyocytes [49]. Human MSCs-conditioned medium can also reduce the size of myocardial infarction in pigs, rats, and mice [50–52], and production of VEGF has been proposed as a critical factor for the MSCs-dependent post-ischemic cardioprotection [44]. Human cardiac-explant-conditioned medium contains soluble factors such as FGF, HGF, IGF-1, and others that can induce MSCs proliferation and a partial transdifferentiation into cardiomyocytes [53]. Kubal et al. [54] has shown that MSCs derived from bone marrow display a potent anti-ischemic effect against cardiomyocyte cell death in humans.

Proangiogenic factors may also be responsible for increased vascular density observed in ischemic tissues of animals treated with stem cells. MSCs represent a source of paracrine proangiogenic factors [55–57], and their in vivo local delivery augments the conductance of vessels through paracrine mechanisms, without incorporating into vessels [58]. Transplantation of MSCs in experimentally infarcted hearts produced an increase in capillary density compared with controls [59–61]. Although not formally proven, a proangiogenic paracrine mechanism seems most plausible.

Paracrine factors likely play an important role in the myocardial protection exerted by CSCs. These rare cells were originally isolated by fluorescence-activated cell sorting on the base of c-kit, MDR1, and Sca-1-reactive protein cell surface markers. In humans, CSCs can amplify and likely commit to the myocyte lineage in response to increased workload, thereby contributing to cell resistance to the stress-induced hypertrophy [62]. During aging the beneficial potential of CSCs (discussed further below) decreases.

In presence of an extensive oxidative and ischemic injury, such as during an infarction, increase in cardiomyocyte cell survival is not sufficient to preserve heart function. De novo formation of blood vessels (neovascularization) and repair of damaged cells/regeneration are also required to maintain cardiac contractility in this circumstance. In this respect, paracrine factors produced by MSCs have proven beneficial, whereas the proangiogenic effects of paracrine factors secreted by CSCs have not been studied in detail. Despite the fact that stem cells incorporate into vascular structures, there is evidence that only a small number of vessels contain donor cells [47, 63].

Since MSCs release paracrine factors, an intriguing working hypothesis is that MSCs injected into ischemic hearts may attract and activate resident CSCs, enabling cardiac regeneration. Evidence supporting this hypothesis has been documented in a study in which MSCs were injected into infarcted pig hearts [64]. Analyses performed 10 days later revealed the appearance, only in MSC-treated animals, of newly formed cardiomyocytes, some of which stained positive for the CSCs marker c-kit, and others for the proliferation marker ki67. Although co-staining for c-Kit and ki67 was not shown, it is plausible that endogenous cardiac regeneration was present.

Finally, survival and self-renewal in many stem cell types appear to be mediated also by autocrine mechanisms [47, 65, 66]. The paracrine/autocrine paradigma embraces the influence of stem cell released factors on the microenvironment modulating stem cell function and cardiac tissue response (Fig. 17.1).

17.4 IGF-1 as a Cardioprotective Factor

In its role as an autocrine/paracrine factor, IGF-1 has been the subject of numerous studies on stem cell function. IGF-1 is a peptide that affects multiple signaling cascades, resulting in a potent proliferative signal that stimulates growth in many different cells and organs and blocks apoptosis. In mammals, the IGF-1 gene spans more than 70 kb and has six exons, giving rise to multiple splice variants. These variants share a common core peptide, flanked by varying termini (Class 1 and 2N-terminal peptides, and Ea, Eb and Ec C-terminal peptides). IGF-1 is both a systemic growth factor produced mainly by the liver in response to GH and a local growth factor acting in an autocrine/paracrine manner in organs such as the heart [67].

One important branch of IGF-1 signaling is the PI3K/Akt/mTOR/p70S6 signaling cascade, which is similar to insulin signaling [67, 68]. MSCs genetically engineered to overexpress AKT (AKT-MSCs) displayed further enhanced myocardial protective properties. Injection of concentrated medium from autologous MSCs and AKT-MSCs into the rat heart at the infarct border immediately after left coronary occlusion led 3 days later to a modest or dramatic, respectively, reduction in apoptosis and infarct size. These results were confirmed in other animal models of infarct and cellular tracing experiments with green fluorescent protein demonstrated that the protective role of MSC does not depend on a transdifferentiation of MSCs into cardiomyocytes [69–72]. The activation of AKT can enhance the protective effects of MSCs, likely through an increase in the production/release of VEGF, FGF, HGF, and IGF-1 [49, 55].

The paracrine action of IGF-1 is illustrated by studies in which MSCs engineered to overexpress IGF-1 were implanted into an infarcted rat heart-activated paracrine mechanisms involving stromal cell-derived factor (SDF)-1alpha that mobilized stem cells of different origins, reduced infarction size, and dramatically improved the indices of LV function, including ejection fraction and fractional shortening, and culminated in an extensive neovascularization [73].

Exogenous administered paracrine factors can also activate endogenous CSCs and their induced neovascularization. In dogs, circulating IGF-1 was injected after infarction to stimulate resident CSCs: this intervention led to the formation of cardiomyocytes and coronary vessels within the infarct and resulted in a marked recovery of contractility of the infarcted heart [74]. Similar results were obtained in mice [75]. A recent study on humans showed that IGF-1 receptor is a reliable marker to identify a young pool of CSCs able to potently modulating stem cell proliferation, commitment to the cardiomyocyte lineage and differentiation [76]. The latter reports highlight the potential of IGF-1 in promoting CSC recruitment and CVU survival and integrity.

IGF-1 may also play a role in stem cell aging. With aging, CSCs undergo telomere shortening, generating a differentiated progeny that acquire a senescent phenotype. CSC aging can be induced by attenuation of the IGF-1 system in rats [77]. Conversely, IGF-1 signaling induces division of CSCs by upregulating telomerase activity and hindering replicative senescence, thereby preserving the pool of functionally competent CSC [74, 75, 77, 78]. IGF-1 also acts as a proangiogenic factor. Activation by IGF-1 injection led CSCs to migrate to the regions of damage, reversing partly the aging cardiomyopathy and inducing vessel regeneration [77]. In fact, c-kit positive CSCs activated with IGF-1 and HGF before their injection in proximity of the site of occlusion of the left coronary artery in rats generated new vessels connected with the primary coronary circulation, and this increase in vascularization more than doubled myocardial blood flow in the infarcted myocardium [79]; as in the case of IGF-1 engineered MSCs [73], SDF-1alpha was critical for these effects [79].

Our lab has recently shown that mIGF-1 (comprising a Class 1 signal peptide and a C-terminal Ea extension peptide) overexpression is able to recover heart functionality after HF-inducing injuries (infarct induced by ligation of the left coronary artery, or after cardiotoxin-induced injury) [80], demonstrating a restoration of cardiac function in post-infarct and injury-challenged mIGF-1 transgenic mice that is facilitated by modulation of the inflammatory response. Local expression of mIGF-1 repairs the heart from injury through production of paracrine signals such as VEGF and MCP-1, cross-talk with the bone marrow, and recruitment of $CD34^+$ stem cells for de novo vascularization of the myocardium [81]. Moreover, mIGF-1 displays cardioprotective features from oxidative and hypertrophic stresses when expressed as a transgene in cardiomyocytes [67, 82, 83]. Of note, in the skeletal muscle, mIGF-1 has an established role in stem cell activation and tissue regeneration by modulation of paracrine factors [84, 85]. Although a comparative study of the diverse IGF-1 isoforms on adult stem cells recruitment during cardiac injury has never been undertaken, it is likely that adult stem cells, such as MSCs and CSCs, in synergy with IGF-1 and other proangiogenic paracrine factors may increase neovascularization and repair from ischemic damage (Fig. 17.1).

17.5 Conclusions and Perspectives

The evidence that adult stem cells exert their protective and repair properties in the heart in synergy with endogenously produced or exogenously administered natural/chemical paracrine factors has important pharmacological implications. These studies can lead to the identification of single factors or of multicomponent combination of drugs able to hit multiple targets involved in cardiac repair and regeneration. The use of adult stem cells, like MSCs and CSCs, has been proposed as advantageous over pluripotent ESCs because, although they can differentiate in beating cardiomyocytes, their employment for therapeutical purposes raised concerns about: (i) ethical issues in several countries [86, 87]; (ii) tumorigenesis (teratoma) observed in a number of studies upon ESCs transplantation [88, 89]; and (iii) immune rejection by the host [90, 91]. The "paracrine paradigma" could overcome the disadvantages associated to ESCs employment. Transgenic overexpression of the cytokine tumor necrosis factor alpha in the recipient mouse myocardium hampered tumorigenic processes upon ESCs implantation and enhanced the efficacy of paracrine cardiogenic signals [92]; Crisostomo et al. [93] have shown that perfusion of ischemic rat hearts with ESCs-conditioned medium produced a post-ischemic functional recovery and anti-inflammatory effect even greater compared to perfusion with MSCs-conditioned medium, without any cell transplantation.

Characterization of the paracrine mediators involved in stem cell homeostasis and function may lead to the replacement of stem cell-based therapy with growth factor-based therapy to prevent HF. Recent encouraging data in mice and in pigs support this strategy [94, 95]. This is certainly an easier approach; however, the main limitation for the clinical translation is the difficulty in finding the right "cocktail" of multiple growth factors contained in the adult stem cells conditioned media [96], also because many of these peptides induce distinct actions at different concentrations and at specific phases post myocardial infarction. A deeper knowledge of the interactions between paracrine factors and their roles in distinct phases of the disease in time and in space is required. Progresses in protein and peptide therapeutics research and in efficient delivery methods will certainly pave the way for the development of cell-free protein-based treatment for cardiac repair.

References

1. Frey N, Olson EN (2003) Cardiac hypertrophy: the good, the bad, and the ugly. Annu Rev Physiol 65:45–79
2. Takimoto E, Kass DA (2007) Role of oxidative stress in cardiac hypertrophy and remodeling. Hypertension 49(2):241–248
3. Seddon M, Looi YH, Shah AM (2007) Oxidative stress and redox signalling in cardiac hypertrophy and heart failure. Heart 93(8):903–907
4. Tendera M (2005) Epidemiology, treatment, and guidelines for the treatment of heart failure in Europe. Eur Heart J 7(J):J5–J9
5. Zannad F, Briancon S, Juilliere Y, Mertes PM, Villemot JP, Alla F, Virion JM (1999) Incidence, clinical and etiologic features, and outcomes of advanced chronic heart failure: the EPICAL Study. Epidemiologie de l'Insuffisance Cardiaque Avancee en Lorraine. J Am Coll Cardiol 33(3):734–742
6. Haldeman GA, Croft JB, Giles WH, Rashidee A (1999) Hospitalization of patients with heart failure: National Hospital Discharge Survey, 1985 to 1995. Am Heart J 137(2):352–360
7. Malek M (1999) Health economics of heart failure. Heart 82(Suppl 4):IV11–IV13
8. Heineke J, Molkentin JD (2006) Regulation of cardiac hypertrophy by intracellular signalling pathways. Nat Rev Mol Cell Biol 7(8):589–600
9. Kajstura J, Urbanek K, Perl S, Hosoda T, Zheng H, Ogorek B, Ferreira-Martins J, Goichberg P, Rondon-Clavo C, Sanada F, D'Amario D, Rota M, Del Monte F, Orlic D, Tisdale J, Leri A, Anversa P (2010) Cardiomyogenesis in the adult human heart. Circ Res 107(2): 305–315
10. Bergmann O, Bhardwaj RD, Bernard S, Zdunek S, Barnabe-Heider F, Walsh S, Zupicich J, Alkass K, Buchholz BA, Druid H, Jovinge S, Frisen J (2009) Evidence for cardiomyocyte renewal in humans. Science 324(5923):98–102
11. Laflamme MA, Murry CE (2011) Heart regeneration. Nature 473(7347):326–335
12. Krause K, Schneider C, Kuck KH, Jaquet K (2011) Stem cell therapy in cardiovascular disorders. Cardiovasc Ther 28(5):e101–e110
13. Sharif F, Bartunek J, Vanderheyden M (2011) Adult stem cells in the treatment of acute myocardial infarction. Catheter Cardiovasc Interv 77(1):72–83
14. Takahashi K, Yamanaka S (2006) Induction of pluripotent stem cells from mouse embryonic and adult fibroblast cultures by defined factors. Cell 126(4):663–676
15. Takahashi K, Tanabe K, Ohnuki M, Narita M, Ichisaka T, Tomoda K, Yamanaka S (2007) Induction of pluripotent stem cells from adult human fibroblasts by defined factors. Cell 131(5):861–872
16. Okita K, Ichisaka T, Yamanaka S (2007) Generation of germline-competent induced pluripotent stem cells. Nature 448(7151):313–317
17. Wong SS, Bernstein HS (2010) Cardiac regeneration using human embryonic stem cells: producing cells for future therapy. Regen Med 5(5):763–775
18. Nelson TJ, Martinez-Fernandez A, Terzic A (2010) Induced pluripotent stem cells: developmental biology to regenerative medicine. Nat Rev Cardiol 7(12):700–710

19. Leeper NJ, Hunter AL, Cooke JP (2010) Stem cell therapy for vascular regeneration: adult, embryonic, and induced pluripotent stem cells. Circulation 122(5):517–526
20. Ausoni S, Sartore S (2009) The cardiovascular unit as a dynamic player in disease and regeneration. Trends Mol Med 15(12):543–552
21. Beltrami AP, Barlucchi L, Torella D, Baker M, Limana F, Chimenti S, Kasahara H, Rota M, Musso E, Urbanek K, Leri A, Kajstura J, Nadal-Ginard B, Anversa P (2003) Adult cardiac stem cells are multipotent and support myocardial regeneration. Cell 114(6):763–776
22. Smart N, Bollini S, Dube KN, Vieira JM, Zhou B, Davidson S, Yellon D, Riegler J, Price AN, Lythgoe MF, Pu WT, Riley PR (2011) De novo cardiomyocytes from within the activated adult heart after injury. Nature 474(7353):640–644
23. Avitabile D, Crespi A, Brioschi C, Parente V, Toietta G, Devanna P, Baruscotti M, Truffa S, Scavone A, Rusconi F, Biondi A, D'Alessandra Y, Vigna E, Difrancesco D, Pesce M, Capogrossi MC, Barbuti A (2011) Human cord blood CD34+ progenitor cells acquire functional cardiac properties through a cell fusion process. Am J Physiol Heart Circ Physiol 300(5):H1875–H1884
24. Fortini C, Toffoletto B, Fucili A, Puppato E, Olivares A, Beltrami AP, Fiorelli V, Bergamin N, Cesselli D, Morelli C, Francolini G, Ferrari R, Beltrami CA (2011) Circulating stem cell vary with NYHA stage in heart failure patients. J Cell Mol Med 15(8):1726–1736
25. Cesselli D, Beltrami AP, D'Aurizio F, Marcon P, Bergamin N, Toffoletto B, Pandolfi M, Puppato E, Marino L, Signore S, Livi U, Verardo R, Piazza S, Marchionni L, Fiorini C, Schneider C, Hosoda T, Rota M, Kajstura J, Anversa P, Beltrami CA, Leri A (2011) Effects of age and heart failure on human cardiac stem cell function. Am J Pathol 179(1):349–366
26. Lionetti V, Cecchini M, Ventura C (2011) Nanomechanics to drive stem cells in injured tissues: insights from current research and future perspectives. Stem Cells Dev 20(4):561–568
27. Ventura C, Cavallini C, Bianchi F, Cantoni S (2008) Stem cells and cardiovascular repair: a role for natural and synthetic molecules harboring differentiating and paracrine logics. Cardiovasc Hematol Agents Med Chem 6(1):60–68
28. Lisi A, Ledda M, de Carlo F, Pozzi D, Messina E, Gaetani R, Chimenti I, Barile L, Giacomello A, D'Emilia E, Giuliani L, Foletti A, Patti A, Vulcano A, Grimaldi S (2008) Ion cyclotron resonance as a tool in regenerative medicine. Electromagn Biol Med 27(2):127–133
29. Ventura C, Maioli M, Asara Y, Santoni D, Mesirca P, Remondini D, Bersani F (2005) Turning on stem cell cardiogenesis with extremely low frequency magnetic fields. Faseb J 19(1):155–157
30. Lionetti V, Fittipaldi A, Agostini S, Giacca M, Recchia FA, Picano E (2009) Enhanced caveolae-mediated endocytosis by diagnostic ultrasound in vitro. Ultrasound Med Biol 35(1):136–143
31. Martino CF, Perea H, Hopfner U, Ferguson VL, Wintermantel E (2010) Effects of weak static magnetic fields on endothelial cells. Bioelectromagnetics 31(4):296–301
32. Okano H, Onmori R, Tomita N, Ikada Y (2006) Effects of a moderate-intensity static magnetic field on VEGF-A stimulated endothelial capillary tubule formation in vitro. Bioelectromagnetics 27(8):628–640
33. Murray JC, Farndale RW (1985) Modulation of collagen production in cultured fibroblasts by a low-frequency, pulsed magnetic field. Biochim Biophys Acta 838(1):98–105
34. Lionetti V, Cantoni S, Cavallini C, Bianchi F, Valente S, Frascari I, Olivi E, Aquaro GD, Bonavita F, Scarlata I, Maioli M, Vaccari V, Tassinari R, Bartoli A, Recchia FA, Pasquinelli G, Ventura C (2010) Hyaluronan mixed esters of butyric and retinoic acid affording myocardial survival and repair without stem cell transplantation. J Biol Chem 285(13):9949–9961
35. Ventura C, Maioli M, Asara Y, Santoni D, Scarlata I, Cantoni S, Perbellini A (2004) Butyric and retinoic mixed ester of hyaluronan: a novel differentiating glycoconjugate affording a high throughput of cardiogenesis in embryonic stem cells. J Biol Chem 279(22):23574–23579
36. Maioli M, Santaniello S, Montella A, Bandiera P, Cantoni S, Cavallini C, Bianchi F, Lionetti V, Rizzolio F, Marchesi I, Bagella L, Ventura C (2010) Hyaluronan esters drive Smad gene expression and signaling enhancing cardiogenesis in mouse embryonic and human mesenchymal stem cells. PLoS ONE 5(11):e15151

37. Ventura C, Cantoni S, Bianchi F, Lionetti V, Cavallini C, Scarlata I, Foroni L, Maioli M, Bonsi L, Alviano F, Fossati V, Bagnara GP, Pasquinelli G, Recchia FA, Perbellini A (2007) Hyaluronan mixed esters of butyric and retinoic acid drive cardiac and endothelial fate in term placenta human mesenchymal stem cells and enhance cardiac repair in infarcted rat hearts. J Biol Chem 282(19):14243–14252
38. Simioniuc A, Campan M, Lionetti V, Marinelli M, Aquaro GD, Cavallini C, Valente S, Di Silvestre D, Cantoni S, Bernini F, Simi C, Pardini S, Mauri P, Neglia D, Ventura C, Pasquinelli G, Recchia FA (2011) Placental stem cells pre-treated with a hyaluronan mixed ester of butyric and retinoic acid to cure infarcted pig hearts: a multimodal study. Cardiovasc Res 90(3):546–556
39. Limana F, Esposito G, D'Arcangelo D, Di Carlo A, Romani S, Melillo G, Mangoni A, Bertolami C, Pompilio G, Germani A, Capogrossi MC (2011) HMGB1 attenuates cardiac remodelling in the failing heart via enhanced cardiac regeneration and miR-206-mediated inhibition of TIMP-3. PLoS ONE 6(6):e19845
40. Forini F, Lionetti V, Ardehali H, Pucci A, Cecchetti F, Ghanefar M, Nicolini G, Ichikawa Y, Nannipieri M, Recchia FA, Iervasi G (2011) Early long-term L-T3 replacement rescues mitochondria and prevents ischemic cardiac remodelling in rats. J Cell Mol Med 15(3):514–524
41. Nagaya N, Kangawa K, Itoh T, Iwase T, Murakami S, Miyahara Y, Fujii T, Uematsu M, Ohgushi H, Yamagishi M, Tokudome T, Mori H, Miyatake K, Kitamura S (2005) Transplantation of mesenchymal stem cells improves cardiac function in a rat model of dilated cardiomyopathy. Circulation 112(8):1128–1135
42. Yoon YS, Wecker A, Heyd L, Park JS, Tkebuchava T, Kusano K, Hanley A, Scadova H, Qin G, Cha DH, Johnson KL, Aikawa R, Asahara T, Losordo DW (2005) Clonally expanded novel multipotent stem cells from human bone marrow regenerate myocardium after myocardial infarction. J Clin Invest 115(2):326–338
43. Caplan AI, Dennis JE (2006) Mesenchymal stem cells as trophic mediators. J Cell Biochem 98(5):1076–1084
44. Markel TA, Wang Y, Herrmann JL, Crisostomo PR, Wang M, Novotny NM, Herring CM, Tan J, Lahm T, Meldrum DR (2008) VEGF is critical for stem cell-mediated cardioprotection and a crucial paracrine factor for defining the age threshold in adult and neonatal stem cell function. Am J Physiol Heart Circ Physiol 295(6):H2308–H2314
45. Chimenti I, Smith RR, Li TS, Gerstenblith G, Messina E, Giacomello A, Marban E (2010) Relative roles of direct regeneration versus paracrine effects of human cardiosphere-derived cells transplanted into infarcted mice. Circ Res 106(5):971–980
46. Shabbir A, Zisa D, Lin H, Mastri M, Roloff G, Suzuki G, Lee T (2010) Activation of host tissue trophic factors through JAK-STAT3 signaling: a mechanism of mesenchymal stem cell-mediated cardiac repair. Am J Physiol Heart Circ Physiol 299(5):H1428–H1438
47. Gnecchi M, Zhang Z, Ni A, Dzau VJ (2008) Paracrine mechanisms in adult stem cell signaling and therapy. Circ Res 103(11):1204–1219
48. Mirotsou M, Jayawardena TM, Schmeckpeper J, Gnecchi M, Dzau VJ (2011) Paracrine mechanisms of stem cell reparative and regenerative actions in the heart. J Mol Cell Cardiol 50(2):280–289
49. Gnecchi M, He H, Liang OD, Melo LG, Morello F, Mu H, Noiseux N, Zhang L, Pratt RE, Ingwall JS, Dzau VJ (2005) Paracrine action accounts for marked protection of ischemic heart by AKT-modified mesenchymal stem cells. Nat Med 11(4):367–368
50. Simpson DL, Boyd NL, Kaushal S, Stice SL, Dudley SC Jr (2012) Use of human embryonic stem cell derived-mesenchymal cells for cardiac repair. Biotechnol Bioeng 109(1):274–283
51. Timmers L, Lim SK, Hoefer IE, Arslan F, Lai RC, van Oorschot AA, Goumans MJ, Strijder C, Sze SK, Choo A, Piek JJ, Doevendans PA, Pasterkamp G, de Kleijn DP (2011) Human mesenchymal stem cell-conditioned medium improves cardiac function following myocardial infarction. Stem Cell Res 6(3):206–214
52. See F, Seki T, Psaltis PJ, Sondermeijer HP, Gronthos S, Zannettino AC, Govaert KM, Schuster MD, Kurlansky PA, Kelly DJ, Krum H, Itescu S (2011) Therapeutic effects of human STRO-3-selected mesenchymal precursor cells and their soluble factors in experimental myocardial ischemia. J Cell Mol Med 25:1533–1542

53. Schittini AV, Celedon PF, Stimamiglio MA, Krieger M, Hansen P, da Costa FD, Goldenberg S, Dallagiovanna B, Correa A (2010) Human cardiac explant-conditioned medium: soluble factors and cardiomyogenic effect on mesenchymal stem cells. Exp Biol Med (Maywood) 235(8):1015–1024
54. Kubal C, Sheth K, Nadal-Ginard B, Galinanes M (2006) Bone marrow cells have a potent anti-ischemic effect against myocardial cell death in humans. J Thorac Cardiovasc Surg 132(5):1112–1118
55. Gnecchi M, He H, Noiseux N, Liang OD, Zhang L, Morello F, Mu H, Melo LG, Pratt RE, Ingwall JS, Dzau VJ (2006) Evidence supporting paracrine hypothesis for AKT-modified mesenchymal stem cell-mediated cardiac protection and functional improvement. Faseb J 20(6):661–669
56. Kinnaird T, Stabile E, Burnett MS, Lee CW, Barr S, Fuchs S, Epstein SE (2004) Marrow-derived stromal cells express genes encoding a broad spectrum of arteriogenic cytokines and promote in vitro and in vivo arteriogenesis through paracrine mechanisms. Circ Res 94(5):678–685
57. Ohnishi S, Yasuda T, Kitamura S, Nagaya N (2007) Effect of hypoxia on gene expression of bone marrow-derived mesenchymal stem cells and mononuclear cells. Stem Cells 25(5):1166–1177
58. Kinnaird T, Stabile E, Burnett MS, Shou M, Lee CW, Barr S, Fuchs S, Epstein SE (2004) Local delivery of marrow-derived stromal cells augments collateral perfusion through paracrine mechanisms. Circulation 109(12):1543–1549
59. Jiang S, Haider H, Idris NM, Salim A, Ashraf M (2006) Supportive interaction between cell survival signaling and angiocompetent factors enhances donor cell survival and promotes angiomyogenesis for cardiac repair. Circ Res 99(7):776–784
60. Nagaya N, Fujii T, Iwase T, Ohgushi H, Itoh T, Uematsu M, Yamagishi M, Mori H, Kangawa K, Kitamura S (2004) Intravenous administration of mesenchymal stem cells improves cardiac function in rats with acute myocardial infarction through angiogenesis and myogenesis. Am J Physiol Heart Circ Physiol 287(6):H2670–H2676
61. Tomita S, Mickle DA, Weisel RD, Jia ZQ, Tumiati LC, Allidina Y, Liu P, Li RK (2002) Improved heart function with myogenesis and angiogenesis after autologous porcine bone marrow stromal cell transplantation. J Thorac Cardiovasc Surg 123(6):1132–1140
62. Urbanek K, Quaini F, Tasca G, Torella D, Castaldo C, Nadal-Ginard B, Leri A, Kajstura J, Quaini E, Anversa P (2003) Intense myocyte formation from cardiac stem cells in human cardiac hypertrophy. Proc Natl Acad Sci U S A 100(18):10440–10445
63. Cossu G, Bianco P (2003) Mesoangioblasts—vascular progenitors for extravascular mesodermal tissues. Curr Opin Genet Dev 13(5):537–542
64. Amado LC, Saliaris AP, Schuleri KH, St John M, Xie JS, Cattaneo S, Durand DJ, Fitton T, Kuang JQ, Stewart G, Lehrke S, Baumgartner WW, Martin BJ, Heldman AW, Hare JM (2005) Cardiac repair with intramyocardial injection of allogeneic mesenchymal stem cells after myocardial infarction. Proc Natl Acad Sci U S A 102(32):11474–11479
65. Kolf CM, Cho E, Tuan RS (2007) Mesenchymal stromal cells. Biology of adult mesenchymal stem cells: regulation of niche, self-renewal and differentiation. Arthritis Res Ther 9(1):204
66. Alfaro MP, Pagni M, Vincent A, Atkinson J, Hill MF, Cates J, Davidson JM, Rottman J, Lee E, Young PP (2008) The Wnt modulator sFRP2 enhances mesenchymal stem cell engraftment, granulation tissue formation and myocardial repair. Proc Natl Acad Sci U S A 105(47): 18366–18371
67. Vinciguerra M, Musaro A, Rosenthal N (2010) Regulation of muscle atrophy in aging and disease. Adv Exp Med Biol 694:211–233
68. Foncea R, Andersson M, Ketterman A, Blakesley V, Sapag-Hagar M, Sugden PH, LeRoith D, Lavandero S (1997) Insulin-like growth factor-I rapidly activates multiple signal transduction pathways in cultured rat cardiac myocytes. J Biol Chem 272(31):19115–19124
69. Takahashi M, Li TS, Suzuki R, Kobayashi T, Ito H, Ikeda Y, Matsuzaki M, Hamano K (2006) Cytokines produced by bone marrow cells can contribute to functional improvement of the infarcted heart by protecting cardiomyocytes from ischemic injury. Am J Physiol Heart Circ Physiol 291(2):H886–H893

70. Lim SY, Kim YS, Ahn Y, Jeong MH, Hong MH, Joo SY, Nam KI, Cho JG, Kang PM, Park JC (2006) The effects of mesenchymal stem cells transduced with AKT in a porcine myocardial infarction model. Cardiovasc Res 70(3):530–542
71. Xu M, Uemura R, Dai Y, Wang Y, Pasha Z, Ashraf M (2007) In vitro and in vivo effects of bone marrow stem cells on cardiac structure and function. J Mol Cell Cardiol 42(2):441–448
72. Uemura R, Xu M, Ahmad N, Ashraf M (2006) Bone marrow stem cells prevent left ventricular remodeling of ischemic heart through paracrine signaling. Circ Res 98(11):1414–1421
73. Haider H, Jiang S, Idris NM, Ashraf M (2008) IGF-1-overexpressing mesenchymal stem cells accelerate bone marrow stem cell mobilization via paracrine activation of SDF-1alpha/CXCR4 signaling to promote myocardial repair. Circ Res 103(11):1300–1308
74. Linke A, Muller P, Nurzynska D, Casarsa C, Torella D, Nascimbene A, Castaldo C, Cascapera S, Bohm M, Quaini F, Urbanek K, Leri A, Hintze TH, Kajstura J, Anversa P (2005) Stem cells in the dog heart are self-renewing, clonogenic, and multipotent and regenerate infarcted myocardium, improving cardiac function. Proc Natl Acad Sci U S A 102(25):8966–8971
75. Urbanek K, Rota M, Cascapera S, Bearzi C, Nascimbene A, De Angelis A, Hosoda T, Chimenti S, Baker M, Limana F, Nurzynska D, Torella D, Rotatori F, Rastaldo R, Musso E, Quaini F, Leri A, Kajstura J, Anversa P (2005) Cardiac stem cells possess growth factor-receptor systems that after activation regenerate the infarcted myocardium, improving ventricular function and long-term survival. Circ Res 97(7):663–673
76. D'Amario D, Cabral-Da-Silva MC, Zheng H, Fiorini C, Goichberg P, Steadman E, Ferreira-Martins J, Sanada F, Piccoli M, Cappetta D, D'Alessandro DA, Michler RE, Hosoda T, Anastasia L, Rota M, Leri A, Anversa P, Kajstura J (2011) Insulin-like growth factor-1 receptor identifies a pool of human cardiac stem cells with superior therapeutic potential for myocardial regeneration. Circ Res 108(12):1467–1481
77. Gonzalez A, Rota M, Nurzynska D, Misao Y, Tillmanns J, Ojaimi C, Padin-Iruegas ME, Muller P, Esposito G, Bearzi C, Vitale S, Dawn B, Sanganalmath SK, Baker M, Hintze TH, Bolli R, Urbanek K, Hosoda T, Anversa P, Kajstura J, Leri A (2008) Activation of cardiac progenitor cells reverses the failing heart senescent phenotype and prolongs lifespan. Circ Res 102(5):597–606
78. Torella D, Rota M, Nurzynska D, Musso E, Monsen A, Shiraishi I, Zias E, Walsh K, Rosenzweig A, Sussman MA, Urbanek K, Nadal-Ginard B, Kajstura J, Anversa P, Leri A (2004) Cardiac stem cell and myocyte aging, heart failure, and insulin-like growth factor-1 overexpression. Circ Res 94(4):514–524
79. Tillmanns J, Rota M, Hosoda T, Misao Y, Esposito G, Gonzalez A, Vitale S, Parolin C, Yasuzawa-Amano S, Muraski J, De Angelis A, Lecapitaine N, Siggins RW, Loredo M, Bearzi C, Bolli R, Urbanek K, Leri A, Kajstura J, Anversa P (2008) Formation of large coronary arteries by cardiac progenitor cells. Proc Natl Acad Sci U S A 105(5):1668–1673
80. Santini MP, Tsao L, Monassier L, Theodoropoulos C, Carter J, Lara-Pezzi E, Slonimsky E, Salimova E, Delafontaine P, Song YH, Bergmann M, Freund C, Suzuki K, Rosenthal N (2007) Enhancing repair of the mammalian heart. Circ Res 100(12):1732–1740
81. Santini MP, Lexow J, Borsellino G, Slonimski E, Zarrinpashneh E, Poggioli T, Rosenthal N (2011) IGF-1Ea induces vessel formation after injury and mediates bone marrow and heart cross-talk through the expression of specific cytokines. Biochem Biophys Res Commun 410(2):201–207
82. Vinciguerra M, Santini MP, Claycomb WC, Ladurner AG, Rosenthal N (2009) Local IGF-1 isoform protects cardiomyocytes from hypertrophic and oxidative stresses via SirT1 activity. Aging (Albany N Y) 2(1):43–62
83. Vinciguerra M, Santini MP, Martinez C, Pazienza V, Claycomb WC, Giuliani A, Rosenthal N (2012) mIGF-1/JNK1/SirT1 signaling confers protection against oxidative stress in the heart. Aging Cell 11:139–149
84. Pelosi L, Giacinti C, Nardis C, Borsellino G, Rizzuto E, Nicoletti C, Wannenes F, Battistini L, Rosenthal N, Molinaro M, Musaro A (2007) Local expression of IGF-1 accelerates muscle regeneration by rapidly modulating inflammatory cytokines and chemokines. Faseb J 21(7): 1393–1402

85. Musarò A, Giacinti C, Borsellino G, Dobrowolny G, Pelosi L, Cairns L, Ottolenghi S, Cossu G, Bernardi G, Battistini L, Molinaro M, Rosenthal N (2004) Stem cell-mediated muscle regeneration is enhanced by local isoform of insulin-like growth factor 1. Proc Natl Acad Sci U S A 101(5):1206–1210
86. de Wert G, Mummery C (2003) Human embryonic stem cells: research, ethics and policy. Hum Reprod 18(4):672–682
87. Juengst E, Fossel M (2000) The ethics of embryonic stem cells—now and forever, cells without end. JAMA 284(24):3180–3184
88. Cao F, Lin S, Xie X, Ray P, Patel M, Zhang X, Drukker M, Dylla SJ, Connolly AJ, Chen X, Weissman IL, Gambhir SS, Wu JC (2006) In vivo visualization of embryonic stem cell survival, proliferation, and migration after cardiac delivery. Circulation 113(7):1005–1014
89. Chaudhry GR, Fecek C, Lai MM, Wu WC, Chang M, Vasquez A, Pasierb M, Trese MT (2009) Fate of embryonic stem cell derivatives implanted into the vitreous of a slow retinal degenerative mouse model. Stem Cells Dev 18(2):247–258
90. Draper JS, Pigott C, Thomson JA, Andrews PW (2002) Surface antigens of human embryonic stem cells: changes upon differentiation in culture. J Anat 200(Pt 3):249–258
91. Zimmermann WH (2011) Embryonic and embryonic-like stem cells in heart muscle engineering. J Mol Cell Cardiol 50(2):320–326
92. Behfar A, Perez-Terzic C, Faustino RS, Arrell DK, Hodgson DM, Yamada S, Puceat M, Niederlander N, Alekseev AE, Zingman LV, Terzic A (2007) Cardiopoietic programming of embryonic stem cells for tumor-free heart repair. J Exp Med 204(2):405–420
93. Crisostomo PR, Abarbanell AM, Wang M, Lahm T, Wang Y, Meldrum DR (2008) Embryonic stem cells attenuate myocardial dysfunction and inflammation after surgical global ischemia via paracrine actions. Am J Physiol Heart Circ Physiol 295(4):H1726–H1735
94. Banquet S, Gomez E, Nicol L, Edwards-Levy F, Henry JP, Cao R, Schapman D, Dautreaux B, Lallemand F, Bauer F, Cao Y, Thuillez C, Mulder P, Richard V, Brakenhielm E (2011) Arteriogenic therapy by intramyocardial sustained delivery of a novel growth factor combination prevents chronic heart failure. Circulation 124(9):1059–1069
95. Ellison GM, Torella D, Dellegrottaglie S, Perez-Martinez C, Perez de Prado A, Vicinanza C, Purushothaman S, Galuppo V, Iaconetti C, Waring CD, Smith A, Torella M, Cuellas Ramon C, Gonzalo-Orden JM, Agosti V, Indolfi C, Galinanes M, Fernandez-Vazquez F, Nadal-Ginard B (2011) Endogenous cardiac stem cell activation by insulin-like growth factor-1/hepatocyte growth factor intracoronary injection fosters survival and regeneration of the infarcted pig heart. J Am Coll Cardiol 58(9):977–986
96. Timmers L, Lim SK, Arslan F, Armstrong JS, Hoefer IE, Doevendans PA, Piek JJ, El Oakley RM, Choo A, Lee CN, Pasterkamp G, de Kleijn DP (2007) Reduction of myocardial infarct size by human mesenchymal stem cell conditioned medium. Stem Cell Res 1(2):129–137

Chapter 18
Adult Neurogenesis in Alzheimer's Disease and Therapies

Philippe Taupin

Abstract Neural stem cells (NSCs) hold the potential to treat and cure a broad range of neurological diseases and injuries. The confirmation that neurogenesis occurs in the adult brain and NSCs reside in the adult central nervous system of mammals, opens new avenues and opportunities for our understanding of development and for cellular therapy. Adult NSCs may contribute to physio- and pathological processes of the nervous system and may be used to repair and restore the damaged or injured nerve pathways. Neurogenesis is enhanced in the brain of patients with Alzheimer's disease (AD). Enhanced neurogenesis in AD may contribute to a regenerative attempt. Neurogenesis in the adult brain may also lead to the generation of aneuploid neuronal cells, particularly in the hippocampus. Hence, adult neurogenesis may be involved in processes both beneficial and detrimental in AD. Systemic administration may provide a strategy of choice for delivering adult-derived neural progenitor and stem cells to treat AD. Adult NSCs offer therefore a promising model for understanding the pathogenesis and pathology of AD and for cellular therapy for the treatment of AD. The elucidation of the involvement and contribution of adult neurogenesis and NSCs in AD will lead to the development of novel strategies for treating and curing the disease.

18.1 Introduction

Neural stem cells (NSCs) are the self-renewing multipotent cells that generate the main phenotypes of the nervous system, nerve cells, astrocytes, and oligodendrocytes. NSCs hold the potential to treat and cure a broad range of neurological diseases

P. Taupin (✉)
School of Biotechnology, Dublin City University, Glasnevin, Dublin, Ireland
e-mail: philippe.taupin@dcu.ie

H. Baharvand and N. Aghdami (eds.), *Advances in Stem Cell Research*,
Stem Cell Biology and Regenerative Medicine, DOI: 10.1007/978-1-61779-940-2_18,

and injuries, ranging from neurodegenerative diseases, such as Alzheimer's disease (AD), Huntington's disease (HD), and Parkinson's disease, cerebral strokes, traumatic brain, and spinal cord injuries [1]. Neurogenesis occurs in the adult brain and NSCs reside in the adult central nervous system (CNS) of mammals, in various species including humans [2, 3]. Hence, the adult brain may be amenable to repair and adult NSCs offer new avenues and opportunities for cellular therapy, particularly for the treatment of neurological diseases and injuries, and particularly AD. AD is the most common form of dementia among elderly. It affects 35 million individuals worldwide [4]. Reports indicate that neurogenesis is enhanced in the brain of patients with AD [5]. This suggests that adult neurogenesis and newly generated neuronal cells of the adult brain may be involved in the pathogenesis and pathology of AD. The understanding of which will lead to a better understanding of AD and to the development of new treatments for the disease. Neural progenitor and stem cells administered systemically migrate to damaged and injured sites in the CNS [6, 7], providing a strategy particularly suitable for treating AD, a disease with widespread neurodegeneration. In this chapter, we will review and discuss the potential role and contribution of adult neurogenesis and newly generated neuronal cells to the pathogenesis and pathology of AD and the potential of adult neurogenesis and NSCs for treating AD.

18.2 Alzheimer's Disease

AD is a neurodegenerative disease. It is characterized by widespread neurodegeneration, amyloid deposits, neurofibrillary tangles, aneuploidy, memory and cognitive deficits, and anosmia [8–11]. The early onset form of AD (EOAD) is a rare form of the disease, primarily inherited. It is diagnosed before age 65. The late onset form of AD (LOAD) is the most common form of the disease, accounting for over 93% of all cases of AD. It is diagnosed after the age of 65. Most cases of LOAD are sporadic forms of the disease. Age is the principal risk factor for AD. The disease affects about 3% of individuals ages 65–74 and about 50% of those 85-year-old and older [12]. The causes for AD include genetic mutations and risk factors. There is no cure for AD which leads to death within 3–9 years after being diagnosed [4, 8]. The disease affects more than 35 million of individuals worldwide.

Three genetic mutations causative for inherited forms of AD have been identified: mutations in the gene of amyloid precursor protein (APP), of presenilin-1 (PSEN1), and of presenilin-2 (PSEN2) [13]. APP is a 695–770 amino acid protein coding for the protein beta-amyloid. Mutations in APP result in an excessive production of the 42 amino acid beta-amyloid peptide and the formation of amyloid deposits [14]. Mutations in PSEN1 and PSEN2, that plays a role in the maturation of APP into protein beta-amyloid, result in the formation of amyloid deposits [15, 16]. Some of the genetic, acquired, and environmental risk factors, causative for the sporadic forms of AD, have been identified, such as the presence of certain alleles in the genetic makeup of the individuals, particularly the presence of the apolipoprotein E varepsilon 4 allele (ApoE4), hypertension, diabetes, and oxidative stress [17–20].

AD is associated with the loss of nerve cells in areas of the brain that are vital to memory and other cognitive abilities, such as the entorhinal cortex and the hippocampus. As the disease progresses, other regions of the brain are affected, leading to severe incapacity and death [21]. Amyloid plaques and neurofibrillary tangles are the histopathological hallmarks of AD. Amyloid plaques are composed primarily of extracellular deposits of amyloid fibrils or protein beta-amyloid [22]. Protein beta-amyloid originates from the post-transcriptional maturation of the APP. Under certain conditions, such as the presence of specific gene mutations or certain risk factors, the 42 amino acid form of protein beta-amyloid aggregates to form deposits of amyloid fibrils or amyloid plaques. Amyloid plaques are distributed throughout the brain of patients with AD, particularly in the regions of degeneration, like the entorhinal cortex and hippocampus [23]. Their density increases as the disease advances. Amyloid plaques are thought to be the first histological change to occur in the brain of patients with AD and may be a causative factor of AD [24]. Neurofibrillary tangles are intracellular deposits of hyperphosphorylated tau proteins, a microtubule-associated phosphoprotein, distributed throughout the brain of patients with AD [25]. The hyperphosphorylation of tau proteins results in their aggregation and in the break down of microtubules [26]. This leads to the formation of neurofibrillary tangles and cell death [27].

Aneuploidy is landmark of the pathology of AD. A substantial percentage, 4–10%, of nerve cells in regions of neurodegeneration of the brain, such as the hippocampus, of patients with AD express proteins of the cell cycle, particularly cyclin B, and are aneuploid [28, 29]. Cell cycle re-entry and DNA duplication, without cell division, is at the origin of aneuploidy in those cells in the brain of patients with AD [30, 31]. Aneuploid nerve cells originating from the re-expression of proteins of the cell cycle are fated to die; they may live in this state for months undergoing a slow death process [32, 33]. Cell cycle re-entry and DNA duplication, without cell division, leading to aneuploidy would be an underlying mechanism of the neurodegenerative process in AD. The genetic imbalance in aneuploid cells results in the over expression of genes by the cells. The genes for ApoE, APP, PSEN1, PSEN2, and tau are located on chromosomes 19, 21, 1, 14, and 17, respectively [34–37]. Cells of AD patients elicit an elevation of aneuploidy, particularly for chromosomes 21 [38]. Aneuploidy for genes involved in the pathology of AD would contribute to the pathogenesis of the disease, as a result of the over expression of those genes. Hence, aneuploidy in AD underlies the process of neurodegeneration and contributes to the pathogenesis of AD, by over-expressing genes involved in the disease and triggering a cascade of events amplifying the development of the disease.

In all, AD is the most common form of senile dementia. It is characterized by protein misfolding, genetic instability, and neurodegeneration.

18.3 Adult Neurogenesis and Neural Stem Cells in Mammals

Neurogenesis occurs in discrete regions of the adult mammalian brain, primarily in the subventricular zone (SVZ) along the ventricles and in the dentate gyrus (DG) of the hippocampus, in various species including humans [39, 40]. Newly generated neuronal cells, in the anterior part of the SVZ, migrate through the rostro-migratory stream to the olfactory bulb, where they differentiate into olfactory interneurons [41]. In the DG, newly generated neuronal cells in the subgranular zone, a layer beneath the granular layer, migrate to the granule cell layer, where they differentiate into granule-like cells and extend axonal projections to the CA3 region of the Ammon's horn [42, 43]. Newly generated neuronal cells in the DG establish synaptic contacts with neighboring and target cells [43–45]. They establish MF-like synapses with their target cells of the CA3 region [45]. The number of neuronal cells generated per day in the adult brain is relatively low, particularly in the DG. In mice, the number of newly generated neuronal cells per day in the DG is estimated at 9,000 cells, or about 0.1% of the granule cell population [46].

Neurogenesis is modulated in the adult brain, particularly in the hippocampus. It is modulated by a broad range of environmental stimuli, physio- and patho-logical processes and conditions, trophic factors/cytokines, neurotransmitters and drugs, including enriched environment, learning and memory tasks, physical activity, AD, epilepsy, and drugs used for treating depression and AD, such as fluoxetine and memantine [47]. Environmental enrichment, learning and memory tasks and physical activity stimulate hippocampal neurogenesis in adult rodents [48–51]. Neurogenesis is enhanced in the adult hippocampus of animal models of epilepsy and strokes [52, 53]. It is enhanced in the SVZ and hippocampus in the brain of patients with HD and AD, respectively [5, 54]. This suggests that neurogenesis in the adult brain is involved in various physio- and pathological conditions and processes of the nervous system, including neurological diseases and drug activity [55].

Newly generated neuronal in the adult mammalian brain would originate for a pool of residual stem cells [2]. Self-renewing multipotent NSCs have been isolated and characterized in vitro, from various regions of the adult brain, including the SVZ, hippocampus and spinal cord, and from various mammalian species, including from human biopsies and postmortem tissues [56–59]. Neural and progenitor cells isolated and characterized from the adult brain express molecular markers, such as the intermediate filament nestin, the transcription factors sox-2, oct-3/4, and the RNA binding protein Musashi 1 [60–63]. The isolation and characterization of neural and progenitor cells from the adult brain reveal that NSCs reside in the adult CNS and that it has the potential for self-repair. Adult-derived neural progenitor and stem cells also provide a source of tissue that may be used for cellular therapy, for transplantation.

The contribution of adult neurogenesis and newly generated neuronal cells of the adult brain to the physio- and pathology of the nervous system remains to be elucidated. Newly generated neuronal cells of the adult brain would contribute to the plasticity of the nervous system and regenerative attempts after injuries [64, 65].

18.4 Adult Neurogenesis and Newly Generated Neuronal Cells in AD

Autopsy studies show that the expression of doublecortin, a marker of immature neuronal cells, is enhanced in the hippocampus of the brain of patients with LOAD [5]. This reveals that neurogenesis may be enhanced in the hippocampus of patients with AD. In animal models of AD, neurogenesis is enhanced in the hippocampus of adult transgenic mice that express the Swedish and Indiana APP mutations [66]. It is reduced in the hippocampus of adult mice deficient for APP or PSEN1, of adult transgenic mice over-expressing variants of APP or PSEN1 and of adult PDAPP transgenic mice, a mouse model with age-dependent accumulation of protein beta-amyloid [67–70]. The expression of nestin and Musashi1, markers of neural progenitor and stem cells, is reduced in the SVZ of patients with AD [71]. This shows that neurogenesis may be reduced in the SVZ of patients with AD. These studies report conflicting data on the modulation of adult neurogenesis in AD, particularly in the hippocampus. Enhanced neurogenesis in the hippocampus would represent a regenerative attempt to compensate for the neuronal loss in the AD brain. Reduced neurogenesis in the SVZ of AD brain may underlie the compromised olfaction associated with the disease [72]. Hence, adult neurogenesis and newly generated neuronal cells of the adult brain play a role in the pathology of AD, the contribution of which remains to be elucidated and determined.

Bromodeoxyuridine (BrdU)-labeling and immunohistochemistry for neuronal markers, and immunohistochemistry for markers of the cell cycle are the techniques the most used to study cell proliferation and neurogenesis in the human brain and in animal models [73]. BrdU is a thymidine analog. It is a mutagenic and toxic substance. It is a marker of DNA synthesis, not a marker of cell proliferation. There are limitations and pitfalls over the use of BrdU-labeling and immunohistochemistry for markers of the cell cycle for studying cell proliferation and neurogenesis [74, 75]. Studying cell proliferation and neurogenesis using the BrdU-labeling paradigm and immunohistochemistry for markers of the cell cycle requires distinguishing cell proliferation and neurogenesis from other events involving DNA synthesis, such as DNA repair, abortive cell cycle re-entry, leading to apoptosis, and gene duplication, without cell division, leading to aneuploidy [76, 77]. Hence, the conflicting data reported on adult neurogenesis in the brain of patients with AD and in animal models of AD may originate from the validity of the techniques and protocols used to study adult neurogenesis. It may also originate from (i) the validity of animal models, such as transgenic mice, as representative of a disease, such as AD, and to study adult phenotypes, such as adult neurogenesis. In fact, transgenic mice deficient or over-expressing genes, such as APP and PSEN1, are not representative models of the disease, but rather models to study the genes involved in the diseases [78]. Conflicting data reported on adult neurogenesis in the brain of patients with AD and in animal models of AD may also originate from the fact that there is no consensus on the term neurogenesis. Some studies only present data on cell proliferation, whereas others present only data on cell survival and neuronal differentiation. It may also originate from the fact that in most studies, and in particular in the human post-mortem studies, only one time point along the pathology is analyzed. Neurogenesis might indeed be differentially regulated along the pathogenesis [79]. Hence, studies on adult neurogenesis must be carefully analyzed, discussed, confirmed, and validated, particularly when studying neurodegenerative diseases like AD. Further studies are therefore mandated to characterize the modulation of adult neurogenesis in AD.

Cell division is the most likely process to generate aneuploid cells. The non-disjunction of chromosomes during the process of adult neurogenesis has the potential to generate populations of cells that are aneuploid, particularly in the neurogenic regions [80]. Aneuploid newly generated neuronal cells of the adult brain would contribute to neurodegeneration and amyloid and neurofibrillary tangles formation, particularly in the neurogenic regions of the adult brain, such as the hippocampus. Neurogenesis is a relatively low frequency event in the adult brain of mammals, estimated at 0.1% of the granule cell population generated per day in the DG of adult mice [46, 48]. Nonetheless, the contribution of aneuploid newly generated neuronal cells of the adult brain to the pathogenesis of AD may be critical, as they are generated particularly in the hippocampus, a region involved in learning and memory and particularly affected in patients with the disease. Hence, adult neurogenesis may be involved and contribute to the pathogenesis of AD [80].

The hyperphosphorylation of tau proteins, by kinases, leads to the break down of microtubules and the disruption in the mitotic spindle, the mutated forms of PSEN1 are detected in interphase kinetochores and centrosomes of dividing cells, and oxidative stress is involved in the segregation and migration of chromosomes during cells division and promotes aneuploidy [81–83]. Causative factors of AD may therefore promote the generation of aneuploid newly generated neuronal cells in the adult brain.

In all, adult neurogenesis and newly generated neuronal cells of the adult brain may be involved in regenerative attempts and in the pathology and pathogenesis of AD. However, the contribution of adult neurogenesis to the pathology of AD remains to be further characterized, as well as the modulation of neurogenesis in the AD brain. The generation of aneuploid newly generated neuronal cells in the adult brain to be demonstrated, particularly in the AD brain. Adult neurogenesis would not only be beneficial, it may also be detrimental for patients with AD.

18.5 Cellular Therapy

The stimulation of endogenous neural progenitor or stem cells of the adult brain and the transplantation of adult-derived neural progenitor and stem cells are being considered for repairing and restoring the damaged or injured nerve pathways, and particularly for treating patients with AD.

The microenvironment or niche controls the developmental potential and therefore therapeutic potential of stem cells [84, 85]. Neurogenesis is modulated in the adult brain in the neurogenic regions. The stimulation of neurogenesis in the hippocampus and SVZ may contribute to reverse or compensate deficits associated with neurological diseases and disorders, particularly the deficits associated with the hippocampus and olfactory bulb, such as learning and memory and anosmia, respectively. Alternatively, drugs and compounds stimulating adult neurogenesis, or neurogenic drugs, may be used to stimulate the proliferation and differentiation of endogenous neural progenitor or stem cells to repair and restore the degenerated or injured nerve pathways [86]. However, the microenvironment may limit such strategy and the molecular and cellular mechanisms underlying the development of neural progenitor or stem cells in the adult brain will need to be elucidated to promote endogenous neural progenitor and stem cell repair, as a therapeutic strategy for treating neurological diseases and injuries, particularly AD [87]. Neuronal cells are generated at sites of degeneration in the diseased brain and after CNS injuries, such as in HD and in experimental models of cerebral strokes [88, 89]. These newly generated neuronal cells originate from the SVZ and migrate through the rostro-migratory stream to the sites of degenerations and injuries. Therapeutic strategies for treating neurological diseases and injuries may aim at identifying and characterizing drugs or compounds that promote the proliferation, migration, and differentiation of neural progenitor and stem cells of the SVZ to repair and restore the degenerated or injured nerve pathways, particularly in AD.

The intracerebral transplantation of adult-derived neural progenitor and stem cells may be used primarily for treating neurodegenerative diseases for which the degeneration is not widespread, such as Parkinson's disease [90]. For neurodegenerative diseases for which the degeneration is widespread, such as AD, HD, and multiple sclerosis, this strategy may not be applicable, as it would require multiple sites of transplantation. Neural progenitor and stem cells migrate to sites of degeneration and diseased areas of the CNS, when administered systemically, such as intravenously [7, 91]. Hence, the intravenous injection of adult-derived neural progenitor and stem cells may provide a strategy of choice and a promising strategy for treating neurodegenerative diseases and particularly AD. It provides a non-invasive mean to deliver stem cells to the CNS. Future directions will involve improving the homing and migration of the stem cells to the diseased and injured sites, such as the ex vivo fucosylation of stem cells prior intravenous injection [92, 93].

18.6 Conclusion and Perspectives

AD is a major public health problem and there is a need to both better understand the pathology and pathogenesis of the disease and design and develop new treatments for the disease. Adult neurogenesis and NSCs offer such opportunities. Studies show that neurogenesis may be enhanced in the brain of patients with AD. Adult neurogenesis has the potential to generate aneuploid neuronal cells in the adult brain and causative factors of AD, such as genetic mutations and other risk factors, may increase this latter risk for AD. Hence, adult neurogenesis may be both beneficial and detrimental to AD. The intravenous administration of adult-derived neural progenitor and stem cells offers a promising model for treating patients with AD. Future studies will involve further characterization of the modulation of adult neurogenesis in AD and the characterization of the generation of aneuploid neuronal cells in the AD brain, particularly in the hippocampus. Future investigations will aim at studying the potential of systemic delivery of adult-derived neural progenitor and stem cells in animal models of AD and to design and develop novel drugs that promote the regenerative capacity of the hippocampus, to promote functional recovery in AD.

References

1. Gage GH (2000) Mammalian neural stem cells. Science 287:1433–1438
2. Taupin P, Gage FH (2002) Adult neurogenesis and neural stem cells of the central nervous system in mammals. J Neurosci Res 69:745–749
3. Duan X, Kang E, Liu CY et al (2008) Development of neural stem cell in the adult brain. Curr Opin Neurobiol 18:108–115
4. Querfurth HW, LaFerla FM (2010) Alzheimer's disease. N Engl J Med 362:329–344
5. Jin K, Peel AL, Mao XO et al (2004) Increased hippocampal neurogenesis in Alzheimer's disease. Proc Natl Acad Sci U S A 101:343–347
6. Aboody KS, Brown A, Rainov NG et al (2001) Neural stem cells display extensive tropism for pathology in adult brain: evidence from intracranial gliomas. Proc Natl Acad Sci U S A 97:12846–12851 (Erratum in: Proc Natl Acad Sci U S A 2001;98:777)
7. Brown AB, Yang W, Schmidt NO et al (2003) Intravascular delivery of neural stem cell lines to target intracranial and extracranial tumors of neural and non-neural origin. Hum Gene Ther 14:1777–1785
8. Burns A, Byrne EJ, Maurer K (2002) Alzheimer's disease. Lancet 360:163–165
9. Dubois B, Feldman HH, Jacova C et al (2007) Research criteria for the diagnosis of Alzheimer's disease: revising the NINCDS–ADRDA criteria. Lancet Neurol 6:734–746
10. Fusetti M, Fioretti AB, Silvagni F et al (2010) Smell and preclinical Alzheimer disease: study of 29 patients with amnesic mild cognitive impairment. J Otolaryngol Head Neck Surg 39:175–181
11. Wesson DW, Wilson DA, Nixon RA (2010) Should olfactory dysfunction be used as a biomarker of Alzheimer's disease? Expert Rev Neurother 10:633–635
12. Wang XP, Ding HL (2008) Alzheimer's disease: epidemiology, genetics, and beyond. Neurosci Bull 24:105–109
13. St George-Hyslop PH, Petit A (2005) Molecular biology and genetics of Alzheimer's disease. C R Biol 328:119–130
14. Wang JF, Lu R, Wang YZ (2010) Regulation of β cleavage of amyloid precursor protein. Neurosci Bull 26:417–427
15. Nishimura M, Yu G, St George-Hyslop PH (1999) Biology of presenilins as causative molecules for Alzheimer disease. Clin Genet 55:219–225
16. Newman M, Musgrave FI, Lardelli M (2007) Alzheimer disease: amyloidogenesis, the presenilins and animal models. Biochim Biophys Acta 1772:285–297
17. Fassbender K, Masters C, Beyreuther K (2000) Alzheimer's disease: an inflammatory disease? Neurobiol Aging 21:433–436

18. Prasher VP, Haque MS (2000) Apolipoprotein E, Alzheimer's disease and Down's syndrome. Br J Psychiatry 177:469–470
19. Ferri CP, Prince M, Brayne C et al (2006) The role of oxidative stress in the pathogenesis of Alzheimer's disease. Bratisl Lek Listy 107:384–394
20. Bertram L, Tanzi RE (2008) Thirty years of Alzheimer's disease genetics: the implications of systematic meta-analyses. Nat Rev Neurosci 9:768–778
21. Brun A, Gustafson L (1976) Distribution of cerebral degeneration in Alzheimer's disease: a clinico-pathological study. Arch Psychiatr Nervenkr 223:15–33
22. Kang J, Lemaire HG, Unterbeck A et al (1987) The precursor of Alzheimer's disease amyloid A4 protein resembles a cell-surface receptor. Nature 325:733–736
23. Anderson DH, Talaga KC, Rivest AJ et al (2004) Characterization of beta amyloid assemblies in drusen: the deposits associated with aging and age-related macular degeneration. Exp Eye Res 78:243–256
24. St George-Hyslop PH (2000) Piecing together Alzheimer's. Sci Am 283:76–83
25. Fukutani Y, Kobayashi K, Nakamura I et al (1995) Neurons, intracellular and extra cellular neurofibrillary tangles in subdivisions of the hippocampal cortex in normal ageing and Alzheimer's disease. Neurosci Lett 200:57–60
26. Kim H, Jensen CG, Rebhun LI (1986) The binding of MAP-2 and tau on brain microtubules in vitro: implications for microtubule structure. Ann N Y Acad Sci 466:218–239
27. Iqbal K, Liu F, Gong CX et al (2009) Mechanisms of tau-induced neurodegeneration. Acta Neuropathol 118:53–69
28. Busser J, Geldmacher DS, Herrup K (1998) Ectopic cell cycle proteins predict the sites of neuronal cell death in Alzheimer's disease brain. J Neurosci 18:2801–2807
29. Kingsbury MA, Yung YC, Peterson SE et al (2006) Aneuploidy in the normal and diseased brain. Cell Mol Life Sci 63:2626–2641
30. Yang Y, Geldmacher DS, Herrup K (2001) DNA replication precedes neuronal cell death in Alzheimer's disease. J Neurosci 21:2661–2668
31. Yang Y, Mufson EJ, Herrup K (2003) Neuronal cell death is preceded by cell cycle events at all stages of Alzheimer's disease. J Neurosci 23:2557–2563
32. Herrup K, Arendt T (2002) Re-expression of cell cycle proteins induces neuronal cell death during Alzheimer's disease. J Alzheimers Dis 4:243–247
33. Yang Y, Herrup K (2007) Cell division in the CNS: protective response or lethal event in post-mitotic neurons? Biochim Biophys Acta 1772:457–466
34. Goldgaber D, Lerman MI, McBride OW et al (1987) Characterization and chromosomal localization of a cDNA encoding brain amyloid of Alzheimer's disease. Science 235:877–880
35. Schellenberg GD, Bird TD, Wijsman EM et al (1992) Genetic linkage evidence for a familial Alzheimer's disease locus on chromosome 14. Science 258:668–671
36. Iqbal K, Grundke-Iqbal I, Smith AJ et al (1989) Identification and localization of a tau peptide to paired helical filaments of Alzheimer disease. Proc Natl Acad Sci U S A 86:5646–5650
37. Nishimura M, Yu G, St George-Hyslop PH (1999) Biology of presenilins as causative molecules for Alzheimer disease. Clin Genet 55:219–225
38. Migliore L, Botto N, Scarpato R et al (1999) Preferential occurrence of chromosome 21 malsegregation in peripheral blood lymphocytes of Alzheimer disease patients. Cytogenet Cell Genet 87:41–46
39. Eriksson PS, Perfilieva E, Bjork-Eriksson T et al (1998) Neurogenesis in the adult human hippocampus. Nat Med 4:1313–1317
40. Curtis MA, Kam M, Nannmark U et al (2007) Human neuroblasts migrate to the olfactory bulb via a lateral ventricular extension. Science 315:1243–1249
41. Lois C, Alvarez-Buylla A (1994) Long-distance neuronal migration in the adult mammalian brain. Science 264:1145–1148
42. Cameron HA, Woolley CS, McEwen BS et al (1993) Differentiation of newly born neurons and glia in the dentate gyrus of the adult rat. Neurosci 56:337–344
43. Markakis EA, Gage FH (1999) Adult-generated neurons in the dentate gyrus send axonal projections to field CA3 and are surrounded by synaptic vesicles. J Comp Neurol 406:449–460

44. Toni N, Teng EM, Bushong EA et al (2007) Synapse formation on neurons born in the adult hippocampus. Nat Neurosci 10:727–734
45. Taupin P (2009) Characterization and isolation of synapses of newly generated neuronal cells of the adult hippocampus at early stages of neurogenesis. J Neurodegener Regene 2:9–17
46. Cameron HA, McKay RD (2001) Adult neurogenesis produces a large pool of new granule cells in the dentate gyrus. J Comp Neurol 435:406–417
47. Taupin P (2005) Adult neurogenesis in the mammalian central nervous system: functionality and potential clinical interest. Med Sci Monit 11:RA247–52
48. Kempermann G, Kuhn HG, Gage FH (1997) More hippocampal neurons in adult mice living in an enriched environment. Nature 386:493–495
49. Gould E, Beylin A, Tanapat P et al (1999) Learning enhances adult neurogenesis in the hippocampal formation. Nat Neurosci 2:260–265
50. van Praag H, Kempermann G, Gage GF (1999) Running increases cell proliferation and neurogenesis in the adult mouse dentate gyrus. Nat Neurosci 2:266–270
51. Shors TJ, Miesegaes G, Beylin A et al (2001) Neurogenesis in the adult is involved in the formation of trace memories. Nature 410:372–376 (Erratum in: Nature 2001;414:938)
52. Parent JM, Yu TW, Leibowitz RT et al (1997) Dentate granule cell neurogenesis is increased by seizures and contributes to aberrant network reorganization in the adult rat hippocampus. J Neurosci 17:3727–3738
53. Liu J, Solway K, Messing RO et al (1998) Increased neurogenesis in the dentate gyrus after transient global ischemia in gerbils. J Neurosci 18:7768–7778
54. Curtis MA, Penney EB, Pearson AG et al (2003) Increased cell proliferation and neurogenesis in the adult human Huntington's disease brain. Proc Natl Acad Sci U S A 100:9023–9027
55. Taupin P (2008) Adult neurogenesis pharmacology in neurological diseases and disorders. Exp Rev Neurother 8:311–320
56. Reynolds BA, Weiss S (1992) Generation of neurons and astrocytes from isolated cells of the adult mammalian central nervous system. Science 255:1707–1710
57. Gage FH, Coates PW, Palmer TD et al (1995) Survival and differentiation of adult neuronal progenitor cells transplanted to the adult brain. Proc Natl Acad Sci U S A 92:11879–11883
58. Roy NS, Wang S, Jiang L et al (2000) In vitro neurogenesis by progenitor cells isolated from the adult human hippocampus. Nat Med 6:271–277
59. Palmer TD, Schwartz PH, Taupin P et al (2001) Cell culture: progenitor cells from human brain after death. Nature 411:42–43
60. Lendahl U, Zimmerman LB, McKay RD (1990) CNS stem cells express a new class of intermediate filament protein. Cell 60:585–595
61. Kaneko Y, Sakakibara S, Imai T et al (2000) Musashi1: an evolutionally conserved marker for CNS progenitor cells including neural stem cells. Dev Neurosci 22:139–153
62. Komitova M, Eriksson PS (2004) Sox-2 is expressed by neural progenitors and astroglia in the adult rat brain. Neurosci Lett 369:24–27
63. Okuda T, Tagawa K, Qi ML et al (2004) Oct-3/4 repression accelerates differentiation of neural progenitor cells in vitro and in vivo. Brain Res Mol Brain Res 132:18–30
64. Taupin P (2006) Adult neurogenesis and neuroplasticity. Restor Neurol Neurosci 24:9–15
65. Ma DK, Bonaguidi MA, Ming GL et al (2009) Adult neural stem cells in the mammalian central nervous system. Cell Res 19:672–682
66. Jin K, Galvan V, Xie L et al (2004) Enhanced neurogenesis in Alzheimer's disease transgenic (PDGF-APPSw, Ind) mice. Proc Natl Acad Sci U S A 101:13363–13367
67. Wen PH, Shao X, Shao Z et al (2002) Overexpression of wild type but not an FAD mutant presenilin-1 promotes neurogenesis in the hippocampus of adult mice. Neurobiol Dis 10:8–19
68. Zhang C, McNeil E, Dressler L et al (2007) Long-lasting impairment in hippocampal neurogenesis associated with amyloid deposition in a knock-in mouse model of familial Alzheimer's disease. Exp Neurol 204:77–87
69. Rodríguez JJ, Jones VC, Tabuchi M et al (2008) Impaired adult neurogenesis in the dentate gyrus of a triple transgenic mouse model of Alzheimer's disease. PLoS ONE 3:e2935

70. Yu Y, He J, Zhang Y et al (2009) Increased hippocampal neurogenesis in the progressive stage of Alzheimer's disease phenotype in an APP/PS1 double transgenic mouse model. Hippocampus 19:1247–1253
71. Ziabreva I, Perry E, Perry R et al (2006) Altered neurogenesis in Alzheimer's disease. J Psychosom Res 61:311–316
72. Li W, Howard JD, Gottfried JA (2010) Disruption of odour quality coding in piriform cortex mediates olfactory deficits in Alzheimer's disease. Brain 133:2714–2726
73. Miller MW, Nowakowski RS (1998) Use of bromodeoxyuridine-immunohistochemistry to examine the proliferation, migration and time of origin of cells in the central nervous system. Brain Res 457:44–52
74. Nowakowski RS, Hayes NL (2001) Stem cells: the promises and pitfalls. Neuropsychopharmacology 25:799–804
75. Gould E, Gross CG (2002) Neurogenesis in adult mammals: some progress and problems. J Neurosci 22:619–623
76. Taupin P (2007) Protocols for studying adult neurogenesis: insights and recent developments. Regen Med 2:51–62
77. Taupin P (2007) BrdU Immunohistochemistry for studying adult neurogenesis: paradigms, pitfalls, limitations, and validation. Brain Res Rev 53:198–214
78. German DC, Eisch AJ (2004) Mouse models of Alzheimer's disease: insight into treatment. Rev Neurosci 15:353–369
79. Taupin P (2010) Adult neurogenesis and neural stem cells as a model for the discovery and development of novel drugs. Expert Opin Drug Discov 5:921–925
80. Taupin P (2009) Adult neurogenesis, neural stem cells and Alzheimer's disease: developments, limitations, problems and promises. Curr Alzheimer Res 6:461–470
81. Li J, Xu M, Zhou H et al (1997) Alzheimer presenilins in the nuclear membrane, interphase kinetochores, and centrosomes suggest a role in chromosome segregation. Cell 90:917–927
82. Ramírez MJ, Puerto S, Galofré P et al (2000) Multicolour FISH detection of radioactive iodine-induced 17cen-p53 chromosomal breakage in buccal cells from therapeutically exposed patients. Carcinogenesis 21:1581–1586
83. Boeras DI, Granic A, Padmanabhan J et al (2008) Alzheimer's presenilin 1 causes chromosome missegregation and aneuploidy. Neurobiol Aging 29:319–328
84. Taupin P (2006) Adult neural stem cells, neurogenic niches and cellular therapy. Stem Cell Rev 2:213–220
85. Mitsiadis TA, Barrandon O, Rochat A et al (2007) Stem cell niches in mammals. Exp Cell Res 313:3377–3385
86. Taupin P (2010) Neurogenic drugs and compounds. Recent Pat CNS Drug Discov 5:253–257
87. Kobel S, Lutolf M (2010) High-throughput methods to define complex stem cell niches. Biotechniques 48:ix-xxii
88. Arvidsson A, Collin T, Kirik D et al (2002) Neuronal replacement from endogenous precursors in the adult brain after stroke. Nat Med 8:963–970
89. Jin K, Sun Y, Xie L et al (2003) Directed migration of neuronal precursors into the ischemic cerebral cortex and striatum. Mol Cell Neurosci 24:171–189
90. Brundin P, Barker RA, Parmar M (2010) Neural grafting in Parkinson's disease Problems and possibilities. Prog Brain Res 184:265–294
91. Pluchino S, Quattrini A, Brambilla E et al (2003) Injection of adult neurospheres induces recovery in a chronic model of multiple sclerosis. Nature 422:688–694
92. Taupin P (2010) Ex vivo fucosylation to improve the engraftment capability and therapeutic potential of human cord blood stem cells. Drug Discovery Today 15:698–699
93. Taupin P (2010) Transplantation of cord blood stem cells for treating hematologic diseases and strategies to improve engraftment. Therapy 7:703–715

Author's Biography

Hossein Baharvand, Ph.D.
Head of Department of Stem Cells and Developmental Biology, Cell Science Research Center, Royan Institute for Stem Cell Biology and Technology, ACECR, P.O. Box 19395-4644, Tehran, Iran.
and
Head of Department of Developmental Biology, University of Science and Culture, ACECR, Tehran, Iran
E-mail: Baharvand@RoyanInstitute.org

Born in 1972, Hossein Baharvand received his PhD degree from the University for Teacher Training (TARBIAT MOALEM), Tehran in 2004 in the field of Developmental Biology. He joined Royan institute when he was an MSc student in 1995. He is currently professor and director of Royan Institute for Stem Cell Biology and Technology (2002-now) and head of Department of Developmental Biology at University of Science and Culture (2006-now). He established the first mouse and human embryonic stem cells in Iran (2003). Later his team has generated several mouse and human embryonic and induced pluripotent stem cells (2008). This has enabled them to pursue many avenues of research into

H. Baharvand and N. Aghdami (eds.), *Advances in Stem Cell Research*,
Stem Cell Biology and Regenerative Medicine, DOI: 10.1007/978-1-61779-940-2,

translational research and regenerative medicine. He has published 130 international and 98 national peer-reviewed papers, as well as five chapters in international books, six books in Persian, and translated six English text books into Persian. He was editor of three international books (2009, 2012, Humana press, Springer, USA). He is editorial board member of eight international journals and was as invited speaker in several meetings. He has received 18 international and national prizes including 10th (2004), 12th (2006), and 17th (2012) annual Razi research festival on medical sciences, 27th annual book of the year of the Islamic republic of Iran (2010), Academy of Medical Sciences of Iran (Dr. Hadavi's Award, 2010), Sicence and Technology Prize of the Islamic Educational, Scientific, and Cultural Organization (ISESCO) in the field of Biology (2010).

Nasser Aghdami, M.D., Ph.D.
Head of Department of Regenerative Medicine, Cell Science Research Center, Royan Institute for Stem Cell Biology and Technology, ACECR, P.O. Box 19395-4644, Tehran, Iran.
E-mail: Nasser.Aghdami@royaninstitute.org

Born in 1971 in Iran, Nasser Aghdami received his MD degree from Urumieh Medical University in 1998 and PhD in Immunology in 2007 from Tarbiat Modares University, Iran. He joined the Hematology, Oncology and BMT Research Center in 2000 as a principal investigator and Royan Institute for Stem Cell Biology and Technology in 2006. Dr. Aghdami's research has focused on the development of cardiovascular development, translational research, and the role of stem cells in tissue regeneration. Since 2008, he is the Head of the Department of Regenerative Medicine and Royan Cell Therapy Center. In this regard, clinical trial design and performance is one of his responsibilities. He has served on numerous national grant review committees, advisory panels and editorial boards related to stem cells, clinical trials, and cell therapy. Dr. Aghdami has published 21 international and 10 national peer-reviewed papers, as well as three chapters in international books, two books in Farsi, and translated one book into Farsi.

Index

H. Baharvand and N. Aghdami (eds.), *Advances in Stem Cell Research*,
Stem Cell Biology and Regenerative Medicine, DOI: 10.1007/978-1-61779-940-2,

Printed by Printforce, the Netherlands